LONDON MATHEMATICAL SOCIETY LECTURE NOT‎ ‎ ‎ES

Managing Editor: Professor Endre Süli, Mathematical Institute, University c‎
Woodstock Road, Oxford OX2 6GG, United Kingdom

The titles below are available from booksellers, or from Cambridge Unive‎
www.cambridge.org/mathematics

London Mathematical Society Lecture Note Series: 463

Differential Geometry in the Large

Edited by

OWEN DEARRICOTT
La Trobe University, Australia

WILDERICH TUSCHMANN
Karlsruhe Institute of Technology, Germany

YURI NIKOLAYEVSKY
La Trobe University, Australia

THOMAS LEISTNER
University of Adelaide

DIARMUID CROWLEY
University of Melbourne

CAMBRIDGE
UNIVERSITY PRESS

CAMBRIDGE
UNIVERSITY PRESS

University Printing House, Cambridge CB2 8BS, United Kingdom

One Liberty Plaza, 20th Floor, New York, NY 10006, USA

477 Williamstown Road, Port Melbourne, VIC 3207, Australia

314–321, 3rd Floor, Plot 3, Splendor Forum, Jasola District Centre, New Delhi – 110025, India

79 Anson Road, #06–04/06, Singapore 079906

Cambridge University Press is part of the University of Cambridge.

It furthers the University's mission by disseminating knowledge in the pursuit of education, learning, and research at the highest international levels of excellence.

www.cambridge.org
Information on this title: www.cambridge.org/9781108812818
DOI: 10.1017/9781108884136

First published 2021
Reprinted 2021

Printed in the United Kingdom by Books Limited, Padstow Cornwall

A catalogue record for this publication is available from the British Library.

ISBN 978-1-108-81281-8 Paperback

„Bekanntlich setzt die Geometrie sowohl den Begriff des Raumes, als die ersten Grundbegriffe für die Con-structionen im Raume als etwas Gegebenes voraus. Sie giebt von ihnen nur Nominaldefinitionen, während die wesentlichen Bestimmungen in Form von Axiomen auftreten. Das Verhältniss dieser Voraussetzungen bleibt dabei im Dunkeln ...

Diese Dunkelheit wurde auch von Euklid bis auf Legendre ... weder von den Mathematikern, noch von den Philosophen, welche sich damit beschäftigten, gehoben. Es hatte dies seinen Grund wohl darin, dass der allgemeine Begriff mehrfach ausgedehnter Grössen, unter welchem die Raumgrössen enthalten sind, ganz unbearbeitet blieb."

From the beginning of Bernhard Riemann's Habilitationsschrift „Ueber die Hypothesen, welche der Geometrie zu Grunde liegen" (Abhandl. Königl. Ges. Wiss. Göttingen, Vol. 13, 133–150, 1868)

"It is known that geometry assumes, as things given, both the notion of space and the first principles of constructions in space. She gives definitions of them which are merely nominal, while the true determinations appear in the form of axioms. The relation of these assumptions remains con-sequently in darkness ...

From Euclid to Legendre ... this darkness was cleared up neither by mathematicians nor by such philosophers as concerned themselves with it. The reason of this is doubt-less that the general notion of multiply extended magni-tudes (in which space-magnitudes are included) remained entirely unworked."

Corresponding passage from William Kingdon Clifford's translation 'On the Hypotheses which lie at the Bases of Geometry' (Nature, Vol. 8, 14–17, 1873)

Contents

Contributors

Christoph Böhm   *University of Münster, Einsteinstraße 62, 48149 Münster, Germany*

Theodora Bourni   *Department of Mathematics, University of Tennessee Knoxville, Knoxville, TN, 37996-1320, USA*

Charles P. Boyer   *Department of Mathematics and Statistics, University of New Mexico, Albuquerque, NM 87131, USA*

Paul Bryan   *Department of Mathematics, Macquarie University NSW 2109, Australia*

Timothy Buttsworth   *Department of Mathematics, Cornell University, Ithaca, NY 14853, USA*

Fernando Galaz-García   *Department of Mathematical Sciences, Durham University, Lower Mountjoy, Stockton Road, Durham DH1 3LE, UK*

Sebastian Goette   *Abteilung für Reine Mathematik, Universität Freiburg, Ernst-Zermelo-Straße 1, D-79104 Freiburg, Germany*

A. Rod Gover   *Department of Mathematics, The University of Auckland, Private Bag 92019, Auckland 1142, New Zealand*

Luis Guijarro   *Departamento de Matemáticas, Facultad de Ciencias, Universidad Autónoma de Madrid, 28049 Cantoblanco, Madrid, Spain*

Hongnian Huang   *Department of Mathematics and Statistics, University of New Mexico, Albuquerque, NM 87131, USA*

Mohammad N. Ivaki   *Department of Mathematics, University of Toronto, Ontario, M5S 2E4, Canada*

Martin Kerin   *School of Mathematics, Statistics and Applied Mathematics, National University of Ireland Galway, University Road, Galway, Ireland, H91 TK33*

Stephan Klaus   *Mathematisches Forschungsinstitut Oberwolfach, Schwarzwaldstrasse 9-11, D-77709 Oberwolfach-Walke, Germany*

Klaus Kröncke   *Fachbereich Mathematik, Bereich AD, Bundesstraße 55, 20146 Hamburg, Germany*

Ramiro A. Lafuente   *School of Mathematics and Physics, The University of Queensland, St Lucia, QLD 4072, Australia*

Mat Langford   *Department of Mathematics, University of Tennessee Knoxville, Knoxville, TN 37996-1320, USA*

Claude LeBrun   *Department of Mathematics, Stony Brook University, Stony Brook, NY 11794-3651, USA*

Eveline Legendre   *Institut de Mathématiques de Toulouse, Université Paul Sabatier, 118 route de Narbonne 31062, Toulouse, France*

Nan Li   *Department of Mathematics, The City University of New York - NYC College of Technology, 300 Jay St., Brooklyn, NY 11201, USA*

Matthias Ludewig   *Fakultät für Mathematik, Universität Regensburg, 93040 Regensburg, Germany*

Jesús Núñez-Zimbrón   *Centro de Ciencias Matemáticas UNAM, Antigua Carretera a Pátzcuaro 8701, Col. Ex Hacienda San José de la Huerta, Morelia, Michoacán, C.P. 58089, México*

Jiayin Pan   *Department of Mathematics, University of California Santa Barbara, Santa Barbara, CA 93106, USA*

Artem Pulemotov   *School of Mathematics and Physics, The University of Queensland, St Lucia, QLD 4072, Australia*

Julian Scheuer   *Department of Mathematics, Columbia University New York, NY 10027, USA*

Lorenz Schwachhöfer   *Fakultät für Mathematik, TU Dortmund University, Vogelpothsweg 87, 44221 Dortmund, Germany*

Krishnan Shankar   *Department of Mathematics, University of Oklahoma, 601 Elm Avenue, Room 423, Norman, OK 73019-3103, USA*

Giuseppe Tinaglia   *Department of Mathematics, King's College London, London WC2R 2LS, UK*

Christina W. Tønnesen-Friedman   *Department of Mathematics, Union College, Schenectady, NY 12308, USA*

Boris Vertman   *Carl von Ossietzky Universität Oldenburg, Institut für Mathematik, Carl-von-Ossietzky-Str. 9-11, D-26129 Oldenburg, Germany*

Andrew K. Waldron   *Center for Quantum Mathematics and Physics, Department of Mathematics, University of California Davis, Davis, CA 95616, USA*

Guofang Wei   *Department of Mathematics, University of California Santa Barbara, Santa Barbara, CA 93106, USA*

Valentina-Mira Wheeler   *School of Mathematics and Applied Statistics, University of Wollongong, Northfields Ave Wollongong, NSW 2522, Australia*

Joself A. Wolf   *Department of Mathematics, University of California, Berkeley, CA 94720-3840, USA*

Introduction

The Australian–German Workshop on Differential Geometry in the Large was a two week programme that ran from the 4th of February 2019 through to the 15th of February 2019, under the auspices of the Mathematical Research Institute (MATRIX), at the old Victorian School of Forestry, in the town of Creswick, Australia, that operates now as a campus of the University of Melbourne.

The first week, 4th to the 8th, was a large international conference with a list of prominent keynote speakers from across the globe. These included Ben Andrews and Neil Trudinger from Australia, Rod Gover from New Zealand, Christoph Böhm and Burkhard Wilking from Germany, Robert Bryant, Karsten Grove, Claude LeBrun, Peter Petersen and Guofang Wei from the United States, Dame Frances Kirwan from the United Kingdom and Tom Farrell and Fuquan Fang from the People's Republic of China. In addition to these keynote speakers a breadth of contributed talks were delivered across a full gamut of topics in Differential Geometry and Geometric Analysis.

In the second week, 11th through to the 15th, the meeting became a research symposium with smaller specialist sessions with the themes that included:

- Geometric evolutions equations and curvature flow,
- Structures on manifolds and mathematical physics,
- Recent developments in non-negative sectional curvature.

In the afternoons, small teams of academics teamed up to work on existing or new projects. Some of the work in this volume was conceived at the meeting.

Part I. Geometric Evolution Equations and Curvature Flow

Geometric flows, particularly the Ricci flow and mean curvature flow, have been of great interest in recent decades.

Given a Riemannian manifold, (M, g), one defines the Ricci flow as the family of Riemannian metrics, (M, g_t), that solve the initial value problem,

$$\frac{\partial g_t}{\partial t} = -2\mathrm{Ric}_{g_t},$$

$$g_0 = g,$$

where Ric_{g_t} represents the Ricci curvature of the metric, g_t.

The Ricci flow since its introduction by Hamilton in 1981 has played an important rôle in the solution of several long standing conjectures. Along with Alexandrov geometry (another topic discussed at the meeting), the Ricci flow played a crucial rôle in Perel'man's resolution of Thurston's geometrization conjecture with the celebrated Poincaré conjecture as a corollary. Other interesting applications of the Ricci flow include Schoen and Brendle's solution of the quarter pinched differentiable sphere theorem and Böhm and Wilking's solution to the Hamilton conjecture that a Riemannian manifold with positive curvature operator is diffeomorphic to a positively curved space form. The study of Ricci flow has bearing on the existence of Ricci solitons and Kähler–Einstein metrics.

The mean curvature flow and its generalizations enjoy interesting smoothing properties for families of submanifolds.

Given a Riemannian embedding, $x : X^{n-1} \hookrightarrow M^n$, the mean curvature flow is a family of immersions, $x_t : X^{n-1} \to M^n$, solving the initial value problem,

$$\frac{\partial x_t}{\partial t} = -H_{x_t}\hat{n},$$

$$x_0 = x,$$

where H_{x_t} is the mean curvature of the hypersurface, $X_t = x_t(X)$, and $\hat{n}$ is the normal of the hypersurface, X_t. The mean curvature flow has minimal surfaces as its critical points and is useful in solving the isoperimetric problem.

Part II. Structures on Manifolds and Mathematical Physics

Structures on Riemannian manifolds can take a variety of forms either through the existence of vector fields or tensor fields or curvature properties. For instance, a Riemannian manifold, (M, g), is said to be Kähler if it carries a parallel almost complex structure, J, i.e. an almost complex structure such that $\nabla J = 0$. A Riemannian manifold, (M, g), is said to be conformally Kähler if there is a conformal factor, f, such that the Riemannian manifold, (M, fg), is Kähler. A Riemannian manifold, (S, g), is said to be Sasakian if it carries a unit Killing, ξ, such that

$$R(X, \xi)Y = g(Y, \xi)X - g(X, Y)\xi$$

for any pair of smooth vector fields, X, Y. Equivalently, (S, g) is Sasakian if the Riemannian cone, $(S \times (0, \infty), t^2 g + dt^2)$, is Kähler.

As an example of a structure on a Riemannian manifold defined in terms of a curvature property consider that a Riemannian manifold, (M, g), is said to be Einstein if there is a constant, λ, such that

$$\text{Ric}_g = \lambda g.$$

One may ask if a given manifold carries a metric with a particular curvature property such as an Einstein metric, or more generally, given a symmetric 2 tensor, T, whether there exists a metric, g, with

$$\text{Ric}_g = T,$$

the so called prescibed Ricci curvature problem.

Another problem concerning the existence of metrics with positive scalar curvature was posed by Hidehiko Yamabe in 1960.

Yamabe problem Given a Riemannian metric, (M, g), where N is compact of dimension 3 or greater, is there a smooth function, $f : M \to \mathbb{R}$, such that the $e^{2f} g$ has constant scalar curvature?

The answer to this problem is yes, though the demonstration proved somewhat more difficult and labour intensive than Yamabe had anticipated, fruitfully occupying Trudinger, Schoen and Aubin for years hence.

More obscure perhaps is the Obata problem.

Obata problem Given the standard Riemannian metric on the sphere, (S^d, g), is there a smooth function, $\sigma : S^d \to \mathbb{R}$, such that

$$|d\sigma|^2 - \frac{2\sigma}{d}\left(\Delta\sigma + \frac{\sigma}{2(d-1)}s\right) = 1$$

where $\sigma^{-2}g$ is not Einstein.

Structures on Riemannian manifolds, (M^n, g), are often closely related to constraints on its holonomy group (the group generated by parallel transport along curves in the manifold). A Riemannian manifold is said to possess *special holonomy* if its holonomy is a proper subgroup of the orthogonal group, $O(n)$. For example, (M^n, g) is Kähler if and only if it is even dimensional and its holonomy is contained in $U\left(\frac{n}{2}\right)$.

Metrics with special holonomy are rare and are expected to strongly constrain the topology of the manifold. Indeed, a celebrated theorem of Deligne and Sullivan states that Kähler manifolds have rational homotopy types which are formal. Through the work of Sullivan, the rational homotopy type of a space is equivalent to certain computable algebraic data and formality is a strong constraint on this data. More generally, it is conjectured that all manifolds with special holonomy will be formal. This conjecture is especially enticing in dimension 7, which is both the only dimension in which the exceptional Lie group, G_2, can appear as a holonomy group and the first dimension in which there are simply connected manifolds which are not formal. Despite concerted efforts, it is still unknown whether all manifolds with holonomy group G_2 are formal.

Part III. Recent Developments in Non-Negative Sectional Curvature

The study of manifolds of non-negative and positive sectional curvature has enjoyed some resurgence in the twenty first century. Although the study of sectional curvature is a topic that stretches back to the beginnings of Riemannian geometry, the study of positive sectional curvature in terms of a comprehensive search for examples only really began with Berger in 1960, with his classification of simply connected normal homogeneous spaces with positive curvature, and extended into the 1970s with Aloff and Wallach's discoveries and Berard-Bérgery's classification.

Methods of construction amount to the realization of the manifolds in question as orbit spaces of free isometric actions on Lie groups endowed with non-negatively curved metrics via the non-decreasing property of O'Neill's formula for the curvature of the target of a Riemannian submersion. As such the consideration of manifolds of non-negative curvature in a sense is a natural precursor to that of positive curvature. Indeed, Gromoll and Meyer's construction of a seven manifold homeomorphic, but not diffeomorphic to the standard 7-sphere as the orbit of an isometric action of $Sp(1) \times Sp(1)$ on $Sp(2)$, obtained by multiplication on both the left and right (a so-called biquotient), is a stunning pioneering use of this simple technique. Eschenburg and Bazaikin went on to extend the technique to find new examples of manifolds with positive sectional curvature as biquotients in dimensions 7 and 13.

Developments in the twenty first century have considered cohomogeneity one actions on Riemannian manifolds and hinge on a key Lemma of Grove and Ziller that a cohomogeneity one manifold with codimension two singular orbits carries a metric with non-negative curvature. Essentially this amounts to the fact that as an orbit space the manifold can be obtained via the union of two non-negatively curved pieces that meet in isometric product collars. The examples of interest discussed by Grove and Ziller are the so-called Milnor spheres that occur as three sphere bundles over the 4-sphere. These arise in this context as isometric quotients of principal bundles that carry cohomogeneity one metrics of the above kind. Goette, Kerin and Shankar extended this technique more recently to apply to any exotic sphere in dimension 7. Some of the total spaces of the three sphere bundles also occur in the context of an up to now partially complete classification scheme of Grove, Wilking, Verdiani and Ziller of simply connected cohomogeneity one manifolds with positive sectional curvature, i.e. the so-called P_k family. This family, along with its companion family the Q_k, were also discussed at the meeting.

In both the study of non-negative and positive sectional curvature, as well as that of the more general theory of lower bounds on sectional, Ricci or scalar curvature, convergence techniques (notably ones involving the Gromov–Hausdorff distance) have for decades become an indispensable and ubiquitous tool in (and also way beyond) global differential geometry whenever it comes to studying sequences of metrics and their possible degenerations. This is, perhaps most prominently, illustrated by Gromov's proof of his celebrated theorem on groups of polynomial growth and Perel'man's proof of Thurston's geometrization conjecture.

Gromov–Hausdorff limits of Riemannian manifolds with a uniform lower bound on sectional curvature are known to be so-called Alexandrov spaces, that is, inner metric spaces of finite Hausdorff dimension with a metric notion of lower curvature bound satisfying variants of Toponogov's comparison theorem, and these interesting objects are nowadays studied in their own right as well. The length metric spaces that arise from limits of manifolds with lower bounds on Ricci curvature are so far much less understood. Their investigation constitutes today a very active field of research, involving, more generally, the study of RCD spaces, i.e. certain metric measure spaces which allow for synthetic notions of

Ricci curvature, based on methods of geometric measure theory, optimal transport and the foundational work of Lott–Villani and Sturm.

The meeting was a large and complex event that sponsored eight international keynote speakers' airfares as well as the airfares for many US based participants. The organizers wish to thank the Australian Mathematical Science Institute, the Ian Potter Foundation, the Australian Mathematical Society and La Trobe University for travel sponsorship of keynote speakers and the DFG national priority research scheme "Geometry at Infinity, SPP2026" and the National Science Foundation (NSF) for sponsoring the airfares of German and some US based participants, respectively; in particular, Lee Kennard deserves high praise for his work in successfully obtaining NSF travel funding. The organizers would also like to thank the International Conference Events Network in the New South Wales office of the Australian Department of Home Affairs for their quick work in securing a visa for one of our speakers giving a contributed talk in the first week.

The meeting provided full room and board to almost all participants as well as availing some paying places open to the general public, and the organizers wish to thank SPP2026, the NSF, MATRIX and the University of Melbourne International Research and Research Training Fund (IRRTF) for support with food and board for participants. The organizers would like to thank MATRIX for the provision of coaches from the old VSF to Melbourne Airport and Central Business District and the sponsorship for the provision of classrooms onsite. The organizers would like to thank the IRRTF for supporting the costs for refreshments, the welcoming dinner and farewell drinks at the Farmers' Arms Hotel and outings to Sovereign Hill and the Convent Gallery in Daylesford.

Owen Dearricott, Kyneton, March 2020, on behalf of the organizers,

Owen Dearricott, La Trobe University
Wilderich Tuschmann, Karlsruhe Institute of Technology (KIT)
Yuri Nikolayevsky, La Trobe University
Thomas Leistner, University of Adelaide
Diarmuid Crowley, University of Melbourne

Group photograph of the attendees at the international
conference week, 4th to 8th of February, 2019.

Group photograph of the attendees at the research
symposium week, 11th to 15th of February, 2019.

Part One

Geometric Evolution Equations and Curvature Flow

1

Real Geometric Invariant Theory

Christoph Böhm and Ramiro A. Lafuente

Abstract

For real reductive Lie groups we reprove the Kempf–Ness theorem about closed orbits and the Kirwan–Ness stratification theorem of the null cone using only geometric and analytic methods.

1.1 Introduction

The Kempf–Ness theorem provides a beautiful and simple geometric criterion for the closedness of orbits in a holomorphic representation of a complex reductive Lie group [KN79]. It implies that a non-closed orbit with positive distance to the origin contains a non-trivial closed orbit in its closure. For orbits in the *null cone* – those containing the origin in their closure – this is no longer true. The Kirwan–Ness stratification theorem describes a Morse-type stratification of this null cone into finitely many invariant submanifolds, with respect to a natural energy functional associated to the moment map of the action [Kir84], [Nes84].

In 1990, Richardson and Slodowy extended the Kempf–Ness theorem to the case of real reductive Lie groups acting linearly on Euclidean vector spaces [RS90]; see also [Mar01], [HS07], [EJ09], [BZ16]. Later on, the stratification theorem was also extended to the case of real reductive Lie groups by Lauret [Lau10] in a very special case, and by Heinzner, Schwarz and Stötzel to the much more general setting of group actions on complex spaces [HSS08]. The proofs of many of these results use the corresponding theorems in the complex case, and in particular rely on deep theorems on reductive algebraic groups and their real points

[Mos55], [BHC62], [BT65], [Bir71]. For instance, a key fact is that a vector has a closed orbit with respect to a real reductive Lie group if and only if the same holds for the complexified group [BHC62], [Bir71].

Our main aim in this chapter is to provide self-contained proofs of the Kempf–Ness theorem and the Kirwan–Ness stratification theorem for real reductive Lie group representations. In this particular case, we follow the arguments given in [HS07] and [HSS08], but use essentially elementary geometric and analytic methods.

A first motivation for writing this chapter is to make this theory available to a wider audience, including differential geometers and geometric analysts, who may lack the background in algebraic geometry necessary for approaching geometric invariant theory in its more classical presentation. Secondly, we were also motivated by the recent successful applications of real geometric invariant theory to the study of Riemannian geometry with symmetries: see Section 1.2 and [Lau10], [BL18b], [BL18a] and [BL19].

We now describe more precisely the technical setting and the main results to be addressed. Given a finite-dimensional vector space V over $\mathbb{R}$, we call a closed Lie subgroup $\mathsf{G} \subset \mathsf{GL}(V)$ *real reductive* if there exists a positive-definite scalar product $\langle \cdot, \cdot \rangle$ on V such that

$$\mathsf{G} = \mathsf{K} \cdot \exp(\mathfrak{p}), \tag{1.1}$$

where $\mathsf{K} := \mathsf{G} \cap \mathsf{O}(V, \langle \cdot, \cdot \rangle)$, $\mathfrak{p} := \mathfrak{g} \cap \mathrm{Sym}(V, \langle \cdot, \cdot \rangle)$, $\mathfrak{g}$ denotes the Lie algebra of G, and $\exp : \mathfrak{gl}(V) \to \mathsf{GL}(V)$ denotes the Lie exponential map. Here $\mathsf{O}(V, \langle \cdot, \cdot \rangle)$ denotes the group of orthogonal linear maps in $\mathsf{GL}(V)$ and $\mathrm{Sym}(V, \langle \cdot, \cdot \rangle)$ is the set of symmetric or self-adjoint endomorphisms of V. The maximal compact subgroup of G is K, and at Lie algebra level (1.1) yields a Cartan decomposition $\mathfrak{g} = \mathfrak{k} \oplus \mathfrak{p}$, that is $[\mathfrak{k}, \mathfrak{p}] \subset \mathfrak{p}$ and $[\mathfrak{p}, \mathfrak{p}] \subset \mathfrak{k}$. Note that, for $\mathsf{G} \subset \mathsf{GL}(V)$, (1.1) is equivalent to saying that its Lie algebra is closed under transpose, see [Kna02, Prop. 7.14], provided that G has only finitely many connected components.

More generally, we consider a faithful representation $\rho : \mathsf{G} \to \mathsf{GL}(V)$ of a real Lie group G with $\rho(\mathsf{G})$ a closed, real reductive subgroup of $\mathsf{GL}(V)$. Such representations will be called real reductive representations of G. To simplify notation, we will suppress ρ in what follows, and consider $\rho(\mathsf{G})$ instead of G.

For example, $\mathsf{GL}(V)$ itself is a real reductive Lie group, and so is any faithful, finite-dimensional representation of a real semisimple Lie group G with finitely many connected components [Mos55] (closedness in this case follows from the fact that the identity component of such a group is the identity component of a real linear algebraic group, see [OV90, Chap. 4, §1.2]). The same is true for $\mathsf{GL}_n(\mathbb{R})$, provided the centre acts by semisimple endomorphisms. Notice that, in contrast to these examples, a non-trivial representation of a non-abelian nilpotent Lie group is never real reductive.

We turn now to the Kempf–Ness theorem. A vector $\bar{v} \in \mathsf{G} \cdot v \subset V$ is called a *minimal vector* if it minimizes the distance to $0 \in V$ within the orbit $\mathsf{G} \cdot v$. If $\mathcal{M} \subset V$ denotes the set of all minimal vectors, it is clear that $\mathcal{M}$ intersects all closed orbits. Conversely, we have the following:

Theorem 1.1.1 *For a real reductive Lie group $\mathsf{G} \subset \mathsf{GL}(V)$ the following hold:*

(i) *Any orbit $\mathsf{G} \cdot v$ containing a minimal vector $\bar{v}$ is closed, and $\mathsf{G} \cdot \bar{v} \cap \mathcal{M} = \mathsf{K} \cdot \bar{v}$.*

(ii) *If the orbit $\mathsf{G} \cdot v$ is not closed, there exists $\alpha \in \mathfrak{p}$ such that the limit $w = \lim_{t \to \infty} \exp(t\alpha) \cdot v$ exists, and the orbit $\mathsf{G} \cdot w$ is closed.*

(iii) *The closure of any orbit contains exactly one closed orbit.*

(iv) *The null cone $\mathcal{N} = \{v \in V : 0 \in \overline{\mathsf{G} \cdot v}\}$ is a closed subset of V.*

Part (iii) of the above theorem was first proved in [Lun75], and parts (i) and (ii) in [RS90], under the assumption that the action of G on V is *rational* in the sense of algebraic geometry. One of its main implications is the fact that the set of closed orbits provides a *good quotient* for the G-action, with much better properties than the potentially non-Hausdorff orbit space. In the complex case, part (i) of the above theorem is known as the Kempf–Ness theorem [KN79], part (ii) is related to the Hilbert–Mumford criterion for stability [MFK94], and part (iii) appears in [Lun73]. Let us mention that by [HS10, App. A] the null cone $\mathcal{N}$ is an algebraic subset, thus the open set of *semistable* vectors $V \backslash \mathcal{N}$ is either empty or dense. However, at the moment our methods do not allow us to prove this fact in an elementary way.

Using the decomposition $\mathsf{G} = \mathsf{KTK}$, see Appendix, Section 1.10.1, the proof of Theorem 1.1.1 can be reduced to the abelian case. Here T is a maximal subgroup of G contained in $\exp(\mathfrak{p})$, necessarily abelian and non-compact. The proof of the abelian case relies on two crucial facts: the convexity of the distance function to the origin along one-parameter subgroups, and the separation of any two closed T-invariant sets by continuous T-invariant functions.

We turn now to the stratification theorem. Endow $\mathfrak{g}$ with an $\mathrm{Ad}(\mathsf{K})$-

invariant scalar product, also denoted by $\langle\,\cdot\,,\cdot\,\rangle$ such that $\langle\mathfrak{k},\mathfrak{p}\rangle = 0$ and

$$\mathrm{ad}(\mathfrak{k}) \subset \mathfrak{so}(\mathfrak{g},\langle\,\cdot\,,\cdot\,\rangle), \qquad \mathrm{ad}(\mathfrak{p}) \subset \mathrm{sym}(\mathfrak{g},\langle\,\cdot\,,\cdot\,\rangle). \tag{1.2}$$

For instance, one possible choice for $\langle\,\cdot\,,\cdot\,\rangle$ is the restriction of the usual scalar product on $\mathfrak{gl}(V)$ induced by that on V.

Definition 1.1.2 ((Real) moment map)   The map $\mathrm{m} : V\backslash\{0\} \to \mathfrak{p}$ defined implicitly by

$$\langle\mathrm{m}(v), A\rangle = \tfrac{1}{\|v\|^2} \cdot \langle A \cdot v, v\rangle, \tag{1.3}$$

for all $A \in \mathfrak{p}$, $v \in V\backslash\{0\}$, is called the *moment map* associated to the action of G on V. The corresponding *energy map* is given by

$$\mathrm{F} : V\backslash\{0\} \to \mathbb{R} \; ; \; v \mapsto \|\mathrm{m}(v)\|^2.$$

This scale-invariant moment map describes the infinitesimal change of the norm in V under the group action: if $(\exp(tA))_{t\in\mathbb{R}}$ is the one-parameter subgroup of G associated to $A \in \mathfrak{p}$, then the corresponding smooth action field on V is given by $X_A(v) := A \cdot v$. It follows that minimal vectors are zeroes of m, and in fact $\mathcal{M} = \mathrm{m}^{-1}(0)$, see Lemma 1.5.1. The K-invariance of the involved scalar products implies that the moment map m is K-equivariant, if we consider on $\mathfrak{p}$ the adjoint action of K. That is, we have $\mathrm{m}(k \cdot v) = k\,\mathrm{m}(v)\,k^{-1}$, for all $k \in \mathsf{K}$. The name *moment map* comes from symplectic geometry, see Section 1.3, and some authors also refer to it in the real case as the G-*gradient map*, see e.g. [HS10].

The following real version of the Kirwan–Ness stratification theorem [Kir84], [Nes84] is due to [HSS08].

Theorem 1.1.3   *There exists a finite subset* $\mathcal{B} \subset \mathfrak{p}$ *and a collection of smooth,* G-*invariant submanifolds* $\{\mathcal{S}_\beta\}_{\beta\in\mathcal{B}}$ *of* V, *with the following properties:*

(i) We have $V\backslash\{0\} = \bigcup_{\beta\in\mathcal{B}} \mathcal{S}_\beta$ *and* $\mathcal{S}_\beta \cap \mathcal{S}_{\beta'} = \emptyset$ *for* $\beta \neq \beta'$.

(ii) We have $\overline{\mathcal{S}_\beta} \setminus \mathcal{S}_\beta \subset \bigcup_{\beta'\in\mathcal{B},\,\|\beta'\|>\|\beta\|} \mathcal{S}_{\beta'}$ *(the closure taken in* $V\backslash\{0\}$).

(iii) A vector v *is contained in* $\mathcal{S}_\beta$ *if and only if the negative gradient flow of* F *starting at* v *converges to a critical point* v_C *of* F *with* $\mathrm{m}(v_C) \in \mathsf{K} \cdot \beta$.

The submanifolds $\mathcal{S}_\beta$ are called *strata*. The set of semistable vectors $V\backslash\mathcal{N}$ is nothing but the stratum $\mathcal{S}_0$ (Corollary 1.9.3). The G-invariance (and scale-invariance) of the strata is justified by the formula for the gradient of F given in Lemma 1.7.2.

It is worth mentioning that for some applications in non-Kählerian Riemannian geometry it is interesting to consider representations where $\mathcal{N} = V$. The reason for this is that from the proof of the stratification theorem one can deduce estimates for the associated moment map, which are trivial on $\mathcal{S}_0$ but highly non-trivial on $\mathcal{N}$: see [Lau10], [BL18b] and Lemma 1.9.1.

Concerning the proof of Theorem 1.1.3, we recall that the energy map F is in general not a Morse–Bott function. Nevertheless, it has the following remarkable property: the image of its critical points under the moment map consists of finitely many K-orbits $K \cdot \beta_1, \ldots, K \cdot \beta_N$. As a consequence, $\mathcal{B} = \{\beta_1, \ldots, \beta_N\}$ is a finite set. For $\beta \in \mathcal{B}$ one denotes by $\mathcal{C}_\beta$ the set of critical points of F with $m(\mathcal{C}_\beta) \subset K \cdot \beta$ and by $\mathcal{S}_\beta \subset V$ the unstable manifold of $\mathcal{C}_\beta$ with respect to the negative gradient flow of the real analytic functional F. One then needs to prove that $\mathcal{S}_\beta$ is G-invariant. The proof of the stratification theorem does in fact go the other way around: one first defines certain sets as *candidates* for being strata, and then proves that they are invariant under the negative gradient flow of F. See Section 1.7 for further details.

The chapter is organized as follows. In Section 1.2 we discuss three explicit examples that illustrate the basic concepts. In Section 1.3 we explain how our setting is related to the notion of moment map in symplectic geometry. In Section 1.4, we assume that $G = T$ is abelian, and prove that any two disjoint closed T-orbits can be separated by a single T-invariant continuous function. More generally, in Section 1.5 we show that two disjoint closed T-invariant sets can be separated by a T-invariant continuous function. In Section 1.6 we generalize this to real reductive Lie groups and complete the proof of Theorem 1.1.1. The stratification theorem, Theorem 1.1.3, is proved in Sections 1.7 and 1.8. In Section 1.9 we mention some immediate applications of the stratification theorem. Finally, the two appendices contain some well-known Lie-theoretic properties of real reductive Lie groups and their subgroups, which we prove based solely on our assumption (1.1).

Acknowledgements It is our pleasure to thank Ricardo Mendes and Marco Radeschi for fruitful discussions, and Michael Jablonski, Martin Kerin, Jorge Lauret and an anonymous referee for their helpful comments. RL was supported by the Alexander von Humboldt Foundation and the Australian Research Council.

1.2 Examples

In order to illustrate the content of Theorems 1.1.1 and 1.1.3 we describe in this section three concrete examples.

One of the simplest examples satisfying (1.1) is $\mathsf{G} = \mathbb{R}_{>0}$ acting on $V = \mathbb{R}^2$ via

$$\lambda \cdot (x, y) := (\lambda x, \lambda^{-1} y),$$

where $\lambda \in \mathbb{R}_{>0}$, $x, y \in \mathbb{R}$. The null cone is the union of the origin and four non-closed orbits, two for each axis. All the orbits corresponding to semistable vectors $(x, y) \in \mathbb{R}^2$, $x, y \neq 0$, are closed. Finally, the set of minimal vectors is $\mathcal{M} = \{(x, y) \in \mathbb{R}^2 : |x| = |y|\}$, and the moment map is $\mathrm{m}(x, y) = (x^2 - y^2)/(x^2 + y^2)$. Notice that by changing the action, say by replacing λx by $\lambda^\pi x$ on the righthand side, one obtains a representation whose complexification is not rational in the sense of algebraic geometry.

Next, consider the less trivial example of $\mathsf{G} = \mathsf{SL}_n(\mathbb{R})$ acting by conjugation on the space $V = \mathrm{Mat}(n, \mathbb{R})$ of $(n \times n)$-matrices with real entries:

$$h \cdot A := h \, A \, h^{-1},$$

where $h \in \mathsf{SL}_n(\mathbb{R})$ and $A \in \mathrm{Mat}(n, \mathbb{R})$. Notice that, as mentioned in the introduction, the image of $\mathsf{SL}_n(\mathbb{R})$ in $\mathsf{GL}(V)$ is automatically a closed subgroup.

We endow $\mathfrak{sl}_n(\mathbb{R})$ and $\mathrm{Mat}(n, \mathbb{R})$ with the usual scalar product induced from that of $\mathbb{R}^n$, $\langle A, B \rangle = \mathrm{tr}\, AB^t$. In this case, for any $A \in \mathrm{Mat}(n, \mathbb{R})$ the orbit $\mathsf{SL}_n(\mathbb{R}) \cdot A$ is closed if and only if A is *semisimple* (i.e. diagonalizable over $\mathbb{C}$). More generally, let $A = S + N$ denote the Jordan decomposition, that is $S, N \in \mathrm{Mat}(n, \mathbb{R})$, S semisimple, N nilpotent and $[S, N] = 0$. Then A is semistable if and only if $S \neq 0$. Thus the null cone consists of the set of nilpotent matrices. If $\mathfrak{p}_0 \subset \mathfrak{sl}_n(\mathbb{R})$ denotes the subset of traceless symmetric matrices, the moment map is given by

$$\mathrm{m} : \mathrm{Mat}(n, \mathbb{R}) \backslash \{0\} \to \mathfrak{p}_0, \qquad \mathrm{m}(A) = \tfrac{1}{\|A\|^2} \cdot [A, A^t].$$

The minimal vectors are the normal matrices. Moreover, the strata $\mathcal{S}_\beta$, $\beta \neq 0$, are in one-to-one correspondence with Jordan canonical forms for a nilpotent matrix N. They are parameterized by ordered partitions $n = n_1 + \cdots + n_r$, $n_1 \leq \cdots \leq n_r$. For explicit computations of the corresponding stratum labels and critical points of F for this example we refer the reader to [Lau02, §4].

Our last example, which was in fact our main motivation for writing this chapter, is as follows: let $V_n := \Lambda^2(\mathbb{R}^n)^* \otimes \mathbb{R}^n$ denote the vector space of skew-symmetric, bilinear maps $\mu : \mathbb{R}^n \times \mathbb{R}^n \to \mathbb{R}^n$, and consider the *change of basis* action of $\mathsf{G} = \mathsf{GL}_n(\mathbb{R})$ on V_n, given by

$$h \cdot \mu(\cdot, \cdot) := h\mu(h^{-1}\cdot, h^{-1}\cdot),$$

where $h \in \mathsf{GL}_n(\mathbb{R})$ and $\mu \in V_n$. This embeds $\mathsf{GL}_n(\mathbb{R})$ as a closed subgroup of $\mathsf{GL}(V_n)$, since the centre acts by multiples of the identity in V_n. Let $\{e_i\}_{i=1}^n$ denote the canonical basis of $\mathbb{R}^n$ and $\{e_i^*\}_{i=1}^n$ its dual. We endow $\mathfrak{gl}_n(\mathbb{R}) \simeq (\mathbb{R}^n)^* \otimes \mathbb{R}^n$ and V_n with the scalar products making the respective bases $\{e_i^* \otimes e_j\}_{i,j}$ and $\{e_i^* \wedge e_j^* \otimes e_k\}_{i<j;k}$ orthonormal. Notice that this scalar product is *not* the one induced from $\mathfrak{gl}(V_n)$ via the above action.

Observe now that V_n contains as an algebraic subset the so-called *variety of Lie algebras*

$$\mathcal{L}_n := \{\mu \in V_n : \mu \text{ satisfies the Jacobi identity}\}.$$

The set $\mathcal{L}_n$ is $\mathsf{GL}_n(\mathbb{R})$-invariant, and an orbit $\mathsf{GL}_n(\mathbb{R}) \cdot \mu \subset \mathcal{L}_n$ consists precisely of those Lie brackets $\tilde{\mu}$ which are isomorphic to μ. The null cone in this case is everything, since $\mathrm{Id} \in \mathfrak{gl}_n(\mathbb{R})$ acts as $-\mathrm{Id}_{V_n}$, thus any orbit contains 0 in its closure. However, by restricting to the action of $\mathsf{SL}_n(\mathbb{R})$ there exist closed orbits, and those in $\mathcal{L}_n$ correspond precisely to the semisimple Lie algebras: see [Lau03a, §8]. Furthermore, within the subset of nilpotent Lie algebras, the critical points for F correspond to *Ricci soliton* nilmanifolds [Lau01].

The key feature is now the fact that the variety $\mathcal{L}_n$ can be thought of as a parameterization of the space of left-invariant Riemannian metrics on n-dimensional Lie groups: see [Lau03b] and e.g. [BL18b]. To this end, suppose that $M^n = \mathsf{L}$ is a (connected) Lie group L with Lie algebra $\mathfrak{l} = \mathbb{R}^n$, scalar product $\langle \cdot, \cdot \rangle$ on $\mathbb{R}^n$ (representing a fixed left-invariant background metric $\bar{g}$ on L) and Lie bracket $\mu_{\mathfrak{l}} \in V_n$. Notice that $\langle \cdot, \cdot \rangle$ induces a scalar product on V_n, again denoted by $\langle \cdot, \cdot \rangle$. Then, the space of left-invariant metrics on L can be canonically identified with the orbit $\mathsf{GL}_n(\mathbb{R}) \cdot \mu_{\mathfrak{l}}$; see [BL18b, Prop. 2.3]. In particular, for $\mu_{\mathfrak{l}} \neq 0$, by Theorem 1.1.3, there exists a stratum label $\beta = \beta_{\mathfrak{l}}$ with $\mathsf{GL}_n(\mathbb{R}) \cdot \mu_{\mathfrak{l}} \subset \mathcal{S}_\beta$.

Moreover, as originally noted by J. Lauret in [Lau06], the moment map for the above action appears naturally in the formula for the Ricci endomorphism associated to $\mu \in \mathsf{GL}_n(\mathbb{R}) \cdot \mu_{\mathfrak{l}}$:

$$\mathrm{Ric}_\mu = \mathsf{M}_\mu - \tfrac{1}{2}\mathsf{B}_\mu - S(\mathrm{ad}_\mu \mathsf{H}_\mu) = \mathrm{Ric}_\mu^\star - S(\mathrm{ad}_\mu \mathsf{H}_\mu),$$

see e.g. [BL18b, eq. (18)]. Here, $\mathsf{M}_\mu = \tfrac{1}{4} \cdot \mathrm{m}(\mu) \cdot \|\mu\|^2$, $\mathrm{m} : V_n \backslash \{0\} \to \mathfrak{p}$ is the moment map and $\mathfrak{p} \subset \mathfrak{gl}_n(\mathbb{R})$ denotes the subset of symmetric matrices. The term B_μ in the above formula denotes the Killing form of $(\mathfrak{l}, \mu)$ and the third summand measures to some extent the unimodularity of the Lie group L: $H_\mu = 0$ if and only if $(\mathfrak{l}, \mu)$ is unimodular.

If the bracket $\mu \in GL_n(\mathbb{R}) \cdot \mu_\mathfrak{l}$ is gauged correctly, that is if μ is contained in the $O(n)$-slice $U_{\beta+}^{\geq 0} \subset V_{\beta+}^{\geq 0} \backslash \{0\}$ of $\mathcal{S}_\beta$, see Lemma 1.7.13, then, by the very definition of the stratum $\mathcal{S}_\beta$, one obtains by Lemma 1.9.1 the curvature estimate

$$\langle M(\mu), \beta_+ \rangle \geq 0 \,,$$

where $\beta_+ = \beta + \|\beta\|^2 \cdot Id_n$, since $tr\, m_\mu = \langle m_\mu, Id_n \rangle = -1$. This in turn implies the key curvature estimate

$$\langle Ric_\mu^\star, \beta_+ \rangle \geq 0 \,.$$

Finally let us remark, that the geometric meaning of being gauged correctly is simply a correct choice of an orthonormal basis of $(\mathfrak{l}, \langle \cdot, \cdot \rangle)$.

1.3 Comparison with Complex and Symplectic Case

In this section we connect our setting to the complex setting, and explain how this relates to the notion of the moment map from symplectic geometry.

We first recall that, despite the fact that the definition of real reductive groups used by Richardson and Slodowy relies on that of a complex reductive Lie group, it follows from [RS90, Sect. 2.2] that they all satisfy our assumption (1.1).

Now let $G \subset GL(V)$ be a closed subgroup satisfying (1.1), and let $V^\mathbb{C} = V \otimes_\mathbb{R} \mathbb{C}$ be the complexified vector space. For simplicity let us assume for the rest of this section that G is connected. After considering the natural inclusion $GL(V) \subset GL(V^\mathbb{C})$, the complexification $\mathfrak{g}^\mathbb{C}$ of the Lie algebra $\mathfrak{g}$ of G can be viewed as a Lie subalgebra of $\mathfrak{gl}(V^\mathbb{C})$. Let $G^\mathbb{C}$ be the connected Lie subgroup of $GL(V^\mathbb{C})$ with Lie algebra $\mathfrak{g}^\mathbb{C}$. We call $G^\mathbb{C}$ the *complexification* of the Lie group G: it is a complex Lie group containing G as a closed subgroup. In order to further simplify the presentation we will assume in this section that $G^\mathbb{C}$ is a closed subgroup of $GL(V^\mathbb{C})$. This is indeed the case for a semisimple G, and moreover is a true assumption in general, as the example of a one-dimensional torus action shows (the weights must be integers).

The inner product $\langle \cdot, \cdot \rangle$ on V induces as usual a Hermitian inner product $\langle\!\langle \cdot, \cdot \rangle\!\rangle$ on $V^\mathbb{C}$. After fixing a unitary basis we may identify $V^\mathbb{C} \simeq \mathbb{C}^{N+1}$

as complex vector spaces, so that $\langle\!\langle\cdot,\cdot\rangle\!\rangle$ becomes the canonical Hermitian inner product on $\mathbb{C}^{N+1}$. Since $\mathsf{G}^{\mathbb{C}}$ acts linearly on $\mathbb{C}^{N+1}$, it also acts on the corresponding projective space $\mathbb{C}P^N$. Moreover, the condition (1.1) implies that the maximal compact subgroup U of $\mathsf{G}^{\mathbb{C}}$ (whose Lie algebra is given by $\mathfrak{u} = \mathfrak{k} \oplus i\mathfrak{p}$) acts by unitary transformations on $(V^{\mathbb{C}}, \langle\!\langle\cdot,\cdot\rangle\!\rangle)$, and in particular its corresponding action on $\mathbb{C}P^N$ preserves the Fubini–Study metric and its associated fundamental 2-form ω_{FS}. For example, when $\mathsf{G} = \mathsf{GL}(V)$ we have $\mathsf{U} = \mathsf{U}(N+1)$, $N+1 = \dim_{\mathbb{C}} V^{\mathbb{C}}$. It is well known (see e.g. [Kir84, Lem. 2.5]) that for this symplectic action of the compact Lie group U on $\mathbb{C}P^N$ there exists a moment map in the sense of symplectic geometry,

$$m_{\mathsf{U}} : \mathbb{C}P^N \to \mathfrak{u}^*, \qquad m_{\mathsf{U}}(x)(A) = \tfrac{v^t A v}{2\pi\|v\|^2},$$

where $v \in \mathbb{C}^{N+1}$ is any vector over $x \in \mathbb{C}P^N$ and $A \in \mathfrak{u} \subset \mathfrak{u}(N+1)$. The inclusion $i\mathfrak{p} \subset \mathfrak{u}$ induces a restriction map $r_{\mathfrak{p}} : \mathfrak{u}^* \to (i\mathfrak{p})^* \simeq \mathfrak{p}$, the last identification being made using the scalar product on $\mathfrak{p} \subset \mathfrak{g}$. Up to a constant scalar multiple, the moment map for the action of G on V (Definition 1.1.2) satisfies

$$m = r_{\mathfrak{p}} \circ m_{\mathsf{U}} \circ \pi \big|_{V\setminus\{0\}},$$

where $\pi : V^{\mathbb{C}}\setminus\{0\} \simeq \mathbb{C}^{N+1}\setminus\{0\} \to \mathbb{C}P^N$ is the usual projection.

1.4 The Abelian Case

In this section we assume that $\mathsf{K} = \{e\}$ and that $\mathsf{G} = \mathsf{T} = \exp(\mathfrak{t})$ is an abelian group of positive-definite matrices, $\mathfrak{t} \subset \mathrm{Sym}(V, \langle\cdot,\cdot\rangle)$. Since commuting symmetric matrices can be diagonalized simultaneously, there exists an orthonormal basis $\{e_1,\ldots,e_N\}$ for V which diagonalizes the action of T. Let $\alpha_1,\ldots,\alpha_N \in \mathfrak{t}$ be the corresponding "weights", that is, the action of T on V is given by

$$\exp(\lambda) \cdot v = \left(e^{\langle\lambda,\alpha_1\rangle}v_1,\ldots,e^{\langle\lambda,\alpha_N\rangle}v_N\right), \tag{1.4}$$

where $\lambda \in \mathfrak{t}$ and $v = (v_1,\ldots,v_N) \in V$, the coordinates being with respect to the chosen basis. The scalar product $\langle\cdot,\cdot\rangle$ on $\mathfrak{t}$ is simply given by restricting the one on $\mathfrak{g}$ (see the paragraph before Definition 1.1.2).

For any subset $I \subset \mathbb{I}_N := \{1,\ldots,N\}$ we set

$$\mathfrak{t}_I := \mathrm{span}_{\mathbb{R}}\{\alpha_i : i \in I\} \subset \mathfrak{t}$$

with the convention that $t_\emptyset = \{0\}$. Moreover we define the vector sub-space $V_I := \{v \in V : v_i = 0 \text{ for all } i \notin I\}$ of V and the open subsets

$$U_I := \{v \in V_I : v_i \neq 0 \text{ for all } i \in I\},$$
$$U_I^+ := \{v \in U_I : v_i > 0 \text{ for all } i \in I\}$$

of V_I. Clearly, U_I is a dense subset of V_I, disconnected if $I \neq \emptyset$, and U_I^+ is one of its connected components. Notice that $V = \cup_{I \subset \mathbb{I}_N} U_I$ as a disjoint union.

Lemma 1.4.1 (Hilbert–Mumford criterion for abelian groups)   *Let $v \in V$ and suppose that $\mathsf{T} \cdot v$ is a non-closed orbit. Then, for any $\bar v \in \overline{\mathsf{T} \cdot v} \setminus \mathsf{T} \cdot v$ there exists $\alpha \in t$ and $g \in \mathsf{T}$ such that $\lim_{t \to \infty} \exp(t\alpha) \cdot v = g \cdot \bar v$.*

Proof   Let $I, J \subset \mathbb{I}_N$ be such that $v \in U_I$, $\bar v \in U_J$. The assumptions imply that $J \subset I$. But if $J = I$ then $\bar v \in \mathsf{T} \cdot v$. Thus $J \subsetneq I$, from which $J^C := I \backslash J \neq \emptyset$.

Let $(\lambda^{(k)}) \subset t$ be a sequence with $\lim_{k \to \infty} \exp(\lambda^{(k)}) \cdot v = \bar v$. From (1.4) we deduce that for all $j \in J$ and all $i \in J^C$ it holds that

$$\lim_{k \to \infty} \langle \lambda^{(k)}, \alpha_j \rangle = \lambda_j^\infty \in \mathbb{R} \quad \text{and} \quad \lim_{k \to \infty} \langle \lambda^{(k)}, \alpha_i \rangle = -\infty. \quad (1.5)$$

In particular, the projection of $\lambda^{(k)}$ onto t_J converges to some $\lambda^\infty \in t_J$ as $k \to \infty$. We decompose $t_I = t_J \oplus t_J^\perp$ orthogonally and for each $i \in J^C$ denote by $\alpha_i^\perp \neq 0$ the orthogonal projection of α_i onto $t_J^\perp$. We claim that 0 is not contained in the convex hull $\mathcal{C} = CH\{\alpha_i^\perp : i \in J^C\}$. Indeed, if $0 \in \mathcal{C}$, then for some $c_i > 0$ we would have that $\gamma := \sum_{i \in J^C} c_i \alpha_i \in t_J$, and by (1.5) the sequence $(\langle \lambda^{(k)}, \gamma \rangle)$ would be bounded and unbounded simultaneously, a contradiction.

Thus, let $\beta \in \mathcal{C}$ be the element of minimal (positive) norm in $\mathcal{C}$. We have that $\beta \perp \alpha_j$ for all $j \in J$, and convexity implies that $-\langle \beta, \alpha_i \rangle = -\langle \beta, \alpha_i^\perp \rangle \leq -\|\beta\|^2 < 0$ for all $i \in J^C$. We obtain $g \cdot \bar v = \lim_{t \to \infty} \exp(-t\beta) \cdot v$, where $g = \exp(-\lambda^\infty)$. $\qquad\square$

Corollary 1.4.2   *Any T-orbit has a closed T-orbit in its closure.*

Proof   By Lemma 1.4.1, an orbit in the closure has strictly smaller dimension because the direction α defining the one-parameter subgroup is a new element in the isotropy subalgebra, easily seen by applying $\exp(s\alpha)$ to $g \cdot \bar v = \lim_{t \to \infty} \exp(t\alpha) \cdot v$. The claim follows now by picking an orbit of minimal dimension. $\qquad\square$

For any subset $I \subset \mathbb{I}_N$ we denote now by

$$\Delta_I := \left\{ \sum_{i \in I} c_i \alpha_i : c_i \geq 0, \sum c_i = 1 \right\} \subset \mathfrak{t}_I$$

the convex hull of the set of weights $\{\alpha_i\}_{i \in I}$, with the convention that $\Delta_\emptyset = \{0\}$. Notice that $\dim \Delta_I = \dim \mathfrak{t}_I$ and that $\Delta_I \neq \mathfrak{t}_I$, provided that $\mathfrak{t}_I \neq \{0\}$. Since the relative interior of a point is that point, we obtain the following characterization of closed T-orbits:

Lemma 1.4.3 *For $v \in V$ let $I \subset \mathbb{I}_N$ with $v \in U_I$. Then, the orbit $\mathsf{T} \cdot v$ is closed if and only if $0 \in \mathfrak{t}$ is in the relative interior $(\Delta_I)^o$ of Δ_I.*

Proof If 0 is not in the interior of Δ_I then $\Delta_I \neq \{0\}$ and there exists a hyperplane $H \subset \mathfrak{t}_I$, such that H does not contain Δ_I and such that Δ_I does not intersect one of the two open half-spaces defined by H. Thus for one of the two unit normal vectors $\beta \in \mathfrak{t}_I$ to H we have that $\langle \beta, \alpha_i \rangle \geq 0$ for all $i \in I$, and the inequality is strict for some $i_0 \in I$. Hence $\bar{v} = \lim_{t \to \infty} \exp(-t\beta) \cdot v$ exists by (1.4), and we have that $\bar{v} \notin \mathsf{T} \cdot v$ since $\mathsf{T} \cdot v \subset U_I$ and $\bar{v} \notin U_I$ because $(\bar{v})_{i_0} = 0$.

Conversely, if the orbit is not closed then the proof of Lemma 1.4.1 implies the existence of a $\beta \in \mathfrak{t}$ with the same properties, from which it follows that 0 is not in the interior of Δ_I. $\square$

Recall that $0 \in V_\emptyset$ and that then $\Delta_\emptyset = \{0\}$. If $v \neq 0$, then there exists a non-empty $I \subset \mathbb{I}_N$ with $v \in U_I$. The above motivates now the following definition.

Definition 1.4.4 We call a non-empty subset $I \subset \mathbb{I}_N$ *admissible*, if $0 \in (\Delta_I)^o$.

Notice that if $0 \in (\Delta_I)^o$, then there exist positive coefficients $c_i > 0$, $i \in I$, such that $0 = \sum_{i \in I} c_i \alpha_i$.

In what follows we aim to show that, given two different closed orbits, there exists a continuous T-invariant function which separates them, and that moreover we can pick finitely many such functions to separate any two closed orbits. To this end, we consider the linear map

$$\phi : \mathfrak{t} \to V \; ; \; \lambda \mapsto \left(\langle \lambda, \alpha_1 \rangle, \ldots, \langle \lambda, \alpha_N \rangle \right).$$

Lemma 1.4.5 *A subset $I \subset \mathbb{I}_N$ is admissible if and only if $U_I^+ \cap \phi(\mathfrak{t})^\perp \neq \emptyset$.*

Proof Notice that $0 \in (\Delta_I)^o$ is equivalent to the existence of $w \in U_I^+$ such that $0 = \sum_{i \in I} w_i \alpha_i$. But this is equivalent to $0 = \sum_{i \in I} \langle \lambda, w_i \alpha_i \rangle = \langle \phi(\lambda), w \rangle$ for all $\lambda \in \mathfrak{t}$ from which the lemma follows.   □

Let us now fix an admissible subset $I \subset \mathbb{I}_N$, and let

$$\mathcal{B}_I := \{w^{(1)}, \dots, w^{(r)}\} \subset U_I^+ \cap \phi(\mathfrak{t})^\perp$$

be a basis for $V_I \cap \phi(\mathfrak{t})^\perp$ consisting of elements with positive entries for $i \in I$. Moreover, we scale the basis elements so that for each $w \in \mathcal{B}_I$ the sum of its entries is 1. For each $w \in \mathcal{B}_I$ consider the real-valued function

$$f_w^+ : V \to \mathbb{R} \; ; \; v \mapsto \begin{cases} \prod_{i=1}^{N} v_i^{w_i}, & \text{if } v_i > 0 \text{ for all } i \in I; \\ 0, & \text{otherwise.} \end{cases}$$

Notice that if $\mathrm{pr}_I : V \to V_I$ denotes the orthogonal projection onto V_I, then

$$f_w^+(v) = f_w^+(\mathrm{pr}_I(v)) \quad \text{and} \quad \mathrm{supp}(f_w^+) = \overline{(\mathrm{pr}_I)^{-1}(U_I^+)}.$$

Lemma 1.4.6 *The function f_w^+ is continuous and T-invariant, and $f_w^+(c \cdot v) = c \cdot f_w^+(v)$ for all $c > 0$ and $v \in V$.*

Proof Continuity is clear, since $w_i \geq 0$ for all $i \in \mathbb{I}_N$. To prove T-invariance, first observe that $V_I \backslash U_I^+$ is a T-invariant set. On the other hand, for $\tilde{v} = \mathrm{pr}_I(v) \in U_I^+$ we compute directly using (1.4) and the fact that $w \in \phi(\mathfrak{t})^\perp$:

$$f_w^+(\exp(\lambda) \cdot v) = \prod_{i=1}^{N} (e^{\langle \lambda, \alpha_i \rangle} v_i)^{w_i} = e^{\langle \lambda, \sum w_i \alpha_i \rangle} f_w^+(v)$$
$$= e^{\langle \phi(\lambda), w \rangle} f_w^+(v) = f_w^+(v).$$

This shows the claim.   □

Lemma 1.4.7 *If $\mathcal{O}_1 \neq \mathcal{O}_2$ are two closed T-orbits in U_I^+ then there exists $w \in \mathcal{B}_I$ such that $f_w^+(\mathcal{O}_1) \neq f_w^+(\mathcal{O}_2)$.*

Proof Assume that this is not the case. Let $v \in \mathcal{O}_1$, $\bar{v} \in \mathcal{O}_2$, and consider the map $\log_I : U_I^+ \to V_I$, assigning to each vector $v \in U_I^+$ the vector $\log_I(v) \in V_I$ whose ith entry is $\log(v_i)$, for all $i \in I$. For each $w \in \mathcal{B}_I$ we have that

$$\langle w, \log_I(v) \rangle = \log f_w^+(v) = \log f_w^+(\bar{v}) = \langle w, \log_I(\bar{v}) \rangle.$$

Hence $\log_I(v) - \log_I(\bar{v}) \perp V_I \cap \phi(\mathsf{t})^\perp$. In other words, $\log_I(v) - \log_I(\bar{v}) \in V_I \cap \phi(\mathsf{t})$, from which it immediately follows that $v \in \mathsf{T} \cdot \bar{v}$. Contradiction.

$\square$

In order to separate orbits that lie in different connected components of U_I, we argue as follows: for each choice of signs $\sigma \in \{\pm 1\}^N$, let $T_\sigma : V \to V$ be the T-equivariant linear map that changes the sign of each coordinate according to σ. For any connected component U_I^c of U_I there exists $\sigma \in \{\pm 1\}^N$ such that $T_\sigma(U_I^c) = U_I^+$. We then define the functions f_w^σ, $w \in \mathcal{B}_I$, by $f_w^\sigma = f_w^+ \circ T_\sigma$. Clearly, they satisfy Lemma 1.4.6, and they separate orbits in the corresponding connected component U_I^c of U_I. We consider now the finite set of continuous, T-invariant, real-valued functions on V:

$$\mathcal{F} := \left\{ f_w^\sigma \; : \; w \in \mathcal{B}_I, \sigma \in \{\pm 1\}^N, I \subset \mathbb{I}_N \text{ admissible} \right\}.$$

Notice that $\mathrm{supp}(f_w^\sigma) \cap V_I = \overline{U_I^c}$.

Proposition 1.4.8 (Separation of closed orbits) *Let $L = |\mathcal{F}| \in \mathbb{N}$. Then there exists a continuous, T-invariant map $\Phi : V \to \mathbb{R}^L$, such that $\Phi(\mathcal{O}_1) \neq \Phi(\mathcal{O}_2)$ for any two closed orbits $\mathcal{O}_1 \neq \mathcal{O}_2$.*

Proof The coordinate functions of the map Φ are of course just functions f_w^σ in $\mathcal{F}$.

First assume that $\mathcal{O}_1, \mathcal{O}_2 \subset U_I$. If they belong to the same connected component of U_I, which without loss of generality we may assume to be U_I^+, then they are separated by Lemma 1.4.7. On the other hand, if this is not the case then the existence of a separating function follows immediately from $\mathrm{supp}(f_w^\sigma) \cap V_I = \overline{U_I^c}$.

We are left with the case $\mathcal{O}_1 \subset U_I$, $\mathcal{O}_2 \subset U_J$ with $I \neq J$. Suppose that $j \in J \backslash I$. Then, there exists $f_w^\sigma \in \mathcal{F}$ with $w \in \mathcal{B}_J$ such that $f_w^\sigma(\mathcal{O}_2) > 0$ and $f_w^\sigma(\mathcal{O}_1) = 0$.

$\square$

1.5 Separation of Closed T-Invariant Sets

The linear action of the real reductive group $\mathsf{G} \subset \mathsf{GL}(V)$ on $(V, \langle \cdot, \cdot \rangle)$ provides us with a smooth *action field* for any $A \in \mathfrak{g}$:

$$X_A(v) := \frac{d}{dt}\Big|_{t=0} \exp(tA) \cdot v = A \cdot v.$$

Notice that for an initial value $v_0 \in V$ the curve $v(t) := \exp(tA) \cdot v_0$ is the corresponding integral curve of X_A. Recall also that we denoted by

$\mathfrak{g} = \mathfrak{k} \oplus \mathfrak{p}$ the Cartan decomposition of the Lie algebra $\mathfrak{g}$ of G. Then, for $A \in \mathfrak{k}$ the vector fields X_A are Killing fields, meaning that their flows consist of isometries.

For fixed $A \in \mathfrak{p}$ and $v \in V$ we let

$$d(t) := d_{A,v}(t) := \|\exp(tA) \cdot v\|^2$$

denote the square of the distance function to the origin along $\exp(tA) \cdot v$.

Lemma 1.5.1 (Convexity of the distance function) *Let $A \in \mathfrak{p}$ and $v \in V$ be given. Then $d'(0) = 2 \cdot \langle A \cdot v, v \rangle$ and $d''(t) = 4 \cdot \|A \cdot \exp(tA) \cdot v\|^2$.*

Proof We have $d'(t) = 2 \cdot \langle A \cdot \exp(tA) \cdot v, \exp(tA) \cdot v \rangle$. From this the claim follows immediately using $A^t = A$. □

Corollary 1.5.2 *Let $A \in \mathfrak{p}$, $v \in V$ and suppose that $\lim_{t \to \infty} \exp(tA) \cdot v = \bar{v} \neq v$ exists. Then for all $t \in \mathbb{R}$ one has $\|\exp(tA) \cdot v\| > \|\bar{v}\|$ for all $t \in \mathbb{R}$.*

Let us mention that, for a fixed v, the function $\mathsf{G} \to \mathbb{R}$, $\exp(tA) \mapsto d(t)$ is usually called a *Kempf–Ness function* in the literature.

Next, set $\mathsf{T} = \exp(\mathfrak{t})$, with $\mathfrak{t} \subset \mathrm{Sym}(V, \langle \cdot, \cdot \rangle)$ abelian, and let

$$\mathcal{M}_\mathsf{T} := \{v \in V : \|v\| \leq \|t \cdot v\| \text{ for all } t \in \mathsf{T}\}$$

denote the set of minimal vectors for the T-action. For any $v \in \mathcal{M}_\mathsf{T}$ the orbit $\mathsf{T} \cdot v$ is closed by Lemma 1.4.1 and Corollary 1.5.2. Conversely, for a closed T-orbit $\mathcal{O}$, the closest point to the origin in $\mathcal{O}$ belongs to $\mathcal{M}_\mathsf{T}$.

Notice that by Lemma 1.5.1 the condition of $v \in V$ being the closest point to the origin of $\mathsf{T} \cdot v$ is equivalent to $\langle \lambda \cdot v, v \rangle = 0$ for all $\lambda \in \mathfrak{t}$. Since this condition is linear in λ and polynomial in v, $\mathcal{M}_\mathsf{T}$ is a closed subset of V.

Next, we show that the continuous, T-invariant map Φ defined in the proof of Proposition 1.4.8 satisfies the following uniform estimate.

Lemma 1.5.3 *There exists $C > 0$ such that $\|v\| \leq C \cdot \|\Phi(v)\|$ for all $v \in \mathcal{M}_\mathsf{T}$.*

Proof Recall that $\Phi(c \cdot v) = c \cdot \Phi(v)$ for all $c > 0$ and $v \in V$. Assume that there exists a sequence $(v_k)_{k \in \mathbb{N}} \subset \mathcal{M}_\mathsf{T}$, $\|v_k\| \equiv 1$, with $\lim_{k \to \infty} \Phi(v_k) = 0$. For a subsequential limit $\bar{v} \in \mathcal{M}_\mathsf{T}$, $\|\bar{v}\| = 1$, we have $\Phi(\bar{v}) = 0$. But the orbit $\mathsf{T} \cdot \bar{v}$ is closed and non-trivial, hence contained in some U_I. As a consequence, there exists one function $f \in \mathcal{F}$ with $f(\mathsf{T} \cdot \bar{v}) > 0$. But this contradicts $\Phi(\bar{v}) = 0$. □

Corollary 1.5.4 *The null cone $\{v \in V : 0 \in \overline{\mathsf{T} \cdot v}\}$ is a closed subset.*

Proof We have that $0 \in \overline{\mathsf{T} \cdot v}$ if and only if $\Phi(v) = 0$. $\qquad \square$

Corollary 1.5.5 (Separation of closed T-invariant subsets) *Let Z_1, Z_2 be two closed, disjoint, T-invariant subsets of V. Then, there exists a continuous T-invariant function $f : V \to [0, 1]$ such that $f|_{Z_1} \equiv 0$ and $f|_{Z_2} \equiv 1$.*

Proof By Urysohn's lemma it is enough to show that $A_1 := \Phi(Z_1)$, $A_2 := \Phi(Z_2)$ are closed, disjoint subsets, since then we can set $f := d \circ \Phi$, where $d : \mathbb{R}^L \to [0, 1]$ is a continuous function with $d|_{A_1} \equiv 0$ and $d|_{A_2} \equiv 1$. See e.g. [Bre93, Chap. I, Lem. 10.2].

To see that $\Phi(Z_1)$ is closed consider a sequence $(\Phi(v_k))_{k \in \mathbb{N}} \subset \Phi(Z_1)$ converging to some $\Phi_0 \in \mathbb{R}^L$. Since Z_1 is closed and T-invariant, we may assume that $v_k \in \mathcal{M}_{\mathsf{T}}$ for all k (recall that any T-orbit has a closed T-orbit in its closure by Corollary 1.4.2, and that Φ is continuous and T-invariant). By Lemma 1.5.3 we then have that (v_k) is bounded, thus it subconverges to some $\bar{v} \in Z_1$. Now $\Phi_0 = \Phi(\bar{v})$, as we wanted to show. Clearly, $\Phi(Z_2)$ is also closed.

If $v_1 \in Z_1$, $v_2 \in Z_2$ are such that $\Phi(v_1) = \Phi(v_2)$, then as above we may assume that $v_1, v_2 \in \mathcal{M}$, so that the corresponding T-orbits are closed. But this contradicts Proposition 1.4.8. $\qquad \square$

1.6 The General Case of Real Reductive Groups

We now focus on proving Theorem 1.1.1. The idea is to reduce it to the abelian case, already settled above. More precisely, let us fix a maximal abelian subalgebra $\mathfrak{t} \subset \mathfrak{p}$, and let $\mathsf{T} := \exp(\mathfrak{t})$ be the corresponding connected abelian Lie subgroup of G. It will be proved in Corollary 1.10.2 that one has $\mathsf{G} = \mathsf{KTK}$, which in some sense says that the non-compactness in G is abelian.

We aim to prove that orbits containing minimal vectors are closed. Recall that we only consider in V the standard vector space topology. Using the convexity of orbits of one-parameter subgroups (Lemma 1.5.1), as a first step we prove the following

Lemma 1.6.1 *Let $v_{\min} \in \mathcal{M} \subset V$ be a minimal vector with $\|v_{\min}\| = 1$ and assume that $\mathsf{G} \cdot v_{\min}$ is not closed. Then, there exists $\epsilon = \epsilon_{v_{\min}} > 0$, such that $\|v\| \geq 1 + \epsilon$ for any $v \in \overline{\mathsf{G} \cdot v_{\min}} \backslash \mathsf{G} \cdot v_{\min}$.*

Proof As we will show below, $\mathsf{K} \cdot v_{\min}$ admits an open, bounded neighbourhood U in $\mathsf{G} \cdot v_{\min}$ such that the following holds: the closure $\overline{U}$ of U in V satisfies $\overline{U} \subset \mathsf{G} \cdot v_{\min}$ and there exists $\epsilon > 0$, such that for all $v \in \overline{\mathsf{G} \cdot v_{\min}} \backslash U$ we have $\|v\| \geq 1 + \epsilon$. It follows then that any $v \in \overline{\mathsf{G} \cdot v_{\min}} \backslash \mathsf{G} \cdot v_{\min}$ satisfies $\|v\| \geq 1 + \epsilon$.

To show this claim, let $\mathfrak{g}_{v_{\min}} \subset \mathfrak{g}$ denote the isotropy subalgebra of $v_{\min}$ and $\mathfrak{p}^{\perp}_{v_{\min}}$ the orthogonal complement of $\mathfrak{g}_{v_{\min}} \cap \mathfrak{p}$ in $\mathfrak{p}$ with respect to the given scalar product on $\mathfrak{g}$. Then $\psi : \mathfrak{p}^{\perp}_{v_{\min}} \to \mathsf{G} \cdot v_{\min}$; $A \mapsto \exp(A) \cdot v_{\min}$ is a local diffeomorphism close to $0 \in \mathfrak{p}^{\perp}_{v_{\min}}$, such that its image intersects $\mathsf{K} \cdot v_{\min}$ transversally. Most importantly, by Lemma 1.5.1 assuming that $\|A\| = 1$ we know that the function $d(t) = d_{A,v_{\min}}(t)$ along $\exp(t \cdot A) \cdot v_{\min}$ satisfies $d'(0) = 0$ and $d''(0) = \|A \cdot v_{\min}\|^2$. Since $A \in \mathfrak{p}^{\perp}_{v_{\min}}$ and $\|A\| = 1$ there exists $\delta(v_{\min}) > 0$ such that $d''(0) \geq \delta(v_{\min}) > 0$ for all such A. Since $d''(t) = \|A \cdot \exp(t \cdot A) \cdot v_{\min}\|^2$ we deduce furthermore that there exist $t_{\delta(v_{\min})} > 0$ and $\epsilon_{v_{\min}} > 0$ such that for all $t \in \mathbb{R}$ with $|t| \geq t_{\delta(v_{\min})}$ and for all $A \in \mathfrak{p}^{\perp}_{v_{\min}}$ with $\|A\| = 1$ we have $d_{A,v_{\min}}(t) \geq (1 + \epsilon_{v_{\min}})^2$.

We consider now the map $\Psi : \mathfrak{p}^{\perp}_{v_{\min}} \times U^{\mathsf{K} \cdot v_{\min}}_{v_{\min}} \to \mathsf{G} \cdot v_{\min}$; $(A, v) \mapsto \exp(A) \cdot v$ for an open neighbourhood $U^{\mathsf{K} \cdot v_{\min}}_{v_{\min}}$ of $v_{\min}$ in $\mathsf{K} \cdot v_{\min}$. Again, we may assume that Ψ is a local diffeomorphism from $B_{t_{\delta(v_{\min})}}(0) \times U^{\mathsf{K} \cdot v_{\min}}_{v_{\min}}$ to its image $U_{v_{\min}} \subset \mathsf{G} \cdot v$. Precisely as above, we deduce that there exist $t_{\delta(v_{\min})} > 0$ and $\epsilon_{v_{\min}} > 0$ such that for all $t \in \mathbb{R} \backslash (-t_{\delta(v_{\min})}, t_{\delta(v_{\min})})$ and for all $A \in \mathfrak{p}^{\perp}_{v_{\min}}$ with $\|A\| = 1$ and for all $v \in U^{\mathsf{K} \cdot v_{\min}}_{v_{\min}}$ we have that $d_{A,v}(t) \geq (1 + \epsilon_{v_{\min}})^2$.

Recall that $\mathsf{K} \cdot v_{\min} \subset \mathcal{M}$, as K acts isometrically. As $\mathsf{K} \cdot v_{\min}$ is compact, there exist finitely many such open neighbourhoods $U_{k_1 \cdot v_{\min}}, \ldots, U_{k_N \cdot v_{\min}}$, such that $U = \cup_{i=1}^{N} U_{k_i \cdot v_{\min}}$ is an open neighbourhood of $\mathsf{K} \cdot v_{\min}$, $k_1 = e, \ldots, k_N \in \mathsf{K}$. We may of course assume that each of the open subsets $U^{\mathsf{K} \cdot v_{\min}}_{k_i \cdot v_{\min}}$ contains a compact subset $A^{\mathsf{K} \cdot v_{\min}}_{k_i \cdot v_{\min}}$, $i = 1, \ldots, N$, such that the interior of these sets still cover $\mathsf{K} \cdot v_{\min}$, that is $\mathsf{K} \cdot v_{\min} \subset U^{\mathsf{K}} := \cup_{i=1}^{N} (A^{\mathsf{K} \cdot v_{\min}}_{k_i \cdot v_{\min}})^o$. This then shows the above claim. $\square$

In the following lemma we show that a non-closed orbit has a closed, G-invariant subset in its closure, intersecting the orbit trivially.

Lemma 1.6.2 *Let $v \in V$ and suppose that the orbit $\mathsf{G} \cdot v$ is not closed. Then there exists $\bar{v} \in \overline{\mathsf{G} \cdot v} \backslash \mathsf{G} \cdot v$, such that $Y = \overline{\mathsf{G} \cdot \bar{v}}$ satisfies $Y \cap \mathsf{G} \cdot v = \emptyset$.*

Proof Notice first that the set $Z := \overline{\mathsf{G} \cdot v}$ is closed and G-invariant, hence it contains a minimal vector $v_{\min} \in \mathcal{M}$. Since by assumption $\mathsf{G} \cdot v$ is not closed, there are now two cases to be considered: $v_{\min} \in \mathsf{G} \cdot v$ and $v_{\min} \in Z \backslash \mathsf{G} \cdot v$.

If $v_{\min} \in G \cdot v$ by Lemma 1.6.1 there exists $\epsilon > 0$ such that for any $\bar{v} \in Z \backslash G \cdot v$ we have $\|\bar{v}\| \geq \|v_{\min}\| + \epsilon$. In particular, the same estimate holds true on $Y = \overline{G \cdot \bar{v}}$. Since $G \cdot v \cap Y$ is G-invariant, this intersection must be empty, since otherwise it would contain $G \cdot v$, hence $v_{\min}$ contradicts the above estimate.

In the case $v_{\min} \in Z \backslash G \cdot v$, we set $Y := \overline{G \cdot v_{\min}}$. Again, by Lemma 1.6.1 any element in $\overline{G \cdot v_{\min}} \backslash G \cdot v_{\min}$ must satisfy $\|v\| \geq \|v_{\min}\| + \epsilon$ for some $\epsilon > 0$. Thus $G \cdot v \cap Y = \emptyset$, since $v_{\min} \in \overline{G \cdot v}$ implies that $G \cdot v$ must contain vectors of norm $\|v_{\min}\| + \epsilon/2$. $\qquad\square$

We can now provide a proof of the Hilbert–Mumford criterion in this setting. The following argument is due to Richardson (see also [Bir71, Thm. 5.2]).

Lemma 1.6.3 (Hilbert–Mumford criterion for real reductive groups) *Let $v \in V$. If the orbit $G \cdot v$ is not closed, then for some $\alpha \in \mathfrak{p}$ the limit $\lim_{t \to \infty} \exp(t\alpha) \cdot v$ exists.*

Proof Let $T = \exp(\mathfrak{t}) \subset G$ be a maximal abelian subalgebra, $\mathfrak{t} \subset \mathfrak{p}$, and choose $\bar{v} \in \overline{G \cdot v} \backslash G \cdot v$ such that $Y := \overline{G \cdot \bar{v}}$ satisfies $Y \cap G \cdot v = \emptyset$; see Lemma 1.6.2. We will show below that there then exists $g \in G$, $k \in K$ and $\alpha \in \mathfrak{t}$ such that $g \cdot \bar{v} = \lim_{t \to \infty} \exp(t\alpha) \cdot (k \cdot v)$. Notice that the lemma follows, since for $\alpha' = k^{-1} \alpha k$, $\bar{v}' = k^{-1} \cdot g \cdot \bar{v}$ we deduce $\bar{v}' = \lim_{t \to \infty} \exp(t\alpha') \cdot v$.

To prove the above claim, suppose on the contrary that $Y \cap \overline{T \cdot k \cdot v} = \emptyset$ for all $k \in K$. Since Y is closed and T-invariant, by Corollary 1.5.5, for each $k \in K$ there exists a continuous T-invariant function $f_k : V \to \mathbb{R}$ with $f_k(\overline{T \cdot (k \cdot v)}) = 1$ and $f_k(Y) \equiv 0$. By continuity, each k has an open neighbourhood U_k in K such that $f_k(T \cdot U_k \cdot v) > 1/2$. Since K is compact, we may extract a finite number of such functions $f_{k_1}, \ldots, f_{k_R}$ such that for $f = f_{k_1} + \cdots + f_{k_R}$ we have that $f((TK) \cdot v) > 1/2$ and $f(Y) \equiv 0$. Since $K \cdot \bar{v} \subset Y$, we deduce $\overline{TK \cdot v} \cap K \cdot \bar{v} = \emptyset$, thus $\bar{v} \notin K(\overline{TK \cdot v})$. Using $G = KTK$ and $\overline{G \cdot v} \subset K(\overline{TK \cdot v})$, we obtain $\bar{v} \notin \overline{G \cdot v}$, a contradiction. To see why one has $\overline{G \cdot v} \subset K(\overline{TK \cdot v})$, observe that if $w = \lim_{i \to \infty} g_i \cdot v$ with $g_i = k_i t_i k_i' \in G = KTK$, by compactness of K one may assume that $k_i \to k_\infty$ and hence $w = k_\infty \cdot \lim_{i \to \infty} t_i k_i' \cdot v \in K(\overline{TK \cdot v})$. $\qquad\square$

Corollary 1.6.4 *Let $v \in V$. Then, the orbit $G \cdot v$ is closed if and only if there exists $v_m \in G \cdot v$ with $\|v_m\| \leq \|g \cdot v_m\|$ for all $g \in G$.*

Proof If the orbit is closed then the existence of v_m is clear. Conversely, assume that there exists a minimal vector v_m but $G \cdot v_m$ is not closed.

Notice that by continuity we also have that $\|v_m\| \le \|\bar{v}\|$ for all $\bar{v} \in \overline{\mathsf{G} \cdot v_m}$. Lemma 1.6.3 together with Corollary 1.5.2 give a contradiction. $\quad\square$

Lemma 1.6.5   *Any orbit* $\mathsf{G} \cdot v$ *contains exactly one closed orbit in its closure.*

Proof   Suppose that $\bar{v} \in \overline{\mathsf{G} \cdot v} \backslash \mathsf{G} \cdot v$ has minimal norm. Then, by Corollary 1.6.4 the orbit $\mathsf{G} \cdot \bar{v} \subset \overline{\mathsf{G} \cdot v}$ is closed. Suppose furthermore that there is a second closed orbit $\mathsf{G} \cdot \bar{w} \subset \overline{\mathsf{G} \cdot v} \backslash \mathsf{G} \cdot \bar{v}$. Then, by Corollary 1.5.5 there exists a continuous, T-invariant function $f : V \to [0,1]$ with $f(\mathsf{G} \cdot \bar{v}) = 1$ and $f(\mathsf{G} \cdot \bar{w}) = 0$. Now let $(w_i)_{i \in \mathbb{N}} \subset \mathsf{G} \cdot v$ be a sequence with $\lim_{i \to \infty} w_i = \bar{w}$. By the claim in the proof of Lemma 1.6.3, for each i there exist $g_i \in \mathsf{G}$, $k_i \in \mathsf{K}$ and $\alpha_i \in \mathfrak{t}$ such that $\lim_{t \to \infty} \exp(t\alpha_i) \cdot (k_i \cdot w_i) = g_i \cdot \bar{v}$. Since f is T-invariant and continuous, we deduce $f(k_i \cdot w_i) = 1$ for all $i \in \mathbb{N}$. On the other hand, the sequence $(k_i \cdot w_i)$ subconverges to a vector in $\mathsf{G} \cdot \bar{w}$, hence $f(k_i \cdot w_i) \to 0$ along that subsequence. Contradiction. $\quad\square$

Proof of Theorem 1.1.1   For (i): if v is a minimal vector then by Lemma 1.5.1 and the decomposition $\mathsf{G} = \mathsf{KTK}$ the set of minimal vectors in the closed orbit $\mathsf{G} \cdot v$ is precisely $\mathsf{K} \cdot v$. For (ii): let $\mathsf{G} \cdot v$ be a non-closed orbit, and pick $\bar{v} \in \overline{\mathsf{G} \cdot v}$ of minimal norm. Then, by Corollary 1.6.4 the orbit $\mathsf{G} \cdot \bar{v}$ is closed. From the proof of Lemma 1.6.3 we know that there exists a one-parameter subgroup such that $\lim_{t \to \infty} \exp(t\alpha') \cdot v \in \mathsf{G} \cdot \bar{v}$. The third item is precisely Lemma 1.6.5. For (iv): let $(v_i) \subset V$ be a sequence with $0 \in \overline{\mathsf{G} \cdot v_i}$ for all i such that $v_i \to v_\infty$. By (ii), it follows that there exist maximal abelian subalgebras $\mathfrak{t}_i \subset \mathfrak{p}$ such that $0 \in \overline{\mathsf{T}_i \cdot v_i}$, where $\mathsf{T}_i := \exp(\mathfrak{t}_i)$. Since by Proposition 1.10.1 all such T_i are conjugate by elements in K, we may assume (after possibly changing v_∞ by $k \cdot v_\infty$, $k \in \mathsf{K}$) that $\mathsf{T}_i \equiv \mathsf{T}$ is constant. The result now follows from Corollary 1.5.4. $\quad\square$

1.7 Stratification

In this section we provide a proof for Theorem 1.1.3. This will be done by using the energy map $\mathrm{F}(v) = \|\mathrm{m}(v)\|^2$ associated to the moment map $\mathrm{m} : V \backslash \{0\} \to \mathfrak{p}$ (see (1.3)) as a Morse function. The map F has the following remarkable property: its critical points are mapped under the moment map onto finitely many K-orbits $\mathsf{K} \cdot \beta_1, \ldots, \mathsf{K} \cdot \beta_N$ in $\mathfrak{p}$ (see Lemma 1.8.1). We set $\mathcal{B} := \{\beta_1, \ldots, \beta_N\}$, and for $\beta \in \mathcal{B}$ we let $\mathcal{C}_\beta$ denote the set of critical points of F with $m(\mathcal{C}_\beta) \subset K \cdot \beta$. It will turn out that the stratum $\mathcal{S}_\beta \subset V \backslash \{0\}$ is the unstable manifold corresponding to $\mathcal{C}_\beta$.

In order to briefly describe how the strata are constructed, let us fix $\beta \in \mathcal{B}$ and let G_β denote the centralizer of β in G. It turns out that critical points v_C of F with $m(v_C) = \beta$ correspond to minimal vectors for the action of a real reductive subgroup $H_\beta \subset G_\beta$ with Lie algebra $\mathfrak{h}_\beta = \beta^\perp$ on a certain subspace $V_{\beta+}^0 \subset V$. This makes it possible to apply Theorem 1.1.1 for the restricted action. Inspired by the negative directions of the Hessian of F at v_C (Lemma 1.8.2) and the fact that F is K-invariant, one defines the stratum $\mathcal{S}_\beta$ as in Definition 1.7.8. After proving that this is a smooth submanifold (Proposition 1.8.3), it will follow that $\mathcal{S}_\beta$ is invariant under the negative gradient flow of F.

Throughout the proof it will turn out to be extremely convenient to break the K-symmetry and work with a fixed β as opposed to the entire K-orbit $K \cdot \beta$. Thus, the crucial results will be proved on the *slice* $V_{\beta+}^{\geq 0}$ (see (1.7)), and then extended to all of $\mathcal{S}_\beta$ by K-invariance. This forces us to work with a certain *parabolic subgroup* Q_β of G associated to β (Definition 1.7.3) which preserves the subspace $V_{\beta+}^{\geq 0}$. Well-known properties of Q_β and of other subgroups of G adapted to β will be needed throughout this section. They will be proved in the Appendix, Section 1.10.2.

Recall that $G \subset GL(V)$ is a closed subgroup satisfying (1.1) (see also Appendix, Section 1.10.1). In particular, $\mathfrak{g} \subset \mathfrak{gl}(V)$ is a Lie subalgebra, with the property that $A^t \in \mathfrak{g}$ for all $A \in \mathfrak{g}$. Recall also that $\mathfrak{g}$ has a Cartan decomposition $\mathfrak{g} = \mathfrak{k} \oplus \mathfrak{p}$, where $\mathfrak{k} \subset \mathfrak{so}(V)$ and $\mathfrak{p} \subset \mathrm{Sym}(V)$.

Notation 1.7.1 For $\beta \in \mathfrak{p}$ we set $\beta^+ := \beta - \|\beta\|^2 \cdot \mathrm{Id}_V \in \mathrm{Sym}(V)$.

This notation appears naturally in the formula for the gradient of F:

Lemma 1.7.2 *The gradient of the energy map* $F : V \backslash \{0\} \to \mathbb{R}$ *is given by*

$$(\nabla F)_v = \frac{4}{\|v\|^2} \cdot m(v)^+ \cdot v.$$

Proof Since F is scale invariant, we have $(\nabla F)_v \perp v$ for $v \in V \backslash \{0\}$. So let $w \perp v$ with $\|v\| = \|w\|$ and set $v(t) = \cos(t)v + \sin(t)w$. Then by (1.3) for $A \in \mathfrak{p}$ we have

$$\langle (d\,m)_{v(t)} \, v'(t), A \rangle = \frac{2}{\|v\|^2} \cdot \langle A \cdot v(t), v'(t) \rangle. \tag{1.6}$$

Thus $\langle (\nabla F)_v, w \rangle = \frac{4}{\|v\|^2} \cdot \langle m(v) \cdot v, w \rangle$ and the lemma follows since again by (1.3) we have $\frac{1}{\|v\|^2} \langle m(v) \cdot v, v \rangle \cdot v = \|m(v)\|^2 \cdot v$. $\qquad \square$

Since $\beta \in \mathfrak{p}$ is a symmetric endomorphism, we may decompose V as a sum of eigenspaces $V_{\beta^+}^r$ of β^+ corresponding to its eigenvalues $r \in \mathbb{R}$. Of major importance will be $V_{\beta^+}^0$, the kernel of β^+, and the sum of the non-negative eigenspaces

$$V_{\beta^+}^{\geq 0} := \bigoplus_{r \geq 0} V_{\beta^+}^r . \tag{1.7}$$

The reason is that the above-mentioned Hessian is non-negative on $V_{\beta^+}^{\geq 0}$ in every critical point $v_C \in \mathcal{C}_\beta$ of F: see Lemma 1.8.2 below. Let us explicitly mention, though, that at this point β is arbitrary.

Analogous to the subspaces $V_{\beta^+}^0$, $V_{\beta^+}^{>0} := \bigoplus_{r>0} V_{\beta^+}^r$ and $V_{\beta^+}^{\geq 0}$, we have certain special subgroups of G. To define them, consider the symmetric endomorphism

$$\operatorname{ad}(\beta) : \mathfrak{g} \to \mathfrak{g} \; ; \; A \mapsto [\beta, A] .$$

Using the eigenspace decomposition $\mathfrak{g} = \bigoplus_{r \in \mathbb{R}} \mathfrak{g}_r$ of $\operatorname{ad}(\beta)$, we denote $\ker(\operatorname{ad}(\beta))$ by $\mathfrak{g}_\beta := \mathfrak{g}_0$, set $\mathfrak{u}_\beta := \bigoplus_{r>0} \mathfrak{g}_r$ and $\mathfrak{q}_\beta = \mathfrak{g}_\beta \oplus \mathfrak{u}_\beta$.

Definition 1.7.3   We denote by $\mathsf{G}_\beta := \{g \in \mathsf{G} : g\beta g^{-1} = \beta\}$ the centralizer of β in G, by $\mathsf{U}_\beta := \exp(\mathfrak{u}_\beta)$, and we set $\mathsf{Q}_\beta := \mathsf{G}_\beta \mathsf{U}_\beta$.

It turns out that $\mathsf{G}_\beta, \mathsf{U}_\beta$ and Q_β are closed subgroups of G, with Lie algebras $\mathfrak{g}_\beta, \mathfrak{u}_\beta$ and $\mathfrak{q}_\beta$, respectively. We refer the reader to the Appendix, Section 1.10.2, for more details and properties of these groups.

Lemma 1.7.4   *The subspace $V_{\beta^+}^{\geq 0}$ is Q_β-invariant.*

Proof   For $A \in \mathfrak{g}$, if the action field $X_A(v) = A \cdot v$ is tangent to a subspace W of V, then the integral curves of X_A starting tangent to W cannot leave W. Thus it suffices to show that for all $A \in \mathfrak{q}_\beta$ and $v \in V_{\beta^+}^{\geq 0}$ we have that $A \cdot v \in V_{\beta^+}^{\geq 0}$.

By linearity we may assume that $v \in V_{\beta^+}^r$, $r \geq 0$, and that A is an eigenvector of $\operatorname{ad}(\beta) : \mathfrak{g} \to \mathfrak{g}$ with eigenvalue $\lambda_A \geq 0$. Then,

$$\beta^+ \cdot (A \cdot v) = [\beta^+, A] \cdot v + A \cdot \beta^+ \cdot v = \lambda_A(A \cdot v) + r(A \cdot v), \quad (1.8)$$

thus $A \cdot v \in V_{\beta^+}^{r+\lambda_A} \subset V_{\beta^+}^{\geq 0}$ and the lemma follows. $\square$

The linear orthogonal projection $p_\beta : V_{\beta^+}^{\geq 0} \to V_{\beta^+}^0$ will be important later on.

Lemma 1.7.5   *The orthogonal projection $p_\beta : V_{\beta^+}^{\geq 0} \to V_{\beta^+}^0$ satisfies the formula*

$$p_\beta(v) = \lim_{t \to \infty} \exp(-t\beta^+) \cdot v; \tag{1.9}$$

it is G_β*-equivariant, and for each* $v \in V^0_{\beta+}$ *the fibre* $p^{-1}_\beta(v)$ *is* U_β-*invariant.*

Proof Let $v = \sum_{r \geq 0} v_r$ with $v^r \in V^r_{\beta+}$. Since the action of $\exp(-t\beta^+)$ on $V^r_{\beta+}$ is simply given by scalar multiplication by e^{-tr}, we immediately obtain (1.9). To show the G_β-equivariance, let $v \in V^{\geq 0}_{\beta+}$ and $h \in \mathsf{G}_\beta$. Since $[\mathsf{G}_\beta, \exp(\beta^+)] = 0$ we deduce

$$p_\beta(h \cdot v) = \lim_{t \to \infty} \exp(-t\beta^+) \cdot (h \cdot v) = h \cdot \lim_{t \to \infty} \exp(-t\beta^+) \cdot v = h \cdot p_\beta(v).$$

To prove that the above fibre $p^{-1}_\beta(v)$ is U_β-invariant, recall that for $u \in \mathsf{U}_\beta$ we have $\lim_{t \to \infty} \exp(-t\beta^+) \cdot u \cdot \exp(t\beta^+) = e$ by Lemma 1.10.6. It follows that

$$p_\beta(u \cdot v) = \lim_{t \to \infty} \exp(-t\beta^+) \cdot u \cdot \exp(t\beta^+) \cdot \exp(-t\beta^+) \cdot v = p_\beta(v),$$

which shows the claim. □

Remark 1.7.6 From the proof of the previous lemma it also follows that for an arbitrary $v \in V$, the limit in (1.9) exists if and only if $v \in V^{\geq 0}_{\beta+}$.

Before introducing the strata $\mathcal{S}_\beta$ algebraically we need to consider one further group related to the Q_β-action on $V^{\geq 0}_{\beta+}$. Recall that the group G_β is reductive, with Cartan decomposition given by $\mathsf{G}_\beta = \mathsf{K}_\beta \exp(\mathfrak{p}_\beta)$ induced from that of G: $\mathsf{K}_\beta = \mathsf{G}_\beta \cap \mathsf{K}$, $\mathfrak{p}_\beta = \mathfrak{g}_\beta \cap \mathfrak{p}$. Consider the following Lie subalgebra of $\mathfrak{g}_\beta$:

$$\mathfrak{h}_\beta := \{A \in \mathfrak{g}_\beta : \langle A, \beta \rangle = 0\}.$$

Definition 1.7.7 The subgroup $\mathsf{H}_\beta \subset \mathsf{G}_\beta$ is defined by

$$\mathsf{H}_\beta = \mathsf{K}_\beta \exp(\mathfrak{p}_\beta \cap \mathfrak{h}_\beta).$$

The group H_β is real reductive, see (1.1). Its Lie algebra is $\mathfrak{h}_\beta$, and we have $\mathsf{G}_\beta = \exp(\mathbb{R}\beta) \times \mathsf{H}_\beta$ by the explicit description of G_β given after Definition 1.10.4. Moreover, it follows from (1.8) that H_β acts on $V^0_{\beta+}$ and that this action satisfies (1.1) with respect to the induced scalar product on $V^0_{\beta+}$. Thus Theorem 1.1.1 applies in this case.

Definition 1.7.8 We call

$$U^0_{\beta+} := \left\{ v \in V^0_{\beta+} : 0 \notin \overline{\mathsf{H}_\beta \cdot v} \right\}$$

the subset of H_β-*semistable* vectors in $V^0_{\beta+}$. We also define accordingly

$$U^{\geq 0}_{\beta+} := p^{-1}_\beta(U^0_{\beta+}).$$

Then, the *stratum* $\mathcal{S}_\beta$ associated with the orbit $\mathsf{K} \cdot \beta$ is the set defined by

$$\mathcal{S}_\beta := \mathsf{G} \cdot U_{\beta+}^{\geq 0} .$$

It will be made clear later that for most $\beta \in \mathfrak{p}$ the stratum $\mathcal{S}_\beta$ is actually empty. However, if the subset $U_{\beta+}^0$ of semistable vectors is non-empty, then it is an open subset of $V_{\beta+}^0$ by Theorem 1.1.1(iv), applied to the action of H_β on $V_{\beta+}^0$. The same holds of course for $U_{\beta+}^{\geq 0}$ in $V_{\beta+}^{\geq 0}$.

Remark 1.7.9 Notice that the strata $\mathcal{S}_\beta$ are scale invariant. Indeed, a vector $v \in V_{\beta+}^0$ is H_β-semistable if and only if $c \cdot v$ is also scale invariant, for any $c \neq 0$. Thus, $U_{\beta+}^0$ is scale invariant, and the same holds for $U_{\beta+}^{\geq 0}$ since p_β is a linear map.

A second observation is that for a critical point v_C of F we have $v_C \in U_{\beta+}^0 \subset \mathcal{S}_\beta$, where $\beta := \mathrm{m}(v_C)$. This also follows from Theorem 1.1.1 applied to the action of H_β on $V_{\beta+}^0$, since Lemma 1.7.10 will imply that the moment map

$$\mathrm{m}_{\mathsf{H}_\beta} : V_{\beta+}^0 \backslash \{0\} \to \mathfrak{p} \cap \mathfrak{h}_\beta$$

for this action, which a priori is given by the orthogonal projection of $\mathrm{m}(v)$ to $\mathfrak{h}_\beta$, satisfies the formula

$$\mathrm{m}_{\mathsf{H}_\beta}(v) = \mathrm{m}(v) - \beta .$$

Let us mention that up to the K-action on $\mathfrak{p}$ there are only finitely many β of the form $\mathrm{m}(v_C)$, for v_C a critical point of the energy map F (see Lemma 1.8.1).

The two main statements to be proved are the fact that the strata are smooth submanifolds, and that not only the critical point v_C, but also the entire flow lines of the negative gradient flow of F converging to v_C, are contained in the corresponding stratum $\mathcal{S}_\beta$. From this, the other assertions in Theorem 1.1.3 will easily follow.

We first compute the moment map on $V_{\beta+}^0$.

Lemma 1.7.10 For $v \in V_{\beta+}^0 \backslash \{0\}$ we have that $\mathrm{m}(v) = \mathrm{m}_{\mathsf{H}_\beta}(v) + \beta$. Moreover, if $\mathrm{m}(v) = \beta$, then v is a critical point of F.

Proof Let $v \in V_{\beta+}^0 \backslash \{0\}$ with $\|v\| = 1$. Since $\beta \in \mathfrak{p}$ and $\beta^+ \cdot v = 0$, (1.2) implies that for any $A \in \mathfrak{g}$ we have that

$$\langle [\beta, \mathrm{m}(v)], A \rangle = \langle \mathrm{m}(v), [\beta, A] \rangle = \langle [\beta, A] \cdot v, v \rangle = \langle [\beta^+, A] \cdot v, v \rangle = 0 .$$

This shows that $m(v) \in \mathfrak{g}_\beta$. Recall that $\mathfrak{g}_\beta = \mathbb{R}\beta \oplus \mathfrak{h}_\beta$, and observe that

$$\langle m(v), \beta \rangle = \langle \beta \cdot v, v \rangle = \langle \beta^+ \cdot v, v \rangle + \|\beta\|^2 = \langle \beta, \beta \rangle.$$

Hence the β-component of $m(v)$ is precisely β. On the other hand, the orthogonal projection of $m(v)$ to $\mathfrak{h}_\beta$ is $m_{H_\beta}(v)$, since $H_\beta \subset G$.

The assertion follows from Lemma 1.7.2, as $m(v)^+ \cdot v = \beta^+ \cdot v = 0$. $\quad\square$

In the next step, we deduce that $F|_{V_{\beta+}^{\geq 0}}$ attains its minimum precisely at the critical points of F with critical value $\|\beta\|^2$.

Lemma 1.7.11   *For $v \in V_{\beta+}^{\geq 0} \setminus \{0\}$ we have that $\|m(v)\| \geq \|\beta\|$, with equality if and only if $v \in V_{\beta+}^0$ is a critical point of F with $m(v) = \beta$.*

Proof   We write $v = \sum_{r \geq 0} v_r$ with $v_r \in V_{\beta+}^r$ and $\|v\|^2 = 1$. Notice that $\beta \cdot v_r = (r + \|\beta\|^2)v_r$. Then, we have that

$$\langle m(v), \beta \rangle = \sum_{r \geq 0} \langle \beta \cdot v_r, v_r \rangle \geq \|\beta\|^2.$$

It is clear that equality holds if and only if $v \in V_{\beta+}^0$. By Cauchy–Schwarz we deduce that $\|m(v)\|^2 \geq \langle m(v), \beta \rangle \geq \|\beta\|^2$, with equality if and only if $m(v) = \beta$ and $v \in V_{\beta+}^0$. By Lemma 1.7.10 v is a critical point of F. $\quad\square$

The following lemma shows that critical points of F which are mapped to β under the moment map m, correspond to minimal vectors for the H_β-action on $V_{\beta+}^0$.

Lemma 1.7.12   *Let $v \in V_{\beta+}^0$. Then, $v \in U_{\beta+}^0$ if and only if there exists a critical point $v_C \in \overline{H_\beta \cdot v} \setminus \{0\}$ of F with $m(v_C) = \beta$.*

Proof   Clearly, $0 \notin \overline{H_\beta \cdot v}$ is equivalent to the existence of a vector $w \in \overline{H_\beta \cdot v}$ of minimal positive norm. This implies that for all $A \in \mathfrak{h}_\beta$ we have

$$0 = \tfrac{d}{dt}\big|_0 \|\exp(tA) \cdot w\|^2 = 2 \cdot \langle A \cdot w, w \rangle = 2 \cdot \|w\|^2 \cdot \langle m(w), A \rangle.$$

Thus, $m(w) \perp \mathfrak{h}_\beta$. By Lemma 1.7.10 we deduce that $m(w) = \beta$ and that w is a critical point.

Conversely, if $0 \in \overline{H_\beta \cdot v}$ then there would be two closed H_β-orbits in $\overline{H_\beta \cdot v}$: the one corresponding to v_C, and $\{0\}$. This contradicts Theorem 1.1.1. $\quad\square$

In the next lemma we show that, up to the action of K, the stratum equals $U_{\beta+}^{\geq 0}$.

Lemma 1.7.13 *The set $U_{\beta+}^{\geq 0}$ is Q_β-invariant and $\mathcal{S}_\beta = \mathsf{K} \cdot U_{\beta+}^{\geq 0}$.*

Proof The group H_β leaves $U_{\beta+}^0$ invariant by definition. Since $U_{\beta+}^0 \subset V_{\beta+}^0$, and the latter is an eigenspace of $\exp(\beta)$, it follows that $\mathsf{G}_\beta = \exp(\mathbb{R}\beta) \times \mathsf{H}_\beta$ preserves $U_{\beta+}^0$, because an orbit is closed if and only if the scaled orbit is also closed. By the definition of $U_{\beta+}^{\geq 0}$ and Lemma 1.7.5 we have that $U_{\beta+}^{\geq 0}$ is G_β-invariant and U_β-invariant. The last assertion follows from the facts that $\mathcal{S}_\beta = \mathsf{G} \cdot U_{\beta+}^{\geq 0}$ and $\mathsf{G} = \mathsf{KQ}_\beta$; see Lemma 1.10.5. $\square$

Corollary 1.7.14 *For $v \in \overline{\mathcal{S}_\beta} \backslash \{0\}$ we have that $\|\mathrm{m}(v)\| \geq \|\beta\|$. Equality holds if and only if $\mathrm{m}(v) \in \mathsf{K} \cdot \beta$, and in this case $v \in \mathcal{S}_\beta$ is a critical point for F.*

Proof From Lemma 1.7.13 and the compactness of K we have that $\overline{\mathcal{S}_\beta} \subset \mathsf{K} \cdot \overline{U_{\beta+}^{\geq 0}} \subset \mathsf{K} \cdot V_{\beta+}^{\geq 0}$. The claim now follows from Lemma 1.7.11 and the K-equivariance (resp. invariance) of m (resp. F). $\square$

This shows that on the stratum $\mathcal{S}_\beta$ the energy map F is bounded below by the critical value $\mathrm{F}(v_C)$, for $v_C \in \mathrm{m}^{-1}(\beta)$. Moreover, the minimum is attained precisely at those critical points v_C of F with $\mathrm{m}(v_C) \in \mathsf{K} \cdot \beta$.

Corollary 1.7.15 *If $v \in \mathcal{S}_\beta$ then there exists a critical point $v_C \in \overline{(\mathbb{R}_{>0} \cdot \mathsf{G}) \cdot v} \subset \overline{\mathcal{S}_\beta}$ of F with $\mathrm{m}(v_C) = \beta$.*

Proof By Lemma 1.7.13 we may replace v by $k \cdot v \in U_{\beta+}^{\geq 0}$, for some $k \in \mathsf{K}$. Applying Lemma 1.7.12 to $p_\beta(k \cdot v) \in U_{\beta+}^0$ and using Lemma 1.7.5 and (1.9) one obtains

$$v_C \in \overline{\mathsf{H}_\beta \cdot p_\beta(k \cdot v)} = \overline{p_\beta(\mathsf{H}_\beta \cdot k \cdot v)} \subset \overline{(\mathbb{R}_{>0} \cdot \mathsf{G}) \cdot v}$$

as in the statement. $\square$

We are now in a position to prove Theorem 1.1.3. The key analytical property is the negativity of the Hessian of the energy map F restricted to the normal space of a stratum at a critical point; see Lemma 1.8.2.

Proof of Theorem 1.1.3 For each critical point v_C of F we consider the set $\mathcal{S}_\beta$ defined in Definition 1.7.8, where $\beta = \mathrm{m}(v_C)$. Notice first that for all $k \in \mathsf{K}$, $\mathcal{S}_{k \cdot \beta \cdot k^{-1}} = \mathcal{S}_\beta$ by K-equivariance. By picking one representative for each K-orbit, say a diagonal β with eigenvalues in non-decreasing order, the strata may be parameterized by a finite set $\mathcal{B}$ by Lemma 1.8.1.

We now prove one direction in Theorem 1.1.3 (iii). Let $v(t)$ denote a solution to the negative gradient flow of F with $v(0) = v \in V \backslash \{0\}$. Since F is scale invariant we have $\|v(t)\| \equiv \|v\|$. Using that F is real analytic, by Łojasiewicz' theorem [Loj63] there exists a unique limit point $\lim_{t \to \infty} v(t) = v_C$, which is of course a critical point of F. Let $\beta := \mathrm{m}(v_C)$ and notice that $v_C \in U^0_{\beta+} \subset \mathcal{S}_\beta$ by Lemma 1.7.12. Since $\mathcal{S}_\beta$ is a smooth embedded submanifold of V by Proposition 1.8.3, there exists an open neighbourhood $\Omega \subset V$ of v_C which is diffeomorphic to the normal bundle of $\mathcal{S}_\beta$ restricted to $\mathcal{S}_\beta \cap \Omega$, such that $\mathcal{S}_\beta \cap \Omega$ is the zero section. For some $t_0 \in \mathbb{R}$ we have $v(t) \in \Omega$, $\forall t \geq t_0$. Also, since $v_C \in V^{\geq 0}_{\beta+} \cap \mathcal{S}_\beta$, we obtain $T_{v_C} \mathcal{S}_\beta = \mathfrak{k} \cdot v_C + V^{\geq 0}_{\beta+}$. By the Hessian computations from Lemma 1.8.2 and a standard second-order argument, see Remark 1.7.16, we conclude that we must have $v(t) \in \mathcal{S}_\beta \cap \tilde{\Omega}$ for all $t \geq t_0$ and some open subset $\tilde{\Omega} \subset \Omega$. Since the flow lines are, up to scaling, tangent to G-orbits, and since the stratum $\mathcal{S}_\beta$ is G-invariant and scale invariant (Remark 1.7.9), we conclude that $v \in \mathcal{S}_\beta$ is as well. As a consequence $V \backslash \{0\} = \bigcup_{\beta \in \mathcal{B}} \mathcal{S}_\beta$.

To prove (i) it remains to show that $\mathcal{S}_\beta \cap \mathcal{S}_{\beta'} \neq \emptyset$ implies $\mathsf{K} \cdot \beta = \mathsf{K} \cdot \beta'$. Suppose that $v \in \mathcal{S}_\beta \cap \mathcal{S}_{\beta'}$. By Corollary 1.7.15 one obtains critical points $v_C \in \overline{\mathcal{S}_\beta}$ and $v'_C \in \overline{\mathcal{S}_{\beta'}}$ with $\mathrm{m}(v_C) = \beta$ and $\mathrm{m}(v'_C) = \beta'$. Since $v_C, v'_C \in \overline{(\mathbb{R}_{>0} \cdot \mathsf{G}) \cdot v}$, we deduce $v_C, v'_C \in \overline{\mathcal{S}_\beta \cap \mathcal{S}_{\beta'}}$. Applying Corollary 1.7.14 twice we get $\|\beta\| = \|\beta'\|$, thus by the rigidity in the equality case in that result we conclude that $\mathsf{K} \cdot \beta = \mathsf{K} \cdot \beta'$.

We can now prove the other direction in (iii). Let $v \in \mathcal{S}_\beta$ and assume that the limit v_C of the negative gradient flow of F starting at v satisfies $\mathrm{m}(v_C) = \beta'$. By the above we have $v \in \mathcal{S}_{\beta'}$, thus $\mathcal{S}_\beta \cap \mathcal{S}_{\beta'} \neq \emptyset$ and hence $\beta' \in \mathsf{K} \cdot \beta$ and $\mathcal{S}_\beta = \mathcal{S}_{\beta'}$.

Finally, to show (ii) let $v \in \overline{\mathcal{S}_\beta} \backslash \mathcal{S}_\beta$, with say $v \in \mathcal{S}_{\beta'}$. By (iii) we may assume that $\mathrm{m}(v) = \beta'$, thus Corollary 1.7.14 applied to $\overline{\mathcal{S}_\beta}$ yields $\|\beta'\| = \|\mathrm{m}(v)\| \geq \|\beta\|$. Since equality would imply that $\beta' \in \mathsf{K} \cdot \beta$ and $\mathcal{S}_{\beta'} = \mathcal{S}_\beta$, contradicting the fact that $v \notin \mathcal{S}_\beta$, we deduce $\|\beta'\| > \|\beta\|$. $\square$

Remark 1.7.16 We now briefly explain the second-order argument mentioned in the proof above. Let $Z(x, y) = (f(x, y), g(x, y))$ be a smooth vector field on $\mathbb{R}^2$, such that $g(x, 0) = 0$ for all $x \in \mathbb{R}$, $f(0, 0) = 0$, $\frac{\partial g}{\partial y}(0, 0) = 2c > 0$ and $\frac{\partial g}{\partial x}(0, 0) = 0$. Then by Taylor's formula $g(x, y) = 2cy + \eta C_1 xy + \tilde{\eta} C_2 y^2$, where $\eta = \eta_{x,y}, \tilde{\eta} = \tilde{\eta}_{x,y} \in (0, 1)$ and $C_1, C_2 \in \mathbb{R}$. It follows that for $y \neq 0$ we have $yg(x, y) > cy^2 > 0$ for all $(x, y) \in B_\epsilon((0, 0))$, $\epsilon > 0$ small enough. This shows that the vector field Z cannot have an integral curve converging to the origin, unless it is contained in the x-axis.

1.8 Properties of Critical Points of the Energy Map

In this section we prove some properties of the critical points of the energy map F. We first show that they are mapped under m onto finitely many K-orbits.

Lemma 1.8.1 *The moment map* m *maps the set of critical points of* F *onto a finite number of* K*-orbits.*

Proof Fix an orthonormal basis $\{e_i\}_{i=1}^N$ for V which diagonalizes the action of T, and let us adopt the notation from Section 1.4. If $v = \sum v_i e_i \in V$ then

$$\alpha \cdot v = \sum \langle \alpha, \alpha_i \rangle \, v_i e_i, \qquad (1.10)$$

for any $\alpha \in \mathfrak{t}$. By (1.3) we obtain $\langle \mathrm{m}(v), \alpha \rangle = \frac{1}{\|v\|^2} \cdot \sum \langle \alpha, \alpha_i \rangle v_i^2$, thus if $\mathrm{m}(v) \in \mathfrak{t}$ it follows that

$$\mathrm{m}(v) = \sum \frac{v_i^2}{\|v\|^2} \cdot \alpha_i \,. \qquad (1.11)$$

Hence $\mathrm{m}(v) \in \Delta_I$, the convex hull of those α_i for which $v_i \neq 0$.

Now let v_C be a critical point of F such that $\beta := \mathrm{m}(v_C)$ is diagonal. Then $\beta \cdot v = \|\beta\|^2 v$ by Lemma 1.7.2. Thus, by equation (1.10) we must have $\langle \beta, \alpha_i \rangle = \|\beta\|^2$ for all i such that $v_i \neq 0$. Equivalently $\langle \beta, \alpha_i - \beta \rangle = 0$ for all such i. We deduce that $\beta \perp (\alpha - \beta)$, for all $\alpha \in \Delta_I$. Since $\beta \in \Delta_I$ by (1.11), β is the element of minimal norm in $\mathcal{CH}$. Therefore, there are only a finite number of possible β. $\qquad \square$

If v_C is a critical point of the energy map F, then the orbit $K \cdot v_C$ also consists of critical points. By the K-equivariance of the moment map, we may therefore assume that $\beta := \mathrm{m}(v_C)$ is diagonal. Hence $\beta^+ \cdot v_C = 0$ by Lemma 1.7.2, thus $v_C \in V_{\beta^+}^{\geq 0}$. As a consequence β^+ preserves the orthogonal complement of v_C in V.

Lemma 1.8.2 *Let* v_C *be a critical point of* F *with* $\mathrm{m}(v_C) = \beta$ *and let* $w \perp v_C$ *be an eigenvector of* β^+ *with eigenvalue* $\lambda_{\beta^+} \in \mathbb{R}$ *and* $\|w\| = \|v_C\|$. *Then*

$$(\mathrm{Hess}_{v_C} \mathrm{F})(w, w) = 4 \cdot \lambda_{\beta^+} + 2 \cdot \|(\mathrm{dm})_{v_C} \cdot w\|^2 \,.$$

Moreover, on the subspaces $\mathfrak{k} \cdot v_C$, $V_{\beta^+}^{\geq 0}$ *and* $(\mathfrak{k} \cdot v_C + V_{\beta^+}^{\geq 0})^\perp$ *the Hessian of* F *is zero, non-negative and negative, respectively.*

Proof As in the proof of Lemma 1.7.2 we set $v(t) = \cos(t)v_C + \sin(t)w$. This implies for all t that $\frac{d}{dt}F(v(t)) = 2 \cdot \langle m(v(t)), (dm)_{v(t)}(v'(t))\rangle$ and differentiating once more yields

$$\frac{d^2}{dt^2}\Big|_0 F(v(t)) = 2 \cdot \frac{d}{dt}\Big|_0 \langle \beta, (dm)_{v(t)} \cdot v'(t)\rangle + 2 \cdot \|(dm)_{v_C} \cdot w\|^2.$$

From (1.6) we deduce

$$\frac{d}{dt}\Big|_0 \langle \beta, (dm)_{v(t)} \cdot v'(t)\rangle = \frac{2}{\|v_C\|^2} \cdot \langle \beta \cdot w, w\rangle - \frac{2}{\|v_C\|^2} \cdot \langle \beta \cdot v_C, v_C\rangle.$$
(1.12)

Since $\beta^+ \cdot v_C = 0$ and $\beta^+ \cdot w = \lambda_{\beta^+} \cdot w$, we have that $\beta \cdot v_C = \|\beta\|^2 v_C$ and $\beta \cdot w = (\lambda_{\beta^+} - \|\beta\|^2) \cdot w$. The formula now follows from $\|w\| = \|v_C\|$.

From the K-invariance of F it follows that $\mathfrak{k} \cdot v_C$ lies in the kernel of the Hessian. The non-negativity of the Hessian on the subspace $V_{\beta^+}^{\geq 0}$ follows from its definition and the formula in Lemma 1.8.2. To prove negativity of the Hessian on the subspace $(\mathfrak{k} \cdot v_C + V_{\beta^+}^{\geq 0})^\perp$, observe that for any $w \perp \mathfrak{g} \cdot v_C$ one has that $(dm)_{v_C} \cdot w = 0$ by (1.6). By Lemma 1.7.4 we get $\mathfrak{g} \cdot v_C \subset \mathfrak{k} \cdot v_C + V_{\beta^+}^{\geq 0}$ and the lemma follows. $\qquad\square$

Finally, we show that the strata $\mathcal{S}_\beta = G \cdot V_{\beta^+}^{\geq 0} = K \cdot U_{\beta^+}^{\geq 0} \subset V\backslash\{0\}$ corresponding to the images of critical values of the energy map F are smooth, embedded submanifolds.

Let $G \times_{Q_\beta} U_{\beta^+}^{\geq 0}$ be the quotient of $G \times U_{\beta^+}^{\geq 0}$ with respect to the action of Q_β given by $q \cdot (g, v) = (gq^{-1}, q \cdot v)$. Since this action is proper and free, $G \times_{Q_\beta} U_{\beta^+}^{\geq 0}$ is a smooth manifold by a classical result of Koszul (cf. [Pal61, Prop. 2.2.1]). This yields a well-defined, smooth, surjective and G-equivariant map

$$\Psi : G \times_{Q_\beta} U_{\beta^+}^{\geq 0} \to \mathcal{S}_\beta \ ; \ [g, v] \mapsto g \cdot v.$$

Here the action of G on $G \times_{Q_\beta} U_{\beta^+}^{\geq 0}$ is given by left multiplication on the G factor, and it commutes with the action of Q_β mentioned above. Notice also that, by linearity of the G-action, the map Ψ is $\mathbb{R}_{>0}$-equivariant, where $\mathbb{R}_{>0}$ acts by scalar multiplication on the second factor in $G \times U_{\beta^+}^{\geq 0}$ (and the action commutes with the Q_β-action and thus passes to the quotient).

Proposition 1.8.3 *Let $v_C \in \mathcal{S}_\beta$ be a critical point of F with $m(v_C) = \beta$. Then, the map $\Psi : G \times_{Q_\beta} U_{\beta^+}^{\geq 0} \to V\backslash\{0\}$ is an embedding whose image is $\mathcal{S}_\beta$.*

Proof We first prove that Ψ is an immersion at $[e, v_C]$. To that end, it suffices to show that the kernel of the differential of the map $\hat{\Psi}$:

$G \times U_{\beta+}^{\geq 0} \to S_\beta$, $(g, v) \mapsto g \cdot v$, is given by the tangent space to the Q_β-orbit $Q_\beta \cdot (e, v_C)$. By linearity of the action, the latter is given by

$$T_{(e,v_C)}(Q_\beta \cdot (e, v_C)) = \{(A, -A \cdot v_C) : A \in \mathfrak{q}_\beta\}.$$

On the other hand, the differential is given by

$$d\hat\Psi|_{(e,v_C)}(A, w) = A \cdot v_C + w,$$

and this vanishes if and only if $(A, w) = (A, -A \cdot v_C)$, where $w \in V_{\beta+}^{\geq 0}$.

Suppose that there exists $A \in \mathfrak{k}$ with $A \cdot v_C \in V_{\beta+}^{\geq 0}$. Then $\exp(tA) \cdot v_C \in V_{\beta+}^{\geq 0}$ and $\exp(tA) \in K$ for all $t \in \mathbb{R}$. The K-equivariance of the moment map m implies that $\| m(\exp(tA) \cdot v_C)\| \equiv \| m(v_C)\| = \|\beta\|$. Since $\exp(tA) \cdot v_C \in U_{\beta+}^{\geq 0} \subset V_{\beta+}^{\geq 0}$, we deduce from Lemma 1.7.11 that $m(\exp(tA) \cdot v_C) = \beta$ for all $t \in \mathbb{R}$. Thus $m(\exp(tA) \cdot v_C) = \exp(tA) \cdot \beta \cdot \exp(-tA)$ for all $t \in \mathbb{R}$. Differentiating shows that $A \in \mathfrak{k}_\beta \subset \mathfrak{q}_\beta$. This shows the above claim.

In order to show that it is an immersion at any point, we recall that Ψ is G-equivariant and $\mathbb{R}_{>0}$-equivariant, and apply Corollary 1.7.15.

In the second step we show that Ψ is injective. By the above, there exists an open neighbourhood Ω of $[e, v_C]$ in $G \times_{Q_\beta} U_{\beta+}^{\geq 0}$ such that $\Psi|_\Omega$ is injective. If Ψ is not globally injective, then there exists $v \in U_{\beta+}^{\geq 0}$ and $g \in G\backslash Q_\beta$ such that $g \cdot v \in U_{\beta+}^{\geq 0}$. By $G = KQ_\beta$, we may assume that $g = k \in K\backslash K_\beta$. Using Corollary 1.7.15, let $(c_s) \subset \mathbb{R}_{>0}$ and $(g_s) \subset G$ be sequences with $g_s = k_s q_s$, $k_s \in K$, $q_s \in Q_\beta$, such that $c_s g_s \cdot (k \cdot v) \to \overline{v_C}$, where $\overline{v_C} \in V_{\beta+}^{\geq 0}$ is a critical point of F with $m(\overline{v_C}) = \beta$. By Lemma 1.7.11 we deduce $\overline{v_C} \in V_{\beta+}^0$ and by Lemma 1.7.12 even $\overline{v_C} \in U_{\beta+}^0$.

By compactness of K we may assume that

$$v_s := c_s q_s \cdot (k \cdot v) \to \overline{v_C}',$$

where $\overline{v_C}' \in K \cdot \overline{v_C}$ is another critical point. Since $k \cdot v \in V_{\beta+}^{\geq 0}$, by Lemma 1.7.4 $v_s \in V_{\beta+}^{\geq 0}$ for all s, thus $\overline{v_C}' \in V_{\beta+}^{\geq 0}$. It follows that $m(\overline{v_C}') = \beta$ by Lemma 1.7.11 and $\overline{v_C}' \in U_{\beta+}^0$, as above. We now write $q_s k = \tilde{k}_s \tilde{q}_s$ with $\tilde{k}_s \in K$, $\tilde{q}_s \in Q_\beta$, and assume that $\tilde{k}_s \to \tilde{k}_\infty \in K$ as $s \to \infty$. By setting $\hat{v}_C := \tilde{k}_\infty^{-1} \cdot \overline{v_C}'$, also a critical point, we have that $c_s \tilde{q}_s \cdot v \to \hat{v}_C \in V_{\beta+}^{\geq 0}$ and $m(\hat{v}_C) = \beta$ again by Lemma 1.7.11. In particular, $\tilde{k}_\infty^{-1} \in K_\beta = G_\beta \cap K$ by K-equivariance of m and the very definition of G_β. Now, if we put $\hat{k}_s := \tilde{k}_\infty \tilde{k}_s^{-1} \notin K_\beta$, it satisfies $\hat{k}_s \to \mathrm{Id}$ and $\hat{k}_s \cdot v_s \to \overline{v_C}'$, or, in other words, $[\hat{k}_s, v_s] \to [e, \overline{v_C}']$. Since $\Psi([\hat{k}_s, v_s]) = \Psi([e, \hat{k}_s \cdot v_s])$ and $\hat{k}_s \notin Q_\beta$, this contradicts the injectivity of Ψ near $[e, \overline{v_C}'] = \lim_{s \to \infty} \Psi([e, v_s])$.

In order to show that Ψ is an embedding it remains to show that Ψ is a proper map. So let $([g_i, v_i])_{i\in\mathbb{N}}$ be a sequence in $\mathsf{G}\times_{\mathsf{Q}_\beta} U_{\beta+}^{\geq 0}$ with $\Psi([g_i, v_i]) \to \Psi([g, v]) = gv$ for $i\to\infty$. We write $g_i = k_i q_i$ with $k_i \in \mathsf{K}$ and $q_i \in \mathsf{Q}_\beta$. Assuming that $k_i \to k$ for $i\to\infty$, we deduce $q_i v_i \to k^{-1} gv$ for $i\to\infty$. Thus $[q_i, v_i] = [k_i, q_i v_i] \to [k, k^{-1}gv]$ for $i\to\infty$. This shows the claim. □

An immediate consequence is the following characterization of the parabolic subgroup Q_β (cf. Lemma 1.7.13).

Corollary 1.8.4 *For $v \in U_{\beta+}^{\geq 0}$, $g \in \mathsf{G}$ we have that $g \cdot v \in U_{\beta+}^{\geq 0}$ if and only if $g \in \mathsf{Q}_\beta$.*

Proof If $g \cdot v \in U_{\beta+}^{\geq 0}$ then $[g, v]$, $[e, g \cdot v] \in \mathsf{G}\times_{\mathsf{Q}_\beta} U_{\beta+}^{\geq 0}$ and $\Psi([g, v]) = \Psi([e, g\cdot v])$, thus $g \in \mathsf{Q}_\beta$ by injectivity. The converse is clear by Lemma 1.7.13. □

Another application is the fact that the stratum fibres over the compact homogeneous space $\mathsf{K}/\mathsf{K}_\beta$:

Corollary 1.8.5 *The map $\Psi_\mathsf{K} : \mathsf{K}\times_{\mathsf{K}_\beta} U_{\beta+}^{\geq 0} \to \mathcal{S}_\beta$, $[k, v] \mapsto k\cdot v$, is a diffeomorphism.*

1.9 Applications

We collect some applications of the above results. They will be fundamental in our study of homogeneous Ricci flows [BL18b], [BL19].

The first one, while only a restatement of Lemma 1.7.11, is very important in applications since it gives non-trivial estimates on strata $\mathcal{S}_\beta$ with $\beta \neq 0$.

Lemma 1.9.1 *For $v \in V_{\beta+}^{\geq 0}\setminus\{0\}$ we have that*

$$\|\mathrm{m}(v)\|^2 \geq \langle \mathrm{m}(v), \beta\rangle \geq \|\beta\|^2$$

with equality if and only if $v \in V_{\beta+}^0$ is a critical point of F with $\mathrm{m}(v) = \beta$.

The second application gives information about the isotropy subgroups

$$\mathsf{G}_v := \{g \in \mathsf{G} : g\cdot v = v\}.$$

Corollary 1.9.2 *For any $v \in U_{\beta+}^{\geq 0}$ we have that both $\mathsf{G}_v \subset \mathsf{H}_\beta \mathsf{U}_\beta$ and $\mathsf{G}_v \cap \mathsf{K} \subset \mathsf{K}_\beta$.*

Proof Let $\varphi \in G_v$. Then $\Psi([\varphi, v]) = v = \Psi([e, v])$. By the injectivity of Ψ (see Proposition 1.8.3) we deduce $[\varphi, v] = [e, v]$, thus $\varphi \in Q_\beta = G_\beta U_\beta$. Hence $\varphi = \varphi_\beta \varphi_u$ with $\varphi_\beta \in G_\beta$ and $\varphi_u \in U_\beta$. Now let $w := p_\beta(v)$, where $p_\beta : V_{\beta+}^{\geq 0} \to V_{\beta+}^0$ denotes the orthogonal projection. Then we have $w \in U_{\beta+}^0$, since $U_{\beta+}^{\geq 0} = p_\beta^{-1}(U_{\beta+}^0)$, and $\varphi_\beta \in G_w$ by Lemma 1.7.5. We write $\varphi_\beta = \varphi_H \exp(a\beta)$ with $\varphi_H \in H_\beta$ and $a \in \mathbb{R}$. Recall now that $U_{\beta+}^0 \subset V_{\beta+}^0$ by Definition 1.7.8 and that $\exp(-t\beta^+)$ acts trivially on $V_{\beta+}^0$ for any $t \in \mathbb{R}$. Thus,

$$(\varphi_H)^n \cdot w = \exp(-na\beta) \cdot w$$
$$= \big(\exp(na\|\beta\|^2 \mathrm{Id}) \exp(-na\beta^+)\big) \cdot w = e^{na\|\beta\|^2} w,$$

for all $n \in \mathbb{Z}$. It follows that $a = 0$, since otherwise we would have $0 \in \overline{H_\beta \cdot w}$, contradicting $w \in U_{\beta+}^0$ and the very definition of $U_{\beta+}^0$.

Note that $G_v \cap K \subset K_\beta$ follows from the fact that $K \cap Q_\beta = K_\beta$; see Lemma 1.10.5. □

The next application shows that the semistable vectors are precisely the stratum corresponding to $\beta = 0$:

Corollary 1.9.3 *We have that* $V \backslash \mathcal{N} = \mathcal{S}_0$.

Proof If $v \in V \backslash \mathcal{N}$, then by Theorem 1.1.1 there exists a minimal vector $\bar{v} \in \overline{G \cdot v} \cap \mathcal{S}_0$, hence $v \in \mathcal{S}_0$ by part (ii) of Theorem 1.1.3. To see that $\mathcal{S}_0 \subset V \backslash \mathcal{N}$, notice that if $v \in \mathcal{S}_0 \cap \mathcal{N}$ then also the cone $C(G \cdot v)$ over $G \cdot v$ is contained in $\mathcal{N}$, and by part (iv) of Theorem 1.1.1 the same is true for its closure. Using that by Lemma 1.7.2 the gradient of F is tangent to $C(G \cdot v)$, we deduce from part (iii) of Theorem 1.1.3 that for the limit v_C of the negative gradient flow of F starting at v_0 we have that $v_C \in \mathcal{S}_0 \cap \mathcal{N}$. Thus, $\mathrm{m}(v_C) = 0$ and hence $v_C \in V \backslash \mathcal{N}$, a contradiction. □

Finally, our last application of the stratification theorem, Theorem 1.1.3, generalizes the uniqueness (up to K-action) of zeros of the moment map within the closure of an orbit (Theorem 1.1.1(i) and (iii)) to critical points of F of higher energy. To that end, one has to restrict to critical points lying in the same stratum, since in general the closure of an orbit may contain several K-orbits of critical points, lying however in strata of higher energy. Let us also mention that this by no means implies that on each stratum there is a unique K-orbit of critical points: these typically come in families of several continuous parameters.

Corollary 1.9.4 *Let* $v \in \mathcal{S}_\beta$. *Then, there is a critical point* $v_C \in \overline{\mathbb{R}_{>0} \cdot G \cdot v} \cap \mathcal{S}_\beta$ *of* F *which is unique up to scaling and the action of* K.

Proof By Lemma 1.7.2, the negative gradient flow of F is tangent to the orbits of $\mathbb{R}_{>0}\mathsf{G}$, where the action of $\mathbb{R}_{>0}$ is just multiplication by positive scalars. The existence of v_C thus follows by Theorem 1.1.3(iii) and the K-invariance of F.

Regarding uniqueness, let $v_C, w_C \in \overline{\mathbb{R}_{>0} \cdot \mathsf{G} \cdot v} \cap \mathcal{S}_\beta$ be two critical points of F. By Proposition 1.8.3, up to the action of K we may assume that $v, v_C, w_C \in U_{\beta+}^{\geq 0}$. From Corollary 1.8.4 and the fact that $U_{\beta+}^{\geq 0}$ is an open cone in $V_{\beta+}^{\geq 0}$ we deduce

$$\overline{\mathbb{R}_{>0} \cdot \mathsf{G} \cdot v} \cap U_{\beta+}^{\geq 0} = \overline{\mathbb{R}_{>0} \cdot \mathsf{Q}_\beta \cdot v} \cap U_{\beta+}^{\geq 0}.$$

This implies that $v_C, w_C \in \overline{\mathbb{R}_{>0} \cdot \mathsf{Q}_\beta \cdot v} \cap U_{\beta+}^0$. Using $\mathsf{G}_\beta = \exp(\mathbb{R}\beta)\mathsf{H}_\beta$ and Lemma 1.7.5 we obtain that

$$v_C = p_\beta(v_C) \in p_\beta\left(\overline{\mathbb{R}_{>0} \cdot \mathsf{Q}_\beta \cdot v}\right) \subset \overline{p_\beta(\mathbb{R}_{>0} \cdot \mathsf{Q}_\beta \cdot v)} = \overline{\mathbb{R}_{>0} \cdot \mathsf{H}_\beta \cdot v_0},$$

where $v_0 = p_\beta(v)$, and the same holds for w_C. We now claim that in fact $v_C, w_C \in \mathbb{R}_{>0} \cdot \overline{\mathsf{H}_\beta \cdot v_0}$. Up to scaling this would allow us to assume that $v_C, w_C \in \overline{\mathsf{H}_\beta \cdot v_0}$, from which the result follows after using Theorem 1.1.1(i) applied to the H_β-action on $V_{\beta+}^0$ (see also Lemma 1.7.10). It remains to justify the above claim.

To that end, let $(\lambda_n v_n)_{n \in \mathbb{N}}$ be a sequence with $\lambda_n \in \mathbb{R}_{>0}$, $v_n \in \mathsf{H}_\beta \cdot v_0$, such that $\lambda_n v_n \to v_C$ as $n \to \infty$. Since $v_0 = p_\beta(v) \in U_{\beta+}^0$ we know that $0 \notin \overline{\mathsf{H}_\beta \cdot v_0}$, or, in other words, there must be a uniform upper bound $|\lambda_n| \leq C$ for all $n \in \mathbb{N}$. On the other hand, Theorem 1.1.1(iii) applied to the H_β-action on $V_{\beta+}^0$ yields the existence of a unique (up to K_β-action) $\tilde{v}_C \in \overline{\mathsf{H}_\beta \cdot v_0}$ of minimal norm $\delta = \|\tilde{v}_C\| > 0$. We also have $\mathrm{m}(\tilde{v}_C) = \beta$; see Lemma 1.7.10. Moreover, uniqueness and that lemma also imply that if $\|v\| \geq 2 \cdot \delta$ then $\|\mathrm{m}(v)\| \geq \|\beta\| + \epsilon$ for some $\epsilon > 0$, for all $v \in \mathsf{H}_\beta \cdot v_0$. Since $\mathrm{m}(v_n) = \mathrm{m}(\lambda_n v_n) \to \mathrm{m}(v_C) = \beta$, we must have that $\|v_n\| < 2 \cdot \delta$. This yields a uniform lower bound $|\lambda_n| \geq c > 0$, and after passing to a convergent subsequence $\lambda_n \to \lambda$ we deduce that $v_C \in \lambda \cdot \overline{\mathsf{H}_\beta \cdot v_0}$ as claimed. $\qquad\square$

1.10 Appendices

1.10.1 Real Reductive Lie Groups

In this section we collect useful properties of closed subgroups $\mathsf{G} \subset \mathsf{GL}(V)$ that satisfy condition (1.1), i.e. those we call *real reductive Lie groups*.

First, let us mention that there exist in the literature several non-equivalent definitions of this concept. To the best of our knowledge, we can mention at least four: those groups in the Harish-Chandra class [HC75, §3]; Knapp's slightly more general version [Kna02, Chap. VII, §2]; Wallach's definition [Wal88]; and Borel's definition [Bor06, §6]. Usually the common aim of these definitions is to enlarge the class of real semisimple Lie groups to allow for inductive arguments, since most structural results remain valid for the more robust classes of real reductive groups. Of all of them perhaps the most succinct is that of Borel: a real reductive Lie group is one with finitely many connected components, and whose Lie algebra is reductive (the direct sum of a semisimple Lie subalgebra and its centre).

Our interest in linear representations allows us to reduce ourselves to the case of linear groups (i.e. those contained in some $\mathsf{GL}(V)$ for a vector space V), thus avoiding lots of technicalities of the general case. Let us also mention that linear groups satisfying Knapp's definition automatically satisfy ours. Conversely, recall that $\mathsf{O}(n)$ for n even is not real reductive in the sense of Knapp. A linear group satisfying Borel's definition will satisfy (1.1) provided the centre acts on V by semisimple endomorphisms (i.e. diagonalizable over $\mathbb{C}$).

A first immediate property of a group G satisfying (1.1) is that both G and $\mathfrak{g}$ are closed under transpose (i.e. they are *self-adjoint*, cf. [Mos55]). The following further property, whose proof is based on that given in [Hel01], shows that maximal subalgebras in $\mathfrak{p}$ are all conjugate by an element in K.

Proposition 1.10.1   *Let $\mathfrak{t}_1, \mathfrak{t}_2 \subset \mathfrak{p}$ be two maximal abelian subalgebras. Then there exists $k \in \mathsf{K}$ such that $k\mathfrak{t}_1 k^{-1} = \mathfrak{t}_2$. In particular, for any maximal abelian subalgebra $\mathfrak{t} \subset \mathfrak{p}$ we have that*

$$\mathfrak{p} = \bigcup_{k \in \mathsf{K}} k\,\mathfrak{t}\,k^{-1}. \tag{1.13}$$

Proof   Recall that we have by assumption an $\mathsf{Ad}(\mathsf{K})$-invariant scalar product $\langle\,\cdot\,,\cdot\,\rangle$ on $\mathfrak{g}$. Let $H_i \in \mathfrak{t}_i$ be generic, so that $Z_{\mathfrak{p}}(H_i) := \{A \in \mathfrak{p} : [A, H_i] = 0\} = \mathfrak{t}_i$, $i = 1, 2$. (Generic vectors do exist, since for a fixed maximal abelian subalgebra the set of non-generic elements is a finite union of hyperspaces, the kernels of the roots of the adjoint representation; see the description following Corollary 1.10.2 below.) Then there exist $k \in \mathsf{K}$, which minimizes $d : \mathsf{K} \to \mathbb{R}$; $k \mapsto \langle kH_1 k^{-1}, H_2 \rangle$, since K is compact. At the infinitesimal level this implies that

$$\langle [Z, kH_1k^{-1}], H_2 \rangle = 0,$$

for all $Z \in \mathfrak{k}$. We deduce $\langle [H_2, kH_1k^{-1}], Z \rangle = 0$ for all $Z \in \mathfrak{k}$, using (1.2) and that $H_i \in \mathfrak{p} \subset \operatorname{Sym}(V)$, $Z \in \mathfrak{k} \subset \mathfrak{so}(V)$. Since $[\mathfrak{p}, \mathfrak{p}] \subset \mathfrak{k}$, we have $[H_2, kH_1k^{-1}] \in \mathfrak{k}$, thus $[H_2, kH_1k^{-1}] = 0$. By definition of H_2 this yields $kH_1k^{-1} \in \mathfrak{t}_2$. For any $A_2 \in \mathfrak{t}_2$ we deduce $[k^{-1}A_2k, H_1] = 0$, from which $A_2 \in \mathfrak{t}_1$ by definition of H_1. Therefore, $k^{-1}\mathfrak{t}_2k \subset \mathfrak{t}_1$, which also reads as $\mathfrak{t}_2 \subset k\mathfrak{t}_1k^{-1}$. By maximality of $\mathfrak{t}_2$ this implies that $\mathfrak{t}_2 = k\mathfrak{t}_1k^{-1}$.

Given a maximal abelian subalgebra $\mathfrak{t}$ of $\mathfrak{p}$, any $A \in \mathfrak{p}$ is contained in a maximal abelian subalgebra $\mathfrak{a}$ by extending $\mathbb{R}A$ to a maximal abelian subalgebra, and $\mathfrak{a} = k\mathfrak{t}k^{-1}$ for some $k \in \mathsf{K}$ by the last paragraph. Hence (1.13). □

Corollary 1.10.2 *For any maximal abelian subalgebra $\mathfrak{t} \subset \mathfrak{p}$ we have that $\mathsf{G} = \mathsf{K}\mathsf{T}\mathsf{K}$, where $\mathsf{T} := \exp(\mathfrak{t})$.*

Proof We have $\mathsf{G} = \mathsf{K}\exp(\mathfrak{p})$, and Proposition 1.10.1 implies that $\exp(\mathfrak{p}) \subset \mathsf{K}\mathsf{T}\mathsf{K}$. □

Let us now fix a maximal abelian subalgebra $\mathfrak{t} \subset \mathfrak{p}$. Choose an orthonormal basis for $(V, \langle \cdot, \cdot \rangle)$ such that $\mathfrak{t}$ is contained in $\mathfrak{d}$, the set of diagonal matrices in $\mathfrak{gl}(V)$. Moreover, by maximality of $\mathfrak{t}$ it also holds that $\mathfrak{t} = \mathfrak{g} \cap \mathfrak{d}$. Denote by $\mathfrak{u}$ (resp. $\mathfrak{u}^t$) the nilpotent subalgebra of $\mathfrak{gl}(V)$ of strictly lower (resp. upper) triangular matrices, and by $\mathsf{U} := \exp(\mathfrak{u})$ the corresponding analytic subgroup of $\mathsf{GL}(V)$.

We now look at the root space decomposition of $\mathfrak{gl}(V)$ with respect to $\mathfrak{t}$. More precisely, $\operatorname{ad}(\mathfrak{t})$ is a commuting family of symmetric endomorphisms of $\mathfrak{gl}(V)$, hence there exists a finite subset $\Sigma \subset \mathfrak{t}^*$ and a decomposition into common eigenspaces

$$\mathfrak{gl}(V) = \bigoplus_{\lambda \in \Sigma} \mathfrak{gl}(V)_\lambda,$$

where for $E \in \mathfrak{t}$ we have $\operatorname{ad}(E)|_{\mathfrak{gl}(V)_\lambda} = \lambda(E) \cdot \operatorname{Id}_{\mathfrak{gl}(V)_\lambda}$. Notice now that $\operatorname{ad}(\mathfrak{t})$ preserves the subspaces $\mathfrak{d}$, $\mathfrak{u}$ and $\mathfrak{u}^t$. This implies that for $\lambda \in \Sigma$ we can decompose $\mathfrak{gl}(V)_\lambda = (\mathfrak{gl}(V)_\lambda \cap \mathfrak{d}) \oplus (\mathfrak{gl}(V)_\lambda \cap \mathfrak{u}) \oplus (\mathfrak{gl}(V)_\lambda \cap \mathfrak{u}^t)$. Moreover, for each $\lambda \neq 0$ at most one of the three subspaces on the right-hand side is non-zero, since for any $E \in \mathfrak{t}$ we have that $\operatorname{ad}(E)|_{\mathfrak{u}} = -\operatorname{ad}(E)|_{\mathfrak{u}^t}$. Thus, Σ can be decomposed as a disjoint union $\Sigma = \{0\} \cup \Sigma^+ \cup \Sigma^-$, such that

$$\mathfrak{t} \subset \mathfrak{gl}(V)_0, \qquad \bigoplus_{\lambda \in \Sigma^+} \mathfrak{gl}(V)_\lambda \subset \mathfrak{u}, \qquad \bigoplus_{\lambda \in \Sigma^-} \mathfrak{gl}(V)_\lambda \subset \mathfrak{u}^t.$$

Consider now the corresponding subalgebras of $\mathfrak{g}$, $\mathfrak{n} := \mathfrak{g} \cap \mathfrak{u}$, $\mathfrak{n}^t := \mathfrak{g} \cap \mathfrak{u}^t$, and the analytic subgroups $T := \exp(\mathfrak{t})$, $N := \exp(\mathfrak{n}) = (G \cap U)_0$, corresponding to $\mathfrak{t}, \mathfrak{n}$, respectively. Since $\mathrm{ad}(\mathfrak{t})$ preserves $\mathfrak{g}$, we have the induced root space decomposition

$$\mathfrak{g} = \bigoplus_{\lambda \in \Sigma} \mathfrak{g}_\lambda,$$

where now some $\mathfrak{g}_\lambda$ might be trivial. The above discussion implies that

$$\mathfrak{n} = \bigoplus_{\lambda \in \Sigma^+} \mathfrak{g}_\lambda, \quad \mathfrak{n}^t = \bigoplus_{\lambda \in \Sigma^-} \mathfrak{g}_\lambda \quad \text{and} \quad \mathfrak{t} \subset \mathfrak{g}_0,$$

and thus $\mathfrak{g} = \mathfrak{g}_0 \oplus \mathfrak{n} \oplus \mathfrak{n}^t$.

Proposition 1.10.3 *If $G \subset GL(V)$ is closed and satisfies* (1.1), *then* $G = KTN$.

Proof Assume first that G is connected. Since $\mathfrak{t} \oplus \mathfrak{n}$ is a subalgebra of $\mathfrak{g}$, it follows that TN is a closed connected subgroup of G, intersecting K trivially: the only lower triangular orthogonal matrix with positive eigenvalues is the identity. The result in this case would follow provided we show that $\mathfrak{g} = \mathfrak{k} \oplus \mathfrak{t} \oplus \mathfrak{n}$. To that end, let $E \in \mathfrak{g} = \bigoplus_{\lambda \in \Sigma} \mathfrak{g}_\lambda$, say $E \in \mathfrak{g}_\lambda$. If $\lambda = 0$ then $E \in \mathfrak{g}_0$. If $\lambda \in \Sigma^+$ then $E \in \mathfrak{n}$. And finally, for $\lambda \in \Sigma^-$, we have $E^t \in \mathfrak{n}$ thus

$$E = (E - E^t) + E^t \in \mathfrak{k} \oplus \mathfrak{n}.$$

The general case follows from the previous one: we know that $G_0 = K_0 TN$. On the other hand, the connected component of the identity G_0 is given by $G_0 = K_0 \exp(\mathfrak{p})$ thanks to (1.1), thus $G/G_0 \simeq K/K_0$ and hence $G = KTN$. $\qquad\square$

1.10.2 The Parabolic Subgroup Q_β

Let $G \subset GL(V)$ be a closed subgroup satisfying (1.1) with Lie algebra $\mathfrak{g} \subset \mathfrak{gl}(V)$ and Cartan decomposition $\mathfrak{g} = \mathfrak{k} \oplus \mathfrak{p}$, $\mathfrak{k} = \mathfrak{g} \cap \mathfrak{so}(V)$, $\mathfrak{p} = \mathfrak{g} \cap \mathrm{Sym}(V)$. We will describe in this section some important subgroups of G associated with a fixed element $\beta \in \mathfrak{p}$.

Consider for such a fixed $\beta \in \mathfrak{p}$ the adjoint map

$$\mathrm{ad}(\beta) : \mathfrak{g} \to \mathfrak{g}, \qquad A \mapsto [\beta, A].$$

Assuming (1.2), the map $\text{ad}(\beta) : \mathfrak{g} \to \mathfrak{g}$ is a symmetric endomorphism with respect to the chosen scalar product on $\mathfrak{g}$. If $\mathfrak{g}_r$ is the eigenspace of $\text{ad}(\beta)$ with eigenvalue $r \in \mathbb{R}$ then $\mathfrak{g} = \bigoplus_{r \in \mathbb{R}} \mathfrak{g}_r$, and we set

$$\mathfrak{g}_\beta := \mathfrak{g}_0 = \ker(\text{ad}(\beta)), \qquad \mathfrak{u}_\beta := \bigoplus_{r>0} \mathfrak{g}_r, \qquad \mathfrak{q}_\beta := \mathfrak{g}_\beta \oplus \mathfrak{u}_\beta.$$

Definition 1.10.4 We denote by

$$\mathsf{G}_\beta := \{g \in \mathsf{G} : g\beta g^{-1} = \beta\}, \quad \mathsf{U}_\beta := \exp(\mathfrak{u}_\beta) \quad \text{and} \quad \mathsf{Q}_\beta := \mathsf{G}_\beta \mathsf{U}_\beta$$

the centralizer of β in G, the unipotent subgroup associated with β, and the parabolic subgroup associated with β, respectively.

To describe these groups more explicitly, let us decompose $V = V_1 \oplus \cdots \oplus V_m$ as a sum of β-eigenspaces corresponding to the real eigenvalues $\lambda_1 < \cdots < \lambda_m$ with multiplicities $n_1, \ldots, n_m \in \mathbb{N}$. Corresponding to this decomposition of V, we write any $g \in \mathsf{GL}(V)$ as

$$g = \begin{pmatrix} g_{11} & \cdots & g_{1m} \\ \vdots & \ddots & \vdots \\ g_{m1} & \cdots & g_{mm} \end{pmatrix}$$

with $g_{ii} \in \mathsf{GL}(V_i)$ and $g_{ij} \in \text{End}(V_j, V_i)$ for $i \neq j$. With respect to a suitable orthonormal basis of V the map $\beta = (\lambda_1 \text{Id}_1, \cdots, \lambda_m \text{Id}_m)$ is diagonal; moreover,

$$\mathsf{G}_\beta = \mathsf{G} \cap \left\{ \begin{pmatrix} g_1 & & \\ & \ddots & \\ & & g_m \end{pmatrix} : g_i \in \mathsf{GL}(V_i) \right\},$$

$$\mathsf{U}_\beta = \mathsf{G} \cap \left\{ \begin{pmatrix} \text{Id}_1 & & & \\ g_{21} & \text{Id}_2 & & \\ \vdots & & \ddots & \\ g_{m1} & \cdots & g_{m,m-1} & \text{Id}_m \end{pmatrix} : g_{ij} \in \text{End}(V_j, V_i), \ i > j \right\},$$

$$\mathsf{Q}_\beta = \mathsf{G} \cap \left\{ \begin{pmatrix} g_{11} & & \\ \vdots & \ddots & \\ g_{m1} & \cdots & g_{mm} \end{pmatrix} : g_{ij} \in \text{End}(V_j, V_i), \ g_{ii} \in \mathsf{GL}(V_i) \right\}.$$

Lemma 1.10.5 *The groups* $\mathsf{G}_\beta, \mathsf{U}_\beta, \mathsf{Q}_\beta$ *are closed in* G, *and their Lie algebras are given respectively by* $\mathfrak{g}_\beta, \mathfrak{u}_\beta, \mathfrak{q}_\beta$. *We have that*

$$\mathsf{G} = \mathsf{K}\mathsf{Q}_\beta, \qquad \mathsf{K} \cap \mathsf{Q}_\beta = \mathsf{K} \cap \mathsf{G}_\beta =: \mathsf{K}_\beta.$$

Moreover, U_β *is connected and normal in* Q_β, G_β *is reductive and sat-isfies* (1.1), *and we have* $U_\beta \cap G_\beta = \{Id_V\}$.

Proof   A simple computation shows that $[\mathfrak{g}_r, \mathfrak{g}_s] \subset \mathfrak{g}_{r+s}$, thus $\mathfrak{u}_\beta$ is a Lie subalgebra of $\mathfrak{g}$ and an ideal in $\mathfrak{q}_\beta$. Hence U_β is a closed subgroup of G, which is normal in Q_β. It is clear that the centralizer subgroup G_β is closed in G.

The fact that $G = KQ_\beta$ follows at once from Proposition 1.10.3, ap-plied to a maximal abelian subalgebra $\mathfrak{t} \subset \mathfrak{p}$ containing β: in this case, we have that $TN \subset Q_\beta$.

If $g \in G_\beta \cap U_\beta$, say $g = \exp(N)$ with $N \in \mathfrak{u}_\beta$, then g preserves the eigenspaces of β, thus so does N. Hence $N \in \mathfrak{g}_\beta$, therefore $N = 0$ and $g = Id_V$.

Now observe that $G_\beta^t = G_\beta$, so in particular we also have $G_\beta \cap U_\beta^t = \{Id_V\}$. Also, since $\mathfrak{g}_r^t = \mathfrak{g}_{-r}$, we obtain in the same way as above that $U_\beta^t \cap Q_\beta = \{Id_V\}$. Since U_β is normal in Q_β we may write $Q_\beta = G_\beta U_\beta = U_\beta G_\beta$, and in particular $Q_\beta^t = G_\beta U_\beta^t$. From this observation it follows that $Q_\beta \cap Q_\beta^t = G_\beta$, and it is now clear that $K \cap Q_\beta = K_\beta$.

Finally, to show that G_β satisfies (1.1), set $\mathfrak{p}_\beta := \mathfrak{g}_\beta \cap \mathrm{Sym}(V) = \mathfrak{g}_\beta \cap \mathfrak{p}$, $\mathfrak{k}_\beta := \mathfrak{g}_\beta \cap \mathfrak{k} = \mathrm{Lie}(K_\beta)$, and let $g = k\exp(A) \in G_\beta$, $A \in \mathfrak{p}$, $k \in K$. Since $k^t = k^{-1}$ we have that $\exp(2A) = gg^t \in G_\beta$ for the symmetric endomorphism $A \in \mathfrak{p}$. Thus, $[A, \beta] = 0$ and hence $[k, \beta] = 0$. This implies that $g \in K_\beta \exp(\mathfrak{p}_\beta)$. On the other hand, it is clear that $K_\beta \exp(\mathfrak{p}_\beta) \subset G_\beta$, hence equality must hold and G_β is real reductive.   □

The following characterization turns out to be very useful in the ap-plications.

Lemma 1.10.6   *For* $\beta \in \mathfrak{p}$ *we have that*

$$G_\beta = \{g \in G : g\exp(\beta)g^{-1} = \exp(\beta)\},$$
$$U_\beta = \{g \in G : \lim_{t\to\infty} \exp(-t\beta)g\exp(t\beta) = Id\},$$
$$Q_\beta = \{g \in G : \lim_{t\to\infty} \exp(-t\beta)g\exp(t\beta) \text{ exists}\}.$$

Proof   Notice that $g \in G_\beta$ if and only if g preserves all the eigenspaces of β. Since these coincide with the eigenspaces of $\exp(\beta)$, the first assertion follows. For the other two claims, adopting the same notation as in the paragraph following Definition 1.10.4, for $g = (g_{ij})_{1 \le i,j \le m} \in GL(V)$ we have that the limit

$$\begin{pmatrix} e^{-t\lambda_1} & & \\ & \ddots & \\ & & e^{-t\lambda_m} \end{pmatrix} \cdot \begin{pmatrix} g_{11} & \cdots & g_{1m} \\ \vdots & \ddots & \vdots \\ g_{m1} & \cdots & g_{mm} \end{pmatrix} \cdot \begin{pmatrix} e^{t\lambda_1} & & \\ & \ddots & \\ & & e^{t\lambda_m} \end{pmatrix}$$

$$= \begin{pmatrix} g_{11} & \cdots & e^{-t(\lambda_1-\lambda_m)}g_{1m} \\ \vdots & \ddots & \vdots \\ e^{-t(\lambda_m-\lambda_1)}g_{m1} & \cdots & g_{mm} \end{pmatrix} \longrightarrow \begin{pmatrix} g_{11} & & \\ & \ddots & \\ & & g_{mm} \end{pmatrix}$$

as $t \to \infty$ exists if and only if $g_{ij} = 0$ for $i < j$, and it is Id_V if and only if in addition we have $g_{ii} = \mathrm{Id}_{V_i}$ for $i = 1, \ldots, m$. $\square$

References

[BHC62] Armand Borel and Harish-Chandra, *Arithmetic subgroups of algebraic groups*, Ann. of Math. (2) **75** (1962), 485–535. MR0147566.

[Bir71] David Birkes, *Orbits of linear algebraic groups*, Ann. of Math. (2) **93** (1971), 459–475. MR0296077.

[BL18a] Christoph Böhm and Ramiro A. Lafuente, *Homogeneous Einstein metrics on Euclidean spaces are Einstein solvmanifolds*, arXiv:1811.12594, 2018.

[BL18b] Christoph Böhm and Ramiro A. Lafuente, *Immortal homogeneous Ricci flows*, Invent. Math. **212** (2018), no. 2, 461–529. MR3787832.

[BL19] Christoph Böhm and Ramiro A. Lafuente, *The Ricci flow on solvmanifolds of real type*, Adv. Math. **352** (2019), 516–540.

[Bor06] Armand Borel, *Lie groups and linear algebraic groups. I. Complex and real groups*, Lie groups and automorphic forms, AMS/IP Stud. Adv. Math., vol. 37, Amer. Math. Soc., Providence, RI, 2006, pp. 1–49. MR2272918.

[Bre93] Glen E. Bredon, *Topology and geometry*, Graduate Texts in Mathematics, vol. 139, Springer-Verlag, New York, 1993. MR1224675.

[BT65] Armand Borel and Jacques Tits, *Groupes réductifs*, Inst. Hautes Études Sci. Publ. Math. (1965), no. 27, 55–150. MR0207712.

[BZ16] Leonardo Biliotti and Michela Zedda, *Stability with respect to actions of real reductive Lie groups*, arXiv:1610.05027, 2016.

[EJ09] Patrick Eberlein and Michael Jablonski, *Closed orbits of semisimple group actions and the real Hilbert-Mumford function*, New developments in Lie theory and geometry, Contemporary Mathematics, vol. 491, Amer. Math. Soc., Providence, RI, 2009, pp. 283–321. MR2537062.

[HC75] Harish-Chandra, *Harmonic analysis on real reductive groups. I. The theory of the constant term*, J. Funct. Anal. **19** (1975), 104–204. MR0399356.

[Hel01] Sigurdur Helgason, *Differential geometry, Lie groups, and symmetric spaces*, Graduate Studies in Mathematics, vol. 34, Amer. Math. Soc., Providence, RI, 2001. Corrected reprint of the 1978 original. MR1834454 (2002b:53081).

[HS07] Peter Heinzner and Gerald W. Schwarz, *Cartan decomposition of the moment map*, Math. Ann. **337** (2007), no. 1, 197–232. MR2262782.

[HS10] Peter Heinzner and Patrick Schützdeller, *Convexity properties of gradient maps*, Adv. in Math. **225** (2010), no. 3, 1119–1133.

[HSS08] Peter Heinzner, Gerald W. Schwarz and Henrik Stötzel, *Stratifications with respect to actions of real reductive groups*, Compos. Math. **144** (2008), no. 1, 163–185. MR2388560 (2009a:32030).

[Kir84] Frances Clare Kirwan, *Cohomology of quotients in symplectic and algebraic geometry*, Mathematical Notes, vol. 31, Princeton University Press, Princeton, NJ, 1984.

[KN79] George Kempf and Linda Ness, *The length of vectors in representation spaces*. In Lønsted, K. (ed.) Algebraic geometry (Proc. Summer Meeting, Univ. Copenhagen, Copenhagen, 1978), Lecture Notes in Mathematics, vol. 732, Springer, Berlin, 1979, pp. 233–243. MR555701.

[Kna02] Anthony W. Knapp, *Lie groups beyond an introduction*, second ed., Progress in Mathematics, vol. 140, Birkhäuser Boston, Inc., Boston, MA, 2002. MR1920389 (2003c:22001).

[Lau01] Jorge Lauret, *Ricci soliton homogeneous nilmanifolds*, Math. Ann. **319** (2001), no. 4, 715–733.

[Lau02] Jorge Lauret, *Finding Einstein solvmanifolds by a variational method*, Math. Z. **241** (2002), 83–99.

[Lau03a] Jorge Lauret, *Degenerations of Lie algebras and geometry of Lie groups*, Differ. Geom. Appl. **18** (2003), no. 2, 177–194. MR1958155.

[Lau03b] Jorge Lauret, *On the moment map for the variety of Lie algebras*, J. Funct. Anal. **202** (2003), no. 2, 392–423. MR1990531.

[Lau06] Jorge Lauret, *A canonical compatible metric for geometric structures on nilmanifolds*, Ann. Global Anal. Geom. **30** (2006), no. 2, 107–138.

[Lau10] Jorge Lauret, *Einstein solvmanifolds are standard*, Ann. of Math. (2) **172** (2010), no. 3, 1859–1877.

[Loj63] Stanislaw Lojasiewicz, *Une propriété topologique des sous-ensembles analytiques réels*, Les équations aux dérivées partielles **117** (1963), 87–89.

[Lun73] Domingo Luna, *Slices étales*, Sur les groupes algébriques, Mém. Soc. Math. Fr. (1973) no. 33, 81–105, MR0342523.

[Lun75] Domingo Luna, *Sur certaines opérations différentiables des groupes de Lie*, Amer. J. Math. **97** (1975), 172–181. MR0364272.

[Mar01] Alina Marian, *On the real moment map*, Math. Res. Lett. **8** (2001), no. 5-6, 779–788. MR1879820.

[MFK94] David Mumford, John Fogarty and Frances Kirwan, *Geometric invariant theory*, third ed., Ergebnisse der Mathematik und ihrer Grenzgebiete (2) [Results in Mathematics and Related Areas (2)], vol. 34, Springer-Verlag, Berlin, 1994. MR1304906 (95m:14012).

[Mos55] George D. Mostow, *Self-adjoint groups*, Ann. of Math. (2) **62** (1955), 44–55. MR0069830.

[Nes84] Linda Ness, *A stratification of the null cone via the moment map*, Amer. J. Math. **106** (1984), no. 6, 1281–1329. With an appendix by David Mumford. MR765581.

[OV90] Arkady L. Onishchik and Èrnest B. Vinberg, Lie groups and algebraic groups, Springer Series in Soviet Mathematics, Springer-Verlag, Berlin, 1990. Translated from the Russian and with a preface by D. A. Leites. MR1064110.

[Pal61] Richard S. Palais, *On the existence of slices for actions of non-compact Lie groups*, Ann. of Math. (2) **73** (1961), 295–323. MR0126506.

[RS90] Roger Wolcott Richardson and Peter Slodowy, *Minimum vectors for real reductive algebraic groups*, J. London Math. Soc. (2) **42** (1990), no. 3, 409–429. MR1087217 (92a:14055).

[Wal88] Nolan R. Wallach, *Real reductive groups. I*, Pure and Applied Mathematics, vol. 132, Academic Press, Inc., Boston, MA, 1988. MR929683.

2

Convex Ancient Solutions to Mean Curvature Flow

Theodora Bourni, Mat Langford and Giuseppe Tinaglia

Abstract

X.-J. Wang [33] proved a series of remarkable results on the structure of (non-compact) convex ancient solutions to mean curvature flow. Some of his results do not appear to be widely known, however, possibly due to the technical nature of his arguments and his exploitation of methods which are not widely used in mean curvature flow. In this expository chapter, we present Wang's structure theory and some of its consequences. We shall simplify some of Wang's analysis by making use of the monotonicity formula and the differential Harnack inequality, and obtain an important additional structure result by exploiting the latter. We conclude by showing that various rigidity results for convex ancient solutions and convex translators follow quite directly from the structure theory, including the new result of Corollary 2.8.3. We recently provided a complete classification of convex ancient solutions to curve shortening flow by exploiting similar arguments [12].

2.1 Introduction

A smooth one-parameter family $\{\mathcal{M}_t^n\}_{t \in I}$ of smoothly immersed hypersurfaces $\mathcal{M}_t^n$ of $\mathbb{R}^{n+1}$ *evolves by mean curvature flow* if there exists a smooth one-parameter family $X : M^n \times I \to \mathbb{R}^{n+1}$ of immersions $X(\cdot, t) : M^n \to \mathbb{R}^{n+1}$ with $\mathcal{M}_t^n = X(M^n, t)$ satisfying

$$\partial_t X(x,t) = \vec{H}(x,t) \quad \text{for all} \quad (x,t) \in M^n \times I \,,$$

where $\vec{H}(\cdot, t)$ is the mean curvature vector field of $X(\cdot, t)$. Unless otherwise stated, we shall *not* assume that M^n is compact.

We will refer to a hypersurface as (*strictly*) *convex* if it bounds a (strictly) convex body, *locally uniformly convex* if its second fundamental form is positive definite and (*strictly*) *mean convex* if its mean curvature (with respect to a consistent choice of unit normal field) is (positive) non-negative. We will refer to a mean curvature flow $\{\mathcal{M}_t^n\}_{t \in I}$ as compact/(strictly) convex/locally uniformly convex/mean convex if each of its timeslices $\mathcal{M}_t^n$ possesses the corresponding property.

An *ancient* solution to mean curvature flow is one which is defined on a time interval of the form $I = (-\infty, T)$, where $T \leq \infty$. Ancient solutions are of interest due to their natural role in the study of high curvature regions of the flow since they arise as limits of rescalings about singularities [20, 26, 27, 29, 30, 34, 35].

A special class of ancient solutions are the *translating solutions*. As the name suggests, these are solutions $\{\mathcal{M}_t^n\}_{t \in (-\infty, \infty)}$ which evolve by translation: $\mathcal{M}_{t+s}^n = \mathcal{M}_t^n + se$ for some fixed vector $e \in \mathbb{R}^{n+1}$. The timeslices $\mathcal{M}_t^n$ of a translating solution $\{\mathcal{M}_t^n\}_{t \in (-\infty, \infty)}$ are all congruent and satisfy the *translator equation*,

$$\vec{H} = e^\perp \,,$$

where $\cdot^\perp$ denotes projection onto the normal bundle. Translating solutions arise as blow-up limits at *type-II* singularities, which are still not completely understood [20, 29, 30] (although there is now a classification of mean convex examples in[1] $\mathbb{R}^3$ [1, 13, 24, 32, 33]). Understanding ancient and translating solutions is therefore of relevance to applications of the flow which require a controlled continuation of the flow through singularities.

Ancient solutions to mean curvature flow also arise in conformal field theory, where, according to Bakas and Sourdis [7], they describe 'to lowest order in perturbation theory ... the ultraviolet regime of the boundary renormalization group equation of Dirichlet sigma models'.

Further interest in ancient and translating solutions to geometric flows arises from their rigidity properties, which are analogous to those of complete minimal/constant mean curvature surfaces, harmonic maps and Einstein metrics; for example, when $n \geq 2$, under certain geometric conditions, the only convex ancient solutions to mean curvature flow are the shrinking spheres [23, 28, 33]. We present a new proof of this fact in Section 2.5.

---

[1] The classification of all proper translating curves in the plane is left as an exercise.

Convex ancient solutions to mean curvature flow are closely related to convex translating solutions.

(i) Consider a *complete* solution $u : \Omega \to \mathbb{R}$, $\Omega \subset \mathbb{R}^n$, to the PDE

$$\operatorname{div}\left(\frac{Du}{\sqrt{\sigma^2 + |Du|^2}}\right) = -\frac{1}{\sqrt{\sigma^2 + |Du|^2}}. \tag{2.1}$$

When $\sigma = 0$, u is the *arrival time*[2] of a mean convex ancient solution to mean curvature flow in $\mathbb{R}^n$ (i.e. the level sets of u form a mean convex ancient solution). In this case, (2.1) is called the *level set flow*. When $\sigma = 1$, the graph of u is a mean convex translator in $\mathbb{R}^{n+1}$ with bulk velocity $-e_{n+1}$. In this case, (2.1) is called the *graphical translator equation*. The converse is also true: every mean convex ancient solution to mean curvature flow (resp. proper, mean convex translating solution) gives rise to a complete solution to (2.1) with $\sigma = 0$ (resp. $\sigma = 1$). This connection was exploited by X.-J. Wang in [33]. We will not make use of it here.

(ii) If $\{\mathcal{M}_t^n\}_{t \in (-\infty, \infty)}$ is a convex translating solution, then the family of rescaled solutions $\{\lambda \mathcal{M}_{\lambda^{-2}t}^n\}_{t \in (-\infty, 0)}$ converges to a (self-similarly shrinking) convex ancient solution as $\lambda \to 0$. See Lemma 2.6.1 below.

(iii) If $\{\mathcal{M}_t^n\}_{t \in (-\infty, T)}$ is a convex ancient solution and $X_j \in \mathcal{M}_{t_j}^n$ denotes a sequence of points such that $t_j \to -\infty$ and $|X_j| \to \infty$ as $j \to \infty$, then a subsequence of the translated family $\{\mathcal{M}_{t+t_j}^n - X_j\}_{t \in (-\infty, T-t_j)}$ converges to a convex translating solution as $j \to \infty$. See Lemma 2.2.1 below.

2.2 Asymptotics for Convex Ancient Solutions

The following lemma describes the asymptotic properties of convex ancient solutions. It will be convenient to make use of the inverse P of the Gauss map, which is defined on a strictly convex solution by

$$\nu(P(e, t), t) = e.$$

---

[2] In this case, the equation is degenerate at critical points of u. At such points, u satisfies (2.1) in an appropriate *viscosity* sense but is a priori only continuous. When the level sets of u are convex and compact, results of Huisken [25, 27] imply that u has a single critical point, where it is C^2.

Lemma 2.2.1 (Asymptotic shape of convex ancient solutions)   *Suppose that $\{\mathcal{M}_t^n\}_{t \in (-\infty, 0)}$ is a convex ancient solution to mean curvature flow in $\mathbb{R}^{n+1}$.*

(i) (Blow-up)

    (a) *If $\{\mathcal{M}_t^n\}_{t \in (-\infty, 0)}$ is compact and 0 is its singular time, there exists a point $p \in \mathbb{R}^{n+1}$ such that the family of rescaled solutions $\{\lambda(\mathcal{M}_{\lambda^{-2}t}^n - p)\}_{t \in (-\infty, 0)}$ converges uniformly in the smooth topology to the shrinking sphere $\{S_{\sqrt{-2nt}}^n\}_{t \in (-\infty, 0)}$ as $\lambda \to \infty$.*

    (b) *If 0 is a regular time of $\{\mathcal{M}_t^n\}_{t \in (-\infty, 0)}$ and $0 \in \mathcal{M}_0^n$, then the family of rescaled solutions $\{\lambda(\mathcal{M}_{\lambda^{-2}t}^n)\}_{t \in (-\infty, 0)}$ converges locally uniformly in the smooth topology to the stationary hyperplane $\{X : \langle X, \nu(0) \rangle = 0\}_{t \in (-\infty, 0)}$ as $\lambda \to \infty$.*

(ii) (Blow-down) *There is a rotation $R \in \mathrm{SO}(n+1)$ such that the rescaled solutions $\{\lambda R \cdot \mathcal{M}_{\lambda^{-2}t}^n\}_{t \in (-\infty, 0)}$ converge locally uniformly in the smooth topology as $\lambda \to 0$ to either*

- *the shrinking sphere $\{S_{\sqrt{-2nt}}^n\}_{t \in (-\infty, 0)}$,*
- *the shrinking cylinder $\{\mathbb{R}^k \times S_{\sqrt{-2(n-k)t}}^{n-k}\}_{t \in (-\infty, 0)}$ for some $k \in \{1, \ldots, n-1\}$, or*
- *the stationary hyperplane $\{\mathbb{R}^n \times \{0\}\}_{t \in (-\infty, 0)}$ of multiplicity either one or two.*

Furthermore:

    (a) *If the limit is the shrinking sphere, then $\{\mathcal{M}_t^n\}_{t \in (-\infty, 0)}$ is a shrinking sphere.*

    (b) *If the limit is the stationary hyperplane of multiplicity one, then $\{\mathcal{M}_t^n\}_{t \in (-\infty, 0)}$ is a stationary hyperplane of multiplicity one.*

(iii) (Asymptotic translators) *Suppose that $\sup_{\mathcal{M}_t^n} |A| < \infty$ for each t. If the second fundamental form of the solution $\{\mathcal{M}_t^n\}_{t \in (-\infty, 0)}$ is bounded on each time interval $(-\infty, T]$, $T < 0$, then, given any sequence of times $s_j \to -\infty$ and a normal direction $e \in \cap_{s \in (-\infty, 0)} \nu(\mathcal{M}_s^n)$, the translated solutions $\{\mathcal{M}_{t+s_j}^n - P_j\}_{t \in (-\infty, -s_j)}$, where $P_j := P(e, s_j)$, converge locally uniformly in C^∞, along some subsequence, to a convex translating solution which translates in the direction $-e$ with speed $\lim\limits_{j \to -\infty} H(P_j, s_j)$.*

Proof   (i) (Blow-up) (a) is an immediate consequence of Huisken's theorem [25]. (b) is straightforward.

(ii) (Blow-down) This was proved by Wang through a delicate analysis of the arrival time [33]. We will derive it as a simple consequence of the monotonicity formula: after a finite translation, we can arrange that the solution reaches the origin at time zero. Since the shrinking sphere also reaches the origin at time zero, it must intersect the solution at all negative times by the avoidance principle. Thus,

$$\min_{q \in \mathcal{M}^n_t} |q| \le \sqrt{-2nt}\,.$$

Since the speed, H, of the solution is bounded in any compact subset of $\mathbb{R}^{n+1} \times (-\infty, 0)$ after the rescaling, given any sequence $\lambda_i \searrow 0$, we can find a subsequence along which $\{\lambda_i \mathcal{M}^n_{\lambda_i^{-2}t}\}_{t \in (-\infty, 0)}$ converges locally uniformly in the smooth topology to a non-empty limit flow.

We will show that the subsequential limit flow is a self-similarly shrinking solution using Huisken's monotonicity formula [26]:

$$\frac{d}{dt} \Theta(t) = - \int_{\mathcal{M}^n_t} \left| \vec{H}(p) + \frac{p^\perp}{-2nt} \right|^2 d\mathcal{H}^1(p)\,, \qquad (2.2)$$

where $\Theta(t)$ is the Gaussian area of $\mathcal{M}^n_t$ (based at the spacetime origin):

$$\Theta(t) := (-4\pi t)^{-\frac{n}{2}} \int_{\mathcal{M}^n_t} e^{-\frac{|X|^2}{-4t}} d\mathcal{H}^n(X)\,.$$

We claim that $\Theta(t)$ is bounded *uniformly*.

Claim 2.2.2   *There exists $C = C(n) < \infty$ such that*

$$\sup_{\tau > 0} \frac{1}{\tau^{n/2}} \int_{\mathcal{M}^n} e^{\frac{-|y|^2}{\tau}} d\mathcal{H}^n(y) < C$$

for any convex hypersurface $\mathcal{M}^n$ of $\mathbb{R}^{n+1}$ and any $\tau > 0$.

Proof   Let $A_i = B_{(i+1)\sqrt{k}}(0) \setminus B_{i\sqrt{k}}(0)$ for $i \in \mathbb{N}$. Since $\mathcal{M}^n$ is convex,

$$|\mathcal{M}^n \cap A_i| \le |B_{(i+1)\sqrt{k}}(0)| = c_n (i+1)^n k^{n/2}$$

and hence

$$\frac{1}{\tau^{n/2}} \int_{\mathcal{M}^n} e^{\frac{-|y|^2}{\tau}} d\mathcal{H}^n(y) = \frac{1}{\tau^{n/2}} \sum_{i=0}^{\infty} \int_{\mathcal{M}^n \cap A_i} e^{\frac{-|y|^2}{\tau}} d\mathcal{H}^n(y)$$

$$\le \frac{1}{\tau^{n/2}} \sum_{i=0}^{\infty} c_n (i+1)^n \tau^{n/2} e^{-i^2} = c_n \sum_{i=0}^{\infty} (i+1)^n e^{-i^2}\,,$$

where the constant c_n depends only on n. This proves the claim.   □

It follows that Θ converges to some limit as $t \to -\infty$. But then

$$-\int_a^b \int_{\lambda \mathcal{M}^n_{\lambda^{-2}t}} \left| \vec{H}(p) + \frac{p^\perp}{-2nt} \right|^2 d\mathcal{H}^1(p) = \Theta_\lambda(b) - \Theta_\lambda(a)$$

$$= \Theta(\lambda^{-2}b) - \Theta(\lambda^{-2}a) \to 0,$$

as $\lambda \to 0$ for any $a < b < 0$, where Θ_λ is the Gaussian area of the λ-rescaled flow. We conclude that the integrand vanishes identically in the limit, and hence any limit of $\{\lambda \mathcal{M}^n_{\lambda^{-2}}\}_{t \in (-\infty,0)}$ along a sequence of scales $\lambda_i \to 0$ is a self-similarly shrinking solution to mean curvature flow.

A well-known result of Huisken [27] (see also Colding and Minicozzi [18]) implies that the only convex examples which can arise are the shrinking spheres, the shrinking cylinders and the stationary hyperplanes of multiplicity either one or two. If the limit flow is the shrinking sphere, then the solution is compact and hence, by part (i), its 'blow-up' is also the shrinking sphere. The monotonicity formula then implies that the Gaussian area is constant, and we conclude that the solution is the shrinking sphere. Similarly, if the limit is the hyperplane of multiplicity one then the solution is the hyperplane of multiplicity one, since the blow-up about any regular point (without loss of generality, the space-time origin) is also a hyperplane of multiplicity one (see [36, Prop. 2.10]). Finally, we note that the limit is unique since convexity ensures that the limiting convex region enclosed by any subsequential limit is contained in the limiting convex region enclosed by any other subsequential limit.

(iii) (Asymptotic translators) This is a consequence of Hamilton's Harnack inequality [21]. (We will in fact make use of Andrews' interpretation of the differential Harnack inequality using the Gauss map parametrization [2].) It implies that the mean curvature of a convex ancient solution to mean curvature flow is pointwise non-decreasing in the Gauss map parametrization. In particular, the limit

$$H_\infty(e) := \lim_{s \to -\infty} H(P(e,s),s)$$

exists. Moreover, the curvature of the family of translated flows is uniformly bounded on any compact time interval. It follows that some subsequence converges locally uniformly in C^∞ to a weakly convex eternal limit mean curvature flow. Since the curvature of the limit is constant in time with respect to the Gauss map parametrization, the rigidity case of the Harnack inequality [21] implies that it moves by translation, with velocity $-H_\infty(e)e$. $\qquad\square$

Remark 2.2.3   In case (iii) of Lemma 2.2.1, the limit will simply be a stationary hyperplane if $\lim\limits_{s\to-\infty} H(P(e,s),s) = 0$. We refer to the 'forwards' limit in part (i) as the *blow-up* of the solution, the 'backwards' limit in part (ii) as its *blow-down* and the translating limits in part (iii) as *asymptotic translators*. In certain cases, the asymptotic translator is unique, and hence the convergence holds for all $s \to -\infty$.

2.3 X.-J. Wang's Dichotomy for Convex Ancient Solutions

Recall that the arrival time $u : \cup_{t\in I}\mathcal{M}_t^n \to \mathbb{R}$ of a mean curvature flow $\{\mathcal{M}_t^n\}_{t\in I}$ of mean convex boundaries $\mathcal{M}_t^n = \partial\Omega_t$ is defined by

$$u(X) = t \iff X \in \mathcal{M}_t^n.$$

Lemma 2.3.1   *The arrival time of a convex ancient solution to mean curvature flow with bounded curvature on each timeslice is locally concave.*

Proof   Straightforward calculations show that the arrival time u of a convex ancient solution $\{\mathcal{M}_t^n\}_{t\in(-\infty,T)}$ to mean curvature flow satisfies

$$Du = -\frac{1}{H}\nu$$

and

$$D^2u = \begin{pmatrix} -A/H & \nabla H/H^2 \\ \nabla H/H^2 & -\partial_t H/H^3 \end{pmatrix}.$$

Fix a point $p \in \mathcal{M}_t^n$ and any vector $V = V^\top + \alpha\nu(p,t) \in T_p\mathbb{R}^{n+1}$. If $\alpha = 0$, then

$$-HD^2u(V,V) = A(V^\top, V^\top) \geq 0.$$

Otherwise, we can scale V so that $\alpha = -H$, in which case the differential Harnack inequality [21] yields

$$-HD^2u(V,V) = A(V^\top, V^\top) + 2\nabla_{V^\top}H + \partial_t H \geq 0.$$

$\square$

In the compact case, concavity of the arrival time can also be obtained from the concavity maximum principle [33, Lem. 4.1 and 4.4].

Definition 2.3.2 An ancient mean curvature flow $\{\mathcal{M}_t^n\}_{t\in(-\infty,T)}$ of mean convex boundaries $\mathcal{M}_t^n = \partial\Omega_t$ shall be called *entire* if it *sweeps out all of space*, in the sense that $\cup_{t\in(-\infty,T)}\mathcal{M}_t^n = \mathbb{R}^{n+1}$. Equivalently, its arrival time is an entire function.

Observe that the shrinking sphere and the bowl soliton are entire ancient solutions, whereas the paperclip and grim reaper are not entire (they only sweep out 'strip' regions between two parallel lines).

Any convex ancient solution which is not entire necessarily lies at all times inside some stationary halfspace. The following remarkable theorem of X.-J. Wang states that, in fact, such a solution must lie at all times in a stationary *slab* region (the region between two parallel hyperplanes).

Theorem 2.3.3 (X.-J. Wang's dichotomy for convex ancient solutions [33, Cor. 2.2]) *Let $\{\mathcal{M}_t^n\}_{t\in(-\infty,0)}$ be a convex ancient solution to mean curvature flow in $\mathbb{R}^{n+1}$ with $\sup_{\mathcal{M}_t^n}|A| < \infty$ for each t. If its blow-down is a stationary hyperplane of multiplicity two, then $\{\mathcal{M}_t^n\}_{t\in(-\infty,0)}$ lies in a stationary slab region. In particular, every convex ancient solution to mean curvature flow with bounded curvature on each timeslice is either entire or lies in a stationary slab region.*

Proof Let $\{\mathcal{M}_t^n\}_{t\in(-\infty,0)}$ be a convex ancient solution which is not entire. After a rotation, we can assume, by Lemma 2.2.1, that the rescaled flows $\{\lambda\mathcal{M}_{\lambda^{-2}t}^n\}_{t\in(-\infty,0)}$ converge as $\lambda \to 0$ to the stationary hyperplane $L := \{x \in \mathbb{R}^{n+1} : x_1 = 0\}$ with multiplicity 2. Since $\mathcal{M}_t^n$ is convex, we may represent each timeslice $\mathcal{M}_t^n$ as the union of the graphs of two functions $v^{\pm}(\,\cdot\,,t)$ over a domain $V_t \subset L$. That is,

$$\mathcal{M}_t^n = \{(v^+(y,t),y) : y \in V_t\} \cup \{(v^-(y,t),y) : y \in V_t\},$$

with $v^+(\,\cdot\,,t) : V_t \to \mathbb{R}$ concave, $v^- : V_t \to \mathbb{R}$ convex, and v^+ lying above v^-. Thus, the function

$$v := v^+ - v^-$$

is positive and concave in its space variable at each time. Note also that $v(p,t) = 0$ for each $p \in \partial V_t$. We need to show that $v(\,\cdot\,,t)$ stays uniformly bounded as $t \to -\infty$.

Since their graphs move by mean curvature flow, the functions v^+ and v^- satisfy

$$\frac{\partial v^{\pm}}{\partial t} = \sqrt{1 + |Dv^{\pm}|^2}\,\mathrm{div}\left(\frac{Dv^{\pm}}{\sqrt{1 + |Dv^{\pm}|^2}}\right). \tag{2.3}$$

Since the blow-down of the solution is the hyperplane $\{X \in \mathbb{R}^{n+1} : \langle X, e_1 \rangle = 0\}$ of multiplicity two, there exists, for every $\varepsilon > 0$, some $t_\varepsilon > -\infty$ such that

$$|p| \geq \varepsilon^{-1} \sqrt{-t} \text{ for all } p \in \partial V_t \qquad (2.4a)$$

and

$$v(0, t) \leq \varepsilon \sqrt{-t} \qquad (2.4b)$$

for all $t < t_\varepsilon$.

Since the solution is convex, it suffices, by (2.4a), to show that $v(0, t)$ stays bounded as $t \to -\infty$. We achieve this with the following two claims.

Claim 2.3.4 ([33, Claim 1 in Lem. 2.1 and 2.2, and Lem. 2.6]) *There exists $t_0 < 0$ and $\alpha > 0$ such that*

$$|p| \, v(0, t) \geq -\alpha t \, (> 0)$$

for each $p \in \partial V_t$ and every $t \leq t_0$.

Proof We will use equations (2.4a) and (2.4b) to show that the tangent planes to v^+ and v^- at the origin are almost horizontal for $t \ll 0$, which will allow us to estimate $|p| \, v(0, t)$ by the area in the plane $e_1 \wedge p$ enclosed by $\mathcal{M}_t$ and $\{\langle X, p \rangle = 0\}$. The argument is quite simple when $n = 1$ and the solution is compact. Removing these hypotheses introduces some technical difficulties, but the idea is essentially the same.

Fix $t < t_\varepsilon$ and $p \in \partial V_t$. By rotating about the x_1-axis, we can arrange that $p = de_2$, where $d > 0$. Set

$$q^\pm(t) := v^\pm(0, t) \, e_1 \, .$$

By the convexity/concavity of the respective graphs, the segment connecting $(p, v^+(p))$ to $q^+(t)$ lies below the graph of $v^+(\cdot, t)$, and the segment connecting (p, v^-) to $q^-(t)$ lies above the graph of $v^-(\cdot, t)$. Thus, comparing their slopes with the slope of the tangents to $v^\pm(\cdot, t)$ at 0 and applying (2.4), we find that

$$\partial_2 v^+(0, t) - \partial_2 v^-(0, t) \geq -\frac{v^+(0, t) - v^-(0, t)}{d} \geq -2\varepsilon^2 \, . \qquad (2.5)$$

If the ray $\{re_2 : r < 0\}$ does not intersect ∂V_t, then, since $\mathcal{M}_t^n$ is convex,

$$\partial_2 v^+(0, t) - \partial_2 v^-(0, t) \leq 0 \, ,$$

Otherwise, applying the same argument to the point $p' := \{re_2 : r < 0\} \cap \partial V_t$, we obtain

$$|\partial_2 v^+(0, t) - \partial_2 v^-(0, t)| \leq 2\varepsilon^2 \, .$$

Denote by $\widehat{\mathcal{M}}_t$ the intersection of $\mathcal{M}_t$ with the plane $e_1 \wedge e_2$. Since the normal velocity of $\widehat{\mathcal{M}}_t$ in $e_1 \wedge e_2$ is the projection onto $e_1 \wedge e_2$ of $-H\nu$, the inward normal speed $\widehat{H}$ of $\widehat{\mathcal{M}}_t$ is given by

$$\widehat{H} = H \frac{|D\widehat{u}|}{|Du|},$$

where $\widehat{u}$ is the restriction of the arrival time u to $e_1 \wedge e_2$. On the other hand, since u is locally concave, the curvature $\widehat{\kappa}$ of $\widehat{\mathcal{M}}_t$ can be estimated by

$$\begin{aligned}
\widehat{\kappa} &= -\frac{1}{|D\widehat{u}|}\left(\Delta\widehat{u} - \frac{D^2\widehat{u}\,(D\widehat{u}, D\widehat{u})}{|D\widehat{u}|^2}\right) \\
&= -\frac{D^2\widehat{u}\,(\widehat{\tau}, \widehat{\tau})}{|D\widehat{u}|} \\
&= -\frac{1}{|D\widehat{u}|}\left(\Delta u - \frac{D^2 u\,(Du, Du)}{|Du|^2} - \mathrm{tr}_{\widehat{\tau}^\perp}D^2 u\right) \\
&\leq \frac{|Du|}{|D\widehat{u}|}H,
\end{aligned}$$

where $\widehat{\tau}$ is a unit tangent vector field to $\widehat{\mathcal{M}}_t$ and where $\widehat{\tau}^\perp$ is its orthogonal compliment in $T\mathcal{M}_t$. Thus,

$$\widehat{H} \geq \widehat{\kappa}\,\frac{|D\widehat{u}|^2}{|Du|^2}.$$

We claim that $\frac{|D\widehat{u}|^2}{|Du|^2}$ is close to 1. Fix $x \in \widehat{\mathcal{M}}_t$. For each $k = 3, \dots, n+1$, let $z_k e_k$, $z_k \in \mathbb{R}$, be the point at which the tangent plane $T_x\mathcal{M}_t$ intersects the e_k-axis (if it exists). Since $\langle \nu, z_k e_k - x\rangle = 0$, we find, when $\nu_k \neq 0$,

$$z_k \nu_k = x_1 \nu_1 + x_2 \nu_2.$$

Without loss of generality, $x_1 \leq x_2 \leq z_k$ for each k. Then

$$x_1^2 + x_2^2 \leq 2z_k^2$$

and hence

$$\nu_k^2 \leq 2(\nu_1^2 + \nu_2^2)$$

for each $k = 3, \dots, n+1$. Since $|\nu|^2 = 1$, this implies that

$$\nu_1^2 + \nu_2^2 \geq \frac{1}{2n-1},$$

and we conclude that

$$|D\widehat{u}|^2 = \langle Du, D\widehat{u} \rangle \geq \frac{1}{2n-1} |Du||D\widehat{u}|.$$

That is,

$$\frac{|D\widehat{u}|^2}{|Du|^2} \geq \frac{1}{(2n-1)^2}.$$

We can now estimate

$$\frac{d}{dt} \mathscr{H}^2(\widehat{\Omega}_t \cap \{\langle X, e_2 \rangle > 0\}) = -\int_{\widehat{\mathcal{M}}_t \cap \{\langle X, e_2 \rangle > 0\}} \widehat{H}\, ds$$

$$\leq -\frac{1}{(2n-1)^2} \int_{\widehat{\mathcal{M}}_t \cap \{\langle X, e_2 \rangle > 0\}} \widehat{\kappa}\, ds$$

$$\leq -\frac{1}{(2n-1)^2} \left(\pi - 2\varepsilon^2\right),$$

where $\widehat{\Omega}_t$ is the convex region in $e_1 \wedge e_2$ bounded by $\widehat{\mathcal{M}}_t$. Integrating this between $t < 4t_\varepsilon$ and t_ε yields, upon choosing $\varepsilon = \frac{\sqrt{\pi}}{2}$,

$$\mathscr{H}^2(\widehat{\Omega}_t^2 \cap \{\langle X, e_2 \rangle > 0\}) \geq -\frac{3(\pi - 2\varepsilon^2)}{4(2n-1)^2} t \geq -\frac{3\pi}{8(2n-1)^2} t.$$

The convex region $\widehat{\Omega}_t$ lies between the tangent hyperplanes to $v^\pm(\,\cdot\,, t)$ at the origin. By (2.5), these tangent hyperplanes intersect the line $\langle X, e_2 \rangle = d$ in $e_1 \wedge e$ at two points with distance at most $2(v^+(0,t) - v^-(0,t))$. Comparing $\mathscr{H}^2(\widehat{\Omega}_t^2 \cap \{\langle X, e_2 \rangle > 0\})$ with the area of this enclosing trapezium yields

$$\frac{3}{2} v(0,t)\, |p| \geq -\frac{3\pi}{8(2n-1)^2} t,$$

which completes the proof of the claim.   □

For each $k \in \mathbb{N}$, set

$$t_k := 2^k t_\varepsilon, \quad v_k := v(\,\cdot\,, t_k), \quad \text{and} \quad d_k := \min_{p \in \partial V_{t_k}} |p|,$$

where $\varepsilon > 0$ is to be determined and t_ε is chosen as in (2.4).

Claim 2.3.5 [33, Claim 2 in Lem. 2.1 and 2.2, and Lem. 2.7] *There exists $\varepsilon > 0$ such that*

$$v_k(0) \leq v_{k-1}(0) + 2^{-\frac{k}{4n}} \sqrt{-t_\varepsilon} \tag{2.6}$$

for all $k \geq 1$.

Proof Since the arrival time u of $\{\mathcal{M}_t^n\}_{t\in(-\infty,0)}$ is locally concave, and since

$$u(v^{\pm}(y,t),y) = t$$

for all $t < 0$ and $y \in V_t$, we find that the functions $t \mapsto v^+(y,t)$ and $t \mapsto -v^-(y,t)$, and hence also $t \mapsto v(y,t)$, are concave for each fixed y.

It follows that

$$\frac{d}{dt}\frac{v(y,t)}{-t} \geq 0 \text{ for fixed } y \in V_{t_0} \text{ and every } t < t_0. \tag{2.7}$$

In particular,

$$\frac{v_k(0)}{-t_k} \leq \frac{v_0(0)}{-t_0}$$

and hence, by (2.4b),

$$v_k(0) \leq 2^k v_0(0) \leq 2^{k+1}\varepsilon\sqrt{-t_\varepsilon}.$$

Given $k_0 \geq 8n$ (to be determined momentarily), we choose $\varepsilon = \varepsilon(k_0)$ small enough that $2^{k_0+1}\varepsilon \leq 2^{-\frac{k_0}{4n}}$, so that

$$v_k(0) \leq 2^{-\frac{k_0}{4n}}\sqrt{-t_\varepsilon} \leq \frac{1}{4}\sqrt{-t_\varepsilon} \text{ for all } k \leq k_0. \tag{2.8}$$

In particular, (2.6) holds for each $k \leq k_0$. We will prove the claim by induction on k.

So suppose that (2.6) holds up to some $k \geq k_0$. Then

$$v_k(0) \leq v_{k_0}(0) + \sqrt{-t_\varepsilon}\sum_{i=k_0+1}^{k}2^{-\frac{i}{4n}},$$

where the second term on the right hand side is taken to be zero if $k = k_0$. Since $\sum_{j=1}^{\infty}2^{-\frac{j}{4n}} < \infty$, we can choose k_0 so that $\sum_{j=k_0+1}^{\infty}2^{-\frac{j}{4n}} < 1/4$. Applying (2.8), we then obtain

$$v_k(0) \leq \frac{1}{2}\sqrt{-t_\varepsilon}. \tag{2.9}$$

By (2.7),

$$\frac{v_{k+1}(0)}{-t_{k+1}} \leq \frac{v_k(0)}{-t_k}$$

and hence

$$v_{k+1}(0) \leq 2v_k(0) \leq \sqrt{-t_\varepsilon}.$$

Since $t \mapsto v(0,t)$ is decreasing, we conclude that

$$v(0,t) \le \sqrt{-t_\varepsilon} \quad \text{for every } t \ge t_{k+1}.$$

Since $d(t) := \min_{p \in \partial V_t} |p|$ is decreasing in t, Claim 2.3.4 implies that there exists $\alpha > 0$ such that

$$d_{k+1} \ge d_k \ge \alpha \frac{-t_k}{v_k(0)} \ge \alpha \frac{-t_k}{\sqrt{-t_\varepsilon}} \tag{2.10a}$$

and

$$d_{k-1} \ge \alpha \frac{-t_{k-1}}{\sqrt{-t_\varepsilon}} = \frac{\alpha}{2} \frac{-t_k}{\sqrt{-t_\varepsilon}}. \tag{2.10b}$$

Now define

$$D_k := \left\{ y \in \mathbb{R}^n : |y| < \frac{\alpha}{2} \frac{-t_k}{\sqrt{-t_\varepsilon}} \right\}. \tag{2.11}$$

Since $y \mapsto v(y,t)$ is concave for fixed t,

$$v(y + se, t) - v(y,t) \le s D_e v(y,t)$$

for any $y \in V_t$, any unit vector e and any s such that $y + se \in V_t$. If $(y,t) \in D_k \times [t_{k+1}, t_k]$ and the ray $\{y + re : r > 0\}$ intersects ∂V_t, we can choose s so that $y + se \in \partial V_t$ and hence

$$D_e v(y,t) \ge -\frac{v(y,t)}{s} \ge -\frac{v(y,t)}{d(t) - |y|}.$$

If, however, the ray $\{y + re : r > 0\}$ does not intersect ∂V_t, then, since $\mathcal{M}_t^n$ is convex, $D_e v(y,t) \ge 0$. Applying the same reasoning with e replaced by $-e$ yields

$$D_e v(y,t) \le \frac{v(y,t)}{d(t) - |y|}.$$

We conclude the gradient estimate

$$|Dv(y,t)| \le \frac{v(y,t)}{d(t) - |y|} \tag{2.12}$$

for all $(y,t) \in D_k \times [t_{k+1}, t_k]$.

Since $t \mapsto v(y,t)$ is concave for fixed y, (2.12) implies

$$v(y,t) \le v(0,t) + |y| |Dv(0,t)|$$

$$\le v(0,t) \left(1 + \frac{|y|}{d(t)} \right)$$

$$\le 2v(0,t).$$

These estimates, (2.10) and the concavity of $t \mapsto v(y,t)$ yield, for all $(y,t) \in D_k \times [t_{k+1}, t_k]$,

$$|Dv(y,t)| \leq \frac{v(y,t)}{d_k - |y|} \leq 4\alpha^{-1} \frac{-t_\varepsilon}{-t_k} \qquad (2.13)$$

and

$$0 \leq -\partial_t v(y,t) \leq \frac{v(y,t)}{-t} \leq \frac{v_k(y)}{-t_k} \leq \frac{2\sqrt{-t_\varepsilon}}{-t_k}. \qquad (2.14)$$

We will use the gradient estimate to bound $\Delta v(y,t)$ in $D_k \times [t_{k+1}, t_k]$ in terms of t_k. Given a positive function $f : \mathbb{N} \to \mathbb{R}$, to be determined momentarily, define

$$\chi_k := \{(y,t) \in D_k \times [t_{k+1}, t_k] : -\Delta v(y,h) \geq f(k)\}.$$

Recalling (2.13) we find, for any $t \in (t_{k+1}, t_k)$,

$$\mathcal{H}^n(\{y \in D_k : (y,t) \in \chi_k\}) f(k) \leq -\int_{D_k} \Delta v(\,\cdot\,,t)\, d\mathcal{H}^n$$

$$\leq \int_{\partial D_k} |Dv(\,\cdot\,,t)|\, d\mathcal{H}^{n-1}$$

$$\leq \sup_{D_k} |Dv(\,\cdot\,,t)|\, \mathcal{H}^{n-1}(D_k)$$

$$\leq C \frac{t_\varepsilon}{t_k} \left(\frac{-t_k}{\sqrt{-t_\varepsilon}} \right)^{n-1}$$

$$= C \frac{2^{k(n-1)}}{-t_k} (-t_\varepsilon)^{\frac{n+1}{2}},$$

where C is a constant which depends only on n. Integrating between t_{k+1} and t_k then yields

$$\mathcal{H}^{n+1}(\chi_k) \leq C \frac{2^{k(n-1)}}{f(k)} (-t_\varepsilon)^{\frac{n+1}{2}} \cdot \frac{t_k - t_{k+1}}{-t_k} = 2C \frac{2^{k(n-1)}}{f(k)} (-t_\varepsilon)^{\frac{n+1}{2}}.$$

Consider now another positive function $g : \mathbb{N} \to \mathbb{R}$, to be determined later. Since

$$\int_{D_k} \mathcal{H}^1(\chi_k \cap \{(y,t) : y = z\})\, d\mathcal{H}^n(z) = \mathcal{H}^{n+1}(\chi_k)$$

$$\leq 2C \frac{2^{k(n-1)}}{f(k)} (-t_\varepsilon)^{\frac{n+1}{2}},$$

there exists $\widehat{D}_k \subset D_k$ with

$$\mathcal{H}^n(\widehat{D}_k) \leq 2C \frac{2^{k(n-1)}}{f(k)g(k)} (-t_\varepsilon)^{\frac{n+1}{2}}$$

such that

$$\mathscr{H}^1(\chi_k \cap \{(y,t) : y = z\}) \le g(k) \text{ for all } z \in D_k \setminus \widehat{D}_k. \qquad (2.15)$$

Now, for any $y \in D_k \setminus \widehat{D}_k$,

$$v_{k+1}(y) - v_k(y) = -\int_{t_{k+1}}^{t_k} \partial_t v(y,s)\, ds$$

$$= -\int_{[t_{k+1},t_k] \setminus I_k(y)} \partial_t v(y,s)\, ds - \int_{I_k(y)} \partial_t v(y,s)\, ds, \qquad (2.16)$$

where $I_k(y) := \{t : (y,t) \in \widehat{D}_k\}$. Using (2.14) and (2.15) to bound the first integral on the right hand side of (2.16) and the graphical mean curvature flow equation (2.3) to bound the second, we find, for any $y \in D_k \setminus \widehat{D}_k$,

$$v_{k+1}(y) - v_k(y) \le \frac{2\sqrt{-t_\varepsilon}}{-t_k} g(k) - f(k) t_k. \qquad (2.17)$$

We now choose $f(k) = \frac{2^{-k(1+\beta/2)}}{\sqrt{-t_\varepsilon}}$ and $g(k) = 2^{k(1-\beta/2)}(-t_\varepsilon)$, where $\beta \in (0,1)$ will be chosen explicitly below. Then

$$\mathscr{H}^n(D_k \setminus \widehat{D}_k) \ge \omega_n \left(\frac{-\pi t_k}{8\sqrt{-t_\varepsilon}} \right)^n - 2^{k\beta+1} C \sqrt{-t_\varepsilon} \left(\frac{-t_k}{\sqrt{-t_\varepsilon}} \right)^{n-1}$$

$$= \left(\omega_n \left(\frac{\pi}{8} \right)^n 2^k - 2^{k\beta+1} C \right) \sqrt{-t_\varepsilon} \left(\frac{-t_k}{\sqrt{-t_\varepsilon}} \right)^{n-1}.$$

Since

$$\mathscr{H}^n(\widehat{D}_k) \le 2^{(\beta+n-1)k+1} C(-t_\varepsilon)^{\frac{n}{2}},$$

there is a point $y_0 \in D_k \setminus \widehat{D}_k$ with

$$0 < |y_0| \le \left(\frac{2C}{\omega_n} \right)^{\frac{1}{n}} 2^{k\left(\frac{\beta-1}{n}+1 \right)} \sqrt{-t_\varepsilon} \quad \text{and} \quad v_k(y_0) \le v_k(0). \qquad (2.18)$$

Since $y \mapsto v(y, t_{k+1})$ is concave and zero on ∂V_t, we find

$$v_{k+1}(0) \le \frac{d_{k+1}}{d_{k+1} - |y|} v_{k+1}(y_0).$$

Set $C' := \left(\frac{2C}{\omega_n}\right)^{\frac{1}{n}}$. Then (2.10a) implies that

$$
\frac{d_{k+1}}{d_{k+1} - |y|} \leq \frac{d_{k+1}}{d_{k+1} - 2^{\left(\frac{\beta-1}{n}+1\right)k}C'\sqrt{-t_\varepsilon}}
$$

$$
= 1 + \frac{2^{\left(\frac{\beta-1}{n}+1\right)k}C'\sqrt{-t_\varepsilon}}{d_{k+1} - 2^{\left(\frac{\beta-1}{n}+1\right)k}C'\sqrt{-t_\varepsilon}}
$$

$$
\leq 1 + 2C' \cdot 2^{\frac{\beta-1}{n}k}
$$

provided k_0 is chosen sufficiently large (depending on β). Thus,

$$
v_{k+1}(0) \leq \left(1 + 2C' \cdot 2^{\frac{\beta-1}{n}k}\right) v_{k+1}(y_0) .
$$

Applying (2.17), estimating $v_k(y_0) \leq v_k(0)$ and recalling (2.9), we find

$$
v_{k+1}(0) \leq (1 + 2C' \cdot 2^{\frac{\beta-1}{n}k})(v_k(0) + 3\sqrt{-t_\varepsilon}\, 2^{-\frac{\beta}{2}k})
$$

$$
\leq v_k(0) + \sqrt{-t_\varepsilon}\left(C' \cdot 2^{\frac{\beta-1}{n}k} + 3 \cdot 2^{-\frac{\beta}{2}k} + 6C' \cdot 2^{\left(\frac{\beta-1}{n}-\frac{\beta}{2}\right)k}\right) .
$$

Choosing now $\beta := 5/8 \in (1/2n, 3/4)$, we obtain, provided k_0 is sufficiently large,

$$
v_{k+1}(0) \leq v_k(0) + \sqrt{-t_\varepsilon}\, 2^{-\frac{k}{4n}} .
$$

This completes the proof of the claim. $\square$

We conclude that

$$
v(0, 2^{k+1}t_\varepsilon) = v_{k+1}(0) \leq v_0(0) + C \sum_{j=1}^{k} 2^{-\frac{j}{4n}} ,
$$

where C is independent of k. Since the sum $\sum_{j=1}^{\infty} 2^{-\frac{j}{4n}}$ is finite and $v(0,t)$ is monotone, we conclude that $v(0,t)$ is bounded uniformly in time.

This completes the proof of Theorem 2.3.3. $\square$

We note that F. Chini and N. Moller [16] have recently obtained a classification of the *convex hulls* of (non-convex) ancient solutions to mean curvature flow more generally.

Theorem 2.3.3 motivates the following definitions.

Definition 2.3.6   A convex, compact ancient solution to mean curvature flow is called

- an *ancient ovaloid* if it is entire;
- an *ancient pancake* if it lies in a stationary slab region.

The Angenent oval is a well-known example of an ancient pancake, while the shrinking sphere is an example of an ancient ovaloid. Further ancient ovaloids were constructed by White [35]. Further ancient pancakes and ancient ovaloids were constructed by Wang [33]. Haslhofer and Hershkovits gave an explicit construction of a family of bi-symmetric ovaloids [23], and Angenent, Daskalopoulos and Šešum proved that there is only one (other than the shrinking sphere) which is uniformly two-convex and non-collapsing[3] [5, 6]. This is related to work of Brendle and Choi, who proved that the bowl is the only non-compact ancient solution which is uniformly two convex and non-collapsing [14, 15]. We gave an explicit construction of a rotationally symmetric ancient pancake and (by analyzing the asymptotic translators) proved that it is the only rotationally symmetric example [9]. A survey of these results can be found in [11].

2.4 Convex Ancient Solutions to Curve Shortening Flow

Lemma 2.2.1(ii) and Theorem 2.3.3 immediately imply that the shrinking spheres are the only convex ancient solutions to curve shortening flow which do not lie in slab regions [33]. By making use of the differential Harnack inequality (as in Lemma 2.2.1(iii)), an elementary enclosed area analysis and Alexandrov's method of moving planes, we were able to obtain a complete classification of convex ancient solutions to curve shortening flow [12].

Theorem 2.4.1 *The only convex ancient solutions to curve shortening flow are the stationary lines, the shrinking circles, the Angenent ovals and the grim reapers.*

The classification of *compact* convex ancient solutions to curve shortening flow (shrinking circles and Angenent ovals) was obtained, using different methods, by Daskalopoulos, Hamilton and Šešum in 2010 [19].

Sections 2.5 and 2.8 illustrate how some of these ideas work in higher dimensions.

---

[3] In the sense of Sheng and Wang [31] (see also [3, 4]).

2.5 Rigidity of the Shrinking Sphere

We will use the asymptotics of the previous section to give a new proof of the following result of Huisken and Sinestrari [28] (see also Haslhofer and Hershkovits [23] and Wang [33, Rem. 3.1]).

Corollary 2.5.1 *Let $\{\mathcal{M}_t^n\}_{t\in(-\infty,0)}$, $n \geq 2$, be a compact, convex ancient solution to mean curvature flow. The following are equivalent:*

1 $\{\mathcal{M}_t^n\}_{t\in(-\infty,0)}$ is a shrinking sphere $\{S_{\sqrt{-2nt}}^n(p)\}_{t\in(-\infty,0)}$, $p \in \mathbb{R}^{n+1}$.

2 $\{\mathcal{M}_t^n\}_{t\in(-\infty,0)}$ is uniformly pinched:

$$\liminf_{t\to-\infty} \min_{M^n\times\{t\}} \frac{\kappa_1}{H} > 0\,.$$

3 $\{\mathcal{M}_t^n\}_{t\in(-\infty,0)}$ has bounded rescaled (extrinsic) diameter:

$$\limsup_{t\to-\infty} \frac{\mathrm{diam}(\mathcal{M}_t)}{\sqrt{-t}} < \infty\,.$$

4 $\{\mathcal{M}_t^n\}_{t\in(-\infty,0)}$ has bounded eccentricity:

$$\limsup_{t\to-\infty} \frac{\rho_+(t)}{\rho_-(t)} < \infty\,,$$

where $\rho_+(t)$ and $\rho_-(t)$ denote, respectively, the circum- and in-radii of $\mathcal{M}_t^n$.

5 $\{\mathcal{M}_t^n\}_{t\in(-\infty,0)}$ has Type-I curvature decay:

$$\limsup_{t\to-\infty} \sqrt{-t} \max_{M^n\times\{t\}} H < \infty\,.$$

6 $\{\mathcal{M}_t^n\}_{t\in(-\infty,0)}$ has bounded speed ratios:

$$\limsup_{t\to-\infty} \frac{\max_{M^n\times\{t\}} H}{\min_{M^n\times\{t\}} H} < \infty\,.$$

7 $\{\mathcal{M}_t^n\}_{t\in(-\infty,0)}$ satisfies a reverse isoperimetric inequality:

$$\limsup_{t\to-\infty} \frac{\mu_t(\mathcal{M})^{n+1}}{|\Omega_t|^n} < \infty\,.$$

Proof Let $\{\mathcal{M}_t^n\}_{t\in(-\infty,0)}$ be a compact, convex ancient solution satisfying one of the conditions 2–7. By Lemma 2.2.1(ii)(a), it suffices to prove that the blow-down of $\{\mathcal{M}_t^n\}_{t\in(-\infty,0)}$ is the shrinking sphere.

Condition 2 (Uniform pinching). Since the pinching condition is scale invariant and violated on any shrinking cylinder $\mathbb{R}^k \times S_{\sqrt{-2(n-k)t}}^{n-k}$,

$k \in \{1, \ldots, n-1\}$, it suffices, by Lemma 2.2.1(ii), to rule out the hyperplane of multiplicity two.

So suppose, to the contrary, that the blow-down is the hyperplane of multiplicity two. theorem 2.3.3 then implies that $\{\mathcal{M}_t^n\}_{t\in(-\infty,0)}$ lies in a slab. By translating parallel to the slab, we obtain, by part (iii) of Lemma 2.2.1, an asymptotic translator Σ^n which is non-trivial since it lies in a slab orthogonal to its translation vector. Thus, by the strong maximum principle, it must satisfy $H > 0$. By the pinching condition, a theorem of Hamilton [20] then implies that Σ^n is compact, which is impossible.

Condition 3 (Bounded rescaled diameter) implies that the blow-down is compact. The theorem in this case then follows immediately from Lemma 2.2.1(ii)(a).

Condition 4 (Bounded eccentricity). This case is almost identical to case 2

Condition 5 (Type-I curvature decay) implies (by integration) that the displacement is bounded by $\sqrt{-t}$. It follows that the blow-down is compact and the theorem again follows immediately from Lemma 2.2.1(ii)(a).

Condition 6 (Bounded speed ratios) implies condition 5.

Condition 7 (Reverse isoperimetric inequality) implies condition 4 (see [28, Lem. 4.4]).

$\square$

2.6 Asymptotics for Convex Translators

The following lemma describes the asymptotic geometry of convex translators.

Lemma 2.6.1 (Asymptotic shape of convex translators)   *Let Σ^n be a convex translator in $\mathbb{R}^{n+1}$ and $\{\Sigma_t^n\}_{t\in(-\infty,\infty)}$, where $\Sigma_t^n := \Sigma^n + te_{n+1}$, the corresponding translating solution to mean curvature flow.*

(i) (Asymptotic shrinker) *There is a rotation $R \in \mathrm{SO}(\mathbb{R}^n \times \{0\})$ about the x_{n+1}-axis such that the family of rescaled solutions $\{\lambda R \cdot \Sigma_{\lambda^{-2}t}^n\}_{t\in(-\infty,0)}$ converge locally uniformly in the smooth topology as $\lambda \to 0$ to either*

  • *the shrinking cylinder $\{S^{n-k}_{\sqrt{-2(n-k)t}} \times \mathbb{R}^k\}_{t\in(-\infty,0)}$ for some $k \in \{1, \ldots, n-1\}$, or*

- the stationary hyperplane $\{\{0\} \times \mathbb{R}^n\}_{t \in (-\infty, 0)}$ of multiplicity either one or two.

  Moreover, if the limit is the stationary hyperplane of multiplicity one, then Σ^n is a vertical hyperplane.

(ii) (Asymptotic translators[4]) *Given any direction* $e \in \partial \nu(\Sigma^n)$ *in the boundary of the Gauss image* $\nu(\Sigma^n)$ *of* Σ^n *and any sequence of points* $X_j \in \Sigma^n$ *with* $\nu(X_j) \to e$, *a subsequence of the translated solutions* $\Sigma^n - X_j$ *converges locally uniformly in the smooth topology to a convex translator which splits off a line (in the direction* $w := \lim_{j \to \infty} \frac{X_j}{|X_j|}$).

Proof The first claim is a straightforward consequence of Lemma 2.2.1 (i). It was proved directly by Wang using a different argument [33, Thm. 1.3].

So, consider the second claim. Up to a translation, we may assume that $X_1 = 0$. After passing to a subsequence, we can arrange that $w_j := X_j / \|X_j\| \to w$ for some $w \in S^n$. Since each Σ_j^n is convex and satisfies the translator equation, the sequence admits the uniform curvature bound

$$|A_j| \leq H_j = -\langle \nu_j, e_{n+1} \rangle \leq 1.$$

It follows, after passing to a further subsequence, that the sequence converges locally uniformly in the smooth topology to a convex translator Σ_∞^n. We claim that Σ_∞^n contains the line

$$L := \{sw : s \in \mathbb{R}\}.$$

First note that the closed convex region $\overline{\Omega}$ bounded by Σ^n contains the ray $\{sw : s \geq 0\}$, since it contains each of the segments $\{sw_j : 0 \leq s \leq s_j\}$, where $s_j := \|X_j\|$. By convexity, $\overline{\Omega}$ also contains the set $\{rsw + (1-r)X_j : s \geq 0, 0 \leq r \leq 1\}$ for each j. It follows that the closed convex region $\overline{\Omega}_j$ bounded by Σ_j^n contains the set $\{rsw - rs_jw_j : s > 0, 0 \leq r \leq 1\}$ since $s_jw_j = X_j$. In particular, choosing $s = 2s_j$, $\{\vartheta(w - w_j) + \vartheta w : 0 \leq \vartheta \leq s_j\} \subset \overline{\Omega}_j$ and, choosing $s = s_j/2$,

$$\{\vartheta w_j - \vartheta(w - w_j) : -s_j/2 \leq \vartheta \leq 0\} \subset \overline{\Omega}_j.$$

Taking $j \to \infty$, we find $\{sw : s \in \mathbb{R}\} \subset \overline{\Omega}_\infty$. It now follows from convexity of $\overline{\Omega}_\infty$ that $\{sw : s \in \mathbb{R}\} \subset \Sigma^n$. We conclude that κ_1 reaches zero somewhere on Σ_∞^n and the lemma follows from the splitting theorem. $\square$

---

[4] See [22, Lem. 2.1] and [8, Lem. 3.1].

In some cases, the asymptotic translators can be shown to depend uniquely on $e \in \partial\nu(\Sigma^n)$, in which case the convergence is independent of the subsequence.

Note that the asymptotic translator will simply be a vertical hyperplane if $e \perp e_{n+1}$.

2.7 X.-J. Wang's Dichotomy for Convex Translators

The following theorem is an immediate consequence of Theorem 2.3.3.

Theorem 2.7.1 (X.-J. Wang's dichotomy for convex translators [33, Cor. 2.2]) *Let Σ be a proper, convex translator. If Σ is not entire, then it lies in a vertical slab region (the region between two parallel vertical hyperplanes).*

We note that F. Chini and N. Moller [17] have recently obtained a classification of the *convex hulls* of the projection onto the subspace orthogonal to the translation direction for general translators.

theorem 2.7.1 motivates the following definitions.

Definition 2.7.2 A proper, locally uniformly convex translator is called

- a *bowloid* if it is entire.
- a *flying wing* if it lies in a slab region.

The grim reaper curve is a well-known example of a flying wing, while the bowl soliton is an example of a bowloid. Further flying wings and bowloids were constructed by Wang [33], providing a counterexample to a conjecture of White [35, Conj. 2]. We proved that each slab of width at least π in $\mathbb{R}^{n+1}$ admits a bi-symmetric[5] flying wing. (Note that no slab of width less than π admits a convex translator: the grim hyperplane is a barrier.) Making use of the differential Harnack inequality (see Lemma 2.6.1(iii)), barrier arguments, and Alexandrov's method of moving planes, we also obtained unique asymptotics and reflection symmetry for such translators [13]. Hoffman *et al.* constructed an $(n-1)$-parameter family of non-entire translating graphs and an $(n-2)$-parameter family of entire translating graphs in $\mathbb{R}^{n+1}$ for each $n \geq 2$ [24]. It is not yet clear whether or not these examples are convex, however (except when

---

[5] Namely, $O(1) \times O(n-1)$-symmetric.

they coincide with the examples in [13]). A survey of these results can be found in [10].

2.8 Rigidity of the Bowl Soliton

We will use the asymptotics of Lemma 2.6.1 in conjunction with the following important result of Haslhofer to obtain various rigidity results for the bowl soliton.

Theorem 2.8.1 (R. Haslhofer [22]) *Let Σ^n, $n \geq 2$, be a convex, locally uniformly convex translator in $\mathbb{R}^{n+1}$. Suppose that the blow-down of the corresponding translating solution $\{\Sigma_t^n\}_{t \in (-\infty, \infty)}$, $\Sigma_t := \Sigma^n + te_{n+1}$, is the shrinking cylinder $S^{n-1}_{\sqrt{-2(n-1)t}} \times \mathbb{R}$. Then Σ^n is rotationally symmetric about a vertial axis (and hence, up to a translation, the bowl soliton).*

Proof See [22]. □

Corollary 2.8.2 (X.-J. Wang [33, Thm. 1.1], J. Spruck and L. Xiao [32]) *Modulo translation, the bowl soliton is the only mean convex, entire translator Σ^2 in $\mathbb{R}^3$. In particular, it is the only translator which arises as a singularity model for a compact, embedded, mean convex mean curvature flow in $\mathbb{R}^3$.*

Proof By the Spruck–Xiao convexity estimate [32], Σ^2 is actually convex. In fact, it must also be locally uniformly convex: else, by the strong maximum principle, it would split off a line; by uniqueness of the grim reaper, the result would either be a vertical plane or a grim plane, neither of which are entire.

Since Σ^2 is entire, Lemma 2.6.1 and Wang's dichotomy (Theorem 2.7.1) imply that its blow-down is the shrinking cylinder, $\{S^1_{\sqrt{-2t}} \times \mathbb{R}\}_{t \in (-\infty, 0)}$. The claim now follows from theorem 2.8.1. □

Corollary 2.8.2 was proved by X.-J. Wang (in the convex case) by a different argument: making use of the fact that the blow-down is the shrinking cylinder, he was able to obtain, by an iteration argument, the estimate

$$|u(x) - u_0(x)| = o(|x|)$$

as $|x| \to \infty$, where graph(u_0) is the bowl soliton whose tip coincides with that of u. A classical theorem of Bernstein then implies that $u - u_0$ is constant, which yields the claim. See [33].

Corollary 2.8.3   *Modulo translation, the bowl soliton is the only convex, non-planar translator Σ^n in $\mathbb{R}^{n+1}$, $n \geq 3$, satisfying*

$$\inf_{\Sigma^n} \frac{\kappa_1 + \kappa_2}{H} > 0 . \tag{2.19}$$

In particular, it is the only translator which arises as a singularity model for a compact, two-convex (immersed) mean curvature flow in $\mathbb{R}^{n+1}$ when $n \geq 3$.

Proof   First note that any convex, non-planar translator satisfying (2.19) is automatically locally uniformly convex. Indeed, if this were not the case, then it would split off a line; but the complimentary subspace $\hat{\Sigma}^{n-1}$ cannot be compact and must satisfy $\inf_{\hat{\Sigma}^{n-1}} \kappa_1/H > 0$, contradicting Hamilton's compactness theorem [20].

We claim that Σ^n is entire. Suppose, to the contrary, that this is not the case. Then, by Wang's dichotomy (theorem 2.7.1), it lies in a slab. Choose $e \in \partial \nu(\Sigma)$ with $\langle e, e_1 \rangle = 0$. By Lemma 2.6.1(ii), we can find a sequence of points $X_j \in \Sigma^n$ such that the translates $\Sigma^n - X_j$ converge locally uniformly in the smooth topology to a convex translator which splits off a line parallel to the slab and has normal e at the origin. Since $\langle e_1, e \rangle = 0$, the limit cannot be a vertical hyperplane. It follows that the limit has positive mean curvature. Moreover, since $\langle e, e_{n+1} \rangle < 0$, the cross section of the splitting cannot be compact. But this contradicts the two-convexity hypothesis (2.19) by Hamilton's compactness theorem [20]. The claim now follows from Haslhofer's analysis (theorem 2.8.1). $\square$

To our knowledge, Corollary 2.8.3 was only previously known under additional hypotheses, such as non-collapsing [22] or certain cylindrical and gradient estimates [8].

References

[1] Altschuler, S. J. and Wu, L. F. *Translating surfaces of the nonparametric mean curvature flow with prescribed contact angle.* Calc. Var. Partial Differ. Equ. **2**(1) (1994), 101–111.

[2] Andrews, B. *Harnack inequalities for evolving hypersurfaces.* Math. Z. **17**(2) (1994), 179–197.

[3] Andrews, B. *Noncollapsing in mean-convex mean curvature flow.* Geom. Topol. **16**(3) (2012), 1413–1418.

[4] Andrews, B., Langford, M. and McCoy, J. *Non-collapsing in fully nonlinear*

curvature flows. Ann. Inst. H. Poincaré Anal. Non Linéaire **30**(1) (2013), 23–32.

[5] Angenent, S., Daskalopoulos, P. and Šešum, N. *Unique asymptotics of ancient convex mean curvature flow solutions.* Preprint, arXiv:1503.01178v3, 2015.

[6] Angenent, S., Daskalopoulos, P. and Šešum, N. *Uniqueness of two-convex closed ancient solutions to the mean curvature flow.* Preprint, arXiv:1804.07230, 2018.

[7] Bakas, I. and Sourdis, C. *Dirichlet sigma models and mean curvature flow.* J. High Energy Phys. **2007**(06) (2007), 057.

[8] Bourni, T. and Langford, M. *Type-II singularities of two-convex immersed mean curvature flow.* Geom. Flows **2** (2017), 1–17.

[9] Bourni, T., Langford, M. and Tinaglia, G. *Collapsing ancient solutions of mean curvature flow.* To appear in J. Differ. Geom. Preprint, arXiv:1705.06981, 2017.

[10] Bourni, T., Langford, M. and Tinaglia, G. *Translating solutions to mean curvature flow.* Preprint, http://www.math.utk.edu/~langford/research.html, 2018.

[11] Bourni, T., Langford, M. and Tinaglia, G. *Ancient solutions to mean curvature flow.* In Emmanouil, I., Fellouris, A., Giannopoulos, A. and Lambropoulou, S. (eds.) Proc. First Congress of Greek Mathematicians, Athens, Greece. June 25--30, 2018. Walter de Gruyter, Berlin, 2020.

[12] Bourni, T., Langford, M. and Tinaglia, G. *Convex ancient solutions to curve shortening flow.* Calc. Var. Partial Differ. Equ. **59**, 133 (2020). https://doi.org/10.1007/s00526-020-01784-8.

[13] Bourni, T., Langford, M. and Tinaglia, G. *On the existence of translating solutions of mean curvature flow in slab regions.* Anal. PDE **13**(4) (2020), 1051–1072.

[14] Brendle, S. and Choi, K. *Uniqueness of convex ancient solutions to mean curvature flow in $\mathbb{R}^3$.* Preprint, arXiv:1711.00823, 2017.

[15] Brendle, S. and Choi, K. *Uniqueness of convex ancient solutions to mean curvature flow in higher dimensions.* Preprint, arXiv:1804.00018, 2018.

[16] Chini, F. and Moller, N. M. *Ancient mean curvature flows and their space-time tracks.* Preprint, arXiv:1901.05481, 2019.

[17] Chini, F. and Moller, N. M. *Bi-halfspace and convex hull theorems for translating solitons.* Preprint, arXiv:1809.01069, 2019.

[18] Colding, T. H. and Minicozzi, II, W. P. *Generic mean curvature flow I: generic singularities.* Ann. of Math. (2) **175**(2) (2012), 755–833.

[19] Daskalopoulos, P., Hamilton, R. and Šešum, N. *Classification of compact ancient solutions to the curve shortening flow.* J. Differ. Geom. **84**(3) (2010), 455–464.

[20] Hamilton, R. S. *Convex hypersurfaces with pinched second fundamental form.* Commun. Anal. Geom. **2**(1) (1994), 167–172.

[21] Hamilton, R. S. *Harnack estimate for the mean curvature flow.* J. Differ. Geom. **41**(1) (1995), 215–226.

[22] Haslhofer, R. *Uniqueness of the bowl soliton.* Geom. Topol. **19**(4) (2015), 2393–2406.

[23] Haslhofer, R. and Hershkovits, O. *Ancient solutions of the mean curvature flow.* Commun. Anal. Geom. **24**(3) (2016), 593–604.

[24] Hoffman, D., Ilmanen, T., Martin, F. and White, B. *Graphical translators for mean curvature flow.* Calc. Var. Partial Differ. Equ. **58**, 117 (2019). https://doi.org/10.1007/s00526-019-1560-x.

[25] Huisken, G. *Flow by mean curvature of convex surfaces into spheres.* J. Differ. Geom. **20**(1) (1984), 237–266.

[26] Huisken, G. *Asymptotic behavior for singularities of the mean curvature flow.* J. Differ. Geom. **31**(1) (1990), 285–299.

[27] Huisken, G. *Local and global behaviour of hypersurfaces moving by mean curvature.* In Greene, R. and Yau, S. T. (eds.) Differential geometry: partial differential equations on manifolds. Proc. Symp. Pure Math. Amer. Math. Soc. **54** Part I, Providence, RI (1993), pp. 175–191.

[28] Huisken, G. and Sinestrari, C. *Convexity estimates for mean curvature flow and singularities of mean convex surfaces.* Acta Math. **183**(1) (1999), 45–70.

[29] Huisken, G. and Sinestrari, C. *Mean curvature flow singularities for mean convex surfaces.* Calc. Var. Partial Differ. Equ. **8**(1) (1999), 1–14.

[30] Huisken, G. and Sinestrari, C. *Convex ancient solutions of the mean curvature flow.* J. Differ. Geom. **101**(2) (2015), 267–287.

[31] Sheng, W. and Wang, X.-J. *Singularity profile in the mean curvature flow.* Methods Appl. Anal. **16**(2) (2009), 139–155.

[32] Spruck, J. and Xiao, L. *Complete translating solitons to the mean curvature flow in* $\mathbb{R}^3$ *with nonnegative mean curvature.* Preprint, arXiv:1703.01003, 2017.

[33] Wang, X.-J. *Convex solutions to the mean curvature flow.* Ann. of Math. (2) **173**(3) (2011), 1185–1239.

[34] White, B. *The size of the singular set in mean curvature flow of mean-convex sets.* J. Amer. Math. Soc. **13**(3) (2000), 665–695 (electronic).

[35] White, B. *The nature of singularities in mean curvature flow of mean-convex sets.* J. Amer. Math. Soc. **16**(1) (2003), 123–138 (electronic).

[36] White, B. *A local regularity theorem for mean curvature flow.* Ann. of Math. (2) **161**(3) (2005), 1487–1519.

3

Negatively Curved Three-Manifolds, Hyperbolic Metrics, Isometric Embeddings in Minkowski Space and the Cross Curvature Flow

Paul Bryan, Mohammad Ivaki and Julian Scheuer

Abstract

This short chapter is a mostly expository note examining negatively curved three-manifolds. We look at some rigidity properties related to isometric embeddings into Minkowski space. We also review the cross curvature flow (XCF) as a tool to study the space of negatively curved metrics on hyperbolic three-manifolds, the largest and least understood class of model geometries in Thurston's geometrisation. The relationship between integrability and embeddability yields interesting insights, and we show that solutions with fixed Einstein volume are precisely the integrable solutions, answering a question posed by Chow and Hamilton when they introduced the XCF.

3.1 Introduction

In the early 1980's, Thurston announced the geometrisation conjecture [29]. At around the same time, Hamilton introduced the Ricci flow [13]. The geometrisation conjecture claimed that closed three-manifolds could be decomposed into pieces modelled on geometric structures of eight possible types, while the Ricci flow deformed metrics by their Ricci curvature. At that time, Thurston classified the possible geometric structures into the eight types and proved the geometrisation conjecture for Haken manifolds, while Hamilton obtained convergence to a constant sectional curvature metric provided the initial metric has positive Ricci curvature.

Needless to say, both these seminal works sparked off tremendous developments continuing to this day. A crowning achievement was Perelman's resolution of the geometrisation conjecture using the Ricci flow with surgery [27, 26, 25]. The legend goes that at the urging of Yau, Hamilton initiated a program to use the Ricci flow to prove the geometrisation conjecture. Roughly speaking, the Ricci flow tends to smooth out irregularities in curvature, but singularities may occur. Hamilton outlined an approach to cut out the singularities with analytically controlled topological surgeries and continue the flow. Perelman proved that this process does indeed work, with only finitely many such surgeries required, after which the remaining pieces converge to one of Thurston's eight model geometries. Tracing the process back provides the necessary decomposition to resolve the geometrisation conjecture.

Of the eight geometries, essentially only the hyperbolic geometries are not fully understood. All the other seven cases may be enumerated in a similar fashion to the uniformisation of surfaces. Around the time Perelman was completing his work on the Ricci flow, Chow and Hamilton introduced a new flow, the cross curvature flow (XCF) [5]. This flow deforms initially negatively curved metrics (also positive but we won't focus on that case) by a fully nonlinear parabolic equation. The Ricci flow is to the heat equation as the XCF is to a Monge–Ampère equation. The aim of the XCF is to deform negatively curved metrics to hyperbolic metrics, thus illuminating the structure of the eighth and least understood model geometry, namely the hyperbolic geometries.

Here we will examine the status of this program, describing the known results to date and drawing together some observations made in the literature. We will touch on some of the difficulties faced in this program with the hope of reinvigorating the study of the XCF so that fresh insights may yield new results for this fascinating flow.

In [5], evidence was given that the XCF deforms arbitrary negatively curved metrics (after normalisation) to a hyperbolic metric. Further evidence was provided in [16], showing that hyperbolic metrics are asymptotically stable. See Section 3.4.4 below for details. The original short time existence proof based on the Nash–Moser implicit function theorem in [5] was not complete, and a complete proof based on the DeTurck trick was given in [3]. Short time existence and uniqueness is discussed in Section 3.4.2. In the particular case where the universal cover embeds isometrically into Minkowski space, the Gauss curvature flow is equivalent to the XCF (Lemma 3.4.4), and convergence to the hyperbolic metric follows by [1]. This is discussed in Section 3.3, where it is also shown that embeddability is equivalent to an integrability condition (Theorem 3.3.4). Whether a Harnack inequality holds was raised in [5]. The Harnack inequality for integrable solutions, as well the as the rigidity of solitons, is discussed in Section 3.4.5. Another question was to classify those solutions with constant Einstein volume, and this is answered in Section 3.4.6.

There are other results for the XCF contained in the literature. Available space precludes complete discussion here. Briefly, long time existence and expansion to infinity of a square torus bundle is obtained in [20]. Long time existence, convergence results and singularity analysis on locally homogeneous spaces of variable curvature where the XCF reduces to an ODE is examined in [8, 9, 12] as well as the backwards behaviour in [4]. On a solid torus, long time existence and curvature bounds for the XCF starting at the 2π-metric of Gromov and Thurston is obtained in [11]. Finally, uniqueness and backwards uniqueness are addressed in [18, 19].

Acknowledgements   Parts of this work were written while JS enjoyed the hospitality of the Department of Mathematics at Columbia University in New York, a visit funded by the "Deutsche Forschungsgemeinschaft" (DFG, German research foundation) within the research scholarship "Quermassintegral preserving local curvature flows", grant number SCHE 1879/3-1. JS would like to thank the DFG, Columbia University and especially Prof. Simon Brendle for their support. PB was supported by the ARC within the research grant "Analysis of fully non-linear geometric problems and differential equations", number DE180100110. We would also like to thank the referee for helpful comments clarifying the exposition.

3.2 Geometrisation of Three-Manifolds

For surfaces, the uniformisation theorem and Gauss–Bonnet theorem provide a complete picture of the topology and its relation to curvature. The three dimensional case is more complicated than the two dimensional case, but is almost completely understood thanks to Perelman's successful completion [27, 26, 25] of Hamilton's program based on the Ricci flow [13] to solve the Poincaré and Thurston geometrisation conjectures [29]. The remaining piece of the puzzle is the structure of hyperbolic, closed three-manifolds. We include here a brief description and refer the reader to [28] and [30] for in depth discussions of geometrisation and [24, 23, 15] for expositions of the Hamilton–Perelman proof. Unless explicitly stated otherwise, the results described here may be found in those references.

The geometrisation conjecture may be stated as follows:

Theorem 3.2.1 (Thurston geometrisation)   *Any closed three-manifold decomposes as a connected sum of prime manifolds, each of which may be cut along tori so that the interior of the resulting manifolds each admits a unique geometric structure of finite volume from among a possible eight types.*

A prime manifold is simply a manifold that cannot be written as a non-trivial connected sum. The decomposition into prime manifolds was given in [21]. To say that M admits a geometric structure is to say that M is diffeomorphic to X/Γ, where X is a G-manifold for some Lie group G acting transitively on X with compact stabilisers, and Γ is a discrete subgroup of G acting freely on X. The classification into eight types of finite volume geometric structures was given by Thurston. Finally, the remaining part of the theorem, that such a decomposition of prime manifolds exists, was proven by Hamilton and Perelman using the Ricci flow with surgery. The eight geometries are

$$\mathbb{R}^3, \quad S^3, \quad \mathbb{H}^3, \quad S^2 \times \mathbb{R}, \quad \mathbb{H}^2 \times \mathbb{R}, \quad \widetilde{SL}_2(\mathbb{R}), \quad \text{Nil}, \quad \text{Solv}.$$

In the case of Solv these are precisely torus and Klein bottle bundles over S^1 or the union of two twisted 1-bundles over the torus or Klein bottle. The remaining six non-hyperbolic geometries are all Seifert fibre bundles, completely determined by the Euler characteristic of the base space, χ, and the Euler number of the bundle, e.

The only remaining case then is the hyperbolic case and this has no classification yet. As a consequence of geometrisation, we have

Theorem 3.2.2 (Hyperbolisation)   *A closed three-manifold admitting a metric of negative sectional curvature also admits a hyperbolic metric, that is, a metric of constant negative sectional curvature.*

Thurston proved this result for atoroidal Haken manifolds, and the general result is a consequence of geometrisation as proven by Hamilton and Perelman. The route to hyperbolisation via geometrisation is quite indirect since the Ricci flow does not generally preserve negative curvature in dimensions greater than two, and, furthermore, singularities may form along the Ricci flow so that surgery is necessary. The decomposition obtained in Theorem 3.2.1 is not unique so that the Ricci flow with surgery need not produce the hyperbolic geometry in the limit. Theorem

3.2.2 is deduced a posteriori from the geometrisation conjecture rather than as a direct consequence of the Ricci flow.

In the following, Theorem 3.3.2 states that metrics of negative sectional curvature satisfying an integrability condition on a closed manifold may be smoothly deformed to a hyperbolic metric. Conjecture 3.3.5 claims that the integrability condition may be removed, which would give a more direct proof of hyperbolisation, while strengthening the result to the statement that arbitrary negatively curved metrics are homotopic to a hyperbolic metric. Note that by the Mostow rigidity theorem [22], hyperbolic structures are classified by fundamental group and are essentially unique. That is, if the fundamental group $\pi_1(M_1)$ of a closed hyperbolic manifold, (M_1, g_1) is isomorphic to a $\pi_1(M_2)$ for another hyperbolic manifold (M_2, g_2), then in fact (M_1, g_1) and (M_2, g_2) are isometric. Equivalently, any homotopy equivalence of hyperbolic manifolds may in fact be homotopied to an isometry. Then Conjecture 3.3.5 would show that the space of negatively curved metrics on a hyperbolic three-manifold is contractible.

3.3 Embeddability and Hyperbolic Metrics

Let (M, g) be a compact, Riemannian manifold with strictly negative curvature. Let $\pi \colon (\tilde{M}, \tilde{g}) \to (M, g)$ be the Riemannian universal cover so that $\pi : \tilde{M} \to M$ is a covering map with $\tilde{M}$ simply connected and $\tilde{g} = \pi^* g$. Let G denote the deck transformation group of the cover and observe that $\tilde{g}$ is invariant under G. That is, $G \leq \mathrm{Diff}(\tilde{M})$ is a group of diffeomorphisms of $\tilde{M}$ and $\varphi^* \tilde{g} = \tilde{g}$ for all $\varphi \in G$ so that G acts by isometry on $(\tilde{M}, \tilde{g})$. Then $\tilde{g}$ induces a metric $\bar{g}$ on the quotient $\tilde{M}/G$ such that

$$(\tilde{M}/G, \bar{g}) \underset{\simeq}{\to} (M, g)$$

is an isometry and the quotient map $\tilde{M} \to \tilde{M}/G$ is just π under this identification. Then $(\tilde{M}/G, \bar{g})$ is a compact Riemannian quotient and we say $(\tilde{M}, \tilde{g})$ is a co-compact Riemannian manifold.

Now, since (M, g) has strictly negative sectional curvature, so does $(\tilde{M}, \tilde{g})$; hence, by the Cartan–Hadamard theorem, $\tilde{M} \simeq \mathbb{R}^3$ is diffeomorphic to $\mathbb{R}^3$ via the exponential map. In particular we may equip $\tilde{M}$ with the hyperbolic metric $\tilde{g}_{\mathbb{H}}$ of constant, negative sectional curvature equal to -1. Let us write $G_{\mathbb{H}}$ for the isometry group of $(\tilde{M}, \tilde{g}_{\mathbb{H}})$.

On $\tilde{M}$, there is a simple, smooth homotopy from $\tilde{g}$ to $\tilde{g}_{\mathbb{H}}$:

$$\tilde{h}(t) = t\tilde{g} + (1 - t)\tilde{g}_{\mathbb{H}}, \quad t \in [0, 1].$$

This gives rise to the following simple lemma:

Lemma 3.3.1   *Let (M, g) be a compact manifold of strictly negative sectional curvature. Then the following statements are equivalent:*

(i) *M admits a metric of constant, negative sectional curvature.*
(ii) *$\tilde{g}_{\mathbb{H}}$ is invariant under G.*
(iii) *G is a subgroup of $G_{\mathbb{H}}$.*
(iv) *g is smoothly homotopic to a metric of constant, negative sectional curvature.*
(v) *Every G-invariant metric $\tilde{g}$ on $\tilde{M}$ is smoothly homotopic to $\tilde{g}_{\mathbb{H}}$ via a smooth G-invariant homotopy.*

Proof   (i) $\Rightarrow$ (ii) The pullback of a constant curvature metric under the Riemannian covering π is the hyperbolic one, which is thus G-invariant.

(ii) $\Rightarrow$ (iii) Clear, since $G_{\mathbb{H}}$ is the whole isometry group.

(iii) $\Rightarrow$ (iv) Since $\tilde{g}$ and $\tilde{g}_{\mathbb{H}}$ are invariant under G, so is $\tilde{h}(t)$, which in turn descends to a homotopy on $\tilde{M}/G$. This pushes forward to the desired homotopy, h on M.

(iv) $\Rightarrow$ (v) For any given $\tilde{g}$, the push forward h of $\tilde{h}$ defined above is the desired homotopy.

(v) $\Rightarrow$ (i) Apply (v) to the pullback of g and push forward the resulting homotopy to M.   □

The question of whether the conditions of Lemma 3.3.1 are satisfied are not easy to check but the lemma affords us with several possible approaches to the problem. In this section we prove Theorem 3.3.2, which gives a *sufficient* condition for when (M, g) admits a metric of constant, negative sectional curvature.

Before we can state the theorem, let us agree on some notation and conventions. Given a metric g with Levi-Civita connection ∇ on a manifold M, our conventions for the curvature tensor are

$$\mathrm{Rm}(X, Y)Z = \nabla_X \nabla_Y Z - \nabla_Y \nabla_X Z - \nabla_{[X,Y]} Z,$$
$$\mathrm{Rm}(X, Y, Z, W) = g(\mathrm{Rm}(X, Y)Z, W).$$

Then the Ricci and scalar curvature are defined by

$$\mathrm{Ric}(X, Y) = \mathrm{Tr}\,\mathrm{Rm}(\cdot, X)Y,$$
$$\mathrm{R} = \mathrm{Tr}_g \mathrm{Ric}, \tag{3.1}$$

where Tr is the trace of an endomorphism and Tr_g is the trace of a bilinear form with respect to the metric g. We define the Einstein tensor by

$$\mathrm{E} = \mathrm{Ric} - \frac{\mathrm{R}}{2} g. \tag{3.2}$$

Writing $\mathrm{E}(X, Y) = g(\mathcal{E}(X), Y)$, we may write the Ricci decomposition of the curvature tensor in three dimensions in the form

$$\mathrm{Rm} = -\mathrm{E} \owedge g + \frac{\mathrm{Tr}\mathcal{E}}{2} g \owedge g, \tag{3.3}$$

where $\owedge$ denotes the Kulkarni–Nomizu product. The sectional curvatures are

$$K(X \wedge Y) = \frac{\mathrm{Rm}(X, Y, Y, X)}{|X \wedge Y|^2}.$$

The curvature operator $\mathcal{R}\mathrm{m}$ is defined by

$$\mathrm{Rm}(X, Y, Z, W) = g(\mathcal{R}\mathrm{m}(X \wedge Y), W \wedge Z).$$

Then, from the Ricci decomposition, given an orthonormal basis of eigenvectors E_i for E with eigenvalues λ_i, we have

$$g(\mathcal{R}\mathrm{m}(E_i \wedge E_j), E_p \wedge E_q) = \left(-\mathrm{E} \owedge g + \frac{\mathrm{Tr}\mathcal{E}}{2} g \owedge g \right)(E_i \wedge E_j, E_p \wedge E_q)$$

$$= g(\lambda_k E_i \wedge E_j, E_p \wedge E_q).$$

Thus

$$\mathcal{R}\mathrm{m}(E_i \wedge E_j) = \lambda_k E_i \wedge E_j$$

and the eigenvalues of $\mathcal{R}\mathrm{m}$ are precisely the eigenvalues of E. The sectional curvatures are then

$$K(E_i \wedge E_j) = -\frac{g(\mathcal{R}\mathrm{m}(E_i \wedge E_j), E_j \wedge E_i)}{|E_i \wedge E_j|^2} = -\lambda_k. \tag{3.4}$$

Therefore E is positive definite (respectively negatively definite) if and only if the sectional curvatures are negative (respectively positive). In the case of negative sectional curvature, E is hence a metric. Define $\mathrm{E}^{-1}(X, Y) = g(\mathcal{E}^{-1}(X), Y)$.

Now we have the following theorem. We say that a symmetric $(0, 2)$-tensor T is *Codazzi* if the covariant three-tensor ∇T is totally symmetric.

Theorem 3.3.2 (Integrability and constant negative sectional curvature) *Let (M, g) be a closed Riemannian three-manifold of strictly negative sectional curvature with the integrability condition that the tensor*

$O = \sqrt{\det \mathcal{E}} E^{-1}$ *is Codazzi. Then g is smoothly homotopic to a metric of constant, negative sectional curvature, where each metric in the homotopy also has negative sectional curvature.*

Note that, as a special case of the considerably more general Thurston geometrisation of three-manifolds, any closed three-manifold admitting a metric of negative sectional curvature also admits a metric of constant negative sectional curvature. The famous final resolution of Thurston geometrisation by Perelman requires considerable machinery, so the more direct and simpler proof described here in this case is desirable. Moreover, the homotopy in the theorem is not a consequence of geometrisation. See Section 3.2 for a brief discussion on geometrisation.

Theorem 3.3.2 follows from the embeddability Theorem 3.3.4 and [1, Thm. 1.1], which says that $\tilde{M}$ may be deformed to the one-sheeted hyperboloid at infinity by the Gauss curvature flow. Before we can state and prove Theorem 3.3.4, we need some more notation concerning extrinsic geometry.

Let $\langle \cdot, \cdot \rangle$ denote the inner-product on Minkowski space and D the corresponding Levi-Civita connection. For a spacelike immersion $F \colon M^n \to \mathbb{R}^{n,1}$ with M oriented, we define the second fundamental form A with respect to a timelike, unit normal field ν by

$$D_{F_* X} F_* Y = F_* \nabla_X Y + A(X,Y)\nu.$$

We also define the Weingarten map via

$$A(X,Y) = g(\mathcal{W}(X), Y)$$

and write $H = \mathrm{Tr}_g A = \mathrm{Tr}\mathcal{W}$.

The basic equations of hypersurfaces (Gauss equation) in Minkowski space are

$$\mathrm{Rm}(X,Y)Z = A(X,Z)\mathcal{W}(Y) - A(Y,Z)\mathcal{W}(X),$$
$$\mathrm{Ric}(X,Y) = g(\mathcal{W}^2(X) - H\mathcal{W}(X), Y), \tag{3.5}$$
$$\mathrm{R} = \|A\|^2 - H^2.$$

We can also relate the eigenvalues λ_i of $\mathcal{E}$ with the principal curvatures κ_i of the embedding. Namely, for distinct indices i, j, k we calculate, with the help of (3.5),

$$\lambda_k = \kappa_k^2 - \kappa_k \sum_l \kappa_l - \frac{\sum_l \kappa_l^2 - (\sum_l \kappa_l)^2}{2} = \kappa_i \kappa_j. \tag{3.6}$$

Hence there holds

$$\mathcal{W} = \sqrt{\det \mathcal{E}} \mathcal{E}^{-1}, \qquad (3.7)$$

since these endomorphisms are simultaneously diagonalisable and share the same eigenvalues.

Therefore, E > 0 if and only if all the principal curvatures κ_i have the same sign (which may be taken to be positive by swapping ν with $-\nu$ if necessary). That is, g has negative sectional curvature if and only if E > 0 if and only if $F(\tilde{M})$ is a locally convex, spacelike hypersurface.

Remark 3.3.3   We see a strong rigidity statement that the *extrinsic geometry* of embedded, spacelike hypersurfaces is completely determined by the *intrinsic geometry*. The extrinsic condition of local convexity is equivalent to the intrinsic condition of negative sectional curvature.

Now we may give an intrinsic characterisation of when $(\tilde{M}, \tilde{g})$ embeds isometrically into Minkowski space as precisely when O is Codazzi.

Theorem 3.3.4 (Integrability implies isometric embeddability)   *Let (M, g) be a closed Riemannian three-manifold of strictly negative sectional curvature. Then the tensor* O $= \sqrt{\det \mathcal{E}} \mathrm{E}^{-1}$ *is Codazzi if and only if the Riemannian universal cover $(\tilde{M}, \tilde{g})$ embeds isometrically into Minkowski space $\mathbb{R}^{3,1}$ as a locally convex, co-compact, spacelike hypersurface. Moreover the action of the group G of deck transformations of $\pi : (\tilde{M}, \tilde{g}) \rightarrow (M, g)$ extends to an isometry of $\mathbb{R}^{3,1}$ preserving the hypersurface.*

Proof   First suppose $(\tilde{M}, \tilde{g})$ embeds isometrically in $\mathbb{R}^{3,1}$. As Minkowski space is flat, the second fundamental form A is Codazzi. Then (3.7) gives $A = $ O, hence O is Codazzi.

Conversely, suppose O is Codazzi. The result is a consequence of the fundamental theorem of hypersurfaces: a simple direct computation diagonalising E shows that O solves the contracted Gauss equation (second equation in (3.5)). The Ricci decomposition (3.3) for $n = 3$ then implies the full Gauss equation (first equation in (3.5)). But the Gauss and Codazzi equations are precisely the integrability conditions required to locally integrate the over-determined system,

$$F^* \langle \cdot, \cdot \rangle = g, \qquad A(F) = \mathrm{O}$$

for F. See for example [17, Thm. 7.1].

Since $(\tilde{M}, \tilde{g})$ is the universal cover of (M, g) with strictly negative sectional curvature, $\tilde{M}$ is diffeomorphic to $\mathbb{R}^3$ by the Cartan–Hadamard theorem and we can globally integrate to obtain F.

That G acts on $\mathbb{R}^{3,1}$ by isometry is a consequence of the rigidity part of the fundamental theorem of hypersurfaces [17, Thm. 7.2]: write $F : \tilde{M} \to \mathbb{R}^{3,1}$ for the embedding, fix a point $x_0 \in \tilde{M}$ and let $\varphi \in G$. Then there is a unique isometry Φ of $\mathbb{R}^{3,1}$ taking $F(x_0)$ to $F(\varphi(x_0))$ and taking the tangent plane $T_{x_0} F(\tilde{M})$ to $T_{\varphi(x_0)} F(\tilde{M})$ while taking the normal $N(x_0)$ to $N(\varphi(x_0))$ such that both normals have the same timelike orientation (e.g both are future pointing). But both the metric $\tilde{g}$ and the second fundamental form h are preserved by φ (recalling that h is intrinsic) and by Φ. Uniqueness of the embedding given a tangent plane and orientation then ensure φ agrees with Φ. Since x_0 and g are arbitrary and G acts transitively on $\tilde{M}$, the result follows. $\qquad\square$

Proof of Theorem 3.3.2 By Theorem 3.3.4 we may embed $(\tilde{M}, \tilde{g})$ into Minkowski space as a locally convex, co-compact (by ambient isometry), spacelike hypersurface. By [1, Thm. 1.1], the rescaled Gauss curvature flow deforms $(\tilde{M}, \tilde{g})$ smoothly to the hyperboloid at infinity with constant negative sectional curvature. Thus the flow provides a smooth homotopy from $(\tilde{M}, \tilde{g})$ to $(\tilde{M}, \tilde{g}_{\mathbb{H}})$.

According to [1, Sect. 12] (see also Lemma 3.4.4), the induced metric $\tilde{g}_t$ on $\tilde{M}$ evolves by the cross curvature flow introduced in [5] (see also Lemma 3.4.4 below):

$$\partial_t \tilde{g}_t = 2 \det \mathcal{E}(\tilde{g}_t) E^{-1}(\tilde{g}_t),$$
$$\tilde{g}_0 = \tilde{g}.$$

At the initial time, we have $\varphi^* \tilde{g}_0 = \tilde{g}_0$ for every $\varphi \in G$. Then, given any $\varphi \in G$, $\bar{g}_t = \varphi^* \tilde{g}_t$ is also a solution to the cross curvature flow with the same initial condition. Hence by uniqueness of solutions [5, 3], Theorem 3.4.5 and Section 3.4.2 below), $\tilde{g}_t = \bar{g}_t = \varphi^* \tilde{g}_t$ and the flow is invariant under the action of G. Lemma 3.3.1 then gives the result. $\qquad\square$

The following conjecture suggests that the integrability assumption in Theorem 3.3.2 could be dropped.

Conjecture 3.3.5 ([5]) The XCF deforms arbitrary negatively curved metrics to a hyperbolic metric.

Evidence for this conjecture includes convergence in the integrable case from Theorem 3.3.2, asymptotic stability of the hyperbolic metric under XCF ([16] and Theorem 3.4.12 below) and monotonicity of an integral quantity measuring the deviation from constant curvature ([5] and Theorem 3.4.11 below).

3.4 The Cross Curvature Flow

3.4.1 Definition and Basic Properties of the Flow

Let (M, g) be a closed, Riemannian manifold and define

$$\text{adjE}(X, Y) - g(X, \text{adj}\mathcal{E}(Y)), \tag{3.8}$$

where adj is the *adjugate* of an endomorphism, sometimes referred to as the *classical adjoint*. The cross curvature flow (XCF) is the evolution equation,

$$\partial_t g_t = 2\text{adjE}(g_t),$$
$$g_0 = g, \tag{3.9}$$

where g_0 has negative sectional curvature. When g_0 has positive sectional curvature, we take instead $\partial_t g = -2\text{adjE}$, though in this chapter we will not be concerned with this case.

Remark 3.4.1   Let $\pi : \tilde{M} \to M$ be the universal cover and $\tilde{g}_t = \pi^* g_t$. Similarly to the proof of Theorem 3.3.2, we then have

$$\partial_t \tilde{g}_t = \pi^* \partial_t \tilde{g}_t = \pi^* \text{adjE}(g_t) = \text{adjE}(\tilde{g}_t)$$

and $\tilde{g}_t$ solves the XCF with initial condition $\tilde{g}_0 = \pi^* g_0$. Conversely, if $\tilde{g}_t$ is a G-invariant solution of the XCF on $\tilde{M}$, then there is a unique solution, g_t, of the XCF on M such that $\tilde{g}_t = \pi^* g_t$.

The definition here makes sense in any dimension. If $\mathcal{E}$ is invertible, we may also write

$$\text{adjE} = \det \mathcal{E} \text{E}^{-1} = \det \mathcal{E} g(\mathcal{E}^{-1} \cdot, \cdot).$$

In three dimensions, g_t has negative sectional curvature if and only if E is positive definite (hence $\mathcal{E}$ is invertible). Then in an orthonormal basis of eigenvectors E_1, E_2, E_3 for $\mathcal{E}$, with eigenvalues $\lambda_1, \lambda_2, \lambda_3$, we have for distinct indices, i, k, ℓ,

$$\text{adj}\mathcal{E}(E_i) = \det \mathcal{E}\mathcal{E}^{-1}(E_i) = \lambda_i \lambda_k \lambda_\ell \frac{1}{\lambda_i} E_i = \lambda_k \lambda_\ell E_i,$$

where i, k, ℓ are distinct indices. Thus

$$\text{adjE}(E_i, E_j) = g(\lambda_k \lambda_\ell E_i, E_j) = \lambda_k \lambda_\ell \delta_{ij}. \tag{3.10}$$

The tensor adjE is referred to as the *cross curvature tensor*. The origin of the name is that the i'th eigenvalue of adjE is the "cross term" $\lambda_k \lambda_\ell$ of the remaining eigenvalues.

There is an equivalent way to write adjE in three dimensions. In fact, both these definitions make sense in any dimension, however it is only in three dimensions that they coincide.

Lemma 3.4.2 *In three dimensions, we have*

$$\mathrm{adjE}(X,Y) = -\frac{1}{2}\mathrm{Ric_E}(X,Y) := -\frac{1}{2}\mathrm{Tr}\left(Z \mapsto \mathrm{Rm}(\mathcal{E}(Z),X)Y\right).$$

Proof As noted above,

$$\mathrm{adjE}(E_i, E_j) = \lambda_k \lambda_\ell \delta_{ij}.$$

There holds

$$\mathrm{Ric_E}(E_i, E_j) = \sum_{m=1}^{3} \mathrm{Rm}(\mathcal{E}(E_m), E_i, E_j, E_m)$$

$$= \sum_{m=1}^{3} \lambda_m \mathrm{Rm}(E_m, E_i, E_j, E_m)$$

$$= -\sum_{m=1}^{3} \lambda_m \hat{\lambda}_{mi} \delta_{ij},$$

where $\hat{\lambda}_{mi} = \lambda_k$ if m, i, k are distinct indices and is zero if $m = i$. Now, if $i = j$, and i, k, ℓ are distinct indices, the sum is over $m = k, \ell$, giving

$$\mathrm{Ric_E}(E_i, E_j) = -(\lambda_k \lambda_\ell + \lambda_\ell \lambda_k) = -2\lambda_k \lambda_\ell.$$

Hence

$$\frac{1}{2}\mathrm{Ric_E}(E_i, E_j) = -\lambda_k \lambda_\ell \delta_{ij} = -\mathrm{adjE}(E_i, E_j).$$

$\square$

Remark 3.4.3 In [5, Lem. 3] and [3, eq. (3)], essentially the same result is obtained by contracting with the measure μ.

In the case of integrable (and hence isometrically embeddable) solutions of XCF, we have the following observation of Ben Andrews.

Lemma 3.4.4 ([1, Sect. 12]) *The induced metric under the Gauss curvature flow of convex, spacelike, co-compact hypersurfaces in Minkowski space evolves by XCF.*

Proof The Gauss curvature flow of hypersurfaces in Minkowksi space is the evolution equation

$$\partial_t F = K\nu,$$

where $K = \det \mathcal{W}$ is the Gauss curvature. Under this equation the metric evolves by

$$\partial_t g = 2KA.$$

Equation (3.7) gives $A = \sqrt{\det \mathcal{E}} E^{-1}$ and

$$K = \det \mathcal{W} = \det(\sqrt{\det \mathcal{E}} \mathcal{E}^{-1}) = (\det \mathcal{E})^{3/2}(\det \mathcal{E})^{-1} = \sqrt{\det \mathcal{E}}$$

so that

$$\partial_t g = 2KA = 2\det \mathcal{E} E^{-1} = 2\mathrm{adj}E.$$

$\square$

3.4.2 Short Time Existence and Uniqueness

Both the question of short time existence and the uniqueness of solutions to geometric, parabolic equations on tensor bundles is complicated by the diffeomorphism invariance of the problem leading to degeneracies in the principal symbol. For the Ricci flow, DeTurck described a method to deal with this degeneracy by breaking the diffeomorphism invariance (i.e. fixing a gauge) to obtain an equivalent, strictly parabolic flow referred to as the DeTurck flow [10]. In [14, Sect. 6] a further simplification of DeTurck's method was given. Buckland then adapted this approach to the XCF, while also pointing out a gap in the original proof of short time existence and uniqueness for XCF [3] (see Remark 3.4.7 below).

Theorem 3.4.5 ([5, Lem. 4]; [3, Thm. 1]) *Given any initial smooth metric g of negative sectional curvature on a closed three-manifold N, there exists a unique solution g_t to the XCF,*

$$\partial_t g_t = 2\mathrm{adj}E(g_t),$$

$$g_0 = g,$$

defined on a maximal time interval $[0, T)$ for some $T > 0$ or $T = \infty$.

To prove the theorem, we need the principal symbol of the cross curvature tensor. Recall that for a nonlinear, second order differential operator $D : E \to F$ acting on vector bundles E, F over M, we define

$$\sigma_\xi[D_s] = \sigma_\xi[D'_s], \tag{3.11}$$

where $s \in \Gamma(E)$ is a section of E and

$$D'_s(u) = \partial_w|_{w=0} D(s + wu)$$

is the linearisation of D around s acting on sections $u \in \Gamma(E)$. Recall that the principal symbol $\sigma_\xi[D'_s]$ of D'_s is obtained by replacing second order derivatives in $D'_s(u)$ by components of ξ.

Recall that E is positive definite when g has negative sectional curvature, hence E is itself a metric. We may thus raise and lower indices using E as well as take traces with E. However, rather than using E directly to raise indices we use the metric raising of $E^\sharp$ defined on one-forms by $E^\sharp(\alpha, \beta) = E(\alpha^\sharp, \beta^\sharp)$. The symbol may be conveniently computed using the equivalent formulation of the cross curvature tensor in Lemma 3.4.2.

Lemma 3.4.6 ([5, Lem. 4]; [3, Thm. 1]) *The principal symbol of the cross curvature tensor is*

$$\sigma_\xi[\mathrm{adjE}_g](V) = |\xi|_E^2\, V - 2\mathrm{Sym}\xi \otimes V(\sharp_E\xi, \cdot) + \mathrm{Tr}_E V \xi \otimes \xi.$$

The degeneracy of the XCF is manifested in the second two terms. If they were absent, then the symbol would simply be $V \mapsto |\xi|_E^2\, V$, which is clearly elliptic. However, the presence of the second two terms means that the symbol has a kernel and is hence not elliptic. This is what the DeTurck approach aims to address.

The proof of Theorem 3.4.5 makes use of the DeTurck XCF,

$$\partial_t g = \mathrm{adjE}(g) + \mathcal{L}_W g =: \sigma_{\mathcal{DT}}, \qquad (3.12)$$

where

$$W = g_0^{-1}\mathrm{div}_g \left(g_0 - \frac{1}{2}\mathrm{Tr}_g g_0\right)^\sharp.$$

Then the symbol is

$$\sigma_\xi[\sigma_{\mathcal{DT}_g}](V) = |\xi|_E^2\, V + 2\mathrm{Sym}\xi \otimes V(\sharp\xi - \sharp_E\xi, \cdot) + (\mathrm{Tr}_E V - \mathrm{Tr}_g V)\xi \otimes \xi. \qquad (3.13)$$

The computation of the symbol of $\mathcal{L}_W g$ is well known and is the same as that for the Ricci flow. See [3], [7, Sect. 3.3, 3.4] or [31, Chap. 5].

That $\sigma_\xi[\sigma_{\mathcal{DT}_g}]$ is uniformly elliptic is not readily apparent, but this is what Buckland shows [3]. Existence now follows easily since the DeTurck XCF (3.12) is uniformly parabolic hence unique solutions exist given any smooth initial data. Then if $\bar{g}$ denotes the solution of the DeTurck XCF, $g = \varphi_t^* \bar{g}$ solves the XCF, where φ_t is the flow of W.

For uniqueness, more work is required, but note that the vector field W is the same as that used for Ricci flow. Thus the same proof as for Ricci flow applies and we may appeal to [14, Sect. 6]. In the formulation used here, φ_t solves the harmonic map heat flow $(M, \bar{g}) \to (M, g_0)$. See [31, Sect. 5.2] and [7, Sect. 3.3, 3.4] for details.

Remark 3.4.7   For the Ricci flow, the DeTurck Ricci flow is defined to be

$$\partial_t g = -2\mathrm{Ric}(g) + \mathcal{L}_W g =: \mathrm{Ric}_{\mathcal{DT}}(g)$$

with the same W as (3.12). In fact, W is adapted to the Ricci flow rather than the XCF since the symbol for the Ricci flow is

$$\sigma_\xi[-2\mathrm{Ric}_g](V) = |\xi|_g^2 V - 2\mathrm{Sym}\xi \otimes V(\sharp\xi, \cdot) + \mathrm{Tr}_g V \xi \otimes \xi.$$

Then for the DeTurck Ricci flow, the last two terms cancel and the symbol for the DeTurck Ricci flow becomes

$$\sigma_\xi[\mathrm{Ric}_{\mathcal{DT}_g}](V) = |\xi|_g^2 V,$$

which is clearly uniformly elliptic. The choice of W comes from the fact that $\mathrm{div}_g \mathrm{E} = 0$ so, defining $L_g(T) = \mathrm{div}_g(T - \frac{1}{2}\mathrm{Tr}_g T)$, we have the integrability condition

$$L_g(\mathrm{Ric}(g)) = \mathrm{div}_g \mathrm{E}(g) = 0.$$

Here $g \mapsto L_g$ depends on g to first order, and computing the symbol of $g \mapsto L_g(\mathrm{Ric}(g))$ also shows the degeneracy in the operator Ric. The original proof of short time existence and uniqueness for Ricci flow made use of this fact by employing the Nash–Moser implicit function theorem. See [13, Sect. 4–6].

One might try a similar approach for XCF, defining an appropriate W to cancel terms so that the symbol becomes $V \mapsto |\xi|_\mathrm{E}^2 V$. Equation (3.14) below suggests the choice

$$W = g_0^{-1}\left(\mathrm{E}^{ij}\nabla_i \mathrm{adjG}(g_0)_{jk} - \frac{1}{2}\mathrm{E}^{ij}\nabla_k \mathrm{adjG}(g_0)_{ij}\right)^\sharp,$$

where $\mathrm{G}(g_0) = g_0 - \frac{1}{2}\mathrm{Tr}_g g_0$ is the Einstein gravity tensor. Presumably such an approach works, with the appropriate replacement for the harmonic map heat flow to obtain uniqueness. Note that this equation also leads to the integrability equation

$$L_g(\mathrm{adjE}(g)) = 0,$$

where $L_g(T) = \mathrm{E}^{ij}\nabla_i \mathrm{adjG}(T)_{jk} - \frac{1}{2}\mathrm{E}^{ij}\nabla_k \mathrm{adjG}(T)_{ij}$.

Buckland [3] observed that $g \mapsto L_g$ is second order in g so that the argument in [13, Sect. 4–6] cannot be applied. However, as we have seen, even though the terms in (3.13) do not cancel, the DeTurck flow using W from the Ricci flow is still uniformly parabolic. Existence for XCF follows immediately, and uniqueness is obtained by the known result using the harmonic map heat flow as in the Ricci flow.

3.4.3 Basic Identities and Evolution Equations

Now we give some fundamental identities and evolution equations for the study of the XCF. Since there are many traces and swapping slots in what follows, it is convenient to use index notation. For convenience, we define $V = E^{-1}$ (that is $V(X,Y) = g(\mathcal{E}^{-1}(X), Y)$).

The first basic identity is an analogue for the cross curvature tensor of the fact that according to the contracted second Bianchi identity, $\mathrm{div}_g E = 0$. It is fundamental in deriving various identities and is closely related to short time existence and uniqueness as discussed above. According to [5, Lem. 1(b)],

$$E^{ij} \nabla_i \mathrm{adj} E_{jk} = \frac{1}{2} E^{ij} \nabla_k \mathrm{adj} E_{ij}. \tag{3.14}$$

The general, non-integrable case poses a number of difficulties. At the heart of these difficulties is the *devil tensor* (3.16) that vanishes if and only if the solution is integrable (Lemma 3.4.8). Let us define

$$\mathsf{T}^{kij} = E^{kl} \nabla_l E^{ij}, \quad \mathsf{T}^i = V_{jk} \mathsf{T}^{ijk} = E^{ij} \nabla_j \ln \det \mathcal{E}. \tag{3.15}$$

From the irreducible decomposition of T^{ijk} under the action of $O(3)$ via the metric V we have ([5, p. 6]),

$$\mathsf{T}^{ijk} - \mathsf{T}^{jik} = \mathsf{L}^{ijk} - \mathsf{L}^{jik} + \frac{1}{2}\left(\mathsf{T}^i E^{jk} - \mathsf{T}^j E^{ik}\right),$$

where the traces of L^{ijk} with respect to V are zero. The devil tensor is defined by

$$\mathsf{D}^{ijk} = \mathsf{L}^{ijk} - \mathsf{L}^{jik} = \mathsf{T}^{ijk} - \mathsf{T}^{jik} - \frac{1}{2}\left(\mathsf{T}^i E^{jk} - \mathsf{T}^j E^{ik}\right). \tag{3.16}$$

It satisfies the curvature-like identities,

$$\mathsf{D}^{ijk} = -\mathsf{D}^{jik}, \quad \mathsf{D}^{ijk} + \mathsf{D}^{kij} + \mathsf{D}^{jki} = 0,$$

and trace identities,

$$V_{ij} \mathsf{D}^{ijk} = V_{ik} \mathsf{D}^{ijk} = V_{jk} \mathsf{D}^{ijk} = 0,$$

$$V_{ij} \nabla_k \mathsf{D}^{kij} = \frac{1}{2} |\mathsf{D}^{ijk}|_V^2.$$

Lemma 3.4.8   *We have the following identity:*

$$|\mathsf{D}^{ijk}|_V^2 = \frac{1}{\det \mathcal{E}} |\nabla_i O_{jk} - \nabla_j O_{ik}|_E^2.$$

In particular, $\mathsf{D} \equiv 0$ if and only if O is Codazzi.

Proof Using the definition we calculate

$$\mathsf{T}^{kij} = -\mathrm{E}^{kl}\mathrm{E}^{im}\mathrm{E}^{jn}\nabla_l V_{mn} = -\mathrm{E}^{kl}\mathrm{E}^{im}\mathrm{E}^{jn}\nabla_l \left(\frac{1}{\sqrt{\det \mathcal{E}}}O_{mn}\right)$$

$$= -\frac{1}{\sqrt{\det \mathcal{E}}}\mathrm{E}^{kl}\mathrm{E}^{im}\mathrm{E}^{jn}\nabla_l O_{mn} + \frac{1}{2}\mathsf{T}^k\mathrm{E}^{ij}.$$

Thus we obtain

$$\mathsf{T}^{kij} - \mathsf{T}^{ikj} = \frac{\mathrm{E}^{kl}\mathrm{E}^{im}\mathrm{E}^{jn}}{\sqrt{\det \mathcal{E}}}(\nabla_m O_{ln} - \nabla_l O_{mn}) + \frac{1}{2}(\mathsf{T}^k\mathrm{E}^{ij} - \mathsf{T}^i\mathrm{E}^{kj}).$$

Then, by definition,

$$\mathsf{D}^{kij} = \frac{\mathrm{E}^{kl}\mathrm{E}^{im}\mathrm{E}^{jn}}{\sqrt{\det \mathcal{E}}}(\nabla_m O_{ln} - \nabla_l O_{mn})$$

and hence

$$|\mathsf{D}^{ijk}|_V^2 \det \mathcal{E} = |\nabla_i O_{jk} - \nabla_j O_{ik}|_\mathrm{E}^2.$$

$\square$

To round out this section, let us point out some evolution equations. Define the operator

$$\square = \mathrm{Tr}_{\mathrm{E}^\sharp}\nabla^2 = \mathrm{E}^{ij}\nabla_{ij}^2. \tag{3.17}$$

It is uniformly elliptic for each fixed t on any time interval $[0, \tau]$ on which $\mathrm{E} > 0$.

We have

$$\partial_t \mathrm{E}^{ij} = \square \mathrm{E}^{ij} - \nabla_\ell \mathrm{E}^{ki}\nabla_k \mathrm{E}^{\ell j} - 4\det \mathcal{E} g^{ij} \tag{3.18}$$

and

$$\partial_t \det \mathcal{E} = \square \det \mathcal{E} - \left(\frac{1}{2(\det \mathcal{E})^2}\left|\mathrm{E}^{ij}\nabla_j \det \mathcal{E}\right|^2 - \frac{1}{2}\left|\mathrm{D}^{ijk}\right|_V^2 + 2\mathrm{H}\right)\det \mathcal{E}. \tag{3.19}$$

Here $\mathrm{H} = \mathrm{Tr}_g \mathrm{adj}\mathrm{E}$. Equation (3.18) is obtained by combining equation (3.14) ([5, Lem. 1(a)]) with [5, Lem. 5] giving the evolution of E. Equation (3.19) is then obtained from (3.18) using (3.15).

The maximum principle cannot be immediately applied to ensure positivity of E (and hence negative sectional curvature) is preserved since the last term of each equation has the wrong sign. However, a slightly less direct approach using the evolution of $\det \mathcal{E}$ effectively leads to the desired conclusion (Proposition 3.4.10).

The XCF expands negatively curved metrics, as can be seen from the evolution of the volume:

$$\frac{d}{dt} \operatorname{Vol}_g(M) = \int_M \mathrm{H} d\mu > 0. \tag{3.20}$$

Such expansion is also apparent in the evolution of hyperbolic metrics.

Example 3.4.9   Let g_0 be a hyperbolic metric with constant sectional curvature $K_0 < 0$. Then the solution of the XCF is

$$g(t) = \sqrt{4K_0^2 t + 1} g_0$$

since scaling g_0 by r results in $\operatorname{adjE}(g_t) = \frac{1}{r} \operatorname{adjE}(g_0) = \frac{K_0^2}{r} g_0$.

We see that, up to scale, hyperbolic metrics are fixed points. In fact, they are the only fixed points up to scale and reparametrisation (i.e. solitons). See Section 3.4.5.

3.4.4 Towards Hyperbolic Convergence

The following results support Conjecture 3.3.5 that the XCF deforms arbitrary negatively curved metrics to a hyperbolic metric. Under the XCF, it is not so easy to prove directly that negative sectional curvature is preserved. However, it is possible to prove that if $\det \mathcal{E} \to 0$, then a singularity must occur and thus negative curvature is effectively preserved since that corresponds to the positivity of E.

Proposition 3.4.10 ([5, p. 8])   *Let T be the maximal time of smooth existence of XCF. Then $\det \mathcal{E} > 0$ on $[0, T)$ and hence the sectional curvatures remain negative as long as the solution is smooth.*

The proof follows from equation (3.19) by writing,

$$\partial_t \ln \det \mathcal{E} = \square \ln \det \mathcal{E} + \frac{1}{2} \left| \mathrm{T}^{ijk} - \mathrm{T}^{jik} \right|_V^2 - 2\mathrm{H}.$$

The last term has the wrong sign to apply the maximum principle. However, $\partial_t \min \ln \det \mathcal{E} \geq -2\mathrm{H}$ at the minimum of $\ln \det \mathcal{E}$. If $t_0 < T$ and $\ln \det \mathcal{E} \to -\infty$ as $t \to t_0$, then $\mathrm{H} \to \infty$. In other words, one eigenvalue of $\mathcal{E}$ goes to zero while another goes to ∞ as t approaches t_0.

Going the other way, namely proving that if $[0, T)$ is the maximal time of existence with $T < \infty$, then necessarily $\inf_{t \in [0,T)} \det \mathcal{E} = 0$ would show that finite time existence implies blow up of E. Such a result would serve as a proxy for more general estimates, such as the smoothing estimates enjoyed by the Ricci flow. The lack of such estimates (even C^2 estimates) is perhaps the major outstanding issue for the XCF. In particular, it could be that the solution blows up in finite time, yet $\det \mathcal{E}$ has a positive lower bound.

The following monotonicity property also supports Conjecture 3.3.5 that the flow evolves negative curvature metrics to a hyperbolic metric.

Theorem 3.4.11 ([5, Thm. 8]) *Under the XCF,*

$$J(M_t) := \int \frac{1}{3}\mathrm{Tr}\mathcal{E} - (\det \mathcal{E})^{\frac{1}{3}} d\mu$$

is non-increasing. Moreover, $\frac{d}{dt}J(M_t) = 0$ if and only if g has constant curvature.

The theorem follows by direct computation and noting that by the arithmetic–geometric mean inequality applied to the eigenvalues of $\mathcal{E}$, $\frac{1}{3}\mathrm{Tr}\mathcal{E} \geq (\det \mathcal{E})^{1/3}$ with equality if and only if $\mathrm{E} = \lambda g$ for some smooth function λ. In the equality case, since $\mathrm{div}_g \mathrm{E} = \mathrm{div}_g g = 0$, we must then also have $\lambda \equiv$ const., and hence equality occurs if and only if g has constant sectional curvature equal to $-\lambda$.

To finish this section, we consider the stability of hyperbolic metrics under the flow. Stability is necessary for Conjecture 3.3.5 to hold, and indeed this is the case. To state the result, some normalisation is necessary. Typically normalising to fix volume is the approach taken, but here following [16] a different normalisation is chosen. For any $K < 0$, note that a constant curvature metric g_K of sectional curvature K satisfies

$$\mathrm{adjE} = K^2 g$$

by (3.4) and (3.10). Then the flow

$$\partial_t g = 2\mathrm{adjE}(g) - 2K^2 g$$

has g_K as a fixed point. Note that, up to scaling and diffeomorphism, solutions of this flow are equivalent to solutions of the XCF by making use of [16, Lem. 1].

The linearisation of the corresponding DeTurck flow around a hyperbolic metric is ([16, Lem. 2])

$$\partial_t V = -K\Delta V - 2K^2 \mathrm{Tr}_g V g + 2K^2 V.$$

Then the spectrum of the right hand side is contained in $(-\infty, -1]$ ([16, Sect. 5]), from which Theorem 3.4.12 follows by standard stable manifold theory.

Theorem 3.4.12 ([16, Thm. 4])   *Any constant curvature metric is asymptotically stable under the XCF after suitable normalisation.*

3.4.5 Harnack Inequality and Solitons

Differential Li–Yau–Hamilton Harnack inequalities have proved an indispensable tool in the study of curvature flows. For the XCF Chow and Hamilton [5, p. 9] remarked that it is hoped such an inequality holds. For integrable solutions of the XCF this is true by proving the analogous Harnack inequality for the corresponding solution of the embedded Gauss curvature flow. In general, the non-vanishing of the devil tensor (3.16) causes significant difficulties in obtaining a Harnack inequality. The evolution of the devil tensor is very complicated and it's not clear whether or not it is amenable to the maximum principle.

Theorem 3.4.13 ([2, Sect. 6])   *The following Li–Yau–Hamilton Harnack inequality holds for integrable solutions to the XCF:*

$$\partial_t \sqrt{\det \mathcal{E}} - \frac{1}{\sqrt{\det \mathcal{E}}} \left| \nabla \sqrt{\det \mathcal{E}} \right|_{\mathrm{E}}^2 + \frac{3}{4t} \sqrt{\det \mathcal{E}} \geq 0.$$

Solitons are closely related to the Harnack inequality. They are fixed points of the flow modulo scaling and diffeomorphism. As such, they are the expected limits (up to rescaling) of the flow. Unlike other flows, such as the Ricci flow, solitons for the XCF are very rigid. In fact, something stronger is true. Recall that a soliton is a solution of XCF such that $g_t = \lambda(t)\varphi_t^* g_0$, where λ is a positive function and φ_t is a one-parameter family of diffeomorphisms. More generally, a *breather* is a solution such that $g_{t_0} = \lambda \varphi^* g_0$ for some $t_0, \lambda > 0$ and a diffeomorphism φ.

Theorem 3.4.14 ([6])   *The only breathers, hence also solitons to the XCF, are constant curvature metrics.*

This follows by monotonicity and scaling: since volume is increasing by equation (3.20) and scales as $\lambda^{3/2}$, for a breather we must have $\lambda > 1$. On the other hand, since J scales as $\lambda^{1/2}$, if J strictly decreases, then we obtain the contradiction $\lambda < 1$. Therefore, J is constant and hence g_t has constant sectional curvature by the equality case of Theorem 3.4.11.

The theorem further supports Conjecture 3.3.5 in that if a limiting metric exists it should be stationary for the flow (up to rescaling and reparametrisation) and hence a soliton. Thus the only expected limits have constant curvature. A Harnack inequality for general solutions would further support the conjecture since it may be used to obtain improved smoothing estimates for the flow.

3.4.6 Monotonicity of Einstein Volume

Since the Einstein tensor is a metric, it induces a volume form. By [5, Prop. 9], the Einstein volume is monotone non-decreasing along the XCF. We may strengthen this result, characterising solutions to the XCF such that $\frac{d}{dt}I(M_t) = 0$ as precisely the integrable solutions, a question posed in [5, p. 9].

Theorem 3.4.15 *Under the XCF of negative sectional curvature, the Einstein volume,*

$$I(M_t) := \int \sqrt{\det \mathcal{E}} \, d\mu,$$

is non-decreasing. Moreover, $\frac{d}{dt}I(M_t) = 0$ if and only if O is Codazzi if and only if the Riemannian universal cover $(\tilde{M}, \tilde{g})$ embeds isometrically into Minkowski space $\mathbb{R}^{3,1}$ as a locally convex, co-compact spacelike hypersurface.

Proof By [5, Prop. 9], we have

$$(\partial_t - \Box)(\det \mathcal{E})^\eta = \left(\frac{\eta}{2}|D^{ijk}|_V^2 + \frac{\eta}{2}(1 - 2\eta)|T^i|_V^2 - 2\eta H\right)(\det \mathcal{E})^\eta,$$

where $\Box$ is defined in (3.17). Integrating by parts gives the identity

$$\frac{d}{dt}\int(\det \mathcal{E})^\eta d\mu = \int\left(\frac{\eta}{2}|D^{ijk}|_V^2 + \frac{\eta}{2}(1 - 2\eta)|T^i|_V^2 + (1 - 2\eta)H\right)(\det \mathcal{E})^\eta d\mu.$$

In particular, for $\eta = \frac{1}{2}$ we obtain

$$\frac{d}{dt}I(M_t) = \frac{1}{4}\int|D^{ijk}|_V^2\sqrt{\det \mathcal{E}}d\mu$$

and monotonicity follows. Now apply Lemma 3.4.8 and Theorem 3.3.4 to obtain the second part. $\quad\Box$

References

[1] Ben Andrews, Xuzhong Chen, Hanlong Fang, and James McCoy. *Expansion of cocompact convex spacelike hypersurfaces in Minkowski space by their curvature*. Indiana Univ. Math. J. **64**(2) (2015), pp. 635–662. doi: 10.1512/iumj.2015.64.5485.

[2] Paul Bryan, Mohammad N. Ivaki, and Julian Scheuer. *Harnack inequalities for curvature flows in Riemannian and Lorentzian manifolds*. J. Reine Angew. Math., **764** (2020), 71–109. doi: https://doi.org/10.1515/crelle-2019-0006.

[3] John A. Buckland. *Short-time existence of solutions to the cross curvature flow on 3-manifolds*. Proc. Amer. Math. Soc. **134** (2006), pp. 1803–1807. doi: 10.1090/S0002-9939-05-08204-3.

[4] Xiaodong Cao, John Guckenheimer, and Laurent Saloff-Coste. *The backward behavior of the Ricci and cross-curvature flows on* $SL(2; \mathbb{R})$. Commun. Anal. Geom. **17**(4) (2009), pp. 777–796. doi: 10.4310/CAG.2009.v17.n4.a9.

[5] Bennett Chow and Richard S. Hamilton. *The cross curvature flow of 3-manifolds with negative sectional curvature*. Turk. J. Math. **28**(1) (2004), pp. 1–10.

[6] Bennett Chow, Sun-Chin Chu, David Glickenstein *et al.* The Ricci flow: techniques and applications – Part I: Geometric aspects. Mathematical Surveys and Monographs 135, Amer. Math. Soc., Providence, RI, 2007.

[7] Bennett Chow and Dan Knopf. The Ricci flow: an introduction. Mathematical Surveys and Monographs 110, Amer. Math. Soc., Providence, RI, 2004. doi: 10.1090/surv/110.

[8] Xiaodong Cao, Yilong Ni, and Laurent Saloff-Coste. *Cross curvature flow on locally homogeneous three-manifolds. I.* Pacific J. Math. **236**(2) (2008), pp. 263–281. doi: 10.2140/pjm.2008.236.263.

[9] Xiaodong Cao and Laurent Saloff-Coste. *Cross curvature flow on locally homogeneous three-manifolds. II.* Asian J. Math. **13**(4) (2009), pp. 421–458. doi: 10.4310/AJM.2009.v13.n4.a1.

[10] Dennis M. DeTurck. *Deforming metrics in the direction of their Ricci tensors*. J. Differ. Geom. **18**(1) (1983), pp. 157–162. url: http://projecteuclid.org/euclid.jdg/1214509286.

[11] Jason DeBlois, Dan Knopf, and Andrea Young. *Cross curvature flow on a negatively curved solid torus*. Algebr. Geom. Topol. **10**(1) (2010), pp. 343–372. doi: 10.2140/agt.2010.10.343.

[12] David Glickenstein. *Riemannian groupoids and solitons for three-dimensional homogeneous Ricci and cross-curvature flows*. Int. Math. Res. Not. **12** (2008) doi: 10.1093/imrn/rnn034.

[13] Richard Hamilton. *Three-manifolds with positive Ricci curvature*. J. Differ. Geom. **17**(2) (1982), pp. 255–306.

[14] Richard S. Hamilton. *The formation of singularities in the Ricci flow*. In Surveys in differential geometry, Vol. II. Proc. Conf. Geometry and Topology, Harvard University, April 23–25, 1993. Edited by C.-C. Hsiung and Shing-Tung Yau. International Press, Cambridge, MA, pp. 7–136. 1995.

[15] Bruce Kleiner and John Lott. *Notes on Perelman's papers.* Geom. Topol. **12**(5) (2008), pp. 2587–2855. doi: 10.2140/gt.2008.57.2587.

[16] Dan Knopf and Andrea Young. *Asymptotic stability of the cross curvature flow at a hyperbolic metric.* Proc. Amer. Math. Soc. **137**(2) (2009), pp. 699–709. doi: 10.1090/S0002-9939-08-09534-8.

[17] Shoshichi Kobayashi and Katsumi Nomizu. *Foundations of differential geometry,* Vol. II. Wiley Classics Library. Reprint of the 1969 original. A Wiley-Interscience Publication. John Wiley & Sons, Inc., New York, 1996.

[18] Brett Kotschwar. *A short proof of backward uniqueness for some geometric evolution equations.* Int. J. Math. **27**(12) (2016), 165002. doi: 10.1142/S0129167X16501020.

[19] Brett Kotschwar. *An energy approach to uniqueness for higher-order geometric flows.* Geom. Anal. **26**(4) (2016), pp. 3344–3368. doi: 10.1007/s12220-015-9670-y.

[20] Li Ma and Dezhong Chen. *Examples for cross curvature flow on 3-manifolds.* Calc. Var. Partial Differ. Equ. **26**(2) (2006), pp. 227–243. doi: 10.1007/s00526-005-0366-1.

[21] John Milnor. *A unique decomposition theorem for 3-manifolds.* Amer. J. Math. **84** (1962), pp. 1–7. doi: 10.2307 2372800.

[22] George D. Mostow. *Quasi-conformal mappings in n-space and the rigidity of hyperbolic space forms.* Inst. Hautes Etudes Sci. Publ. Math. **34** (1968), pp. 53–104.

[23] John Morgan and Gang Tian. Ricci flow and the Poincaré conjecture. Clay Mathematics Monographs **3**. Amer. Math. Soc., Providence, RI; Clay Mathematics Institute, Cambridge, MA, 2007.

[24] John Morgan and Gang Tian. The geometrization conjecture. Clay Mathematics Monographs **5**. Amer. Math. Soc., Providence, RI; Clay Mathematics Institute, Cambridge, MA, 2014.

[25] Grisha Perelman. *The entropy formula for the Ricci flow and its geometric applications.* arXiv Mathematics e-prints (Nov. 2002).

[26] Grisha Perelman. *Ricci flow with surgery on three-manifolds.* arXiv Mathematics e-prints (Mar. 2003).

[27] Grisha Perelman. *Finite extinction time for the solutions to the Ricci flow on certain three-manifolds.* arXiv Mathematics e-prints (July 2003).

[28] Peter Scott. *The geometries of 3-manifolds.* Bull. London Math. Soc. **15**(5) (1983), pp. 401–487. doi: 10.1112/blms/15.5.401.

[29] William P. Thurston. *Three-dimensional manifolds, Kleinian groups and hyperbolic geometry.* Bull. Amer. Math. Soc. (N.S.) **6**(3) (1982), pp. 357–381. doi: 10.1090/S0273-0979-1982-15003-0.

[30] William P. Thurston (ed. Silvio Levy). Three-dimensional geometry and topology. Vol. 1. Princeton Mathematical Series **35**. Princeton University Press, Princeton, NJ, 1997.

[31] Peter Topping. Lectures on the Ricci flow. London Mathematical Society Lecture Note Series **325**. Cambridge University Press, Cambridge, 2006. doi: 10.1017/CBO9780511721465.

4

A Mean Curvature Flow for Conformally Compact Manifolds

A. Rod Gover and Valentina-Mira Wheeler

Abstract

Using tools from conformal geometry we obtain a natural variant of the usual mean curvature flow. It is a flow that moves the hypersurface manifold at a speed equal to the normal projection of its scale tractor. This produces a curvature flow that is well defined to the boundary on Poincaré–Einstein manifolds and more generally on conformally compact manifolds and their generalisations.

Study of this flow, and its variants, provides information on the existence of minimal surfaces meeting the boundary at infinity of conformally compact manifolds. We treat the special case of hyperbolic space and retrieve the existence of the generalised flow for all times. A next step is the convergence result, left as a conjecture here. Under a curvature bound we conjecture the convergence of the flow to the minimal hypersurface in hyperbolic space with the given boundary at infinity, the existence of which is originally due to Anderson.

4.1 Introduction

A Riemannian $(n+1)$-manifold (M, g) is said to be *conformally compact* if M may be identified with the interior of a smooth compact manifold $\bar{M}$ with boundary, and there exists a *defining function* u for the boundary ∂M so that

$$g = u^{-2}\bar{g} \text{ on } M, \tag{4.1}$$

where $\bar{g}$ is a smooth metric on $\bar{M}$. In calling u a *defining function* for ∂M we mean that u is a smooth real valued function on $\bar{M}$ such that

∂M is exactly the zero locus $\mathcal{Z}(u)$ of u and which furthermore satisfies the condition that du is non-zero at every point of ∂M. In this case the metric g is complete and for this metric the boundary ∂M is at infinity, in that it is an infinite distance from any point of M. A conformally compact manifold is said to be *asymptotically hyperbolic* if the defining function u satisfies $|du|_g^2 = 1$ along the boundary ∂M, as in this case the sectional curvatures of g approach -1 at the boundary. A yet more restrictive case is that of *Poincaré–Einstein* manifolds which satisfy the condition that g is Einstein (then necessarily negative Einstein) with the Ricci cuvature satisfying $\text{Ric}^g = -ng$. These structures all generalise the *Poincaré ball* where hyperbolic space $\mathbb{H}^{n+1}$ is identified with the interior of a unit ball B^{n+1} in $(n+1)$-dimensional Euclidean space $\mathbb{E}^{n+1}$, the interior of which is equipped with the metric

$$g_{\mathbb{H}} = \frac{4}{(1 - |x|_{g_{\mathbb{E}}}^2)^2} g_{\mathbb{E}},$$

where $g_{\mathbb{E}}$ is the Euclidean metric and x denotes the usual coordinates in $\mathbb{R}^{n+1} \cong \mathbb{E}^{n+1}$.

These structures play an extremely important role in linking conformal and Riemannian geometry in both geometric analysis [1, 16, 17] and mathematical physics [20, 32]. In both settings there has been considerable interest in *hypersurfaces* (i.e. embedded codimension-1 submanifolds) Σ that meet the boundary transversely, so that $\Lambda := \Sigma \cap \partial M$ is a smooth conformal hypersurface in ∂M [4, 5, 2]. When such hypersurfaces satisfy a normalising curvature condition, such as minimality, then, for example, the hypersurface may be determined (at least asymptotically to some order) by its boundary Λ. In such cases this provides a link between conformal and Riemannian hypersurface geometry and some very deep invariants are exposed [18, 19].

Given this context it is interesting to consider hypersurface flow problems that are tailored for the treatment of conformally compact manifolds. Here we introduce such a problem.

Conformal geometry is used to provide a modification of the mean curvature flow equation that is naturally adapted to deal with the fact that the metric (4.1) is singular along the boundary, where u is zero. Part of the idea here is that, although the mean curvature H_g is not conformally invariant, for smooth hypersurfaces Σ in M there is a generalising notion $\bar{H}$ of mean curvature that is smooth to the boundary, where the metric is singular, and this extends H_g (i.e. $\bar{H} = H_g$ on M). This is described in Section 4.2.4 below.

For an n-manifold with boundary $\bar{\Sigma}$, with $\bar{\Sigma} = \Sigma \cup \partial\Sigma$, and a smoothly parametrised family of hypersurface embeddings denoted $F(t) : \bar{\Sigma} \to \bar{M}$, we are interested in problems where $F(\partial\Sigma) \subset \partial M$ and the flow equation is

$$\frac{\partial F}{\partial t} = -\bar{H}\nu_{\bar{g}}, \qquad (4.2)$$

where $\nu_{\bar{g}}$ is a unit normal to $\bar{\Sigma}$ in the metric $\bar{g}$. (More generally one could study other problems; see Remark 4.2.5.) There are natural boundary conditions associated with this setup and this is all described in Section 4.2.6.

In 1956 Mullins [33] introduced mean curvature flow as a mathematical tool to describe and study the motion of idealised grain boundaries in two dimensions. Since then mean curvature flow has been studied in many settings and variants. It is not within the scope of this chapter to include a comprehensive list of references, but some key ones can be found in the survey of various aspects of the mean curvature flow of hypersurfaces by Colding, Minicozzi and Pedersen [10], and then in the references therein. In 1978 Brakke [9] created the first computational platform for the mean curvature flow and completed a study from the perspective of geometric measure theory. Smooth compact surfaces evolved by mean curvature flow in Euclidean space were investigated by Huisken in [25] and [28], and on arbitrary ambient manifolds in [26]. The study of the evolution of complete graphs by mean curvature flow in $\mathbb{R}^{n+1}$ was completed by Ecker and Huisken [14], with interior estimates being improved in [15]. Nonparametric graphs evolving by mean curvature flow with Dirichlet and Neumann boundary conditions have been initially studied in the geometric setting by Huisken [27].

Some of the results from Euclidean space generalise to the setting of Riemannian manifolds. One of the first was the generalisation of the convex bodies contraction to spheres of Huisken, as mentioned above [26]. The long time existence of the mean curvature flow in hyperbolic space was first considered by Unterberger [39]. He proved that if the initial surface is a radial graph and has bounded hyperbolic height over $\mathbb{S}^n_+$ then, under mean curvature flow, it converges smoothly to $\mathbb{S}^n_+$, which is minimal with respect to the hyperbolic metric.

The result was generalised to the case of modified mean curvature flow in hyperbolic space by Lin and Xiao [31], where they show the existence, uniqueness and convergence of complete, embedded star-shaped hypersurfaces with prescribed boundary at infinity. Later on their result was improved by Allmann, Lin and Zhou [3] by removing the uniform local ball condition, required in the previous work, at the asymptotic boundary. We, in part, base our development on the last mentioned works.

The parabolic results are motivated by the following results in the elliptic setting. The existence and regularity of hypersurfaces of constant mean curvature in hyperbolic space $\mathbb{H}^{n+1}$ with prescribed asymptotic boundary at infinity, either complete or with different boundary conditions, have been studied for many years. See Anderson [4, 5] for the complete setting, Hardt and Lin [24, 30] for boundary setting and more regularity issues, then Tonegawa [38] and Nelli and Spruck [34] for more existence and uniqueness issues. In 2000, Guan and Spruck [21] proved the existence and uniqueness of smooth complete hypersurfaces of constant mean curvature with values between $(-n, n)$ in hyperbolic space and with prescribed $C^{1,1}$ star-shaped asymptotic boundary at infinity. The result was also recovered using a calculus of variation method by others, for example by De Silva and Spruck [12].

The chapter is structured as follows: in Section 4.2 we describe our conformal manifolds and the tractor calculus terminology and define the geometric curvature flow that naturally arises from the extended notion of mean curvature. In Section 4.3 we treat the flow through its associated partial differential equation and give existence and convergence results.

4.2 Conformal Geometry and Hypersurfaces in Conformally Compact Manifolds

In the following we consider only metrics of definite signature. So (M, g) will mean a Riemannian manifold. To facilitate our later constructions we assume that this is of dimension $n+1 \geq 3$. We write ∇, or sometimes ∇^g, to denote the Levi-Civita connection of g. Then the *curvature* of ∇ will be denoted by $R_{ab}{}^c{}_d \in \Gamma(\wedge^2 T^* M \otimes \mathrm{End}(TM))$, where we adopt the convention $R_{ab}{}^c{}_d \xi^d := (\nabla_a \nabla_b - \nabla_b \nabla_a)\xi^c$ for a vector field ξ^a, and we use abstract indices in the sense of Penrose. The *Ricci tensor* of ∇ is defined by $\mathrm{Ric}_{bd} := R_{cb}{}^c{}_d$ and the *Schouten tensor* P_{ab} is a (metric) trace modification of this determined by

$$\mathrm{Ric}_{ab} = (n-1)P_{ab} + Jg_{ab},$$

where $J := g^{ab} P_{ab}$.

It will be useful to recall the notion of conformal structure.

4.2.1 Conformal Manifolds

A conformal manifold (M, c) is a smooth manifold M equipped with an equivalence class c of Riemannian metrics, with equivalence relation

$$g \sim \hat{g} \quad \text{if} \quad \hat{g} = \Omega^2 g, \quad \text{where} \quad 0 < \Omega \in C^\infty(M).$$

In conformal geometry density bundles, $\mathcal{E}[w]$ play an important role. For these we follow the conventions in e.g. [11]. A section σ of $\mathcal{E}[w]$, $w \in \mathbb{R}$, may be understood as an equivalence class of (metric, function)

$$(g, s) \sim (\hat{g}, \Omega^w s), \qquad \text{where} \qquad g, \hat{g} \in c, \tag{4.3}$$

with $\hat{g} = \Omega^2 g$ as above. Typically, when g is understood we drop that from the pair, and just view the section σ to be represented by the function s. Note that $\mathcal{E}[0]$ means the trivial bundle (whose sections are simply scalar functions).

It follows that each metric $g \in c$ determines a corresponding positive section σ_g of $\mathcal{E}[1]$ (see [11] for further details) and thus a canonical section of $(S^2 T^* M)[2]$ called the *conformal metric*,

$$\boldsymbol{g} := \sigma_g^2 g,$$

that is independent of the choice $g \in c$. This may be used to raise and lower indices, but doing so introduces a weight (adjustment).

4.2.2 The Tractor Connection

On a general conformal manifold (M, c) there is no distinguished connection on the tangent bundle TM, but there is a canonical connection on a closely related bundle of rank, $\mathrm{rank}(TM) + 2$, i.e. rank $n + 3$ given our conventions. This is the *conformal tractor bundle* $\mathcal{T}$ [8, 11]. Given a metric $g \in c$, this decomposes into a direct sum that we often display vertically,

$$
\begin{array}{c}
\mathcal{E}[1] \\
\oplus \\
\mathcal{T} \overset{g}{=} T^* M[1] \\
\oplus \\
\mathcal{E}[-1]
\end{array}
$$

where for any bundle $\mathcal{B}$ we write $\mathcal{B}[w]$ as a shorthand for $\mathcal{B} \otimes \mathcal{E}[w]$ and "$\overset{g}{=}$" should be read as "equals, calculating in the scale g". We also sometimes use an abstract index notation in which we write $\mathcal{E}^A$ as an alternative notation for $\mathcal{T}$.

In terms of the metric $g \in c$ the tractor connection is then given by

$$\nabla_a^{\mathcal{T}} \begin{pmatrix} \sigma \\ \mu_b \\ \rho \end{pmatrix} \overset{g}{:=} \begin{pmatrix} \nabla_a \sigma - \mu_a \\ \nabla_a \mu_b + \mathbf{g}_{ab} \rho + P_{ab} \sigma \\ \nabla_a \rho - P_{ab} \mu^b \end{pmatrix},$$

where on the right hand side ∇ denotes the Levi-Civita connection for g and P_{ab} is the Schouten tensor. But actually the tractor connection is conformally invariant and so the result is a well-defined section of $T^*M \otimes \mathcal{T}$, see [11].

For a conformal manifold $(M, \mathbf{c})$, the formula

$$V_A \overset{g}{=} (\sigma, \mu_a, \rho) \mapsto 2\sigma\rho + \mathbf{g}^{ab} \mu_a \mu_b =: h(V, V)$$

defines, by polarisation, a signature $(n + 2, 1)$ metric on $\mathcal{T}$. It is easily verified that this is conformally invariant and is preserved by $\nabla^{\mathcal{T}}$. We call h the *tractor metric*. As a symmetric bilinear form field on the bundle $[\mathcal{T}]_g$ this takes the form

$$h(V', V) \overset{\mathbf{g}}{=} \begin{pmatrix} \sigma' & \mu' & \rho' \end{pmatrix} \begin{pmatrix} 0 & 0 & 1 \\ 0 & \mathbf{g}^{-1} & 0 \\ 1 & 0 & 0 \end{pmatrix} \begin{pmatrix} \sigma \\ \mu \\ \rho \end{pmatrix}.$$

Note that by construction $h(V', V)$ has weight 0.

4.2.3 Conformally Compact Manifolds

Conformally compact manifolds were defined in the introduction. These have a convenient interpretation using conformal geometry and the tractor connection.

Given a conformal manifold with boundary $(\bar{M}, \mathbf{c})$, a metric $\bar{g}$ corresponds to a scale $\tau \in \Gamma(\mathcal{E}_+[1])$. Then if u is a defining function (chosen positive on M) for the boundary ∂M, we have that

$$\sigma := u\tau \tag{4.4}$$

is a *defining density* for the boundary, meaning ∂M is exactly the zero locus of σ and $\nabla^{\bar{g}}\sigma$ is non-zero at all points of ∂M. Conversely given a background scale τ such a defining density determines a defining function. A conformally compact manifold does not fix the choice of $\bar{g} \in \mathbf{c}$, equivalently τ; in the setting of the introduction we may change u to any $f \cdot u$, where f is a smooth positive function. Changing this does not affect the condition that $\nabla^{\bar{g}}\sigma$ is non-zero at all points of ∂M. So we have a convenient definition that avoids any such choices, as follows.

Definition 4.2.1   A conformally compact manifold is a conformal manifold with boundary $(\bar{M}, \mathbf{c})$ equipped with a boundary defining density $\sigma \in \Gamma(\mathcal{E}[1])$.

From this it follows immediately that a conformally compact manifold has a conformal structure on ∂M (namely $\mathbf{c}|_{\partial M}$) and the metric $g = \sigma^{-2}\mathbf{g}$ on the interior M. Then, with the additional choice of a scale $\tau \in \Gamma(\mathcal{E}_+[1])$, equivalently $\bar{g} \in \mathbf{c}$, one comes to the defining function $u = \sigma \cdot \tau^{-1}$ so that in summary

$$\bar{g} = \tau^{-1}\mathbf{g} \quad \text{and} \quad g = \sigma^{-2}\mathbf{g}$$

on $\bar{M}$ and M, respectively, and so

$$g = u^{-2}\bar{g} \quad \text{on} \quad M.$$

We now introduce a tool that we will require later. Canonically associated to the tractor connection there is a second order differential $\Gamma(\mathcal{E}[1]) \to \Gamma(\mathcal{T})$ given by the formula

$$D_B \sigma \overset{\bar{g}}{=} \begin{pmatrix} \sigma \\ \nabla_b^{\bar{g}} \sigma \\ -\frac{1}{n+1}(\Delta^{\bar{g}}\sigma + J^{\bar{g}}\sigma) \end{pmatrix},$$

for any metric $\bar{g} \in \mathbf{c}$. Any section of $\mathcal{T}$ that is parallel for the tractor connection is in the image of this operator, and it is easily seen that this condition is sufficient to define this conformally invariant operator [11]. Of course a section of $\mathcal{T}$ in the image of the operator is not necessarily parallel. However, in any case where the image

$$I_A := D_A \sigma \tag{4.5}$$

is nowhere vanishing we term I a *scale tractor*.

In particular on a conformally compact manifold $(M, \mathbf{c}, \sigma)$ we have canonically the scale tractor $I = D\sigma$ corresponding to the defining density σ. On the interior M, $h(I, I)$ gives $-\frac{1}{n(n+1)} \mathrm{Sc}^g$, where $\mathrm{Sc}^g = g^{ab}\mathrm{Ric}_{ab}^g$ is the scalar curvature of the metric $g = \sigma^{-2}\mathbf{g}$. But $h(I, I)$ extends this smoothly to the boundary. Thus the scale tractor satisfies $h(I, I)|_{\partial M} = 1$ if and only if the metric g is asymptotically hyperbolic. The scale tractor I is parallel if and only if g is Einstein. If in addition $\mathrm{Ric}^g = -n(n+1)g$, equivalently $h(I, I) = 1$, then the metric g is Poincaré–Einstein with the usual normalisations [11].

4.2.4 Hypersurfaces

Associated to any hypersurface Σ in a conformal manifold $(\bar{M}, \mathbf{c})$ is a *normal tractor* $N_A \in \Gamma(\mathcal{T}|_N)$ [8]. Given a metric $\bar{g} \in \mathbf{c}$ this (conformally invariant quantity) is given by

$$
N_A \overset{\bar{g}}{=} \begin{pmatrix} 0 \\ \nu_a \\ -\mathsf{H}^g \end{pmatrix},
$$

where ν_a is a weight 1 normal to Σ that is of unit length in that it satisfies $g^{ab}\nu_a\nu_b = 1$ and H^g is the mean curvature of Σ, but viewed as a density of weight -1 [11]. It follows that $h(N, N) = 1$.

From this tractor field and the scale tractor we obtain a natural generalisation of the mean curvature as follows.

Definition 4.2.2   On a hypersurface Σ, in a conformally compact manifold $(M, \mathbf{c}, \sigma)$, the *extended mean curvature* is

$$
\bar{H} := -h(I, N) = -h^{AB} I_A N_B,
$$

where $I = D\sigma$ is the canonical scale tractor and N is the normal tractor along Σ.

Note that this is well defined along any smooth hypersuface Σ in $\bar{M}$. In the following proposition we see that it smoothly extends the mean curvature H^g (of Σ) to the boundary $\Lambda \in \partial M$, where $g = \sigma^{-1}\mathbf{g}$ is singular.

Proposition 4.2.3   *In a conformally compact manifold (M, g, σ), let Σ be a smoothly embedded hypersurface that meets the boundary ∂M transversely in a smooth boundary hypersurface Λ. At any point $p \in M$ the extended mean curvature $\bar{H}$ (of Σ) agrees with the mean curvature H^g of Σ in the metric $g = \sigma^{-2}\mathbf{g}$,*

$$
\bar{H}_p = H^g_p \qquad \text{for} \qquad p \in M.
$$

Along ∂M it measures minus the angle between Σ and ∂M.

Proof   Working at a point $p \in M$ and calculating in the scale of the metric $g = \sigma^{-2}\mathbf{g}$, we have $\nabla^g \sigma = 0$ and so

$$
\bar{H} = -h(I, N) \overset{g}{=} -\left(-\frac{J^{\bar{g}}\sigma}{(n+1)} \quad 0 \quad \sigma \right) \cdot \begin{pmatrix} 0 \\ \nu \\ -\mathsf{H}^g \end{pmatrix} = \sigma \mathsf{H}^g = H^g,
$$

where $\cdot$ indicates the cotractor–tractor contraction. On the other hand, now calculating in the metric $\bar{g} = \tau^{-2}g$ that goes to the boundary we have

$$\bar{H} = -h(I,N) \overset{\bar{g}}{=} - \left(-\frac{\Delta^{\bar{g}}\sigma + J^{\bar{g}}\sigma}{(n+1)} \quad \nabla^{\bar{g}}\sigma \quad \sigma \right) \cdot \begin{pmatrix} 0 \\ \nu \\ -\mathsf{H}^{\bar{g}} \end{pmatrix} = \sigma\mathsf{H}^{\bar{g}} - \nu^a \nabla^{\bar{g}}_a \sigma,$$

(4.6)

and so at $q \in \Lambda$

$$\bar{H} = -h(I,N)_q = -g(\nu, \nu_{\partial M})_q \tag{4.7}$$

as $\sigma = 0$ along ∂M, and $\nu_{\partial M} := g^{ab}\nabla_b\sigma$ is the weight -1 unit (meaning $g(\nu_{\partial M}, \nu_{\partial M}) = 1$) normal to ∂M. □

4.2.5 A Hypersurface Flow for Conformally Compact Manifolds

In a Riemannian manifold with metric g the mean curvature flow equation is $\frac{\partial F}{\partial t} = -H^g \nu_g$, where $F(t)$ is a smoothly parametrised family of hypersurface embeddings and H^g, ν_g are, respectively, the mean curvature and unit normal along the hypersurface at time t, and with respect to the metric g. In the case of a conformally compact manifold the canonical interior metric g is singular at the boundary ∂M.

For an n-manifold with boundary $\bar{\Sigma}$, with $\bar{\Sigma} = \Sigma \cup \partial\Sigma$, and a smoothly parametrised family of hypersurface embeddings denoted $F(t) : \bar{\Sigma} \to \bar{M}$, we are interested in problems where $F(\partial\Sigma) \subset \partial M$. For such hypersurfaces $\bar{\Sigma}$ that meet the boundary ∂M at infinity in conformally compact manifolds, the extended mean curvature suggests a variant of mean curvature flow. First we observe that $\bar{H}$, and hence also $\bar{H}\nu$, extends to the boundary. However, the normal field ν carries a conformal weight of -1 whereas the flow vector $\frac{\partial F}{\partial t}$ is by construction a weight 0 vector field. Thus it is natural to choose a true scale $\tau \in \Gamma(\mathcal{E}_+[1])$ and work instead with the corresponding $\bar{\Sigma}$ normal field $\nu_{\bar{g}}$ that is of weight 0 and unit length with respect to the metric $\bar{g} = \tau^{-2}g$ that extends to the boundary. Thus we arrive at the flow equation (4.2):

$$\frac{\partial F}{\partial t} = -\bar{H}\nu_{\bar{g}}.$$

Although conformally motivated, one can think of this as simply a flow on the Riemannian manifold with boundary $(\bar{M}, \bar{g})$ where we use the flow speed $\bar{H}$ instead of the mean curvature $H^{\bar{g}}$. Since on the interior M the curvature $\bar{H}$ agrees with the mean curvature H^g, stationary solutions to the flow are provided by hypersurfaces that are locally minimal with respect to the complete conformally compact metric g.

To eliminate densities and weights throughout we now use the scale $\tau \in \Gamma(\mathcal{E}_+[1])$ to trivialise density bundles, and recall the associated defining function u of expression (4.4). Then from (4.6) we have

$$\frac{\partial F}{\partial t} = -\bar{H}\nu_{\bar{g}} \quad \Leftrightarrow \quad \frac{\partial F}{\partial t} = (du(\nu_{\bar{g}}) - uH^{\bar{g}})\nu_{\bar{g}}. \tag{4.8}$$

Remark Note that on conformally compact manifolds we may more generally consider flows of the form

$$\frac{\partial F}{\partial t} = -f(\bar{H})\nu_{\bar{g}}, \tag{4.9}$$

where $f : \mathbb{R} \to \mathbb{R}$ is potentially any smooth real function depending on the generalised mean curvature.

4.2.6 Boundary Conditions

Given (4.7), we see that a natural Neumann boundary condition is to require $g(\nu, \nu_{\partial M})_q = 0$ at all points $q \in \Lambda$. In fact, applying the properties of the tractor normal and tractor scale to this Neumann boundary condition transforms it into a fixed Dirichlet condition as follows:

$$F(p,t) = F_0(p) \in \Lambda \quad \forall (p,t) \in \partial\Sigma \times [0,T). \tag{4.10}$$

More generally we can show that either of the Neumann or Dirichlet boundary condition will imply the other one

At Σ we have, by (4.7), that, for all times of existence and everywhere on the boundary $\partial\Sigma$, the speed of the flow vanishes, i.e. $-\bar{H} = -g(\nu, \nu_{\partial M}) = 0$, and thus keeps the hypersurface boundary fixed. That is for all times of existence $\partial\Sigma_t = \Sigma_t \cap \partial M = \Lambda$.

Conversely, let us also note that if we have defined a flow with a fixed Dirichlet boundary condition, as in (4.10), the compatibility condition at the boundary would be that the speed of the flow vanishes there to ensure that the boundary stays fixed. That will imply that $\bar{H}$ vanishes at the boundary, and so again by (4.7) ν is orthogonal to $\nu^{\partial M}$ (equivalently I is orthogonal to N^{Σ_t}) for all times at the boundary. This is exactly the Neumann condition that we consider.

4.2.7 The Flow Problem

Putting the above together and using the scale τ to trivialise the density bundles we come to the flow problem that we treat.

The family of hypersurfaces $(\Sigma_t)_{t \in [0,T)}$ (for T some maximal time), where $\Sigma_t = F_t(\Sigma)$, is said to be evolving by *generalised mean curvature flow with Neumann boundary condition* on ∂M if

$$\frac{\partial F}{\partial t}(p,t) = -\bar{H}\nu^{\Sigma_t}(p,t)$$

$$= h(I, N^{\Sigma_t})\nu^{\Sigma_t}(p,t), \quad \forall (p,t) \in \Sigma \times [0,T), \text{ flow equation, } \quad (4.11)$$

$$F(p,0) = F_0(p), \qquad \forall p \in \Sigma, \qquad\qquad \text{initial condition,}$$

$$\left\langle \nu^{\Sigma_t}, \nu^{\partial M} \right\rangle = h(N^{\Sigma_t}, I) = 0, \quad \forall (p,t) \in \partial\Sigma \times [0,T), \quad \text{Neumann condition,}$$

$$F(\partial\Sigma) \subset \partial M, \qquad\qquad \forall t \in [0,T), \qquad\qquad \text{Neumann condition.}$$

Above we have introduced the notation $\langle \cdot, \cdot \rangle$ for the metric $\bar{g}$. Then $\bar{H} = -h(I, N^{\Sigma_t})$ denotes the generalised mean curvature. On the interior this agrees with the mean curvature H^g of Σ_t with respect to the ambient metric $g = u^{-2}\bar{g}$ of M. Also I denotes the scale tractor, and its inner product with N^{Σ_t} is taken using the tractor metric h. Once again ν^{Σ_t} is the normal to Σ_t in the $\bar{g}$ metric such that $|\nu^{\Sigma_t}|_{\bar{g}} = 1$. As a final remark here, in the following we will often discuss the generalisation of the above where we use $f(\bar{H})$, as in expression (4.9), in place of $\bar{H}$ in (4.11). We will still call this the generalised mean curvature flow and refer to (4.11).

4.3 The Flow Problem

4.3.1 Treating the Flow as a Nonlinear Partial Differential Equation

4.3.1.1 Generalised Gaussian Coordinates

The first step in the study of (4.2) or the more general flow as in (4.9), with speed given by a smooth function of the generalised mean curvature, $f(\bar{H})$, is a short time existence result. The mean curvature flow of hypersurfaces is equivalent to a system of weakly parabolic quasilinear partial differential equations of second order. To see this one defines a scalar quantity associated with the flow movement for a positive time, see for example [29].

The first obstacle in the study of many geometrically defined partial differential equations, including the class of mean curvature flow equations that we consider, is the degeneracy of the operator given by the zeroes appearing in their principal symbol. The solution for the mean curvature flow uses the well-known DeTurck trick to show that the partial differential equation associated with the mean curvature flow is equivalent, up to a family of tangential diffeomorphisms, to a related strongly parabolic equation. Solving the latter we then recover a solution to the mean curvature flow. The DeTurck trick, [13], was first invented for the Ricci flow, however the method applies to many other geometric flows [22, 23].

In [7] Andrews gave a general short time existence for the case of fully nonlinear curvature flows moving in ambient Riemannian manifolds with speeds given by a function of the principal curvatures of the hypersurfaces in ambient Riemannian manifolds. This treats all second order flows and it is sufficient for our purposes. The results for Euclidean ambient space is treated at large in Andrews' notes [6].

For the case of the boundary value problems, Huisken [27] treated the nonparametric mean curvature flow of graphs with either Neumann or Dirichlet boundary conditions, proving a long time existence result and regularity estimates up to and including the boundary. In the case of curved boundaries, as is the case of the free boundary mean curvature flow, the graph function to be considered for a short time has to be chosen such that the curvature of the contact hypersurface is taken into account. Short time existence for the free boundary mean curvature of hypersurfaces has been completed by Stahl in [36] using a combination of the DeTurck trick and a parabolic theory of boundary partial differential equations developed in Lieberman [29].

Following [37] we can define the associated partial differential equation as follows. Begin by defining generalised Gaussian coordinates that will account for the curved boundary. Let $\Sigma_0 = F_0(\Sigma)$ be a smooth immersion of Σ into M, and let $x = (x_1, \ldots, x_n)$ be coordinates on Σ. In a tubular neighbourhood $\mathcal{U} \subset M$ of Σ_0 we can define a smooth vector field with the following properties:

$$\xi\big|_{\Sigma_0} = \nu^{\Sigma_0}, \quad \xi\big|_{\mathcal{U} \cap \partial M} \in T\partial M, \quad \text{and} \quad |\xi|_{\bar{g}} = 1.$$

Using this vector field define the flux lines $\varphi = \varphi(x, \cdot)$ to the corresponding flow perpendicular to Σ_0 and tangential to ∂M as follows: for any given point $p = \varphi(x, s) \in \mathcal{U}$ we can define $x_{n+1}(p) :=$ length of the flux line through p between p and the intersection point $p_0 = \varphi(x, 0)$ on Σ_0, and $(x_1, \ldots, x_n)(p) :=$ coordinates of p_0 on Σ. These coordinates will be referred to as generalised Gaussian coordinates and, with respect to these coordinates, the set $\mathcal{U} \cap \partial M$ appears as part of an "orthogonal cylinder".

We can now define a scalar initial boundary value problem, for some positive δ, for a scalar function $\omega : \Sigma \times [0, \delta) \to \mathbb{R}$ representing the height in the direction of the above defined flux by

$$\frac{\partial}{\partial t}\omega = -uvf(\cdot, t, \omega, D\omega, D^2\omega), \text{ in } \Sigma \times [0, \delta), \qquad (4.12)$$

$$\omega(\cdot, 0) = 0 \text{ on } \Sigma,$$

$$\omega(\cdot, t) = 0 \text{ in } \partial\Sigma \times [0, \delta),$$

where u is the conformal factor and $f(\bar{H}) = f(\cdot, t, \omega, D\omega, D^2\omega)$ is the smooth function depending on the generalised mean curvature, thus on the sum of the eigenvalues of the second fundamental form as defined in (4.2). The quantity $v = \langle \xi, \nu^{\Sigma_t} \rangle^{-1} (x, \omega(x))$ measures the loss of graphicality in the direction of the flux; ν^{Σ_t} is a choice of unit outer normal to $\Sigma_t = (x, \omega(x, t))$, described as a graph over the initial hypersurface Σ_0.

As described in Section 4.2.6, we see that our problem can be seen as a Dirichlet boundary value problem. Thus we omit here the Neumann boundary condition, which is a linear combination between components of the first derivative of ω and the components of the vector field defining the flux.

Following a similar construction to that of Ecker and Huisken [15] and Stahl [37] one can show the existence of a family of tangential diffeomorphisms that can be used to show (4.2) is equivalent to (4.12). Thus the existence of a unique solution for the above scalar initial boundary value problem is equivalent to the existence of a unique solution for the flow problem (4.2).

4.3.1.2 The ε-Flows

The partial differential equation associated with the flow is degenerate at the boundary where the conformal factor vanishes. Thus we work $\varepsilon > 0$ away from the boundary by defining the family of ε-flows such that when $\varepsilon \to 0$ we recover our initial generalised mean curvature flow (4.2).

Let us define $\Sigma^\varepsilon = \{x \in \Sigma \mid dist(x, \partial\Sigma) \geq \varepsilon\}$ for $\varepsilon > 0$. For example in the case of the hyperbolic ambient space this means we are working in the upper half plane of distance $(\geq) \varepsilon$ away from the $\{x_{n+1} = 0\}$

hyperplane. The distance here is the geodesic distance function taken with respect to the metric $\bar{g}$.

We define the family of hypersurfaces $(\Sigma_t^\varepsilon)_{t \in [0,T)}$ (for T some maximal time), where $\Sigma_t^\varepsilon = F_t(\Sigma^\varepsilon)$ are said to be evolving by a generalised mean curvature ε-flow with Neumann boundary condition on Σ if

$$\frac{\partial F}{\partial t}(p,t) = -f(\bar{H})\nu^{\Sigma_t^\varepsilon}(p,t)$$

$$= f(h(I, N^{\Sigma_t^\varepsilon}))\nu^{\Sigma_t^\varepsilon}(p,t), \quad \forall(p,t) \in \Sigma^\varepsilon \times [0,T), \quad \text{flow equation,}$$
$$(4.13)$$

$$\begin{aligned} F(p,0) &= F_0(p), &\forall p \in \Sigma^\varepsilon, &\quad \text{initial condition,} \\ F(p,t) &= F_0(p) &\forall(p,t) \in \partial\Sigma^\varepsilon \times [0,T), &\quad \text{Dirichlet.} \end{aligned}$$

First we note that in the interior the flow is the same as (4.2) (or rather its generalisation as in (4.9)). We impose a Dirichlet condition on the boundary due to the restriction of our points being at a distance of at least ε away from the boundary $\partial\Sigma$.

Having all ε-flows satisfy a Dirichlet boundary condition and taking the limit (after showing existence) as $\varepsilon \to 0$, we obtain again a Dirichlet boundary condition, which we have already seen is equivalent to a Neumann boundary condition at $\partial\Sigma \subset \partial M$.

4.3.1.3 Existence for a Short Time

There are two conditions on the speed of the flow that are necessary to be able to guarantee a short time existence. We state them as from Andrews [7] and discuss our particular case of speed $f(\bar{H})$. Please keep in mind that on the interior $\bar{H}$ agrees with the mean curvature H^g of Σ_t with respect to the ambient metric $g = u^{-2}\bar{g}$ of M, thus it is the actual (mean) sum of the principal curvatures.

Condition 1 $f(\bar{H})$ is a symmetric function of the principal curvatures of the hypersurfaces Σ_t generated by the flow. This ensures that the speed is given by a second order partial differential operator acting on the surface, and depends only on the local geometry of the hypersurface. In our case, on the interior of Σ, $f(\bar{H}) = f(H^g)$ is a function of the sum of the principal curvatures (i.e. the mean curvature) and so this condition is satisfied automatically.

Condition 2 f is an increasing function of the principal curvatures of the hypersurfaces. That is, we have

$$\frac{\partial f}{\partial k_i} > 0, \text{ for all } k_i \text{ principal curvature of } \Sigma_t, \; i = 1, \ldots, n.$$

This condition ensures that the flow is parabolic. In our case this translates to $\frac{df}{dH} > 0$.

Theorem 4.3.1 (Flow existence) *For every positive $\varepsilon > 0$ there exists a unique smooth solution for the ε-flow (4.13) under Condition 1 and Condition 2.*

Proof Use [7, Thm. 3-14], where Condition 1 and Condition 2 are used, and [37, Thm. 2.1] in the setting of the generalised Gaussian coordinates defined in Section 4.3.1.1.   □

4.3.2 Generalised Mean Curvature Flow in Hyperbolic Space

Let us consider the case of our hypersurfaces being embedded in hyperbolic space $\Sigma \subset M = \mathbb{H}^{n+1} \subset \mathbb{R}^{n+1}$, and we specialise henceforth to $f = \mathrm{id}$. We use the model of the half space to represent the hyperbolic space. Here g will be the hyperbolic metric of the interior and $\bar{g}$ will be the Euclidean metric – which of course goes to the boundary. Thus the flow will move points in Euclidean space with speed equal to minus the hyperbolic mean curvature $-\bar{H} = -Hu + \langle e_{n+1}, \nu^{\Sigma_t} \rangle$ in the direction of its Euclidean normal ν^{Σ_t}. Here u is the square root of the conformal factor.

Note that if u corresponds to the conformal compactification for the hyperbolic space we have $u(F) = x_{n+1}$, thus we also have that its full derivative in Euclidean space is $Du = e_{n+1} = (0, \ldots, 0, 1)$. This has been used above in defining the generalised mean curvature.

The hyperbolic case of our flow links nicely with the previous mean curvature flow in hyperbolic space considered by Unterberger [39], followed by Lin and Xiao [31] and later by Allmann, Lin and Zhou [3]. Their results are dependent on an initial radial graphicality being preserved. The speed of our flow is a scaled version of the speed previously considered and the scale factor is given by the conformal factor, which in this case is the $n + 1$ component of the position vector or the distance from the $\mathbb{R}^n$ plane. Due to the higher order degeneracy given by the extra scaling factor vanishing at the infinity boundary, an initial radial graphicality is no longer preserved. Thus, the generalised mean curvature flow we have defined requires a different approach to obtain regularity estimates necessary for the long time existence.

4.3.3 Long Time Existence and Convergence

Mean curvature flow of any compact hypersurface with finite area becomes extinct in finite time in any ambient Riemannian manifold. One can see this by enclosing it with a sphere and using the celebrated result of Huisken [25, 26], which shows that any convex initial hypersurfaces contract to points in finite time. Thus the long time existence of a mean curvature flow solution is highly dependent on an extra initial condition that can obstruct the contraction of the flow. Boundary conditions seem to be a good way to achieve that. Usually they are successful under an extra first order condition which allows the geometric flow to be rewritten in terms of scalar partial differential equations, as for example a graphicality condition. This approach was used by Huisken in [27] for nonparametric mean curvature flow with boundary conditions, and by Ecker and Huisken [14, 15] for entire graphs evolving by mean curvature flow.

We consider here the setting of graphs in the direction of a translation in Euclidean space. Different from Unterberger [39], we consider initial surfaces $F_0(\Sigma) = F(\Sigma, 0)$ which are embedded in the upper half space model of hyperbolic space $\mathbb{H}^{n+1}$ and can be written as entire Euclidean graphs above the hyperplane $\mathbb{R}^n = \{x \in \mathbb{R}^{n+1} | x_{n+1} = 0\}$ in the positive half space $\mathbb{R}^{n+1}_+ = \{x \in \mathbb{R}^{n+1} | x_{n+1} > 0\}$.

The graph direction is e_{n+1}. Unterberger, and then the following work of Lin and Xiao [31] and Allmann, Lin and Zhou [3], considers graphs in the radial direction defined over the half sphere in Euclidean space. In our case the extra multiple of the conformal factor in the speed of the evolution brings a new obstacle in the preservation of the positivity of the radial graph function.

Using an initial graphicality in e_{n+1} we prove the following theorem and state the following conjecture.[1]

Theorem 4.3.2 (Long time existence) *Let $F_0 : \Sigma \to \mathbb{H}^{n+1}$ be such that $\Sigma_0 = F_0(\Sigma)$ is an entire locally Lipschitz continuous graph over $\mathbb{R}^n$ in $\mathbb{R}_+^{n+1}$. Then the flow (4.2) has a smooth solution in $C^\infty(\Sigma \times (0, \infty))$ for all positive times and $F(\Sigma, t)$ is a graph over $\mathbb{R}^n$ for all positive times.*

The long time existence does not require any bounds on the initial height as also noted by Unterberger [39]. If one also includes the condition that the Euclidean height of our initial hypersurface over $\mathbb{R}^n$ is bounded, the following convergence result should be obtained.

Conjecture 4.3.3 (Convergence) *If Σ_0 has bounded hyperbolic height over $\mathbb{R}^n$ in $\mathbb{R}_+^{n+1}$ then, under generalised mean curvature flow (4.2), F_t converges in C^∞ to $\mathbb{S}_+^n$, the half sphere contained in $\mathbb{R}_+^{n+1}$.*

The general plan to show long time existence is based on interior and boundary estimates of quantities associated with the flow that give us the C^0, C^1, C^2 and finally C^∞ estimates. These are the hyperbolic height, the gradient which is the graph function, the curvature and the higher derivatives of curvature, respectively. These quantities and in particular clever combinations of them (noted previously by Huisken, Ecker, Unterberger and so on) have nice interior evolutions for mean curvature flow in Euclidean space. This carries over in some extent for the mean curvature flow in hyperbolic space, but as one approaches the boundary, where the conformal factor vanishes, the evolutions become degenerate. This happens even more for our case where the degeneracy is actually of second order due to a second conformal factor that arises from working with a generalised flow.

To bypass this for mean curvature flow in hyperbolic space, one defines (Unterberger has done this first for the mean curvature flow in this setting) a family of flows on domains ε away from the boundary, as we have done for the short time existence. We call these ε-flows. For these flows one obtains all the estimates independent of ε. These estimates are based on interior estimates by evolutions and boundary estimates by construction of barriers. Thus one obtains a long time solution of the ε-flows with regularity constants independent of ε. Then one takes $\varepsilon \to 0$ to find a solution up to the boundary for the flow.

---

[1] The completed proof of long time existence and the to be completed proof of the conjectured convergence will be contained in a planned paper treating the case of hyperbolic space in detail.

The ε independent estimates (ε being our lower bound for the conformal factor) are sufficient to let us obtain a smooth long term solution up to the boundary for our flow.

4.3.4 Generalised Mean Curvature Flow in Riemannian Manifolds

The same method can also be applied to the case of general Riemannian manifolds where the mean curvature flow with Dirichlet or Neumann boundary conditions has been studied and long time existence and convergence results are achieved with uniform time independent estimates.

One such work are the work of Priwitzer [35] who studied mean curvature flow of hypersurfaces in Riemannian manifolds under a transversality condition. The transversality condition is equivalent to the graphicality condition used by Ecker and Huisken [14, 15] to prove existence of long time solutions for entire graphs in Euclidean space. In a similar way, a star-shaped condition, i.e. radial graphicality, is used by Unterberger [39], Lin and Xiao [31] and Allmann, Lin and Zhu [3] to prove long time existence for mean curvature flow in hyperbolic space.

References

[1] Albin, P. *Renormalizing curvature integrals on Poincaré–Einstein manifolds.* Adv. Math., **221** (1), (2009) 140–169.

[2] Alexakis, S., and Mazzeo, R. *Renormalized area and properly embedded minimal surfaces in hyperbolic 3-manifolds.* Commun. Math. Phys., **297**(3), (2010) 621–651.

[3] Allmann, P., Lin, L., and Zhu, J. *Modified mean curvature flow of entire locally Lipschitz radial graphs in hyperbolic space.* arXiv preprint arXiv:1707.07087, 2017.

[4] Anderson, M. T. *Complete minimal varieties in hyperbolic space.* Invent. Math. **69**(3), (1982) 477–494.

[5] Anderson, M. T. *Complete minimal hypersurfaces in hyperbolic n-manifolds.* Comment. Math. Helv. **58**(1), (1983) 264–290.

[6] Andrews, B. *Flow of hypersurfaces by curvature functions.* In Dziuk, G., Huisken, G. and Hutchison, J. (eds.), Theoretical and Numerical Aspects of Geometric Variational Problems. Proc. Centre for Mathematics and its Applications, vol. 26. Mathematical Sciences Institute, Australian National University. 1991, pp. 1–10.

[7] Andrews, B. *Contraction of convex hypersurfaces in Riemannian spaces.* J. Differ. Geom. **39**(2), (1994) 407–431.

[8] Bailey, T. N., Eastwood, M. G., and Gover, A. R. *Thomas's structure bundle for conformal, projective and related structures.* Rocky Mountain J. Math. **24**(4) (2014) 1191–1217.

[9] Brakke, K. A. The motion of a surface by its mean curvature. Princeton University Press, Princeton, NJ. 1978.

[10] Colding, T., Minicozzi, W., and Pedersen, E. *Mean curvature flow.* Bull. Amer. Math. Soc. **52**(2), (2015) 297–333.

[11] Curry, S. and Gover, A. R. *An introduction to conformal geometry and tractor calculus, with a view to applications in general relativity.* in Daudé, T., Häfner, D. and Nicolas, J. (Eds.) Asymptotic analysis in general relativity, 86—170, London Mathematical Soceity Lecture Note Series, vol. 443, Cambridge University Press, Cambridge, 2018.

[12] De Silva, D. and Spruck, J. *Rearrangements and radial graphs of constant mean curvature in hyperbolic space.* Calc. Var. Partial Differ. Equ. **34**(1), (2009) 73–95.

[13] DeTurck, D. M. *Deforming metrics in the direction of their Ricci tensors.* J. Differ. Geom. **18**(1), (1983) 157–162.

[14] Ecker, K. and Huisken, G. *Mean curvature evolution of entire graphs.* Ann. of Math. (2), **130**(2), (1989) 453–471.

[15] Ecker, K. and Huisken, G. *Interior estimates for hypersurfaces moving by mean curvature.* Invent. Math., **105**(1), (1991) 547--569.

[16] Fefferman, C. and Graham, C. R. The ambient metric. Annals of Mathematics Studies, vol. 178, Princeton University Press, Princeton, NJ. 2012.

[17] Gover, A. R. and Waldron, A. *Conformal hypersurface geometry via a boundary Loewner-Nirenberg-Yamabe problem.* arXiv preprint arXiv:1506.02723. 2015.

[18] Gover, A. R. and Waldron, A. K. *Renormalized volumes with boundary.* Commun. Contemp. Math., **21**(2), (2019) 1850030, 31.

[19] Graham, C. R. and Reichert, N. *Higher-dimensional Willmore energies via minimal submanifold asymptotics.* arXiv preprint arXiv:1704.03852. 2017.

[20] Graham, C. R. and Witten, E. *Conformal anomaly of submanifold observables in AdS/CFT correspondence.* Nucl. Phys. B, **546**(1-2), (1999) 52–64.

[21] Guan, B. and Spruck, J. *Hypersurfaces of constant mean curvature in hyperbolic space with prescribed asymptotic boundary at infinity.* Amer. J. Math. **122**(5), (2000) 1039–1060.

[22] Hamilton, R. S. Harmonic maps of manifolds with boundary. Lecture Notes in Math. **471**, Springer-Verlag, Berlin, pp. 304–318. 1975.

[23] Hamilton, R. S. *Three-manifolds with positive Ricci curvature.* J. Differ. Geom. **17**(2), (1982) 255–306.

[24] Hardt, R. and Lin, F.-H. *Regularity at infinity for area-minimizing hypersurfaces in hyperbolic space.* Invent. Math. **88**(1), (1987) 217–224.

[25] Huisken, G. *Flow by mean curvature of convex surfaces into spheres.* J. Differ. Geom. **20**(1), (1984) 237–266.

[26] Huisken, G. *Contracting convex hypersurfaces in Riemannian manifolds by their mean curvature.* Invent. Math., **84**(3), (1986) 463–480.

[27] Huisken, G. *Non-parametric mean curvature evolution with boundary conditions*. J. Differ. Equ. **77**, (1989) 369–378.

[28] Huisken, G. *Asymptotic behavior for singularities of the mean curvature flow*. J. Differ. Geom. **31**(1), (1990) 285–299.

[29] Lieberman, G. M. Second order parabolic differential equations. World Scientific Publishing Co., River Edge, NJ. 1996.

[30] Lin, F.-H. *On the Dirichlet problem for minimal graphs in hyperbolic space*. Invent. Math. **96**(3), (1989) 593–612.

[31] Lin, L. and Xiao, L. *Modified mean curvature flow of star-shaped hypersurfaces in hyperbolic space*. Commun. Anal. Geom. **20**(5), (2012) 1061–1096.

[32] Maldacena, J. *The large N limit of superconformal field theories and supergravity*. Adv. Theor. Math. Phys. **2**(2), (1998) 231–252.

[33] Mullins, W. W. *Two-dimensional motion of idealized grain boundaries*. J. Appl. Phys. **27**, (1956) 900-904

[34] Nelli, B. and Spruck, J. *On the existence and uniqueness of constant mean curvature hypersurfaces in hyperbolic space*. In Jost, J. (ed.) Geometric analysis and the calculus of variations, International Press, Cambridge, MA, pp. 253–266. 1996.

[35] Priwitzer, B. *Mean curvature flow with Dirichlet boundary conditions in Riemannian manifolds with symmetries*. Ann. Global Anal. Geom. **23**(2), (2003) 157–171.

[36] Stahl, A. *Über den mittleren Krümmungsfluss mit Neumannrandwerten auf glatten Hyperflächen*. Ph.D. thesis, Fachbereich Mathematik, Eberhard-Karls-Universität, Tübingen, Germany. 1994.

[37] Stahl, A. *Regularity estimates for solutions to the mean curvature flow with a Neumann boundary condition*. Calc. Var. Partial Differ. Equ. **4**(4), (1996) 385–407.

[38] Tonegawa, Y. *Existence and regularity of constant mean curvature hypersurfaces in hyperbolic space*. Math. Z. **221**(1), (1996) 591–615.

[39] Unterberger, P. *Evolution of radial graphs in hyperbolic space by their mean curvature*. Commun. Anal. Geom. **11**(4), (2003) 675–698.

5
A Survey on the Ricci Flow on Singular Spaces

Klaus Kröncke and Boris Vertman

Abstract

In this chapter we provide an overview of our recent results concerning Ricci de Turck flow on spaces with isolated conical singularities. The crucial characteristic of the flow is that it preserves the conical singularity. Under certain conditions, Ricci-flat metrics with isolated conical singularities are stable, and positive scalar curvature is preserved under the flow. We also discuss the relation to Perelman's entropies in the singular setting, and outline open questions and future reseach directions.

5.1 Introduction and Geometric Preliminaries

Geometric flows, among them most notably the Ricci flow, provide a powerful tool to attack classification problems in differential geometry and construct Riemannian metrics with prescribed curvature conditions. The interest in this research area has only grown since the Ricci flow was used decisively in the Perelman's proof of Thurston's geometrization and the Poincaré conjectures.

The present chapter summarizes recent results of our research program on the Ricci flow in the setting of singular spaces, obtained in the papers [VER16, KRVE19A, KRVE19B, KMV20], where the latter one is written by the authors of the survey jointly with Tobias Marxen.

Let us mention some related work without claiming to provide a full list of references. The two-dimensional Ricci flow reduces to a scalar equation and has been studied on surfaces with conical singularities by Mazzeo, Rubinstein and Šešum [MRS15] and Yin [YIN10]. The Ricci flow in two dimensions is equivalent to the Yamabe flow, which has been studied in general dimension on spaces with edge singularities by Bahuaud and Vertman [BAVE14] and [BAVE16].

In the setting of Kähler manifolds, Kähler–Ricci flow reduces to a scalar Monge–Ampère equation and has been studied in the case of edge singularities in connection to the recent resolution of the Calabi–Yau conjecture on Fano manifolds by Donaldson [DON11] and Tian [TIA15]; see also Jeffres, Mazzeo and Rubinstein [JMR16]. Kähler–Ricci flow in the case of isolated conical singularities is geometrically, though not analytically, more intricate than edge singularities and has been addressed by Chen and Wang [CHWA15] and Wang [WAN16], as well as Liu and Zhang [LIZH17].

We should point out that in the singular setting, Ricci flow loses its uniqueness and need not preserve the given singularity structure. In fact, Giesen and Topping [GITO10, GITO11] constructed a solution to the Ricci flow on surfaces with singularities, which becomes instantaneously complete. Alternatively, Simon [SIM13] constructed Ricci flow in dimensions two and three that smooths out the singularity.

5.1.1 Isolated Conical Singularities

A compact Riemannian manifold with an isolated conical singularity is illustrated in Figure 5.1. The precise definition is as follows.

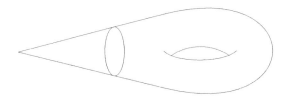

Figure 5.1 A compact manifold with an isolated conical singularity.

Definition 5.1.1 Consider a compact smooth manifold $\overline{M}$ with boundary $\partial M = F$ and open interior denoted by M. Let $\mathscr{C}(F)$ be a tubular neighborhood of the boundary, with open interior $\mathscr{C}(F) = (0,1)_x \times F$, where x is a defining function of the boundary. Consider a smooth Riemannian metric g_F on the boundary F with $n = \dim F$. An incomplete Riemannian metric g on M with an isolated conical singularity is then defined to be smooth away from the boundary and

$$g \restriction \mathscr{C}(F) = dx^2 + x^2 g_F + h,$$

where the higher order term h has the following asymptotics at $x = 0$. Let $\bar{g} = dx^2 + x^2 g_F$ denote the exact conical part of the metric g over $\mathscr{C}(F)$ and $\nabla_{\bar{g}}$ the corresponding Levi-Civita connection. Then we require that for some $\gamma > 0$ and all integer $k \in \mathbb{N}_0$ the pointwise norm satisfies

$$|x^k \nabla_{\bar{g}}^k h|_{\bar{g}} = O(x^\gamma), \quad x \to 0. \tag{5.1}$$

Remark 5.1.2 We emphasize here that we do not assume that the higher order term h is smooth up to $x = 0$ and do not restrict the order $\gamma > 0$ to be integer. In that sense the notion of conical singularities in the present discussion is more general than the classical notion of conical singularities where h is usually assumed to be smooth up to $x = 0$ with $\gamma = 1$. This minor generalization is necessary, since the Ricci de Turck flow, which will be introduced below, preserves a conical singularity only up to a higher order term h as above.

We call (M, g) a compact space with an isolated conical singularity, or a *conical manifold*, and g a conical metric. The definition naturally extends to conical manifolds with finitely many isolated conical singularities. Since the analytic arguments are local in nature, we may assume without loss of generality that M has a single conical singularity only.

There are various examples for such spaces. The simplest case is an orbifold X/G, where X is a smooth compact manifold with a discrete group G acting by isometries, which is not necessarily acting strictly discontinuous and admits finitely many fixed points. The interior of the quotient X/G defines a compact manifold with isolated conical singularities. There exist also examples of compact Ricci-flat manifolds with non-orbifold isolated conical singularities, constructed by Hein and Sun [HESU17].

We now recall elements of b-calculus by Melrose [MEL92, MEL93]. We choose local coordinates (x, z) on the conical neighborhood $\mathscr{C}(F)$, where x is the defining function of the boundary, $n = \dim F$ and $(z) = (z_1, \ldots, z_n)$ are local coordinates on F. We consider the b-vector fields $\mathcal{V}_b$, which by definition are smooth in the interior M and tangent to the boundary F. In local coordinates (x, z), b-vector fields $\mathcal{V}_b$ are locally generated by

$$\left\{ x\frac{\partial}{\partial x}, \partial_z = \left(\frac{\partial}{\partial z_1}, \ldots, \frac{\partial}{\partial z_n} \right) \right\},$$

with coefficients being smooth on $\overline{M}$. The b-vector fields form a spanning set of section for the b-tangent bundle ${}^bTM$, i.e. $\mathcal{V}_b = C^\infty(M, {}^bTM)$. The b-cotangent bundle ${}^bT^*M$ is generated locally by the following one-forms

$$\left\{ \frac{dx}{x}, dz_1, \ldots, dz_n \right\}.$$

These differential forms are singular in the usual sense, but smooth as sections of the b-cotangent bundle ${}^bT^*M$. We extend $x : \mathscr{C}(F) \to (0,1)$ smoothly to a non-vanishing function on M and define the incomplete b-tangent space ${}^{ib}TM$ by the requirement $C^\infty(M, {}^{ib}TM) = x^{-1}C^\infty(M, {}^bTM)$. The dual incomplete b-cotangent bundle ${}^{ib}T^*M$ is related to its complete counterpart by

$$C^\infty(M, {}^{ib}T^*M) = xC^\infty(M, {}^bT^*M), \tag{5.2}$$

with the spanning sections given locally over $\mathscr{C}(F)$ by

$$\{dx, x dz_1, \ldots, x dz_n\}.$$

With respect to the notation just introduced, the conical metric g in Definition 5.1.1 is a smooth section of the vector bundle of the symmetric 2-tensors of the incomplete b-cotangent bundle ${}^{ib}T^*M$, i.e. $g \in C^\infty(\mathrm{Sym}^2({}^{ib}T^*M))$, provided the pointwise norm of the higher order term h is smooth at $x = 0$.

5.1.2 Ricci de Turck Flow and the Lichnerowicz Laplacian

The Ricci flow of an initial metric g is an evolution equation for metrics which reads as $(t \geq 0)$

$$\partial_t g(t) = -2\,\mathrm{Ric}(g(t)), \quad g(0) = g. \tag{5.3}$$

Due to diffeomorphism invariance of the Ricci tensor, this evolution equation fails to be strongly parabolic. One overcomes this problem by adding an additional Lie derivative term to the equation which breaks the diffeomorphism invariance and defines the Ricci de Turck flow via

$$\partial_t g(t) = -2\,\mathrm{Ric}(g(t)) + \mathcal{L}_{W(g(t), \tilde{g})}g(t), \tag{5.4}$$

with the same initial condition as before. Here, $W(t)$ is the de Turck
vector field defined in the usual local index notation in terms of the
Christoffel symbols for the metrics $g(t)$ and a reference metric $\widetilde{g}$:

$$W(g,\widetilde{g})^k = g^{ij}\left(\Gamma_{ij}^k(g) - \Gamma_{ij}^k(\widetilde{g})\right). \tag{5.5}$$

This flow is equivalent to the Ricci flow via diffeomorphisms. The linearization of the right hand side of (5.4) is given by

$$\frac{d}{ds}\left[-2\operatorname{Ric}(\widetilde{g}+sk) + \mathcal{L}_{W(\widetilde{g}+sk,\widetilde{g})}(\widetilde{g}+sk)\right]\Big|_{s=0} = -\Delta_L k, \tag{5.6}$$

where Δ_L is an elliptic operator, which is known as the Lichnerowicz
Laplacian of g, acting on symmetric 2-tensors by

$$\Delta_L \omega_{ij} = \Delta \omega_{ij} - 2g^{pq}\operatorname{Rm}_{qij}^r \omega_{rp} + g^{pq}R_{ip}\omega_{qj} + g^{pq}R_{jp}\omega_{iq}.$$

Here Δ denotes the rough Laplacian, and $\operatorname{Rm}_{qij}^r$ and R_{ij} denote the
components of the $(1,3)$-Riemann curvature tensor and the Ricci tensor,
respectively.

Thus, in contrast to the Ricci flow, the Ricci de Turck flow is parabolic
in the strict sense, and, at least in the smooth compact setting, standard
existence theorems guarantee well-posedness for its initial value problem.
Therefore, from the analytical perspective, the Ricci de Turck flow is
much easier to handle than the standard Ricci flow. As it appears in its
linearization, the Lichnerowicz Laplacian and its spectral properties will
be fundamental for considerations in this chapter.

Acknowledgements   The authors thank MATRIX for hosting the
program *Differential Geometry in the Large*, where this work was initiated. Furthermore, the authors thank the priority programme *Geometry at Infinity* of the German Research Foundation DFG for financial
support and for providing an excellent platform for joint research.

5.2 Existence of the Singular Ricci de Turck Flow

We shall present here the short time existence result for the singular
Ricci de Turck flow obtained by Vertman [VER16] in a simple way,
which is sufficient for the purpose of the present discussion. We consider
a compact conical manifold (M, g_0). We study the Ricci de Turck flow
with g_0 as the initial metric. While the reference metric $\widetilde{g}$ is usually
taken as the initial metric g_0, in the case of $\widetilde{g}$ being Ricci-flat the initial
metric g_0 can be chosen as a sufficiently small perturbation of $\widetilde{g}$.

To get short time existence for the Ricci de Turck flow in the conical setting, we need two more conditions. The first one is that we require subquadratic blowup of the Ricci and scalar curvature close to the singular point. The second is a spectral condition on the Lichnerowicz Laplacian which we will explain in the following subsection.

5.2.1 Tangential Stability

Let (M, h) be a compact conical Ricci-flat manifold. We make the definition $S := \mathrm{Sym}^2({}^{ib}T^*M)$. The Lichnerowicz Laplacian $\Delta_L : C_0^\infty(M, S) \to C_0^\infty(M, S)$ is a differential operator of second order, which can be written in local coordinates near the conical singularity as follows. We choose local coordinates (x, z) over the singular neighborhood $\mathscr{C}(F) = (0, 1)_x \times F$. In [VER16] we introduced a decomposition of compactly supported smooth sections $C_0^\infty(\mathscr{C}(F) : S \upharpoonright \mathscr{C}(F))$ over the conical neighborhood $\mathscr{C}(F) = (0, 1) \times F$:

$$C_0^\infty(\mathscr{C}(F), S \upharpoonright \mathscr{C}(F)) \to C_0^\infty((0,1), C^\infty(F) \times \Omega^1(F) \times \mathrm{Sym}^2(T^*F)),$$
$$\omega \mapsto (\omega(\partial_x, \partial_x), \omega(\partial_x, \cdot), \omega(\cdot, \cdot)),$$
$$(5.7)$$

where $\Omega^1(F)$ denotes differential 1-forms on F. Under such a decomposition, the Lichnerowicz Laplace operator Δ_L associated to the singular Riemannian metric g attains the following form over $\mathscr{C}(F)$:

$$\Delta_L = -\frac{\partial^2}{\partial x^2} - \frac{n}{x}\frac{\partial}{\partial x} + \frac{\square_L}{x^2} + \mathcal{O},\qquad(5.8)$$

where $\square_L$ is a differential operator on $C^\infty(F) \times \Omega^1(F) \times \mathrm{Sym}^2(T^*F)$, called the tangential operator of the Lichnerowicz Laplace operator. The higher order term $\mathcal{O} \in x^{-2+\gamma}\mathcal{V}_b^2$ is a second order differential operator with higher asymptotic behavior at $x = 0$ than the leading part.

Definition 5.2.1 Let (F^n, g_F) be a closed Einstein manifold[1] with Einstein constant $(n - 1)$. Then (F^n, g_F) is called (strictly) tangentially stable if the tangential operator $\square_L$ of the Lichnerowicz Laplacian on its cone restricted to tracefree tensors is non-negative (resp. strictly positive).

---

[1] If (M, g_0) is a conical manifold satisfying condition 2 in Definition 5.2.2, then the cross section (F, g_F) of the cone is automatically Einstein with Einstein constant $(n - 1)$.

5.2.2 The Existence Result

The conditions for short time existence of the Ricci de Turck flow are subsummarized under the notion of admissible metrics. For convenience of the reader, we have moved the definitions of some function spaces and vector bundles to the Appendix, Section 5.7.

Definition 5.2.2 Let (M, g_0) be a compact conical manifold. Then the conical metric g_0 is said to be *admissible*, if it satisfies the following assumptions for $\gamma > 0$ as in (5.1), some $k \in \mathbb{N}$ and $\alpha \in (0, 1)$.

1 The cross section (F, g_F) is assumed to be tangentially stable.
2 The Ricci curvature has subquadratic growth at the singularitiy, i.e. $\mathrm{Ric}(g_0) = O(x^{-2+\gamma})$ as $x \to 0$. More precisely,

$$\mathrm{Ric}(g_0) \in \mathcal{H}^{k,\alpha}_\gamma(M \times [0,T], S). \tag{5.9}$$

In particular, the leading exact part $\bar{g} = dx^2 + x^2 g_F$ of the conical metric g_0 with $g_0 \upharpoonright \mathscr{C}(F) = \bar{g} + h$ is Ricci-flat, and the higher order term h not only satisfies (5.1), but in particular is an element of $\mathcal{C}^{k+3,\alpha}_{\mathrm{ie}}(M, S)_\gamma$.
3 For any $X_1, \ldots, X_4 \in C^\infty(\overline{M}, {}^{ib}TM)$ we have for the curvature $(0, 4)$-tensor

$$Rm(g_0)(X_1, X_2, X_3, X_4) \in x^{-2} \mathcal{C}^{k+1,\alpha}_{\mathrm{ie}}(M).$$

The main result of [VER16, Thm. 4.1], see also [KRVE19A, Thm. 1.2], is then the following theorem.

Theorem 5.2.3 *Let (M, g_0) be a conical manifold with an admissible metric g_0. Let the reference metric $\tilde{g}$ be either equal to g_0 or an admissible conical Ricci-flat metric, in which case g_0 is assumed to be a sufficiently small perturbation of $\tilde{g}$ in $\mathcal{H}^{k+2,\alpha}_\gamma(M, S)$.*

Then there exists some $T > 0$, such that the Ricci de Turck flow (5.4), with the reference metric $\tilde{g}$ starting at g_0, admits a solution $g(t), t \in [0, T]$, which is an admissible perturbation of g_0, i.e. $g(t) \in \mathcal{H}^{k+2,\alpha}_{\gamma'}(M, S)$ for each t, all $k \in \mathbb{N}$ and some $\gamma' \in (0, \gamma)$ sufficiently small.

Remark 5.2.4 A precise definition of the space $\mathcal{H}^{k+2,\alpha}_{\gamma'}(M, S)$ is given in the Appendix, Section 5.7. Let us give a brief description here. Decompose a symmetric 2-tensor h into its pure trace part and its trace-free part as $h = \mathrm{tr} h \cdot g + h_0$. Then $h \in \mathcal{H}^{k+2,\alpha}_{\gamma'}(M, S)$ means that, as we approach the singularity, $\nabla^l h_0 = O(x^{-l+\gamma'})$ for $l \in \{0, \ldots, k+2\}$,

$\nabla^l \text{tr}\, h = O(x^{-l+\gamma'})$ for $l \in \{1, \ldots, k+2\}$ but $\text{tr}\, h = O(1)$. The last condition distinguishes $\mathscr{H}^{k+2,\alpha}_{\gamma'}$ from a standard weighted Hölder space and ensures that multiples of the metric are also contained in this space.

This result is obtained through a careful microlocal analysis of the heat kernel for the Lichnerowicz Laplacian. The heat kernel asymptotics is then used to establish mapping properties of the heat operator on the Hölder spaces $\mathscr{H}^{k+2,\alpha}_{\gamma}$. It is here that tangential stability enters in order to obtain these mapping properties. Short time existence of the Ricci de Turck flow in these spaces is then a consequence of a fixed point argument, which requires the metric to be admissible in the sense above to go through.

Let us now explain in what sense the flow preserves the conical singularity. Given an admissible perturbation g of the conical metric g_0, the pointwise trace of g with respect to g_0, denoted as $\text{tr}_{g_0} g$, is by definition of admissibility an element of the Hölder space $\mathcal{C}^{k,\alpha}_{\text{ie}}(M, S_1)^b_\gamma$, restricting at $x = 0$ to a constant function $(\text{tr}_{g_0} g)(0) = u_0 > 0$. Setting $\widetilde{x} := \sqrt{u_0} \cdot x$, the admissible perturbation $g = g_0 + h$ attains the form

$$g = d\widetilde{x}^2 + \widetilde{x}^2 g_F + \widetilde{h},$$

where $|\widetilde{h}|_g = O(x^\gamma)$ as $x \to 0$. Note that the leading part of the admissible perturbation g near the conical singularity differs from the leading part of the admissible metric g_0 only by scaling.

5.2.3 Characterizing Tangential Stability

A crucial part of [KRVE19A] is devoted to a detailed discussion of the tangential stability. The operator Δ_E appearing in the theorem below denotes the Einstein operator on symmetric 2-tensors over F. It is given by

$$\Delta_E \omega_{ij} = \Delta \omega_{ij} - 2g^{pq} \text{Rm}^r_{qij} \omega_{rp},$$

where Δ is the rough Laplacian. We prove the following general characterization of tangential stability in terms of spectral conditions on (F, g_F). Below, TT denotes the space of symmetric 2-tensors which are trace free and divergence free at each point.

Theorem 5.2.5 *Let (F, g_F), $n \geq 3$, be a compact Einstein manifold with constant $(n-1)$. We write Δ_E for its Einstein operator, and denote the Laplace–Beltrami operator by Δ. Then (F, g_F) is tangentially stable if and only if $\text{Spec}(\Delta_E|_{TT}) \geq 0$ and $\text{Spec}(\Delta) \cap (n, 2(n+1)) = \varnothing$. Similarly,*

(M, g) *is strictly tangentially stable if and only if* $\mathrm{Spec}(\Delta_E|_{TT}) > 0$ *and* $\mathrm{Spec}(\Delta) \cap [n, 2(n+1)] = \varnothing$.

Establishing this result amounts to a careful anaylsis of the Lichnerowicz Laplacian. This analysis relies heavily on a decomposition of symmetric 2-tensors established in [KRÖ17] to understand the spectrum of Δ_L on Ricci-flat cones.

Any spherical space form is tangentially stable because the Lichnerowicz Laplacian of its cone is the rough Laplacian since the cone is flat. However, the spaces $\mathbb{S}^n$ and $\mathbb{RP}^n$ are not strictly tangentially stable since $2(n+1) \in \mathrm{Spec}(\Delta)$ in both cases. This property may also hold for other spherical space forms. In the following theorem, we use Theorem 5.2.5 and eigenvalue computations in [CAHE15] to characerize tangential stability of symmetric spaces.

Theorem 5.2.6 *Let* (F^n, g_F), $n \geq 2$, *be a closed Einstein manifold with constant* $(n-1)$, *which is a symmetric space of compact type. If it is a simple Lie group* G, *it is strictly tangentially stable if* G *is one of the following spaces:*

$$\mathrm{Spin}(p) \ (p \geq 6, p \neq 7), \qquad E_6, \qquad E_7, \qquad E_8, \qquad F_4. \qquad (5.10)$$

If (F, g_F) *is a symmetric space of compact type* G/K, *it is strictly tangentially stable if it is one of the real Grasmannians*

$$\frac{\mathrm{SO}(p+q)}{\mathrm{SO}(p) \times \mathrm{SO}(q)} \quad p \geq 3, \quad q \geq 3, \quad (p, q) \neq (3, 3), \qquad (5.11)$$

or one of the following spaces:

$$E_6/[\mathrm{Sp}(4)/\{\pm I\}], \qquad E_6/\mathrm{SU}(2) \cdot \mathrm{SU}(6), \qquad E_7/[\mathrm{SU}(8)/\{\pm I\}]$$
$$E_7/\mathrm{SO}(12) \cdot \mathrm{SU}(2), \quad E_8/\mathrm{SO}(16), \quad E_8/E_7 \cdot \mathrm{SU}(2), \quad F_4/Sp(3) \cdot \mathrm{SU}(2).$$
$$(5.12)$$

Remark 5.2.7 In our original version of the theorem ([KRVE19A, Thm. 3.2]), the spaces $\mathrm{SU}(2p)/\mathrm{SO}(p)$, $p \geq 6$, also appear in the list. However, a small error occurred there. These spaces are tangentially stable only, not strictly tangentially stable. In fact, the condition

$$\mathrm{Spec}(\Delta) \cap (n, 2(n+1)) = \varnothing$$

for the Laplace–Beltrami operator holds for these spaces, but $\Delta_E|_{TT}$ has a kernel, see e.g. [CAHE15, Table 2].

5.3 Stability of the Singular Ricci de Turck Flow

Our main result in [KRVE19A] establishes long time existence and convergence of the Ricci de Turck flow for sufficiently small perturbations of conical Ricci-flat metrics, assuming linear and tangential stability and integrability. More precisely we consider a compact conical Ricci-flat manifold (M, h_0) and g_0 a sufficiently small perturbation of h_0, not necessarily Ricci flat. We study the Ricci de Turck flow with h_0 as the reference metric, and g_0 as the initial metric,

$$\partial_t g(t) = -2\mathrm{Ric}(g(t)) + \mathcal{L}_{W(t)} g(t), \quad g(0) = g_0, \tag{5.13}$$

where $W(t)$ is the de Turck vector field given by

$$W(t)^k = g(t)^{ij} \left(\Gamma_{ij}^k(g(t)) - \Gamma_{ij}^k(h_0) \right). \tag{5.14}$$

Definition 5.3.1 We say that (M, h_0) is linearly stable if the Lichnerowicz Laplacian Δ_L with domain $C_0^\infty(M, S)$ is non-negative.

Definition 5.3.2 We say that (M, h_0) is *integrable* if for some $\gamma > 0$ there exists a smooth finite-dimensional manifold $\mathscr{F} \subset \mathscr{H}_\gamma^{k,\alpha}(M, S)$ such that

1 $T_{h_0} \mathscr{F} = \ker \Delta_{L,h_0} \subset \mathscr{H}_\gamma^{k,\alpha}(M, S)$,
2 all Riemannian metrics $h \in \mathscr{F}$ are Ricci flat.

Our main result is as follows.

Theorem 5.3.3 *Consider a compact conical Ricci-flat Riemannian manifold (M, h_0). Assume that (M, h_0) satisfies the following three additional assumptions:*

(i) (M, h_0) is tangentially stable in the sense of Definition 5.2.1,
(ii) (M, h_0) is linearly stable in the sense of Definition 5.3.1,
(iii) (M, h_0) is integrable in the sense of Definition 5.3.2.

If h_0 is not strictly tangentially stable, we assume in addition that the singularities are orbifold singularities. Then, for sufficiently small perturbations g_0 of h_0, there exists a Ricci de Turck flow, with a change of reference metric at discrete times, starting at g_0 and converging to a conical Ricci-flat metric h^* as $t \to \infty$.

Remark 5.3.4 If (M, h_0) is a smooth compact manifold, tangential stability is always satisfied. In this case, the statement coincides with the stability results of compact smooth Ricci-flat manifolds obtained in [SES06].

In contrast to the smooth compact case, we cannot work with a priori estimates because the curvature is unbounded. Instead, the mapping properties of the heat kernel of the Lichnerowicz Laplacian on the spaces $\mathscr{H}_\gamma^{k,\alpha}(M,S)$ play a pivotal role in our proof.

We also study examples of compact conical manifolds where the integrability condition is satisfied. This includes flat spaces with orbifold singularities as well as Kähler manifolds. More precisely we establish the following results.

Proposition 5.3.5   *Let (M, h_0) be a flat manifold with an orbifold singularity. Then it is linearly stable and integrable.*

The proof of this result is quite simple. Linear stability follows from the absence of curvature and integrability follows from construction of the submanifold explicitly as an affine space over h_0 modeled over the space of parallel tensors.

Theorem 5.3.6   *Let (M, h_0) be a Ricci-flat Kähler manifold where the cross section is either strictly tangentially stable or a space form. Then h_0 is linearly stable and integrable.*

This result has been obtained in the smooth compact case in the 1980s (see e.g. [TIA87]). The result in our setting follows from the careful adaptation of techniques from the compact case and using the analysis of weighted Sobolev spaces.

5.4 Perelman's Entropies on Singular Spaces

In this section, we review the results obtained in [KRVE19B] on Perelman's entropies on compact conical manifolds. From now on, let (M^m, g) be a compact conical manifold of dimension m and $n = m - 1$. We denote the corresponding integration volume by dV_g. We first introduce the three entropies of interest.

5.4.1 The λ-Functional

Perelman's λ-functional is defined as

$$\lambda(g) = \inf_{\omega \in H_1^1(M)} \left\{ \int_M (\mathrm{scal}(g)\omega^2 + 4|\nabla\omega|_g^2)\, dV_g \,\bigg|\, \omega > 0,\, \int_M \omega^2\, dV_g = 1 \right\}.$$

The corresponding Euler–Lagrange equation is

$$4\Delta_g \omega_g + \text{scal}(g)\omega_g = \lambda(g)\omega_g. \tag{5.15}$$

5.4.2 The Ricci Shrinker Entropy

Consider the functional $\mathcal{W}_-(g, f, \tau)$:

$$\mathcal{W}_-(g, \omega, \tau) := \frac{1}{(4\pi\tau)^{m/2}} \int_M [\tau(\text{scal}(g) \cdot \omega^2 + 4|\nabla\omega|_g^2) - 2\omega^2 \ln \omega - m\omega^2]\, dV_g.$$

The Ricci shrinker entropy is then defined in terms of the weighted Sobolev space $H_1^1(M)$ (see the Appendix, Section 5.7, for the precise definiton) by

$$\mu_-(g, \tau) = \inf\left\{ \mathcal{W}_-(g, \omega, \tau) \,\middle|\, \omega \in H_1^1(M), \omega > 0, \frac{1}{(4\pi\tau)^{m/2}} \int_M \omega^2\, dV_g = 1 \right\}.$$

The corresponding Euler–Lagrange equation is given by

$$\tau(-4\Delta_g\omega_g - \text{scal}(g)\omega_g) + 2\log(\omega_g)\omega_g + (m + \mu_-(g, \tau))\omega_g = 0.$$

It can be shown exactly as in [CCG+07, Cor. 6.34] that, if $\lambda(g) > 0$, the real number $\nu_-(g) = \inf\{\mu_-(g, \tau) \mid \tau > 0\}$ exists and is attained by a parameter τ_g and a minimizing function ω_g.

5.4.3 The Ricci Expander Entropy

Consider the functional $\mathcal{W}_+(g, f, \tau)$:

$$\mathcal{W}_+(g, \omega, \tau) := \frac{1}{(4\pi\tau)^{m/2}} \int_M [\tau(\text{scal}(g) \cdot \omega^2 + 4|\nabla\omega|_g^2) + 2\omega^2 \ln \omega + m\omega^2]\, dV_g.$$

The expander entropy is then defined by

$$\mu_+(g, \tau) = \inf\left\{ \mathcal{W}_+(g, \omega, \tau) \,\middle|\, \omega \in H_1^1(M), \omega > 0, \frac{1}{(4\pi\tau)^{m/2}} \int_M \omega^2\, dV_g = 1 \right\}$$

and the corresponding Euler–Lagrange equation is

$$\tau(-4\Delta_g\omega_g - \text{scal}(g)\omega_g) - 2\log(\omega_g)\omega_g + (-m + \mu_+(g, \tau))\omega_g = 0. \tag{5.16}$$

It is now shown exactly as in [FIN05, p. 10] that, if $\lambda(g) < 0$, the real number $\nu_+(g) = \sup\{\mu_+(g, \tau) \mid \tau > 0\}$ exists and is attained by a parameter τ_g and a minimizing function ω_g.

These functionals are defined almost exactly as in the smooth compact setting. The only difference is that one minimizes in the space $H_1^1(M)$ instead of $C^\infty(M)$. However, the weighted Sobolev space $H_1^1(M)$ is exactly the right space as it covers the blowup of the scalar curvature at the singular points.

The functionals λ and μ_- were already studied by Dai and Wang [DaWa17, DaWa18] in the setting of compact conical manifolds. They showed that they are well defined and posess minimizers provided that the scalar curvature of the cross section satisfies $\mathrm{scal}(g_F) > m - 2$. The minimizers satisfy for any $\varepsilon > 0$ the asymptotics

$$\omega_g(x) = o\left(x^{-\frac{m-2}{2}-\varepsilon}\right), \quad \text{as } x \to 0.$$

By the same methods, we obtained such analogous results for μ_+ under the same assumptions [KrVe19b, Thm. 2.16].

Our main contribution is that these results can be massively improved under the assumption of a subquadratic growth of the scalar curvature at the singularity. Subquadratic growth of the scalar curvature at the conical singularity translates into the condition $\mathrm{scal}(g_F) = n(n-1)$ for the scalar curvature of the cross section. We then have the following result.

Theorem 5.4.1 *Let (M^m, g) be a compact conical Riemannian manifold. Let $n = m-1$ and (F^n, g_F) be the cross section of the conical part of the metric g and assume that $\mathrm{scal}(g_F) = n(n-1)$. Let ω_g be a minimizer in the definition of the λ-functional, shrinker or the expander entropy. Then there exists an $\overline{\gamma} > 0$ such that ω_g admits a partial asymptotic expansion*

$$\omega_g(x, z) = \text{const.} + O(x^{\overline{\gamma}}), \quad \text{as } x \to 0,$$

and moreover, for any integer $k \in \mathbb{N}$,

$$|\nabla_g^k \omega_g|_g(x, z) = O(x^{\overline{\gamma}-k}), \quad \text{as } x \to 0.$$

This result is proved by writing the minimizers in terms of the heat operator, which allows us to use its mapping properties (Schauder estimates) as already employed in the proof of the short time existence of the Ricci de Turck flow. Mapping properties of the heat operator allow us to improve the asymptotics of the minimizers incrementally and the statement is obtained by an iteration argument.

This improvement allows us to study the relation to the Ricci solitons. Ricci solitions are Riemannian metrics g such that for its Ricci curvature $\mathrm{Ric}(g)$, some vector field X, the Lie derivative $\mathcal{L}_X$ and a constant $c \neq 0$, the following equations are satisfied:

$$\mathrm{Ric}(g) + \mathcal{L}_X g = 0 \quad \text{(steady Ricci soliton)},$$
$$\mathrm{Ric}(g) + \mathcal{L}_X g = cg, \quad c > 0 \quad \text{(shrinking Ricci soliton)}, \qquad (5.17)$$
$$\mathrm{Ric}(g) + \mathcal{L}_X g = cg, \quad c < 0 \quad \text{(expanding Ricci soliton)}.$$

We have the following relation between solitons and solutions to (normalized) Ricci flow equations.

1 Any steady Ricci soliton is up to a diffeomorphism a constant solution to the Ricci flow

$$\partial_t g(t) = -2\,\mathrm{Ric}\,(g(t)), \quad g(0) = g. \qquad (5.18)$$

2 Any shrinking or expanding Ricci soliton is up to a diffeomorphism a constant solution to the normalized Ricci flow

$$\partial_t g(t) = -2\,\mathrm{Ric}\,(g(t)) + 2c\,g(t), \quad g(0) = g. \qquad (5.19)$$

Recall also that a Ricci soliton is called *a gradient soliton* if $X = \nabla f$ for some function $f : M \to \mathbb{R}$.

Using improved asymptotics of Theorem 5.4.1 we prove the following results, generalizing well known theorems in the compact smooth setting.

Theorem 5.4.2 *Let (M^m, g) be a compact conical Riemannian manifold. Let $n = m - 1$ and (F^n, g_F) be the cross section of the conical part of the metric g. Then the following statements hold.*

(i) *Suppose that $m \geq 5$ and $\mathrm{scal}(g_F) = n(n-1)$. Then, if (M, g) is a Ricci soliton, it is gradient. Moreover, if (M, g) is steady or expanding, it is Einstein.*

(ii) *In dimension $m = 4$, the assertions of part (i) hold if $\mathrm{Ric}(g_F) = (n-1)g_F$.*

In addition, we prove monotonicity of the entropies along the singular Ricci de Turck flow.

Theorem 5.4.3 *Let (M, g) be a compact conical Riemannian manifold of dimension $\dim M \geq 4$ and a tangentially stable cross section.*

(i) *Then the λ-functional is non-decreasing along the Ricci de Turck flow, preserving conical singularities, and constant only along Ricci-flat metrics.*

(ii) *Whenever defined, the shrinker and the expander entropies are non-decreasing along the (normalized) Ricci de Turck flow, preserving conical singularities, and constant only along shrinking and expanding solitons, respectively.*

5.5 Curvature Quantities Along Singular Ricci de Turck Flow

In this section, we present the main results of our most recent work [KMV20].

5.5.1 Bounded Ricci Curvature Along Singular Ricci de Turck Flow

For a Ricci flow of a smooth compact manifold, the Ricci curvature along the flow can be controlled by the Ricci curvature of the initial metric and the norm of the full curvature tensor along the flow. This follows from the evolution equation for the Ricci curvature along a Ricci flow and the maximum principle. In the singular case, we cannot hope to get such an estimate because the Ricci curvature may be initially unbounded. However, we obtain a very surprising result.

Theorem 5.5.1 *The singular Ricci de Turck flow preserves the initial regularity of the Ricci curvature. In particular, if the initial metric has bounded Ricci curvature, the Ricci curvature remains bounded under the flow.*

The proof studies the evolution equation for the Ricci curvature along the singular Ricci de Turck flow, which is a non-linear parabolic equation for the Lichnerowicz Laplacian Δ_L. Similar to the heat kernel analysis and mapping properties of the heat operator for Δ_L which form a basis for previous works [VER16, KRVE19A, KRVE19B], the recent work [KMV20] establishes regularity results for solutions to such equations.

We are not able to deduce an analoguous result for the scalar curvature. The reason is that the norm of the Ricci tensor appears as the reaction term in the evolution equation of the scalar curvature. Thus, unbounded Ricci curvature at the singularity pushes the scalar curvature to infinity after infinitesimal time. In contrast, the evolution equation on the Ricci curvature is tensorial, which allows more flexibility. However, we are able to prove a different property of the scalar curvature along the Ricci flow which is well known in the smooth compact case.

5.5.2 Positive Scalar Curvature Along Singular Ricci de Turck Flow

It is well known that a Ricci flow on smooth compact manifolds preserves the condition of positive scalar curvature. This is an easy consequence of the maximum principle applied to the evolution equation on the scalar curvature. Moreover, the strong maximum principle implies that in this setting a metric of non-negative scalar curvature that is not Ricci flat will evolve to a metric of positive scalar curvature immediately. In all these cases, the Ricci flow becomes extinct after finite time.

In [KMV20], we establish such results also for the singular Ricci flow on compact conical manifolds. However, we need our tangential stability assumption (which guarantees short time existence for the singular Ricci flow) to be strengthened somewhat further to obtain such results. In fact, we need to assume that $\Box_L > n$ on the orthogonal complement of constant functions. Then we can obtain existence of singular Ricci de Turck flow in $\mathscr{H}_\gamma^{k,\alpha}(M \times [0,T], S)$ with $\gamma > 1$. In this case, the de Turck vector field is sufficiently regular (i.e. it goes to zero as we approach the conical singularity). This allows us to prove the desired statement.

Theorem 5.5.2 *The singular Ricci de Turck flow preserves positivity of scalar curvature along the flow, provided that the de Turck vector field is sufficiently regular. Moreover, if the initial metric has non-negative scalar curvature and is not Ricci flat, the Ricci de Turck flow will become extinct after finite time.*

The stronger assumption on the tangential operator can be characterized explicitly in a similar manner to that in Theorem 5.2.5. For brevity, we provide only a list of symmetric spaces of compact type that satisfy that assumption.

Theorem 5.5.3 *Amongst the symmetric spaces of compact type, only*

$$E_8, \quad E_7/[SU(8)/\{\pm I\}], \quad E_8/SO(16), \quad E_8/E_7 \cdot SU(2)$$

satisfy the conditions $\Box_L > n$.

5.6 Open Questions and Further Research Directions

One obvious but intricate future research direction is clearly an extension of the analysis to non-isolated cones, so-called wedges, and more generally stratified spaces with iterated cone–wedge singularities. Already the existence of the various entropies in the edge setting is an open question.

On the other hand, it is clearly imperative to weaken the conditions of (strict) tangential stability and integrability for more general applications. This might require a setup of the Ricci de Turck flow in L^p based Sobolev spaces instead of Hölder spaces, as the authors have done till now. This approach would also allow us to study the flow of singular metric without a subquadratic blowup of the Ricci curvature.

Another question is whether we can descend to the Ricci flow. This depends on whether the de Turck vector field points toward or away from the singularity.

5.7 Appendix: Sobolev and Hölder Spaces

Let (M, g) be a conical manifold. Let ∇_g denote its Levi-Civita covariant derivative. Let the boundary defining function $x : \mathscr{C}(F) \to (0,1)$ be extended smoothly to $\overline{M}$, nowhere vanishing on M. We consider the space $L^2(M)$ of square-integrable scalar functions with respect to the volume form of g. We define for any $s \in \mathbb{N}$ and any $\delta \in \mathbb{R}$ the weighted Sobolev space $H_\delta^s(M)$ as the closure of compactly supported smooth functions $C_0^\infty(M)$ under

$$\|u\|_{H_\delta^s} := \sum_{k=0}^{s} \|x^{k-\delta} \nabla_g^k u\|_{L^2}. \tag{5.20}$$

Note that $L^2(M, E) = H_0^0(M, E)$ by construction.

Remark 5.7.1 An equivalent norm on the weighted Sobolev space $H_\delta^s(M)$ can be defined for any choice of local bases $\{X_1, \ldots, X_m\}$ of $\mathcal{V}_b$ as follows. We omit the subscript g from the notation of the Levi-Civita covariant derivative and write

$$\|u\|_{H_\delta^s} = \sum_{k=0}^{s} \sum_{(j_1,\ldots,j_k)} \| x^{-\delta} \left(\nabla_{X_{j_1}} \circ \cdots \circ \nabla_{X_{j_k}} \right) u \|_{L^2}. \tag{5.21}$$

Definition 5.7.2 The Hölder space $\mathcal{C}^{\alpha}_{\mathrm{ie}}(M \times [0,T]), \alpha \in [0,1)$ consists of functions $u(p,t)$ that are continuous on $\overline{M} \times [0,T]$ with finite αth Hölder norm

$$\|u\|_{\alpha} := \|u\|_{\infty} + \sup\left(\frac{|u(p,t) - u(p',t')|}{d_M(p,p')^{\alpha} + |t-t'|^{\frac{\alpha}{2}}}\right) < \infty, \qquad (5.22)$$

where the distance function $d_M(p,p')$ between any two points $p,p' \in M$ is defined with respect to the conical metric g, and in terms of the local coordinates (x,z) in the singular neighborhood $\mathscr{C}(F)$ is given equivalently by

$$d_M((x,z),(x',z')) = \left(|x-x'|^2 + (x+x')^2 d_F(z,z')^2\right)^{\frac{1}{2}},$$

where $d_F(z,z')$ is the distance function between any $z,z' \in F$ with respect to g_F. The supremum is taken over all $(p,p',t) \in M^2 \times [0,T]^2$.

We now extend the notion of Hölder spaces to sections of the vector bundle $S = \mathrm{Sym}^2({}^{ib}T^*M)$ of symmetric 2-tensors.

Definition 5.7.3 Denote the fiberwise inner product on S induced by the Riemannian metric g, again by g. The Hölder space $\mathcal{C}^{\alpha}_{\mathrm{ie}}(M \times [0,T], S)$ consists of all sections ω of S which are continuous on $\overline{M} \times [0,T]$, such that, for any local orthonormal frame $\{s_j\}$ of S, the scalar functions $g(\omega, s_j)$ are $\mathcal{C}^{\alpha}_{\mathrm{ie}}(M \times [0,T])$.

The α-th Hölder norm of ω is defined using a partition of unity $\{\varphi_j\}_{j\in J}$ subordinate to a cover of local trivializations of S, with a local orthonormal frame $\{s_{jk}\}$ over $\mathrm{supp}(\varphi_j)$ for each $j \in J$. We put

$$\|\omega\|^{(\varphi,s)}_{\alpha} := \sum_{j\in J}\sum_{k} \|g(\varphi_j\omega, s_{jk})\|_{\alpha}. \qquad (5.23)$$

Norms corresponding to different choices of $(\{\varphi_j\},\{s_{jk}\})$ are equivalent and we may drop the superscript (φ,s) from the notation. The supremum norm $\|\omega\|_{\infty}$ is defined similarly. All the constructions naturally extend to sections in the sub-bundles S_0 and S_1, where the Hölder spaces for S_1 reduce to the usual spaces in Definition 5.7.2.

---

[2] Finiteness of the Hölder norm $\|u\|_{\alpha}$ in particular implies that u is continuous on the closure $\overline{M}$ up to the edge singularity, and the supremum may be taken over $(p,p',t) \in \overline{M}^2 \times [0,T]$. Moreover, as explained in [VER16], we can assume without loss of generality that the tuples (p,p') are always taken from within the same coordinate patch of a given atlas.

We now turn to weighted and higher order Hölder spaces. We extend the boundary defining function $x : \mathscr{C}(F) \to (0,1)$ smoothly to a non-vanishing function on M. The weighted Hölder spaces of higher order are now defined as follows.

Definition 5.7.4   1 The (hybrid) weighted Hölder space for $\gamma \in \mathbb{R}$ is

$$\mathcal{C}^{\alpha}_{\mathrm{ie}}(M \times [0,T], S)_{\gamma} := x^{\gamma} \mathcal{C}^{\alpha}_{\mathrm{ie}}(M \times [0,T], S)$$
$$\cap \; x^{\gamma+\alpha} \mathcal{C}^{0}_{\mathrm{ie}}(M \times [0,T], S)$$

with Hölder norm $\|\omega\|_{\alpha,\gamma} := \|x^{-\gamma}\omega\|_{\alpha} + \|x^{-\gamma-\alpha}\omega\|_{\infty}$.

2 The weighted higher order Hölder spaces are defined by

$$\mathcal{C}^{k,\alpha}_{\mathrm{ie}}(M \times [0,T], S)_{\gamma} := \{\omega \in \mathcal{C}^{\alpha}_{\mathrm{ie},\gamma} \mid \{\mathcal{V}^{j}_{b} \circ (x^2 \partial_t)^l\}\omega \in \mathcal{C}^{\alpha}_{\mathrm{ie},\gamma}$$
$$\text{for all } j + 2l \leq k\},$$

for any $\gamma \in \mathbb{R}$ and $k \in \mathbb{N}$. For any $\gamma > -\alpha$ and $k \in \mathbb{N}$, we also define

$$\mathcal{C}^{k,\alpha}_{\mathrm{ie}}(M \times [0,T], S)^{b}_{\gamma} := \{\omega \in \mathcal{C}^{\alpha}_{\mathrm{ie}} \mid \{\mathcal{V}^{j}_{b} \circ (x^2 \partial_t)^l\}\omega \in \mathcal{C}^{\alpha}_{\mathrm{ie},\gamma}$$
$$\text{for all } j + 2l \leq k\}.$$

The corresponding Hölder norms are defined using a finite cover of coordinate charts trivializing S_0 and a subordinate partition of unity $\{\varphi_j\}_{j \in J}$. By a slight abuse of notation, we identify $\mathcal{V}_b$ with its finite family of generators over each coordinate chart. Writing $\mathscr{D} := \{\mathcal{V}^{j}_{b} \circ (x^2 \partial_t)^l \mid j + 2l \leq k\}$ the Hölder norm on $\mathcal{C}^{k,\alpha}_{\mathrm{ie}}(M \times [0,T], S)_{\gamma}$ is then given by

$$\|\omega\|_{k+\alpha,\gamma} = \sum_{j \in J} \sum_{X \in \mathscr{D}} \|X(\varphi_j \omega)\|_{\alpha,\gamma} + \|\omega\|_{\alpha,\gamma}. \qquad (5.24)$$

For the Hölder norm on $\mathcal{C}^{k,\alpha}_{\mathrm{ie}}(M \times [0,T], S)^{b}_{\gamma}$, replace $\|\omega\|_{\alpha,\gamma}$ by $\|\omega\|_{\alpha}$ in (5.24).

The Hölder norms for different choices of coordinate charts, the subordinate partition of unity or vector fields $\mathcal{V}_b$, are equivalent. Analogously we also consider time-independent Hölder spaces, which are denoted in the same way with $[0,T]$ deleted from notation above.

The vector bundle S decomposes into a direct sum of sub-bundles $S = S_0 \oplus S_1$, where the sub-bundle $S_0 = \mathrm{Sym}^2_0(^{ib}T^*M)$ is the space of trace free (with respect to the fixed metric g) symmetric 2-tensors, and S_1 is the space of pure trace (with respect to the fixed metric g) symmetric 2-tensors. Definition 5.7.4 extends verbatim to sub-bundles S_0 and S_1.

Remark 5.7.5 The spaces presented here are slightly different from the spaces originally introduced in [VER16]. There, in the case of S_1 sections, higher order weighted Hölder spaces were defined in terms of $x^\gamma \mathcal{C}_{ie}^\alpha$ instead of $\mathcal{C}_{ie,\gamma}^\alpha$. Here, we present a more unified definition, which will become much more convenient below. The arguments of [VER16] still carry over to yield regularity statements in these unified Hölder spaces.

Definition 5.7.6 Let (M, g) be a compact conical manifold, $\gamma_0, \gamma_1 \in \mathbb{R}$. In order to simplify notation, we set

$$\mathcal{C}_{ie}^{k,\alpha}(M \times [0,T], S)_\gamma^b := \mathcal{C}_{ie}^{k,\alpha}(M \times [0,T], S)_\gamma$$

for $\gamma \leq -\alpha$, with the same convention for S_1. Then we define the following spaces.

1 If (M, g) is not an orbifold, we set

$$\mathscr{H}_\gamma^{k,\alpha}(M \times [0,T], S) := \mathcal{C}_{ie}^{k,\alpha}(M \times [0,T], S_0)_\gamma \oplus \mathcal{C}_{ie}^{k,\alpha}(M \times [0,T], S_1)_\gamma^b.$$

2 If (M, g) is an orbifold, we set

$$\mathscr{H}_\gamma^{k,\alpha}(M \times [0,T], S) := \mathcal{C}_{ie}^{k,\alpha}(M \times [0,T], S_0)_\gamma^b \oplus \mathcal{C}_{ie}^{k,\alpha}(M \times [0,T], S_1)_\gamma^b.$$

Note that for the non-orbifold case, the different choice of spaces for the S_0 and S_1 components simply ensures that the S_1 component is bounded but not included in $x^\gamma C_{ie}^0$ for some positive weight γ, so that the space contains the metric itself. In the case of orbifold singularities, the condition is imposed on both the S_0 and S_1 components.

References

[BAVE14] Eric Bahuaud and Boris Vertman, *Yamabe flow on manifolds with edges*, Math. Nachr. **287** (2014), no. 23, 127–159.

[BAVE16] Eric Bahuaud and Boris Vertman, *Long time existence of the edge Yamabe flow*, J. Math. Soc. Japan. **71** (2019), no. 2, 651–688.

[CAHE15] Huai-Dong Cao and Chenxu He, *Linear stability of Perelman's ν-entropy on symmetric spaces of compact type*, J. Reine Angew. Math. **709** (2015), no. 1, 229–246.

[CHWA15] Xiuxiong Chen and Yuanqi Wang, *Bessel functions, heat kernel and the conical Kähler - Ricci flow*, J. Funct. Anal. **269**, (2015), no. 2, 551–632.

[CCG+07] Bennett Chow, Sun-Chin Chu, David Glickenstein, *et al.* The Ricci flow: techniques and applications. Part I: Geometric aspects, Mathematical Surveys and Monographs 135. American Mathematical Society Providence, RI, 2007.

[DaWa17] Xianzhe Dai and Changliang Wang, *Perelman's W-functional on manifolds with conical singularities.* Preprint on arXiv:1711.08443 [math.DG] (2017), to appear in Math. Res. Lett.

[DaWa18] Xianzhe Dai and Changliang Wang, *Perelman's lambda-functional on manifolds with conical singularities,* J. Geom. Anal. **28** (2018), no. 4, 3657–3689.

[Don11] Simon K. Donaldson, *Kähler metrics with cone singularities along a divisor.* In Panos Pardalos and Themistocles M. Sassias (eds.), Essays in mathematics and its applications. Springer-Verlag, Heidelberg, 2012, pp. 49–79.

[FIN05] Mikhail Feldman, Tom Ilmanen and Lei Ni, *Entropy and reduced distance for Ricci expanders,* J. Geom. Anal. **15** (2005), no. 1, 49–62.

[GiTo10] Gregor Giesen and Peter M. Topping, *Ricci flow of negatively curved incomplete surfaces,* Calc. Var. Partial Differ. Equ. **38** (2010), no. 3–4, 357–367.

[GiTo11] Gregor Giesen and Peter M. Topping, *Existence of Ricci flows of incomplete surfaces,* Commun. Partial Differ. Equ. **36** (2011), no. 10, 1860–1880.

[HeSu17] Hans-Joachim Hein and Song Sun, *Compact Calabi-Yau manifolds with isolated conical singularities,* Publ. Math. IHES **126** (2017), 73–130.

[JMR16] Thalia Jeffres, Rafe Mazzeo, and Yanir Rubinstein, *Kähler-Einstein metrics with edge singularities,* Ann. of Math. **183** (2016), no. 1, 95–176.

[Krö17] Klaus Kröncke, *Stable and unstable Einstein warped products,* Trans. Amer. Math. Soc. **369** (2017), 6537–6563.

[KrVe19a] Klaus Kröncke and Boris Vertman, *Stability of Ricci de Turck flow on singular spaces,* Calc. Var. Partial Differ. Equ. **58** (2019), no. 2, 74.

[KrVe19b] Klaus Kröncke and Boris Vertman, *Perelman's entropies for manifolds with conical singularities.* Preprint on arXiv:1902.02097 [math.DG] (2019).

[KMV20] Klaus Kröncke, Tobias Marxen and Boris Vertman, *Bounded Ricci curvature and positive scalar curvature under Ricci flow.* Preprint on arXiv:2002.06586 [math.DG] (2020).

[LiZh17] Jiawei Liu and Xi Zhang, *Conical Kähler–Ricci flows on Fano manifolds,* Adv. Math. **307** (2017), 1324–1371.

[MRS15] Rafe Mazzeo, Yanir Rubinstein and Natasha Šešum, *Ricci flow on surfaces with conic singularities,* Anal. PDE **8** (2015), no. 4, 839–882.

[Mel92] Richard B. Melrose, *Calculus of conormal distributions on manifolds with corners,* Intl. Math. Res. Not., no. **3** (1992), 51–61.

[Mel93] Richard B. Melrose, The Atiyah-Patodi-Singer index theorem, Research Notes in Mathematics, vol. 4. A K Peters Ltd., Wellesley, MA, 1993.

[Ses06] Natasa Šešum, *Linear and dynamical stability of Ricci-flat metrics,* Duke Math. J. **133** (2006), no. 1, 1–26.

[Sim13] Miles Simon, *Local smoothing results for the Ricci flow in dimensions two and three,* Geom. Topol. **17** (2013), no. 4, 2263–2287.

[TIA87] Gang Tian, *Smoothness of the universal deformation space of compact Calabi-Yau manifolds and its Petersson–Weil metric.* In S.-T. Yau (ed.), Mathematical aspects of string theory. Advanced Series in Mathematical Physics **1**. World Scientific Publishing, Singapore, 1987, pp. 629–646.

[TIA15] Gang Tian, *K-stability and Kähler–Einstein metrics*, Commun. Pure Appl. Math. **68** (2015), no. 7, 1085–1156.

[VER16] Boris Vertman, *Ricci flow on singular manifolds.* Preprint on arXiv:1603.06545 [math.DG] (2016).

[WAN16] Yuanqi Wang, *Smooth approximations of the conical Kähler–Ricci flows*, Math. Ann. **365** (2016), no. 1–2, 835–856.

[YIN10] Hao Yin, *Ricci flow on surfaces with conical singularities*, J. Geom. Anal. **20** (2010), no. 4, 970–995.

Part Two

Structures on Manifolds and Mathematical Physics

6

Some Open Problems in Sasaki Geometry

Charles P. Boyer, Hongnian Huang, Eveline Legendre
and Christina W. Tønnesen-Friedman

Abstract

We discuss two open problems in Sasaki geometry concerning the existence of extremal and constant scalar curvature Sasaki metrics. Our approach studies two functionals, the Einstein–Hilbert functional and the Sasaki energy functional. Finally, examples that affirmatively satisfy our problems are presented.

6.1 Introduction

The purpose of this chapter is to discuss two open problems in Sasaki geometry. These problems involve the so-called Sasaki cone, which, although different in nature, plays a role in Sasaki geometry similar to that of the Kähler cone in Kähler geometry. In the latter it is well understood by simple examples that constant scalar curvature Kähler (cscK) metrics need not be isolated and that there are complex manifolds whose Kähler cone admits extremal Kähler metrics, but no cscK metric. However, in the case of the Sasaki cone there are no known examples of the analogous phenomenon. This leads to two important open problems:

Problem 6.1.1 *Are constant scalar curvature rays in the Sasaki cone isolated?*

Problem 6.1.2 *If there are extremal rays of Sasaki metrics in the Sasaki cone, is there always at least one constant scalar curvature ray?*

When the contact bundle $\mathcal{D}$ has vanishing first Chern class (or more, generally is a torsion class), constant scalar curvature Sasaki (cscS) metrics turn out to be Sasaki–η-Einstein and, up to a transversal homothety,

Sasaki–Einstein. In this case, by a famous result of Martelli, Sparks and Yau [MSY08], the constant scalar curvature ray, whenever it exists, is not only isolated but also unique in the Sasaki cone. This fact has been used recently by Donaldson and Sun in [DS14, DS17] to study the moduli space of compact Kähler Fano manifolds and more precisely to prove uniqueness of the rescaled pointed Gromov–Hausdorff limits in this setting. An affirmative answer to Problem 6.1.1 would be an extension of this very useful Martelli–Sparks–Yau theorem.

A partial answer to Problem 6.1.1 was given by [BHLTF17, Lem. 4.1] as well as by [BHL18, Cor. 1.7]. In particular, the latter says that if the zero set Z of rays of the Sasaki–Futaki invariant lies in a two-dimensional subcone of the Sasaki cone $\mathfrak{t}^+(\mathcal{D}, J)$, it is a finite set. Moreover, for toric contact structures on a lens space bundle over S^2 the Sasaki cone $\mathfrak{t}^+(\mathcal{D}, J)$ has dimension 3, and in [Leg11] it is proved that all constant scalar curvature rays (cscS) are isolated in this case. In Theorem 6.3.24 below, we give another partial result.

In the case of Problem 6.1.2 involving the so-called $S_{\mathbf{w}}^3$ join $M \star_1 S_{\mathbf{w}}^3$, where M is a Sasaki manifold with constant scalar curvature, it was proven in [BTF16] that $M \star_1 S_{\mathbf{w}}^3$ has a cscS metric. However, this uses the admissible construction of Apostolov, Calderbank, Gauduchon and Tønnesen-Friedman [ACG06, ACGTF04, ACGTF08] for which M itself needs to be cscS. On the other hand, when the contact bundle $\mathcal{D}$ has vanishing first Chern class (or more generally is a torsion class), there are known obstructions to the existence of cscS metrics due to Gauntlett, Martelli, Sparks and Yau [GMSY07]. Moreover, it was shown in [BvC18] that in all known cases the Sasaki cone admits no extremal metrics whatsoever. In terms of K-stability an affirmative answer to Problem 6.1.2 is equivalent to stating that any Sasaki cone that admits a K-semistable polarization relative to a fixed maximal torus $\mathbb{T}$ also admits a K-semistable polarization with respect to an arbitrary torus (for notions of K-stability in the Sasaki context, see [CS18, BHLTF17, BvC18]).

The Sasaki cone $\kappa(\mathcal{D}, J)$ can be thought of as the moduli space of Sasaki metrics with a fixed underlying contact CR structure $(\mathcal{D}, J)$, where $\mathcal{D}$ is the contact bundle and J is a complex structure on $\mathcal{D}$. On a Sasaki manifold M^{2n+1} the dimension k of $\kappa(\mathcal{D}, J)$ satisfies $1 \leq k \leq n + 1$. We are also interested in the moduli space $\mathfrak{e}(\mathcal{D}, J)$ of extremal Sasaki metrics as well as the moduli space $\kappa_{csc}(\mathcal{D}, J)$ of constant scalar curvature Sasaki metrics. A result of [BGS08] says that $\dim \mathfrak{e}(\mathcal{D}, J)$ is either 0 or k, and an affirmative answer to Problems 6.1.1 and 6.1.2 says that if $\dim \mathfrak{e}(\mathcal{D}, J) = k$, the dimension of $\kappa_{csc}(\mathcal{D}, J)$ is exactly 1. The full moduli space, which is described in [Boy19], is obtained by varying J also. However, in this chapter we fix the contact CR structure $(\mathcal{D}, J)$ to study two important functionals on $\kappa(\mathcal{D}, J)$: the Einstein–Hilbert functional $\mathbf{H}$ and the Sasaki energy functional $\mathcal{SE}$. The variational calculus for $\mathbf{H}$ was performed in [BHLTF17], so in this chapter we derive the Euler–Lagrange equations for $\mathcal{SE}$ and compare the critical sets of $\mathbf{H}$ and $\mathcal{SE}$. We end with an application to the case when M is a lens space bundle over a compact Riemann surface.

Acknowledgements   This chapter is roughly based on a talk given by CB at the Australian–German Workshop on Differential Geometry in the Large held at the mathematical research institute MATRIX in Creswick, Victoria, Australia, Feb. 4 to Feb. 15, 2019. He would like to thank MATRIX for its hospitality and support.

CB was partially supported by grant #519432 from the Simons Foundation. EL was partially supported by France ANR project EMARKS No ANR-14-CE25-0010. CT-F was partially supported by grant #422410 from the Simons Foundation.

6.2 Brief Review of Sasaki Geometry

Recall that a Sasakian structure on a contact manifold M^{2n+1} of dimension $2n + 1$ is a special type of contact metric structure $\mathcal{S} = (\xi, \eta, \Phi, g)$ with underlying almost CR structure $(\mathcal{D}, J)$, where η is a contact form such that $\mathcal{D} = \ker \eta$, ξ is its Reeb vector field, $J = \Phi|_{\mathcal{D}}$, and $g = d\eta \circ (\times \Phi) + \eta \otimes \eta$ is a Riemannian metric. $\mathcal{S}$ is a Sasakian structure if ξ is a Killing vector field and the almost CR structure is integrable, i.e. $(\mathcal{D}, J)$ is a CR structure. We refer to [BG08] for the fundamentals of Sasaki geometry. We call $(\mathcal{D}, J)$ a *CR structure of Sasaki type*, and $\mathcal{D}$ a *contact structure of Sasaki type*. We shall always assume that the Sasaki manifold M^{2n+1} is compact and connected.

6.2.1 The Sasaki Cone

Within a fixed contact CR structure $(\mathcal{D}, J)$ there is a conical family of Sasakian structures known as the Sasaki cone. We are also interested in a variation within this family. To describe the Sasaki cone we fix a Sasakian structure $\mathcal{S}_o = (\xi_0, \eta_o, \Phi_o, g_o)$ on M whose underlying CR structure is $(\mathcal{D}, J)$ and let $\mathfrak{t}$ denote the Lie algebra of the maximal torus in the automorphism group of $\mathcal{S}_o$. The *(unreduced) Sasaki cone* [BGS08] is defined by

$$\mathfrak{t}^+(\mathcal{D}, J) = \{\xi \in \mathfrak{t} \mid \eta_o(\xi) > 0 \text{ everywhere on } M\}, \qquad (6.1)$$

which is a cone of dimension $k \geq 1$ in $\mathfrak{t}$. The reduced Sasaki cone $\kappa(\mathcal{D}, J)$ is $\mathfrak{t}^+(\mathcal{D}, J)/\mathcal{W}$, where $\mathcal{W}$ is the Weyl group of the maximal compact subgroup of $\mathfrak{CR}(\mathcal{D}, J)$, which, as mentioned previously, is the moduli space of Sasakian structures with underlying CR structure $(\mathcal{D}, J)$. However, it is more convenient to work with the unreduced Sasaki cone $\mathfrak{t}^+(\mathcal{D}, J)$. It is also clear from the definition that $\mathfrak{t}^+(\mathcal{D}, J)$ is a cone under the transverse scaling defined by

$$\mathcal{S} = (\xi, \eta, \Phi, g) \mapsto \mathcal{S}_a = (a^{-1}\xi, a\eta, g_a),$$
$$\text{where } g_a = ag + (a^2 - a)\eta \otimes \eta, \quad a \in \mathbb{R}^+. \qquad (6.2)$$

So, Sasakian structures in $\mathfrak{t}^+(\mathcal{D}, J)$ come in rays, and, since the Reeb vector field ξ is Killing, $\dim \mathfrak{t}^+(\mathcal{D}, J) \geq 1$, and it follows from contact geometry that $\dim \mathfrak{t}^+(\mathcal{D}, J) \leq n+1$. When $\dim \mathfrak{t}^+(\mathcal{D}, J) = n+1$ we have a toric contact manifold of Reeb type studied in [BM93, BG00, Ler02, Ler04, Leg11, Leg16]. In this case there is a strong connection between the geometry and topology of $(M, \mathcal{S})$ and the combinatorics of $\mathfrak{t}^+(\mathcal{D}, J)$. Much can also be said in the complexity 1 case $(\dim \mathfrak{t}^+(\mathcal{D}, J) = n)$ [AH06].

We let $\mathfrak{R}(\mathcal{D}, J)$ denote the set of rays in $\mathfrak{t}^+(\mathcal{D}, J)$, so that $\mathfrak{t}^+(\mathcal{D}, J)$ is the open cone over the semi-algebraic set $\mathfrak{R}(\mathcal{D}, J)$ [BCR98]. The combinatorial structure of $\mathfrak{R}(\mathcal{D}, J)$ can be involved. For example, let M be a toric Sasaki manifold that is an S^1 bundle over a compact toric Hodge manifold N (or orbifold). A choice of Reeb vector field in $\mathfrak{t}^+(\mathcal{D}, J)$ gives the intersection of the dual moment cone with a hyperplane giving a generalized Delzant polytope[1] P. We think of the interior of the dual polytope P^* as representing the space of rays $\mathfrak{R}(\mathcal{D}, J)$. For example, in the case that N is a Bott manifold [BCTF19] of complex dimension n, the closure $\overline{\mathfrak{R}(\mathcal{D}, J)}$ is a *cross-polytope* or *n-cross* which is dual to P, which here is combinatorially an n-cube. A 3-cross is an octahedron.

---

[1] We use the term generalized here since it is a Delzant polytope only if the Reeb field is regular and the polytope lies on an integral lattice with primitive normal vectors, in which case the quotient is a Hodge manifold. If P lies on the lattice but the normal vectors are not primitive, the Sasakian structure is quasi-regular and the quotient is a Hodge orbifold; and if P does not lie on an integral lattice, the Sasakian structure is irregular and there is no well-defined quotient.

6.2.2 The Transverse Holomorphic Structure

A Sasakian structure $\mathcal{S} = (\xi, \eta, \Phi, g)$ determines not only a CR contact structure $(\mathcal{D}, J)$, but a transverse holomorphic structure $(TM/\mathcal{F}_\xi, \bar{J})$ as well, where $\mathcal{F}_\xi$ is the foliation defined by the Reeb vector field ξ. Here, instead of fixing a contact structure $\mathcal{D}$ we fix the Reeb vector field ξ. This gives a contractible space of Sasakian structures, namely

$$\mathcal{S}(\xi, \bar{J}) = \{\varphi \in C_B^\infty(M) \mid (\eta + d_B^c \varphi) \wedge (d\eta + i\partial_B \bar{\partial}_B \varphi)^n \neq 0, \int_M \varphi \eta \wedge (d\eta)^n = 0\}, \tag{6.3}$$

where $d_B^c = \frac{i}{2}(\bar{\partial} - \partial)$. The space $\mathcal{S}(\xi, \bar{J})$ is an infinite dimensional Fréchet manifold. Each Reeb vector field gives an isotopy class $\mathcal{S}(\xi, \bar{J})$ of contact structures, and determines a basic cohomology class $[d\eta]_B \in H^{1,1}(\mathcal{F}_\xi)$, and each representative determines a transverse Kähler structure with transverse Kähler metric $g^T = d\eta \circ (\times \Phi)$. Note that $d\eta$ is not exact as a basic cohomology class, since η is not a basic 1-form. We want to search for a 'preferred' Sasakian structure $\mathcal{S}_\varphi$ which represents the cohomology class $[d\eta]_B$. This leads to the study [BGS08] of the Calabi functional given by equation (6.8) below.

6.2.3 The Lie Algebra of Killing Potentials

Note that for a CR structure of Sasaki type the group $\mathfrak{CR}(\mathcal{D}, J)$ of CR transformations has dimension at least one. Moreover, if M is compact $\mathfrak{CR}(\mathcal{D}, J)$ is a compact Lie group except for the standard CR structure on the sphere $\mathbb{S}^{2n+1}$, where $\mathfrak{CR}(\mathcal{D}, J) = \mathbb{SU}(n+1, 1)$. We are mainly concerned with reducing things to a maximal torus $\mathbb{T}^k$ in $\mathfrak{CR}(\mathcal{D}, J)$ where $1 \leq k \leq n+1$, and its Lie algebra t. Before doing so we briefly discuss the holomorphic viewpoint which gives rise to an infinite dimensional Lie algebra $\mathfrak{h}^T(\xi, \bar{J})$ of transverse holomorphic vector fields; however, the infinite dimensional part of $\mathfrak{h}^T(\xi, \bar{J})$ is generated by the smooth sections $\Gamma(\xi)$ of line bundle generated by the Reeb vector field, so we have a finite dimensional quotient algebra $\mathfrak{h}^T(\xi, \bar{J})/\Gamma(\xi)$ whose complexification consists of the complexification of the Lie algebra $\mathfrak{aut}(\mathcal{S})$ of the Sasaki automorphism group together with a possible non-reductive part. See [BGS08] for details. Here we concern ourselves with the Abelian Lie algebras $\mathfrak{t}^\mathbb{C} = \mathfrak{t} \otimes \mathbb{C}$ and t associated with the maximal torus action. We refer to the potentials associated with $\mathfrak{t}^\mathbb{C}$ as *holomorphy potentials* since the action of elements of $\mathfrak{t}^\mathbb{C}$ is transversely holomorphic.

Consider the strict contact moment map $\mu : M \to \mathfrak{con}(M, \eta)^*$ with respect to the Fréchet Lie group of strict contact transformations $\mathfrak{Con}(M, \eta)$ defined by

$$\langle \mu(x), X \rangle = \eta(X). \tag{6.4}$$

The function $\eta(X)$ is basic with respect to ξ, and there is a Lie algebra isomorphism between $\mathfrak{con}(\eta)$ and the Lie algebra of smooth ξ invariant functions $C^\infty(M)^\xi$ with Lie algebra structure given by the Jacobi–Poisson bracket defined by

$$\{f, g\} = \eta([X^f, X^g]), \tag{6.5}$$

where $d(\eta(X^f)) = -X^f \lrcorner d\eta$. This isomorphism extends to a Lie algebra isomorphism between $\mathfrak{con}(\mathcal{D})$ and all smooth functions $C^\infty(M)$. Each choice of contact form η or equivalently Reeb vector field ξ defines such an isomorphism. We have

Lemma 6.2.1 *Any $X \in \mathfrak{con}(\eta)$ can be written uniquely[2] as*

$$X = \Phi \circ \mathrm{grad}_T \eta(X) + \eta(X)\xi,$$

where the gradient is taken with respect to the transverse metric g^T.

Proof This follows from

$$0 = \mathscr{L}_X \eta = d(\eta(X)) + \Phi \circ \mathrm{grad}_T \eta(X) \lrcorner d\eta. \qquad \square$$

We now restrict our attention to the finite dimensional Lie subalgebra $\mathfrak{aut}(\mathcal{S})$ of the compact Lie group $\mathfrak{Aut}(\mathcal{S})$ of Sasaki automorphisms. These are Killing vector fields in $\mathfrak{con}(\eta)$ that commute with ξ and leave Φ invariant. They also leave invariant the transverse Kähler structure. Following the Kähler case (see [Gau10]) we say that $\eta(X)$ is a *Killing potential* (for $\mathcal{S}$) when $X \in \mathfrak{aut}(\mathcal{S})$. We denote[3] by $\mathcal{K}^\xi$ the real vector space spanned by all the Killing potentials. $\mathcal{K}^\xi$ forms a Lie subalgebra of $C^\infty(M)^\xi$ isomorphic to the Lie algebra $\mathfrak{aut}(\mathcal{S})$. We actually restrict ourselves to considering a real maximal torus $\mathbb{T}^k$ of $\mathfrak{Aut}(\mathcal{S})$ and its real Lie algebra t. We let $\mathcal{H}^\xi$ denote the subalgebra of $\mathcal{K}^\xi$ that is isomorphic to t. Choosing a maximal torus in $\mathfrak{Aut}(\mathcal{S})$ is equivalent to choosing

[2] The sign $\pm \Phi \circ \mathrm{grad}_T \, \pi_g s_g^T$ in the formula depends on the sign convention for the transverse Kähler form $d\eta$. We use the convention $g^T = d\eta \circ (\times \Phi)$, which gives a plus sign.

[3] We only consider those Sasakian structures in the family given by $\mathfrak{t}^+(\mathcal{D}, J)$, so a choice of $\mathcal{S}$ is equivalent to a choice of Reeb vector field $\xi \in \mathfrak{t}^+(\mathcal{D}, J)$, which is equivalent to specifying the Sasaki metric g of $\mathcal{S}$. We often abuse notation and label objects by ξ, η, g or $\mathcal{S}$ depending on the emphasis.

maximal torus in $\mathfrak{CR}(\mathcal{D}, J)$ [BGS08], so the Lie algebra $\mathfrak{t}$ is independent of the choice of $\mathcal{S} \in \mathfrak{t}^+(\mathcal{D}, J)$; however, we emphasize that $\mathcal{H}^\xi$ depends on $\xi \in \mathfrak{t}^+(\mathcal{D}, J)$, since the isomorphism $\zeta \mapsto \eta(\zeta) \in \mathcal{H}^\xi$ does. We are interested in how $\mathcal{H}^\xi$ changes as ξ varies in $\mathfrak{t}^+(\mathcal{D}, J)$, so we define

$$\mathcal{H} = \bigcup_{\xi \in \mathfrak{t}^+(\mathcal{D}, J)} \mathcal{H}^\xi. \tag{6.6}$$

Lemma 6.2.2 *Let $(M, \mathcal{D}, J, \mathbb{T}^k)$ be a contact CR manifold of Sasaki type with an effective action of a torus $\mathbb{T}^k$ that preserves the CR structure. Then we have the following.*

(1) *The elements of $\mathcal{H}$ are basic with respect to every $\xi \in \mathfrak{t}^+(\mathcal{D}, J)$. Equivalently, $\mathcal{H} \subset C^\infty(M)^{\mathbb{T}}$.*

(2) *The nonconstant elements of $\mathcal{H}^{\xi_1}$ and $\mathcal{H}^{\xi_2}$ are related by*

$$\eta_2(\zeta) = \frac{1}{\eta_1(\xi_2)} \eta_1(\zeta),$$

giving an isomorphism $\mathcal{H}^{\xi_1} \approx \mathcal{H}^{\xi_2}$ of Abelian Lie algebras.

(3) *If $\xi_1, \xi_2 \in \mathfrak{t}^+(\mathcal{D}, J)$ are not colinear, then $\mathcal{H}^{\xi_1} \cap \mathcal{H}^{\xi_2} = \mathbb{R}$, where $\mathbb{R}$ denotes the constants.*

(4) *Each choice of $\xi \in \mathfrak{t}^+(\mathcal{D}, J)$ gives a Lie algebra monomorphism $\iota_\xi : \mathbb{R} \to \mathcal{H}^\xi$ defined by $\iota_\xi(a) = a\eta(\xi) = a$.*

Proof Item (1) is well known. Given $\xi_1, \xi_2 \in \mathfrak{t}^+(\mathcal{D}, J)$, the corresponding contact forms η_1 and η_2 satisfy $\eta_2 = f\eta_1$ for some nowhere vanishing smooth function f. But since ξ_2 is the Reeb field of η_2 this implies $f = \frac{1}{\eta_1(\xi_2)}$, proving the first part of (2). That we have a Lie algebra isomorphism follows from the Abelian nature of $\mathcal{H}^\xi$.

To prove (3) for any $a \in \mathbb{R}$, take $\zeta = a\xi_1$ and $\zeta' = a\xi_2$ then $\eta_1(\zeta) = a = \eta_2(\zeta')$. So $\mathcal{H}^\xi \cap \mathcal{H}^{\xi'}$ contains the constants. Conversely, if $\eta_1(\zeta) = \eta_2(\zeta')$ for some $\zeta, \zeta' \in \mathfrak{t}$, then (2) implies

$$\eta_1(\zeta) = \frac{\eta_1(\zeta')}{\eta_1(\xi_2)}.$$

But then the only way that $\eta_1(\zeta)$ and $\eta_1(\zeta')$ can both be in $\mathcal{H}^{\xi_1}$ is that $\zeta' = a\xi_2$ and $\zeta = a\xi_1$, which implies (3). Item (4) is clear.   □

We call the element $\eta_1(\xi_2) \in \mathcal{H}^{\xi_1}$ the *transfer function* from η_1 to η_2. Note that the smooth functions $\eta_1(\xi_2)$ and $\eta_2(\xi_1)$ are invariant under the same transverse scaling $(\xi_1, \xi_2) \mapsto (a^{-1}\xi_1, a^{-1}\xi_2)$ and satisfy the following relations:

$$\eta_1(\xi_1) = 1, \qquad \eta_2(\xi_2) = 1, \qquad \eta_2(\xi_1)\eta_1(\xi_2) = 1. \tag{6.7}$$

6.3 Extremal Sasaki Geometry

The notion of extremal Kähler metrics was introduced as a variational problem by Calabi in [Cal56] and studied in greater depth in [Cal82]. This was then emulated in [BGS08] for the Sasaki case, namely

$$\mathcal{E}_2(g) = \int_M s_g^2 dv_g, \qquad (6.8)$$

where the variation is taken over the space $\mathcal{S}(\xi, \bar{J})$. As in the Kähler case the Euler–Lagrange equation is a 4th order PDE

$$\mathcal{L}\varphi = (\bar{\partial}\partial^\#)^* \bar{\partial}\partial^\# \varphi$$

$$= \frac{1}{4}\left(\Delta_B^2 \varphi + 4g(\rho^T, i\partial\bar{\partial}\varphi) + 2(\partial s^T) \lrcorner \partial^\# \varphi\right) = 0 \qquad (6.9)$$

whose critical points are those Sasaki metrics whose $(1,0)$ gradient $\partial^\# s_g$ of the scalar curvature s_g is transversely holomorphic. Such Sasaki metrics (structures) are called *extremal*. An important special case is the Sasaki metrics of constant scalar curvature (cscS), in which case $\partial^\# s_g$ is the zero vector field.

Since both the volume functional and the total transverse scalar curvature functional do not depend on the representative in $\mathcal{S}(\xi, \bar{J})$, the functional (6.8) is essentially equivalent to the functional

$$\mathcal{E}_2^T(g) = \int_M (s_g^T)^2 dv_g. \qquad (6.10)$$

This latter functional has the advantage of behaving nicely under transverse scaling (6.2). It is important to realize that if $\mathcal{S}$ is extremal, so is $\mathcal{S}_a$ for all $a \in \mathbb{R}^+$, and if g has constant scalar curvature so does g_a for all $a \in \mathbb{R}^+$. So extremal and cscS structures come in rays.

6.3.1 Transverse Futaki–Mabuchi

The Sasaki version χ of the Futaki–Mabuchi vector field [FM95] was introduced in [BvC18] and used to define the Sasaki version of K-relative stability. Following [FM95] we consider the L^2 inner product $\langle \cdot, \cdot \rangle$ on the polarized Sasaki manifold $(M, \mathcal{S})$, or more generally the inner product on tensors, p-form, functions, etc.,

$$\langle \alpha, \beta \rangle = \int_M g(\alpha, \beta) dv_g. \qquad (6.11)$$

The L^2 inner product on functions induces an inner product on the Lie algebra t that depends on the choice of Reeb vector field $\xi \in \mathfrak{t}^+(\mathcal{D}, J)$,

$$\langle \zeta, \zeta' \rangle_\xi = \langle \eta(\zeta), \eta(\zeta') \rangle = \int_M \eta(\zeta)\eta(\zeta')dv_g. \tag{6.12}$$

Remark 6.3.1   In fact, the inner product (6.12) defines an inner product on the Fréchet Lie algebra $\mathfrak{con}(\eta)$; however, we shall only make use of it on the Abelian subalgebra t.

Then for each $\xi \in \mathfrak{t}^+$ the inner product (6.12) gives an orthogonal splitting

$$\mathfrak{t} = \mathbb{R}\xi \oplus \mathfrak{t}_0, \tag{6.13}$$

and under the isomorphism $\mathfrak{t} \approx \mathcal{H}^\xi$ we have orthogonal splittings

$$\mathcal{H}^\xi \approx \mathbb{R} \oplus \mathcal{H}_0^\xi, \tag{6.14}$$

where $\mathbb{R}$ denotes the constants and

$$\mathcal{H}_0^\xi = \{\eta(\zeta) \mid \int_M \eta(\zeta)dv_\xi = 0, \ \zeta \in \mathfrak{t}\}.$$

Letting $\mathfrak{h}^T(\xi, \bar{J})$ denote the Lie algebra of transverse holomorphic vector fields on $(M, \mathcal{S})$, we recall the Sasaki–Futaki invariant [BGS08] (or transversal Futaki invariant) $\mathbf{F}_\xi : \mathfrak{h}^T(\xi, \bar{J}) \to \mathbb{C}$ defined by

$$\mathbf{F}_\xi(X) = \int X\psi_g dv_\xi,$$

where the basic transverse Ricci potential ψ_g satisfies $\rho^T = \rho_h^T + i\partial\bar{\partial}\psi_g$ and ρ_h^T is the harmonic part of the transverse Ricci form ρ^T. By [BGS08, Prop. 5.1], $\mathbf{F}_\xi$ depends only on the class $\mathcal{S}(\xi, \bar{J})$, and we know that $\mathbf{F}_\xi$ is degenerate on $\mathfrak{h}^T(\xi, \bar{J})$ since it vanishes on the infinite dimensional subalgebra of sections $\Gamma(L_\xi)$ of the line bundle L_ξ. So we restrict attention to the finite dimensional Lie algebra $\mathfrak{h}_\xi^T = \mathfrak{h}^T(\xi, \bar{J})/\Gamma(L_\xi)$. From this we get a map $\mathbf{F} : \mathfrak{t}^+ \times \mathfrak{h}_\xi^T :\to \mathbb{C}$ defined by $\mathbf{F}(\xi, X) = \mathbf{F}_\xi(X)$. By [BGS08, Lem. 4.6] it follows that if $\zeta \in \mathfrak{t}$ then $\Phi\zeta \in \mathfrak{h}_\xi^T$, which gives the map $\mathbf{F} : \mathfrak{t}^+ \times \mathfrak{t} \to \mathbb{R}$ defined by $\mathbf{F}(\xi, a) = \mathbf{F}_\xi(\Phi(a))$. So for each Sasakian structure $\mathcal{S} \in \mathfrak{t}^+$ we have its Sasaki–Futaki invariant $\mathbf{F}_\xi \circ \Phi : \mathfrak{t} \to \mathbb{R}$ on t defined by

$$\mathbf{F}_\xi \circ \Phi(\zeta) = \int \Phi\zeta\psi_g dv_\xi. \tag{6.15}$$

Definition 6.3.2   [BvC18] We define the *Sasaki–Futaki–Mabuchi vector field* χ_ξ to be the dual of $\mathbf{F}_\xi \circ \Phi$ with respect to the inner product (6.12) on t, that is $\mathbf{F}_\xi \circ \Phi(\zeta) = \langle \chi_\xi, \zeta \rangle_\xi$.

So the Sasaki–Futaki invariant becomes

$$\mathbf{F}_\xi \circ \Phi(\zeta) = \int_M \eta(\zeta)\eta(\chi_\xi)dv_\xi. \tag{6.16}$$

The fact that $\mathbf{F}_\xi \circ \Phi(\xi) = 0$ implies

$$\langle \xi, \chi_\xi \rangle_\xi = \int_M \eta(\chi_\xi)dv_\xi = 0, \tag{6.17}$$

or equivalently $\eta(\chi_\xi) \in \mathcal{H}_0^\xi$ and $\chi_\xi \in \mathfrak{t}_0$.

Consider the projection $\pi_g : C^\infty(M)^\xi \longrightarrow \mathcal{H}^\xi$ onto the space $\mathcal{H}^\xi$ of Killing potentials, or equivalently $\pi : \mathfrak{con}(\eta) \longrightarrow \mathfrak{t}$. From the orthogonal decomposition (6.14) we see that the projection of the scalar curvature s_g^T onto the constants $\mathbb{R}$ is just the average scalar curvature $\bar{s}_\xi$ defined by $\bar{s}_\xi = \frac{\mathbf{S}_\xi}{\mathbf{V}_\xi}$, where $\mathbf{S}_\xi$ is the total transverse scalar curvature of $\mathcal{S}$ and $\mathbf{V}_\xi$ is its volume. As in [FM95] we have

Lemma 6.3.3   *The Sasaki–Futaki–Mabuchi vector field χ_ξ can be written uniquely as*

$$\chi_\xi = \Phi \circ \mathrm{grad}_T\, \pi_g s_g^T + (\pi_g s_g^T - \bar{s}_g^T)\xi$$

where the gradient is taken with respect to the transverse metric g^T. Moreover, χ_ξ is independent of the choice of $\mathcal{S} \in \mathcal{S}(\xi, \bar{J})$.

Proof   By Definition 6.3.2, χ_ξ is the unique vector field in $\mathfrak{t}_0$ that is dual to the Sasaki–Futaki invariant. Since $\mathfrak{t}$ is a subalgebra of $\mathfrak{con}(\eta)$ and $\chi_\xi \in \mathfrak{t}$, Lemma 6.2.1 says that χ_ξ takes the form $\Phi \circ \mathrm{grad}_T\, \eta(\chi_\xi) + \eta(\chi_\xi)\xi$. But from [BGS08] $F_\xi \circ \Phi$ is the transverse Futaki invariant with respect to the transverse Kähler metric g^T. Thus, χ_ξ is just the transverse Futaki–Mabuchi vector field with respect to g^T, which implies that $\eta(\chi_\xi) = \pi_g s_g^T$ up to a constant. But then, since $\chi_\xi \in \mathfrak{t}_0$ and $s_g^T - \pi_g s_g^T$ is orthogonal to the constants, we have

$$0 = \int_M \eta(\chi_\xi)dv_g = \int_M \pi_g s_g^T dv_g + c \int_M dv_g$$
$$= \int_M (s_g^T + c)dv_g = (\bar{s}_g^T + c)\mathbf{V}_g,$$

which gives the result. The last statement follows from Definition 6.3.2 and [BGS08, Prop. 5.1].   □

It is easy to obtain the relationship between extremality and Killing potentials.

Lemma 6.3.4 *On a Sasaki manifold the following are equivalent:*

(1) $\mathcal{S}$ is extremal;
(2) $s_g^T \in \mathcal{H}^\xi$;
(3) $\pi_g s_g^T = s_g^T$;
(4) $\pi_g s_g^T - \bar{s}_g^T \in \mathcal{H}_0^\xi$.

Moreover, $\chi_\xi = 0$ if and only if $\pi_g s_g^T = \bar{s}_g^T$ whether it is extremal or not.

As a corollary of Lemma 6.3.3 we have

Proposition 6.3.5 *On the Lie algebra* t, *the inner product $\langle \cdot, \cdot \rangle_\xi$ depends only on the isotopy class $(\mathcal{S}, \bar{J})$.*

In [FM02] Futaki and Mabuchi generalized their bilinear form to a multilinear form. Accordingly, we can do the same, although we make no use of it here. For each $\xi \in \mathfrak{t}^+(\mathcal{D}, J)$ we define the symmetric multilinear form $\Phi_\xi^l : \mathrm{sym}^l(\mathfrak{t}) \longrightarrow \mathbb{R}$ by

$$\Phi_\xi^l(\zeta_1, \ldots, \zeta_l) = \int_M \eta(\zeta_1) \cdots \eta(\zeta_l) dv_g. \tag{6.18}$$

As in [FM02] we have

Proposition 6.3.6 *Φ_ξ^k depends only on the isotopy class $\mathcal{S}(\xi, \bar{J})$.*

6.3.2 The Einstein–Hilbert Functional

We now consider the Einstein–Hilbert functional that was studied in [Leg11, BHLTF17, BHL18],

$$\mathbf{H}_\xi = \frac{\mathbf{S}_\xi^{n+1}}{\mathbf{V}_\xi^n}, \tag{6.19}$$

as a function on the Sasaki cone $\mathfrak{t}^+$. Since both $\mathbf{V}_\xi$ and $\mathbf{S}_\xi$ are independent of the choice of Sasakian structure in $\mathcal{S}(\xi, \bar{J})$, the Einstein–Hilbert functional $\mathbf{H}_\xi$ depends only on the isotopy class of contact structure, and is, moreover, invariant under transverse scaling. We are interested in the set crit $\mathbf{H} \subset \mathfrak{t}^+$ of critical points of $\mathbf{H}$. Since $\mathbf{H}$ is invariant under transverse scaling we can restrict crit $\mathbf{H}$ to an appropriate slice $\Sigma \subset \mathfrak{t}^+$ if desired. It is easy to check the following behavior under transverse scaling.

Lemma 6.3.7 *The following relations hold under the transverse scaling operation $\xi \mapsto a^{-1}\xi$:*

(1) $s^T_{a^{-1}\xi} = a^{-1}s^T_\xi$;

(2) $\bar{s}^T_{a^{-1}\xi} = a^{-1}\bar{s}^T_\xi$;

(3) $\mathbf{S}_{a^{-1}\xi} = a^n \mathbf{S}_\xi$;

(4) $\mathbf{V}_{a^{-1}\xi} = a^{n+1}\mathbf{V}_\xi$;

(5) $\mathbf{H}_{a^{-1}\xi} = \mathbf{H}_\xi$;

(6) $\mathbf{F}_{a^{-1}\xi} = a^{n+1}\mathbf{F}_\xi$;

(7) $\chi_{a^{-1}\xi} = a^{-2}\chi_\xi$;

(8) $\langle \zeta, \zeta' \rangle_{a^{-1}\xi} = a^{n+3}\langle \zeta, \zeta' \rangle_\xi$.

By Lemma 6.3.7 the zeroes and critical points of $\mathbf{V}_\xi, \mathbf{S}_\xi, \mathbf{F}_\xi$ come in rays. We let Z^+ denote the zero set in $\mathfrak{t}^+(\mathcal{D}, J)$ of the Sasaki–Futaki invariant $\mathbf{F}$. Likewise we let $Z \subset \mathfrak{R}(\mathcal{D}, J)$ denote the zero set of rays of the Sasaki–Futaki invariant $\mathbf{F}$. We denote by $\mathfrak{r}_\xi$ the ray through $\xi \in \mathfrak{t}^+(\mathcal{D}, J)$. From Definition 6.3.2 and Lemma 6.3.3 we have

Proposition 6.3.8 *A Reeb vector field $\xi \in \mathfrak{t}^+(\mathcal{D}, J)$ lies in Z^+ if and only if $\pi_g s^T_g = \bar{s}^T_g$. Moreover, $\xi \in Z^+$ is extremal if and only if it is cscS.*

We are interested in the underlying structure of Z and Z^+.

Proposition 6.3.9 *Z and Z^+ are real affine algebraic varieties, and Z^+ is the cone over Z. Hence, the number of connected components of Z is finite, and Z is a compact subset of $\mathfrak{R}(\mathcal{D}, J)$.*

Proof From [BHLTF17, Lem. 3.1] we have

$$d\mathbf{H}_\xi(a) = n(n+1)\bar{s}^n_\xi \mathbf{F}_\xi(\Phi(a)). \qquad (6.20)$$

We first consider the case $\bar{s}^n_\xi \neq 0$ and let $\mathfrak{r}_{\xi_o} \in Z$ be its corresponding ray. Moreover, since $s^T_o \neq 0$ there is a neighborhood U_o of ξ_o such that equation (6.20) holds in U_o with $\mathbf{S}_\xi \neq 0$ for all $\xi \in U_o$. But from [BHL18] we know that both the total transverse scalar curvature $\mathbf{S}$ and the Einstein–Hilbert functional $\mathbf{H}$ are rational functions of ξ. It follows from (6.20) that the Sasaki–Futaki invariant $\mathbf{F}$ is a rational function of ξ on U_o. So its zero set is a real algebraic variety [BCR98]. Now consider the case $\mathbf{S}_{\xi_o} = 0$. The second statement of [BHLTF17, Lem. 3.1] says that when $\mathbf{S}_\xi = 0$ the following holds:

$$d\mathbf{S}_\xi = n\mathbf{F}_\xi \circ \Phi.$$

Thus, $\mathfrak{r}_{\xi_o} \in Z$ if and only if $\mathfrak{r}_{\xi_o}$ is a critical ray of $\mathbf{S}_{\xi_o}$, which is a rational function of ξ_o. So the result follows as above. The compactness of Z follows since $\mathbf{H}$ is a proper function [BHL18]. $\qquad \square$

Remark 6.3.10 Note that if $s^T_{\xi_o} = 0$ then it has $s^T_{a^{-1}\xi_o} = 0$ along the entire ray $\mathfrak{r}_{\xi_o}$ and it is the unique ray in $\mathfrak{t}^+(\mathcal{D}, J)$ with this property by [BHL18].

Since real algebraic varieties are CW complexes, Proposition 6.3.9 implies

Corollary 6.3.11 *Z is locally path connected.*

But we know that a ray $\mathfrak{r}_\xi \in Z$ need not be extremal. Identifying the tangent space of $\mathfrak{t}^+$ at $\xi \in \mathfrak{t}^+$ with the Lie algebra $\mathfrak{t}$ itself, we show here that the gradient vector field grad $\mathbf{H}$ viewed as an element of $\mathfrak{t}$ is proportional to the Futaki–Mabuchi vector field χ_ξ. Specifically we have the following corollary of [BHLTF17, Lem. 3.1]:

Theorem 6.3.12 *For each $\xi \in \mathfrak{t}^+$ the vector field grad $\mathbf{H}_\xi$ satisfies*

(1) grad $\mathbf{H}_\xi = n(n+1)\bar{\mathbf{s}}^n_\xi \chi_\xi$;
(2) $\langle$grad $\mathbf{H}_\xi, \xi\rangle_\xi = 0$;
(3) ξ *is a critical point of* $\mathbf{H}_\xi$ *if and only if* grad $\mathbf{H}_\xi = 0$;
(4) *if* $\mathbf{S}_\xi \neq 0$ *then* ξ *is a critical point of* $\mathbf{H}_\xi$ *if and only if* $\pi_g s^T_g = \bar{\mathbf{s}}^T_g$;
(5) *if* $\mathbf{S}_\xi \neq 0$ *then* ξ *is a critical point of* $\mathbf{H}_\xi$ *if and only if* $\chi_\xi = 0$. *Moreover, in this case* χ_ξ *is a rational function of* ξ.

Proof Taking the dual of equation (6.20) with respect to the Futaki–Mabuchi inner product $\langle \cdot, \cdot \rangle_\xi$ on $\mathfrak{t}$ and using Definition 6.3.2 gives (1), from which (2) follows, and (3) follows by duality. The last two statements follows from (1) and the results of [BHL18]. □

6.3.3 The Sasaki Energy Functional

In [BGS09], a functional, the L^2 norm of the projection $\pi_g s_g$, which provides a lower bound to the Calabi functional (6.8), namely

$$\int_M s^2_g dv_\xi \geq \int_M (\pi_g s_g)^2 dv_g =: \mathcal{SE}_2(\xi), \tag{6.21}$$

was studied. However, this functional does not behave well under transverse scaling, which is desirable when varying in the Sasaki cone. Thus, we consider a related functional

$$\mathcal{SE}^T_2(g) = \int_M (\pi_g s^T_g)^2 dv_g \tag{6.22}$$

which gives a lower bound to the transverse Calabi energy functional (6.10), namely

$$\mathcal{E}_2^T(g) = \int_M (s_g^T)^2 dv_\xi \geq \int_M (\pi_g s_g^T)^2 dv_g, \tag{6.23}$$

as we vary through elements in the Sasaki cone $\mathfrak{t}^+(\mathcal{D}, J)$ with a fixed volume. However, as with the Einstein–Hilbert functional it is convenient to normalize and consider

$$\mathcal{SE}(\xi) := \mathcal{SE}^T(\xi) = \frac{(\int_M (\pi_g s_g^T)^2 dv_g)^{n+1}}{(\int_M dv_g)^{n-1}}, \tag{6.24}$$

which is homogeneous with respect to transverse scaling, i.e. $\mathcal{SE}(a^{-1}\xi) = \mathcal{SE}(\xi)$. So the critical points of $\mathcal{SE}$ are manifestly rays $\mathfrak{r}_\xi \in \mathfrak{R}(\mathcal{D}, J)$. Following [Sim00] for the Kähler case, we call $\mathcal{SE}(\xi)$ the *Sasaki energy functional*.

From the fact that $\chi_\xi \in \mathcal{H}_0^\xi$ we have the equality

$$\int_M \pi_g s_g^T dv_g = \int_M s_g^T dv_g = \mathbf{S}_\xi, \tag{6.25}$$

which suggests a strong relation between the Einstein–Hilbert functional (6.19) and the Sasaki energy functional (6.24).

We now consider the variation of the functionals (6.24) and (6.22). Generally we could consider a path of contact forms

$$\eta_t = \frac{1}{\eta(\xi_t)}\eta + d^c\varphi_t, \tag{6.26}$$

where φ_t is a ξ_t-basic function; however, it is enough to take the variation to lie within a fixed contact CR structure $(\mathcal{D}, J)$ by choosing φ_t to be a constant. We are mainly interested in the scale invariant functional $\mathcal{SE}$, although it is easier to work with the functional $\mathcal{SE}_2^T$. However, in order to obtain rays as critical points of $\mathcal{SE}_2^T$ we need to choose a slice that intersects each ray once. However, from the definition of $\mathcal{SE}$, equation (6.24), we have

$$\frac{d\mathcal{SE}(\xi_t)}{dt} = (n+1)\frac{\mathcal{SE}_2^T(\xi_t)^n}{\mathbf{V}(\xi_t)^{n-1}}\frac{d\mathcal{SE}_2^T(\xi_t)}{dt} - (n-1)\frac{\mathcal{SE}_2^T(\xi_t)^{n+1}}{\mathbf{V}(\xi_t)^n}\frac{d\mathbf{V}(\xi_t)}{dt}. \tag{6.27}$$

Thus, if we choose variations of $\mathcal{SE}_2^T$ with fixed volume, the critical points of $\mathcal{SE}$ and $\mathcal{SE}_2^T$ are essentially the same. Indeed, we have

Lemma 6.3.13 *Under variations of fixed volume a critical point of $\mathcal{SE}_2^T$ is a critical point of $\mathcal{SE}$. Conversely, if s_g^T is not identically zero,* *a critical point of $\mathcal{SE}$ is a critical point of $\mathcal{SE}_2^T$ under variations of fixed volume.*

Remark 6.3.14 Note that

$$\frac{d\mathbf{V}(\xi_t)}{dt} = -(n+1)\int_M \eta(\dot{\xi})dv_g, \qquad (6.28)$$

so the fixed volume constraint is realized by the equation

$$\int_M \eta(\dot{\xi})dv_g = 0. \qquad (6.29)$$

We shall often make use of the following lemma.

Lemma 6.3.15 *For any $f \in C^\infty(M)^\xi$, $\pi_g f$ is the unique element A in $\mathcal{H}^\xi$ such that*

$$\langle A, h \rangle = \langle f, h \rangle$$

for all $h \in \mathcal{H}^\xi$.

Next we give the Euler–Lagrange equations for both functionals $\mathcal{SE}_2^T$ and $\mathcal{SE}$.

Theorem 6.3.16 *For $t \in (-\varepsilon, \varepsilon)$ let ξ_t be a C^1 path of Reeb vector fields compatible with a fixed CR structure $(\mathcal{D}, J)$, then we have*

$$\frac{d\mathcal{SE}(\xi_t)}{dt}\Big|_{t=0} = (n+1)\frac{\mathcal{SE}_2^T(\xi)^n}{\mathbf{V}(\xi)^{n-1}}\int_M F(\xi)\eta(\dot{\xi})dv_g, \qquad (6.30)$$

where $F(\xi)$ is given by

$$F(\xi) = -2ns_g^T \pi_g s_g^T - 2(2n+1)\Delta(\pi_g s_g^T) + (n+1)(\pi_g s_g^T)^2 + (n-1)\frac{\mathcal{SE}_2^T(\xi)}{\mathbf{V}(\xi)}.$$

So the Euler–Lagrange equations of $\mathcal{SE}$ are

$$\pi_g\left(-2ns_g^T \pi_g s_g^T - 2(2n+1)\Delta(\pi_g s_g^T) + (n+1)(\pi_g s_g^T)^2 + (n-1)\frac{\mathcal{SE}_2^T(\xi)}{\mathbf{V}(\xi)}\right) = 0. \qquad (6.31)$$

Remark 6.3.17 Note that the formula for the variation of $\mathcal{SE}_2^T$ is given by (6.30) without the multiplicative factor and the last term in $F(\xi)$, namely

$$\frac{d\mathcal{SE}_2^T(\xi_t)}{dt}\Big|_{t=0} = \int_M \left(-2ns_g^T \pi_g s_g^T - 2(2n+1)\Delta(\pi_g s_g^T) + (n+1)(\pi_g s_g^T)^2\right)\eta(\dot{\xi})dv_g. \qquad (6.32)$$

So the Euler–Lagrange equations for variations of $\mathcal{SE}_2^T$ with fixed volume are

$$\pi_g\left(-2ns_g^T\pi_gs_g^T - 2(2n+1)\Delta(\pi_gs_g^T) + (n+1)(\pi_gs_g^T)^2\right) = c, \quad (6.33)$$

where c is any constant.

In Examples 6.4.2 and 6.4.3 below we give critical points of $\mathcal{SE}$, hence, solutions of (6.31) that are not cscS. Moreover, those in Example 6.4.3 consist of one cscS ray and two extremal rays that are not cscS. So in this case we have two solutions of (6.31) such that $\pi_gs_g^T = s_g^T$ which are not constant.

For the proof of Theorem 6.3.16 we first give some lemmas, the first of which was given in [BGS09] as well as [BHLTF17].

Lemma 6.3.18 *For variations over a C^1 path of the form (6.26) with a fixed contact CR structure we have*

$$\dot{s_t^T} = -(2n+1)\Delta(\eta(\dot{\xi})) + s^T\eta(\dot{\xi}). \quad (6.34)$$

Next we have

Lemma 6.3.19 *Consider a C^1 path of Reeb vector fields ξ_t, $t \in (-\varepsilon, \varepsilon)$, compatible with a fixed CR structure $(\mathcal{D}, J)$. Then for any $h \in \mathcal{H}^\xi$*

$$\langle\dot{\pi_gs_g^T}, h\rangle = -(2n+1)\langle\Delta^g\eta(\dot{\xi}), h\rangle - n\langle s_g^T\eta(\dot{\xi}), h\rangle + (n+1)\langle\pi_gs_g^T\eta(\dot{\xi}), h\rangle. \quad (6.35)$$

Proof Apply Lemma 6.3.15 with $f = s_g^T$ and use the variation of the volume form

$$\left(\frac{d}{dt}dv_t\right)_{t=0} = -(n+1)\eta(\dot{\xi})dv_g \quad (6.36)$$

to give

$$\langle\dot{\pi_gs_g^T}, h\rangle = \langle\dot{s_g^T}, h\rangle - (n+1)\langle s_g^T, h\eta(\dot{\xi})\rangle + (n+1)\langle\pi_gs_g^T, h\eta(\dot{\xi})\rangle.$$

Applying Lemma 6.3.18 to this gives the claim. $\qquad\square$

Proof of Theorem 6.3.16 Using (6.27) we obtain

$$\frac{d\mathcal{SE}(\xi_t)}{dt}\Big|_{t=0} = (n+1)\frac{\mathcal{SE}_2^T(\xi)^n}{\mathbf{V}(\xi)^{n-1}}\int_M\left(2\pi_gs_g^T(\pi_gs_g^T) - (n+1)(\pi_gs_g^T)^2\right)\eta(\dot{\xi})dv_g$$

$$+ (n+1)(n-1)\frac{\mathcal{SE}_2^T(\xi)^{n+1}}{\mathbf{V}(\xi)^n}\int_M\eta(\dot{\xi}). \quad (6.37)$$

Putting $h = \pi_g s_g^T$ in (6.35), the first integral becomes

$$-2(2n+1)\int_M (\pi_g s_g^T)\Delta\eta(\dot\xi)dv_g - 2n\int_M s_g^T(\pi_g s_g^T)\eta(\dot\xi)dv_g$$
$$+(n+1)\int_M (\pi_g s_g^T)^2\eta(\dot\xi)dv_g.$$

Integrating the first term by parts twice and rearranging (6.37) gives (6.30). Then, since $\eta(\dot\xi)$ is an arbitrary element of $\mathcal{H}^\xi$, the Euler–Lagrange equations (6.31) follow from item (6.30) by applying Lemma 6.3.15. $\square$

Proposition 6.3.20 *Let $\mathcal{S}$ be a Sasakian structure. Then*

(1) any ray $\mathfrak{r}_\xi \in Z$ is a critical point of $\mathcal{SE}$;
(2) if $\mathbf{S}_\xi \neq 0$ a critical point of $\mathbf{H}$ is a critical point of $\mathcal{SE}$.

Proof We easily see that $\pi_g s_g^T = \bar{\mathbf{s}}_g^T$ is a solution of (6.31), proving (1). Item (2) then follows from item (4) of Theorem 6.3.12. $\square$

Remark 6.3.21 By Proposition 6.3.8 a Reeb field ξ is in Z^+ if and only if $\pi_g s_g^T = \bar{\mathbf{s}}_g^T$, and we see that this is a solution to the Euler–Lagrange equations (6.31); hence, it is a critical point of $\mathcal{SE}_2^T$ under variations of fixed volume.

Proposition 6.3.22 *Non-extremal critical points of $\mathcal{SE}$ exist.*

Proof The critical points satisfying $\pi_g s_g^T = \bar{\mathbf{s}}_g^T$ need not be extremal. Indeed when $c_1(\mathcal{D}) = 0$ (or a torsion class) the results of Gauntlett, Martelli, Sparks and Yau [GMSY07, MSY08] (cf. [BG08, Thm. 11.3.14]) give critical points that attain an absolute minimum of $\mathbf{H}_\xi$ with $\mathbf{S}_\xi > 0$ that do not have a Sasaki metric of constant scalar curvature. More generally, the results of [BHL18] show that a global minimum of $\mathbf{H}_\xi$ is attained whether ξ is extremal or not, and these are realized when $\pi_g s_g^T = \bar{\mathbf{s}}_g^T$. $\square$

Remark 6.3.23 There are many explicit examples of non-extremal critical points in Proposition 6.3.22 which include homotopy spheres. See for example the Tables in [BvC18]. For all of these examples there are no extremal Sasaki metrics in the entire Sasaki cone.

Thus, we have the following theorem.

Theorem 6.3.24 *Consider the family of Sasakian structures $\mathfrak{t}^+(\mathcal{D}, J)$ and let $\xi \in \mathfrak{t}^+(\mathcal{D}, J)$ have constant scalar curvature $s_g^T \neq 0$. Suppose* *also that there is no eigenfunction of the Laplacian Δ in $\mathcal{K}^\xi$ with eigen-value $\frac{s_g^T}{2n+1}$. Then its ray is isolated in the space of rays with vanishing transversal Futaki invariant.*

Proof Recall that a Sasaki structure has vanishing transversal Futaki invariant if and only if $\pi_g s_g^T$ is a constant. The map $\Pi : \mathfrak{t}^+(\mathcal{D}, J) \longrightarrow C^\infty(M)^{\mathbb{T}}$ defined by

$$\Pi(\xi) = \pi_g s_g^T$$

is a homogeneous map of degree 1 defined on $\mathfrak{t}^+(\mathcal{D}, J)$, a convex open subset of a finite dimensional affine space. Identifying the tangent space of $\mathfrak{t}^+(\mathcal{D}, J)$ with the Lie algebra $\mathfrak{t}$ and the tangent space of the Fréchet manifold $C^\infty(M)^{\mathbb{T}}$ with itself, we see that if the differential $d_\xi\Pi : \mathfrak{t} \to C^\infty(M)^{\mathbb{T}}$ is injective at a point ξ, the constant rank theorem implies that there exists an open neighborhood U of ξ such that the restriction $\Pi : U \to C^\infty(M)^{\mathbb{T}}$ is injective. Assume, moreover, that $\Pi(\xi)$ is a constant, then the pre-image of the constants in $\Pi(U)$ coincides with one isolated ray in U. We show that $d_\xi\Pi$ is injective when ξ is cscS. Now, from Lemma 6.3.19 we have

$$d_\xi\Pi(\eta(\dot{\xi})) = \frac{d}{dt}\left(\pi_{g_{\xi_t}} s_{g_{\xi_t}}^T\right)_{|t=0} = -(2n+1)\Delta^g\eta(\dot{\xi}) - n s_g^T\eta(\dot{\xi}) + (n+1)\pi_g s_g^T\eta(\dot{\xi}).$$

$$(6.38)$$

So if s_g^T is constant (6.38) becomes

$$d_\xi\Pi(\eta(\dot{\xi})) = -(2n+1)\Delta^g\eta(\dot{\xi}) + s_g^T\eta(\dot{\xi}).$$

Therefore, if $\dot{\xi}$ is colinear to ξ, the assumption $s_g^T \neq 0$ implies that $d_\xi\Pi(\eta(\dot{\xi})) \neq 0$. Otherwise, $d_\xi\Pi(\eta(\dot{\xi})) = 0$ gives that $\eta(\dot{\xi}) \in \mathcal{H}^\xi$ is an eigenfunction of Δ^g with eigenvalue $s_g^T/(2n+1)$. That is, the hypothesis of the theorem guarantees that $d_\xi\Pi : \mathfrak{t} \to C^\infty(M)^{\mathbb{T}}$ is injective. This concludes the proof. □

Remark 6.3.25 In the Kähler–Einstein case, the space of Killing potentials coincides with the eigenspace of the first (non-trivial) eigenvalue λ_1^g of the Laplacian thanks to results of Matsushima [Mat57] (see [Gau10, Thm. 3.6.2]) which also provide a lower bound (which would read in our notation $s_g^T/(2n)$) on the eigenvalues of the Laplacian. There is an analogous result in the Sasaki case and this is exactly what we used (i.e $\lambda_1^g \geq s_g^T/2n > s_g^T/(2n+1)$) in the proof of [BHLTF17, Thm. 1.7] to get a local convexity result in the Sasaki–η-Einstein case. In the toric Kähler–Einstein case, the fact that torus invariant Killing potentials (i.e the affine linear functions on the moment polytope) are eigenfunctions of the same eigenvalue even characterizes Kähler–Einstein metrics; see [LSD18, Prop. 1.4]. Therefore, it would be surprising that there would be no constant scalar curvature Kähler metrics having a Killing potential as eigenfunction of the Laplacian; what we can hope for, however, is that if it does the eigenvalue is not as low as $s_g^T/(2n+1)$.

Proposition 6.3.9 and Theorem 6.3.24 imply

Corollary 6.3.26 *Suppose that every cscS metric in the family* $\mathfrak{t}^+(\mathcal{D}, J)$ *satisfies the hypothesis of Theorem 6.3.24. Then the zero set of rays Z of the Sasaki–Futaki invariant is finite. In particular, the number of cscS rays in* $\mathfrak{t}^+(\mathcal{D}, J)$ *is finite.*

6.4 The Functionals H, $\mathcal{SE}$ on Lens Space Bundles Over Riemann Surfaces

In this section we study the extremal Sasakian structures on lens space bundles over Riemann surfaces of genus $\mathcal{G}$ from the point of view of the functionals $\mathbf{H}(b)$ and $\mathcal{SE}(b)$. This is a special case of what we have called an $S_{\mathbf{w}}^3$ join [BTF14, BTF16], which generally represents lens space bundles over a Hodge manifold written as $M \star_{\mathbf{l}} S_{\mathbf{w}}^3$, where $\mathbf{w} = (w_1, w_2)$, $\mathbf{l} = (l_1, l_2)$ and the components of both are relatively prime positive integers. In this case, since the critical points all belong to a two-dimensional subcone $\mathfrak{t}_{\mathbf{w}}^+$ of $\mathfrak{t}^+(\mathcal{D}, J)$, the critical rays are all isolated in the subcone $\mathfrak{t}_{\mathbf{w}}^+$. So Problem 6.1.1 is answered in the affirmative for the $\mathbf{w}$ cone of these Sasaki manifolds. Moreover, [BTF16, Thm. 1.1] says that Problem 6.1.2 is also answered in the affirmative in this case.

For ease of discussion, we take M to be the constant scalar curvature Sasaki structure on an S^1 bundle over a Riemann surface $\Sigma_{\mathcal{G}}$ with its standard Fubini–Study metric. The functional $\mathbf{H}$ for these manifolds was studied in [BHLTF17]. Choosing a Reeb field in $\mathfrak{t}_{\mathbf{w}}^+$ with coordinates (v_1, v_2) gives a ray $\mathfrak{r}_b \in \mathfrak{R}(\mathcal{D}, J)$, where $b = v_2/v_1$. So from [BHLTF17] we have

$$\mathbf{H}(b) = \frac{\left(b^2 l_1 w_1 + 2bl_2(1 - \mathcal{G}) + l_1 w_2\right)^3}{b^2(bw_1 + w_2)^2}. \tag{6.39}$$

It follows from [BHLTF17] that the critical points of $H(b)$ correspond to the points where either $H(b) = 0$ or $F(b) = 0$, where

$$F(b) = b^3 l_1 w_1^2 + b^2 (\mathcal{G} l_2 w_1 + 2 l_1 w_1 w_2 - l_2 w_1)$$
$$- b(\mathcal{G} l_2 w_2 + 2 l_1 w_1 w_2 - l_2 w_2) - l_1 w_2^2. \qquad (6.40)$$

Note that F is just the Sasaki–Futaki invariant up to a constant multiple. So the zero set of F is Z, which consists precisely of those Sasaki metrics in the two-dimensional subcone $\mathfrak{t}_{\mathbf{w}}^+ \subset \mathfrak{t}^+(\mathcal{D}, J)$ that satisfy $\pi_g s_g^T = \bar{s}_g^T$. Indeed, in this case the vanishing of F for a given ray guarantees the existence of an admissible csc metric.[4] Note that for $\mathcal{G} > 0$ we have the equality $\mathfrak{t}_{\mathbf{w}}^+ = \mathfrak{t}^+(\mathcal{D}, J)$. We also mention that it can be seen from equation (6.39) that the zeroes of $\mathbf{H}_\xi$ (also $\mathbf{S}_\xi$) for which the transverse scalar curvature s_g^T does not vanish identically are all inflection points in this case.

We now consider the Sasaki energy functional $\mathcal{SE}$. A straightforward computation using results from [ACGTF08, BTF14, BTF16, BHLTF17] gives

$$\mathcal{SE}(b) = \frac{(g_1(b))^3}{b^4 (w_1 b + w_2) (b^2 w_1^2 + 4 b w_1 w_2 + w_2^2)^3}, \qquad (6.41)$$

where

$$g_1(b) = b^5 l_1^2 w_1^3 + 3 b^4 l_1^2 w_1^2 w_2 + 3 b l_1^2 w_1 w_2^2 + l_1^2 w_2^3$$
$$+ b^3 w_1 \left(l_2^2 (\mathcal{G} - 1)^2 + 2(1 - \mathcal{G}) l_1 l_2 w_2 - l_1^2 w_1 w_2 \right)$$
$$+ 2 b^2 w_2 \left(l_2^2 (\mathcal{G} - 1)^2 + 2(1 - \mathcal{G}) l_1 l_2 w_1 - l_1^2 w_1 w_2 \right). \qquad (6.42)$$

Now we observe that $\lim_{b \to 0} \mathcal{SE}(b) = \lim_{b \to +\infty} \mathcal{SE}(b) = +\infty$, and the derivative equals

$$\mathcal{SE}'(b) = \frac{4 F(b) (g_1(b))^2 g_2(b)}{b^5 (b w_1 + w_2)^2 (b^2 w_1^2 + 4 b w_1 w_2 + w_2^2)^4},$$

where $F(b)$ is given by equation (6.40) and $g_2(b)$ is given by

$$g_2(b) = b^5 l_1 w_1^4 + b^4 (-\mathcal{G} l_2 w_1^3 + 7 l_1 w_1^3 w_2 + l_2 w_1^3)$$
$$+ b^3 (-2 \mathcal{G} l_2 w_1^2 w_2 + 3 l_1 w_1^3 w_2 + 10 l_1 w_1^2 w_2^2 + 2 l_2 w_1^2 w_2)$$
$$+ b^2 (-2 \mathcal{G} l_2 w_1 w_2^2 + 10 l_1 w_1^2 w_2^2 + 3 l_1 w_1 w_2^3 + 2 l_2 w_1 w_2^2)$$
$$+ b(-\mathcal{G} l_2 w_2^3 + 7 l_1 w_1 w_2^3 + l_2 w_2^3) + l_1 w_2^4.$$

---

[4] Without going into details here, this follows essentially from [ACGTF08, Sect. 2.4] together with the discussion in [BTF16, Sect. 5.1]

Clearly, $g_1(b) = 0$ is equivalent to $\mathcal{SE}(b) = 0$, which in turn corresponds to $\pi_g s_g^T$ being constantly zero. In the admissible case at hand this implies that the ray has a csc admissible metric of vanishing constant transverse scalar curvature. In particular, $F'(b)$ would vanish as well. We also mention that if $\pi_g s^T$ is constantly zero, then we must have $\mathcal{G} > 1$. Thus, critical points of $\mathcal{SE}$, other than the ones coming from csc rays, correspond to solutions to $g_2(b) = 0$ for $b > 0$ with $b \neq w_2/w_1$. Note that for $\mathcal{G} \leq 1$ there are no such solutions, but that for $\mathcal{G} \geq 2$ and l_2 sufficiently large we do indeed get solutions to $g_2(b) = 0$ with $b > 0$. Moreover, as the examples below will show, these extra critical points may or may not be extremal, and in general they do not arise in the same way as in the case for $\mathbf{H}(b)$. Note also that if $\mathcal{SE}(b)$ has precisely one critical point, then, due to the limit behavior, this has to be an absolute minimum of $\mathcal{SE}(b)$.

6.4.1 Explicit Examples

Next we give examples explicitly describing the critical points of $\mathcal{SE}$ and $\mathbf{H}$ and their relationship. We know that the critical sets crit of both functionals contain the zero set Z of the Sasaki–Futaki invariant $\mathbf{F}$; however, generally crit($\mathbf{H}$) and crit($\mathcal{SE}$) are different.

Example 6.4.1   Here we take $\mathbf{l} = (1,1)$ and $\mathbf{w} = (w_1, w_2) = (3, 2)$ so that $M = M_3 \star_{1,1} S_{3,2}^3$ is an S^3 bundle over a Riemann surface $\Sigma_{\mathcal{G}}$. One can check that for $\mathcal{G} \leq 3$, both $\mathbf{H}(b)$ and $\mathcal{SE}(b)$ have only one critical point (a global minimum), located at the b-value corresponding to the csc ray. For $\mathcal{G} \geq 4$,

$$\mathbf{H}(b) = \frac{\left(3b^2 - 2(\mathcal{G}-1)b + 2\right)^3}{b^2(3b+2)^2}$$

has three distinct critical points. However, by using Descartes' rule of signs (giving the maximum possible number of positive real roots) on $g_2(b)$ for $\mathcal{G} \leq 15$, supplemented by a manual check for $\mathcal{G} = 16, 17$, we see that $\mathcal{SE}(b)$ is given by

$$\frac{\left(27b^5 + 54b^4 + 6b^3\mathcal{G}^2 - 36b^3\mathcal{G} - 6b^3 + 4b^2\mathcal{G}^2 - 32b^2\mathcal{G} + 4b^2 + 36b + 8\right)^3}{b^4(3b+2)\left(9b^2 + 24b + 4\right)^3}$$

has only one critical point for $\mathcal{G} = 0, 1, \ldots, 17$. For $\mathcal{G} \geq 18$ it can be checked that $g_2(b)$ has two distinct positive real roots, neither of which corresponds to the unique csc ray. These zeroes are outside of the extremal range.

As a more specific example within this example, let us suppose that $\mathcal{G} = 4$. One may check that in this case every ray in the Sasaki cone has an admissible extremal Sasaki metric. The three critical points of $\mathbf{H}(b)$ are the inflection points at $b = \frac{1}{3}\left(3 - \sqrt{3}\right)$ and $b = \frac{1}{3}\left(\sqrt{3} + 3\right)$, plus the location of the global minimum, $b \approx 0.81$, corresponding to the csc ray. The last value is then also the location of the global minimum (and only extremum point) of $\mathcal{SE}$.

Example 6.4.2   The Sasaki manifold M is a lens space bundle over a genus 2 Riemann surface Σ_2 which is represented as a Sasaki join $M = M_3 \star_{1,101} S^3_{3,2}$, where M_3 is the constant sectional curvature -1 Sasaki structure on the primitive S^1 bundle over Σ_2. We refer to [BTF14] for this join construction and to [BTF16, BHLTF18] for the general description of Sasaki joins. The Einstein–Hilbert functional for M is treated in [BHLTF17, Exam. 5.8]. Now, the Sasaki cone $\mathfrak{t}^+(\mathcal{D}, J)$ of M is two dimensional, represented by the first quadrant $v_1 > 0, v_2 > 0$. Then setting $b = \frac{v_2}{v_1}$ in [BHLTF17] we showed that $\mathbf{H}(b)$ has three critical points located at $b \approx 0.099, 0.685, 67.3$. Moreover, the range of admissible extremal[5] structures in the Sasaki cone is the open interval (b_1, b_2) with $b_1 \approx 0.295$ and $b_2 \approx 1.455$. Only one of the critical points, $b \approx 0.685$, lies in this range and it is a local minimum with constant scalar curvature. The two remaining critical points lie outside of the admissible extremal range, and they are inflection points corresponding to $\mathbf{S}_b = 0$.

Now consider the Sasaki energy functional

$$\mathcal{SE}(b) = \frac{\left(27b^5 + 54b^4 + 58746b^3 + 38356b^2 + 36b + 8\right)^3}{b^4(3b + 2)\left(9b^2 + 24b + 4\right)^3}. \tag{6.43}$$

This also has three critical points, $b \approx 0.023, 0.685, 30.3$. Note also that the functional $\mathcal{SE}_2^T$ has the same critical points. However, we now find that the cscS metric $b \approx 0.685$ represents a local maximum of $\mathcal{SE}(b)$, while both $b \approx 0.023$ and $b \approx 30.3$ lie outside of the admissible extremal range and both represent local minima, with the latter being an absolute minimum.

Notice that, although $\mathbf{H}$ and $\mathcal{SE}$ have the same number of critical points, two of them are inflection points of $\mathbf{H}$ with $\mathbf{S}_b = 0$ and s_g^T is not identically zero. They are not critical points of $\mathcal{SE}$; nevertheless, $\mathcal{SE}$ has two critical points that are not critical points of $\mathbf{H}$.

---

[5] Whether there exist extremal Sasaki metrics on M in the same class $(\mathcal{S}, \bar{J})$ that are not admissible is an open question at this time, although there are expected to be none.

Example 6.4.3   As a variant of Example 6.4.2, we can calculate that, for $\mathcal{G} = 2$, $l_1 = 1$, $l_2 = 19$, $w_1 = 3$ and $w_2 = 2$,

$$\mathcal{SE}(b) = \frac{\left(27b^5 + 54b^4 + 1674b^3 + 964b^2 + 36b + 8\right)^3}{b^4(3b+2)\left(9b^2 + 24b + 4\right)^3}.$$

Numeric computer calculations indicate that here $b \approx 0.4466$ and $b \approx 2.497$ are relative minima (with the latter being the absolute minimum), while $b \approx 0.7335$ is a relative maximum corresponding to the csc ray. We can also numerically check that here the range of admissible extremal structures in the Sasaki cone is the open interval (b_1, b_2) with $b_1 \approx 0.0472$ and $b_2 \approx 5.93$. Thus the critical points of $\mathcal{SE}$ all correspond to admissible extremal rays. Comparatively, $b \approx 0.7335$ is the location of a local (global) minimum of $\mathbf{H}(b) = \frac{(3b^2 - 38b + 2)^3}{b^2(3b+2)^2}$, and this function has inflection points at $b \approx 0.05285$ and $b \approx 12.61$ corresponding to $\mathbf{S}_b = 0$. Again, as in Example 6.4.2, $\mathbf{H}$ and $\mathcal{SE}$ have the same number of critical points, but two of them play distinct roles in the two functionals. Moreover, as mentioned in Remark 6.3.17, the critical points of $\mathcal{SE}$ are all extremal.

Example 6.4.4   We consider a similar lens space bundle but now over a Riemann surface of genus $\mathcal{G} = 0$, in which case M is an S^3 bundle over S^2. As before, we put $l_1 = 1$, $l_2 = 101$, $w_1 = 3$ and $w_2 = 2$, which implies that M is the non-trivial S^3 bundle over S^2. In this case the admissible extremal range is the entire first quadrant $v_1 > 0, v_2 > 0$ which is the so-called $\mathbf{w}$ subcone $\mathfrak{t}_{\mathbf{w}}^+$ of the three-dimensional Sasaki cone $\mathfrak{t}^+(\mathcal{D}, J)$. Restricted to $\mathfrak{t}_{\mathbf{w}}^+$, both $\mathbf{H}$ and $\mathcal{SE}$ have precisely three critical points. For $\mathcal{SE}$ we have

$$\mathcal{SE}(b) = \frac{(8 + 36b + 43204b^2 + 63594b^3 + 54b^4 + 27b^5)^3}{(b^4(2 + 3b)(4 + 24b + 9b^2)^3)}.$$

The b values corresponding to csc rays are the only critical points of $\mathcal{SE}$. In this case we get multiple cscS rays corresponding to $b \approx 0.022$, $b \approx 0.644$ and $b \approx 31.67$. The value in the middle corresponds to a local maximum whereas the other two values are locations of relative minima. Note that the existence of multiple cscS rays were first discovered in [Leg11] for precisely these types of Sasaki manifolds. One can check that

$$\mathbf{H}(b) = \frac{(2 + 202b + 3b^2)^3}{b^2(3b + 2)}$$

and that $\mathbf{S}_b > 0$, so all critical points of $\mathbf{H}(b)$ are cscS rays. Thus, in this case the critical points of $\mathbf{H}$ and $\mathcal{SE}$ coincide when restricted to the

$\mathbf{w}$ subcone of $\mathfrak{t}^+(\mathcal{D}, J)$. The numerators of the differentials of $\mathbf{H}$ and $\mathcal{SE}$ have common factors.

References

[ACG06] Vestislav Apostolov, David M. J. Calderbank, and Paul Gauduchon, *Hamiltonian 2-forms in Kähler geometry. I. General theory*, J. Differ. Geom. **73** (2006), no. 3, 359–412.

[ACGTF04] Vestislav Apostolov, David M. J. Calderbank, Paul Gauduchon, and Christina. W. Tønnesen-Friedman, *Hamiltonian 2-forms in Kähler geometry. II. Global classification*, J. Differ. Geom. **68** (2004), no. 2, 277–345.

[ACGTF08] Vestislav Apostolov, David M. J. Calderbank, Paul Gauduchon, and Christina W. Tønnesen-Friedman, *Hamiltonian 2-forms in Kähler geometry. III. Extremal metrics and stability*, Invent. Math. **173** (2008), no. 3, 547–601.

[AH06] Klaus Altmann and Jürgen Hausen, *Polyhedral divisors and algebraic torus actions*, Math. Ann. **334** (2006), no. 3, 557–607.

[BCR98] Jacek Bochnak, Michel Coste, and Marie-Françoise Roy, *Real algebraic geometry*, Ergebnisse der Mathematik und ihrer Grenzgebiete (3) [Results in Mathematics and Related Areas (3)], vol. 36, Springer-Verlag, Berlin, 1998. Translated from the 1987 French original, revised by the authors.

[BCTF19] Charles P. Boyer, David M. J. Calderbank, and Christina W. Tønnesen-Friedman, *The Kähler geometry of Bott manifolds*, Adv. Math. **350** (2019), 1–62.

[BG00] Charles P. Boyer and Krzysztof Galicki, *A note on toric contact geometry*, J. Geom. Phys. **35** (2000), no. 4, 288–298.

[BG08] ———, Sasakian geometry. Oxford Mathematical Monographs, Oxford University Press, Oxford, 2008.

[BGS08] Charles P. Boyer, Krzysztof Galicki, and Santiago R. Simanca, *Canonical Sasakian metrics*, Commun. Math. Phys. **279** (2008), no. 3, 705–733.

[BGS09] ———, *The Sasaki cone and extremal Sasakian metrics*. In Krzysztof Galicki and Santiago R. Simanca (eds.), Riemannian topology and geometric structures on manifolds, Progress in Mathematics, vol. 271, Birkhäuser Boston, Boston, MA, 2009, pp. 263–290.

[BHL18] Charles Boyer, Hongnian Huang, and Eveline Legendre, *An application of the Duistermaat–Heckman theorem and its extensions in Sasaki geometry*, Geom. Topol. **22** (2018), no. 7, 4205–4234.

[BHLTF17] Charles P. Boyer, Hongnian Huang, Eveline Legendre, and Christina W. Tønnesen-Friedman, *The Einstein-Hilbert functional and the Sasaki-Futaki invariant*, Int. Math. Res. Not. **2017** (2017), no. 7, 1942–1974.

[BHLTF18] _____, *Reducibility in Sasakian geometry*, Trans. Amer. Math. Soc. **370** (2018), no. 10, 6825–6869.

[BM93] A. Banyaga and P. Molino, *Géométrie des formes de contact complètement intégrables de type toriques*. Séminaire Gaston Darboux de Géométrie et Topologie Différentielle, 1991–1992 (Montpellier), Université de Montpellier II, Montpellier, 1993, pp. 1–25.

[Boy19] Charles P. Boyer, *Contact structures of Sasaki type and their associated moduli*, Complex Manifolds **6** (2019), no. 1, 1–30.

[BTF14] Charles P. Boyer and Christina W. Tønnesen-Friedman, *Extremal Sasakian geometry on S^3-bundles over Riemann surfaces*, Int. Math. Res. Not. **2014** (2014), no. 20, 5510–5562.

[BTF16] _____, *The Sasaki join, Hamiltonian 2-forms, and constant scalar curvature*, J. Geom. Anal. **26** (2016), no. 2, 1023–1060.

[BvC18] Charles P. Boyer and Craig van Coevering, *Relative K-stability and extremal Sasaki metrics*, Math. Res. Lett. **25** (2018), no. 1, 1–19.

[Cal56] Eugenio Calabi, The space of Kähler metrics, Proc. Int. Congress of Mathematicians, Vol. 2, Amsterdam, 1954, North-Holland, Amsterdam, 1956, pp. 206–207.

[Cal82] _____, *Extremal Kähler metrics*, Seminar on differential geometry, Ann. Math. Stud., vol. 102, Princeton University Press, Princeton, NJ, 1982, pp. 259–290.

[CS18] Tristan C. Collins and Gábor Székelyhidi, *K-semistability for irregular Sasakian manifolds*, J. Differ. Geom. **109** (2018), no. 1, 81–109.

[DS14] Simon Donaldson and Song Sun, *Gromov–Hausdorff limits of Kähler manifolds and algebraic geometry*, Acta Math. **213** (2014), no. 1, 63–106.

[DS17] _____, *Gromov–Hausdorff limits of Kähler manifolds and algebraic geometry, II*, J. Differ. Geom. **107** (2017), no. 2, 327–371.

[FM95] Akito Futaki and Toshiki Mabuchi, *Bilinear forms and extremal Kähler vector fields associated with Kähler classes*, Math. Ann. **301** (1995), no. 2, 199–210.

[FM02] _____, *Moment maps and symmetric multilinear forms associated with symplectic classes*, Asian J. Math. **6** (2002), no. 2, 349–371.

[Gau10] Paul Gauduchon, *Calabi's extremal Kähler metrics*, preliminary version, 2010.

[GMSY07] Jerome P. Gauntlett, Dario Martelli, James Sparks, and Shing-Tung Yau, *Obstructions to the existence of Sasaki-Einstein metrics*, Commun. Math. Phys. **273** (2007), no. 3, 803–827.

[Leg11] Eveline Legendre, *Existence and non-uniqueness of constant scalar curvature toric Sasaki metrics*, Compos. Math. **147** (2011), no. 5, 1613–1634.

[Leg16] _____, *Toric Kähler-Einstein metrics and convex compact polytopes*, J. Geom. Anal. **26** (2016), no. 1, 399–427.

[Ler02] E. Lerman, *Contact toric manifolds*, J. Symplectic Geom. **1** (2002), no. 4, 785–828.

[Ler04] _____, *Homotopy groups of K-contact toric manifolds*, Trans. Amer. Math. Soc. **356** (2004), no. 10, 4075–4083 (electronic).

[LSD18] Eveline Legendre and Rosa Sena-Dias, *Toric aspects of the first eigenvalue*, J. Geom. Anal. **28** (2018), no. 3, 2395–2421.

[Mat57] Y. Matsushima, *Sur la structure du groupe d'homéomorphismes analytiques d'une certaine variété kählérienne*, Nagoya Math. J. **11** (1957), 145–150.

[MSY08] Dario Martelli, James Sparks, and Shing-Tung Yau, *Sasaki–Einstein manifolds and volume minimisation*, Commun. Math. Phys. **280** (2008), no. 3, 611–673.

[Sim00] Santiago R. Simanca, *Strongly extremal Kähler metrics*, Ann. Global Anal. Geom. **18** (2000), no. 1, 29–46.

7

The Prescribed Ricci Curvature Problem for Homogeneous Metrics

Timothy Buttsworth and Artem Pulemotov

Abstract

The prescribed Ricci curvature problem consists in finding a Riemannian metric g on a manifold M such that the Ricci curvature of g equals a given $(0, 2)$-tensor field T. We survey the recent progress on this problem in the case where M is a homogeneous space.

7.1 Introduction

This chapter provides a survey of the recent results on the prescribed Ricci curvature problem with a focus on homogeneous metrics. In Section 7.2, we formulate the problem and briefly review its history. In Section 7.3, we discuss the progress achieved in the framework of compact homogeneous spaces in papers [46, 47, 32, 15]. Also, we comment on applications to the analysis of the Ricci iteration. Section 7.4 raises several open questions and discusses the investigation of the problem on non-compact homogeneous spaces.

Acknowledgements AP's research was supported under Australian Research Council's Discovery Project funding scheme (DP180102185).

7.2 The Prescribed Ricci Curvature Problem

Consider a smooth manifold M and a symmetric $(0,2)$-tensor field T on M. The prescribed Ricci curvature problem consists in finding a Riemannian metric g such that

$$\text{Ric}\, g = T, \tag{7.1}$$

where Ric denotes the Ricci curvature. The investigation of this problem is an important segment of geometric analysis with strong ties to evolution equations. For example, DeTurck's work on (7.1) underlay his subsequent discovery of the famous DeTurck trick for the Ricci flow. There is kinship between (7.1) and the Einstein equation of general relativity.

Several results regarding the local solvability of (7.1) are available in the literature. The first such result appeared in DeTurck's paper [24]. It states that (7.1) has a solution in a neighbourhood of a point $o \in M$ provided that T is non-degenerate at o. Alternative proofs were given in [10, 36]. The local solvability of (7.1) was further pursued in [18, 29, 44, 45].

Even if T is non-degenerate, one may be unable to find g satisfying (7.1) on *all* of M. As DeTurck and Koiso demonstrated in [28], as long as M is compact and T is positive-definite, there is a constant $c_T > 0$ such that $c_T T$ is not the Ricci curvature of any metric; see also [9]. Sufficient conditions for the global solvability of (7.1) were obtained by several authors in a range of contexts. Namely, Hamilton in [34], DeTurck in [27], Delanoë in [22], and Delay in a series of papers (see [23] and references therein) proved the invertability of the Ricci curvature near various 'nice' Einstein metrics. Their results relied on versions of the inverse and implicit function theorems. Pina and his collaborators (see [43] and references therein) studied the solvability of (7.1) in a conformal class. Pulemotov in [44, 45] and Smith in [51] investigated boundary-value problems for (7.1). DeTurck's and Calvaruso's papers [26, 17] provide a sample of the research done in the Lorentzian setting.

Several mathematicians who investigated the prescribed Ricci curvature problem did so in the presence of symmetry. More precisely, they assumed that the metric g and the tensor field T in (7.1) were invariant under the action of a group G on the manifold M. Hamilton in [34] considered several different G with M being the 3-dimensional sphere $\mathbb{S}^3$. Cao and DeTurck in [18] examined the situation where $G = \text{SO}(d)$ and $M = \mathbb{R}^d$. Pulemotov in [44, 45] assumed G acted on M with cohomogeneity 1.

If the manifold M is closed, instead of trying to prove the existence of a metric g with Ric g equal to T, one should look for a metric g and a positive number c such that

$$\mathrm{Ric}\, g = cT. \qquad (7.2)$$

This paradigm was originally proposed by Hamilton in [34] and DeTurck in [27]. To explain it, we consider the problem of finding a metric on the 2-dimensional sphere $\mathbb{S}^2$ with prescribed positive-definite Ricci curvature T. According to the Gauss–Bonnet theorem, if such a metric exists, then the volume of $\mathbb{S}^2$ with respect to T equals 4π. It is relatively easy to prove the converse (see [25, Cor. 2.2] and also [34, Thm. 2.1]). Consequently, it is always possible to find g and c such that (7.2) holds on $\mathbb{S}^2$. In fact, in this setting, c is uniquely determined by T. Hamilton suggests in [34, Sect. 1] that the purpose of c is to compensate for the invariance of the Ricci curvature under scaling of the metric. Note that the shift of focus from (7.1) to (7.2) may be unreasonable on an open manifold or a manifold with non-empty boundary.

For rather detailed surveys of older results on the prescribed Ricci curvature problem, see [10, Chap. 5] and [8, Sect. 6.5].

7.3 Compact Homogeneous Spaces

In [46], Pulemotov initiated the systematic investigation of equation (7.2) for homogeneous metrics. Consider a compact connected Lie group G and a closed connected subgroup $H < G$. Let M be the homogeneous space G/H. It will be convenient for us to assume that the natural action of G on M is effective and that the dimension of M is at least 3. We denote by $\mathcal{M}$ the set of G-invariant Riemannian metrics on M with the manifold structure as in [42, pp. 6318–6319]. The properties of $\mathcal{M}$ are discussed in [11, Sect. 4.1] in great detail.

Let T be a G-invariant tensor field on M. Since M is compact, the prescribed Ricci curvature problem for homogeneous metrics on M consists in finding $g \in \mathcal{M}$ and $c > 0$ that satisfy equation (7.2). Our main objective in this section is to state a number of theorems concerning the existence of such g and c. The proofs of most of these theorems rely on the variational interpretation of (7.2) proposed in [46] and given by Lemma 7.3.1 below.

7.3.1 The Variational Interpretation

The scalar curvature $S(g)$ of a metric $g \in \mathcal{M}$ is constant on M. Therefore, we may interpret $S(g)$ as the result of applying a functional $S : \mathcal{M} \to \mathbb{R}$ to $g \in \mathcal{M}$. Consider the space

$$\mathcal{M}_T = \{g \in \mathcal{M} \mid \mathrm{tr}_g T = 1\}$$

with the manifold structure inherited from $\mathcal{M}$.

Lemma 7.3.1   *Given $g \in \mathcal{M}_T$, formula (7.2) holds for some $c \in \mathbb{R}$ if and only if g is a critical point of the restriction of the functional S to $\mathcal{M}_T$.*

In most cases, $S|_{\mathcal{M}_T}$ is unbounded below; see [32, Rem. 2.2]. Therefore, one typically searches for maxima, local minima and saddle points of $S|_{\mathcal{M}_T}$ rather than global minima. For a discussion of the properness of $S|_{\mathcal{M}_T}$, see [32, Rem. 2.11].

Lemma 7.3.1 parallels the well-known variational interpretation of the Einstein equation. Indeed, a metric $g \in \mathcal{M}$ satisfies

$$\mathrm{Ric}\, g = \lambda g \qquad\qquad\qquad (7.3)$$

for some $\lambda \in \mathbb{R}$ if and only if it is (up to scaling) a critical point of S on

$$\mathcal{M}_1 = \{g \in \mathcal{M} \mid M \text{ has volume 1 with respect to } g\};$$

see, e.g., [52, Sect. 1]. The restrictions $S|_{\mathcal{M}_T}$ and $S|_{\mathcal{M}_1}$ have substantially different properties. For instance, one can easily deduce from formula (7.7) below that $S|_{\mathcal{M}_T}$ is necessarily bounded above if T is positive-definite. This fact is essential to many of the results surveyed here. However, by the same token, $S|_{\mathcal{M}_1}$ is unbounded above in most situations; see [52, Thm. 2.4] and [11, Thm. 1.2].

The contrast in the properties of $S|_{\mathcal{M}_T}$ and $S|_{\mathcal{M}_1}$ is hardly surprising. Indeed, despite their apparent similarity, equations (7.2) and (7.3) are very different in nature. In particular, the former is not diffeomorphism-invariant for general T. Intuitively, one may think of the relationship between (7.2) and (7.3) as the relationship between the Poisson equation $\Delta u = f$ and the Helmholtz equation $\Delta u = u$.

It is worth noting that (7.1), as well as (7.2), admits a variational interpretation; see [34, Thm. 3.1]. However, there seem to be no existence results for (7.1) to date that arise from this variational interpretation.

7.3.2 Maximal Isotropy

If the tensor field T is negative-definite, no metric $g \in \mathcal{M}$ can satisfy (7.2) with $c > 0$. Indeed, the isometry group of (M, g) would have to be finite by Bochner's theorem (see [10, Thm. 1.84]), which is clearly impossible. If T is negative-semidefinite and (7.2) holds for some $g \in \mathcal{M}$ with $c > 0$, then G must be abelian, by the same token. In this case, every G-invariant metric on M is Ricci flat. If T has mixed signature, the techniques underlying the results in Sections 7.3.2, 7.3.5 and 7.3.6 appear to be ineffective. Thus, we will focus on the situation where T is positive-semidefinite and non-zero. In [46], Pulemotov proved the following result.

Theorem 7.3.2 *Suppose H is a maximal connected Lie subgroup of G. If T is positive-semidefinite and non-zero, then there exists a Riemannian metric $g \in \mathcal{M}_T$ such that $S(g) \geq S(g')$ for all $g' \in \mathcal{M}_T$. The Ricci curvature of g coincides with cT for some $c > 0$.*

Clearly, H is a maximal connected Lie subgroup of G if the isotropy representation of G/H is irreducible. In this case, the result is obvious. In [52], Wang and Ziller use variational methods to show that homogeneous spaces satisfying the conditions of Theorem 7.3.2 necessarily support Einstein metrics. They discuss numerous examples of such spaces. Theorem 7.3.2 is similar in spirit to their result. The functionals involved exhibit the same asymptotic behaviour.

Our next goal is to state two sufficient conditions for the solvability of (7.2) in the case where the maximality assumption of Theorem 7.3.2 does not hold. In order to do so, we need to introduce structure constants of the homogeneous space G/H and an extension of the scalar curvature functional S.

7.3.3 The Structure Constants

Denote by $\mathfrak{g}$ and $\mathfrak{h}$ the Lie algebras of G and H, respectively. Choose a scalar product Q on $\mathfrak{g}$ induced by a bi-invariant Riemannian metric on G. In what follows, $\oplus$ stands for the Q-orthogonal sum. Clearly,

$$\mathfrak{g} = \mathfrak{m} \oplus \mathfrak{h}$$

for some $\mathrm{Ad}(H)$-invariant space $\mathfrak{m}$. The representation $\mathrm{Ad}(H)|_{\mathfrak{m}}$ is equivalent to the isotropy representation of G/H. We standardly identify $\mathfrak{m}$ with the tangent space $T_H M$.

Choose a Q-orthogonal $\mathrm{Ad}(H)$-invariant decomposition

$$\mathfrak{m} = \mathfrak{m}_1 \oplus \cdots \oplus \mathfrak{m}_s \tag{7.4}$$

such that $\mathfrak{m}_i \neq \{0\}$ and $\mathrm{Ad}(H)|_{\mathfrak{m}_i}$ is irreducible for each $i = 1, \ldots, s$. The space $\mathfrak{m}$ may admit more than one decomposition of this form. However, by Schur's lemma, the summands $\mathfrak{m}_1, \ldots, \mathfrak{m}_s$ are determined uniquely up to order if $\mathrm{Ad}(H)|_{\mathfrak{m}_i}$ and $\mathrm{Ad}(H)|_{\mathfrak{m}_j}$ are inequivalent whenever $i \neq j$. Let d_i be the dimension of $\mathfrak{m}_i$. It is easy to show that the number s and the multiset $\{d_1, \ldots, d_s\}$ are independent of the chosen decomposition (7.4).

Denote by B the Killing form of $\mathfrak{g}$. For every $i = 1, \ldots, s$, because $\mathrm{Ad}(H)|_{\mathfrak{m}_i}$ is irreducible, there exists $b_i \geq 0$ such that

$$B|_{\mathfrak{m}_i} = -b_i Q|_{\mathfrak{m}_i}.$$

Given $\mathrm{Ad}(H)$-invariant subspaces $\mathfrak{u}_1$, $\mathfrak{u}_2$ and $\mathfrak{u}_3$ of $\mathfrak{m}$, define a tensor $\Delta(\mathfrak{u}_1, \mathfrak{u}_2, \mathfrak{u}_3) \in \mathfrak{u}_1 \otimes \mathfrak{u}_2^* \otimes \mathfrak{u}_3^*$ by setting

$$\Delta(\mathfrak{u}_1, \mathfrak{u}_2, \mathfrak{u}_3)(X, Y) = \pi_{\mathfrak{u}_1}[X, Y], \qquad X \in \mathfrak{u}_2, \ Y \in \mathfrak{u}_3,$$

where $\pi_{\mathfrak{u}_1}$ stands for the Q-orthogonal projection onto $\mathfrak{u}_1$. Let $\langle \mathfrak{u}_1 \mathfrak{u}_2 \mathfrak{u}_3 \rangle$ be the squared norm of $\Delta(\mathfrak{u}_1, \mathfrak{u}_2, \mathfrak{u}_3)$ with respect to the scalar product on $\mathfrak{u}_1 \otimes \mathfrak{u}_2^* \otimes \mathfrak{u}_3^*$ induced by $Q|_{\mathfrak{u}_1}$, $Q|_{\mathfrak{u}_2}$ and $Q|_{\mathfrak{u}_3}$. The fact that Q comes from a bi-invariant metric on G implies

$$\langle \mathfrak{u}_1 \mathfrak{u}_2 \mathfrak{u}_3 \rangle = \langle \mathfrak{u}_{\rho(1)} \mathfrak{u}_{\rho(2)} \mathfrak{u}_{\rho(3)} \rangle$$

for any permutation ρ of the set $\{1, 2, 3\}$. Given $i, j, k \in \{1, \ldots, s\}$, denote

$$\langle \mathfrak{m}_i \mathfrak{m}_j \mathfrak{m}_k \rangle = [ijk].$$

The numbers $([ijk])_{i,j,k=1}^s$ are called the *structure constants* of the homogeneous space M; see [52, Sect.1].

7.3.4 The Scalar Curvature Functional and its Extension

Let us write down a convenient formula for the scalar curvature functional S and define an extension of this functional. In what follows, we implicitly identify $g \in \mathcal{M}$ with the $\mathrm{Ad}(H)$-invariant scalar product induced by g on $\mathfrak{m}$ via the identification of $T_H M$ and $\mathfrak{m}$. Given a flag $\mathfrak{v} \subset \mathfrak{u} \subset \mathfrak{m}$, denote by $\mathfrak{u} \ominus \mathfrak{v}$ the Q-orthogonal complement of $\mathfrak{v}$ in $\mathfrak{u}$. The scalar curvature of $g \in \mathcal{M}$ is given by the equality

$$S(g) = -\frac{1}{2} \mathrm{tr}_g B|_{\mathfrak{m}} - \frac{1}{4} |\Delta(\mathfrak{m}, \mathfrak{m}, \mathfrak{m})|_g^2, \tag{7.5}$$

where $|\cdot|_g$ is the norm induced by g on $\mathfrak{m} \otimes \mathfrak{m}^* \otimes \mathfrak{m}^*$. If the decomposition (7.4) is such that

$$g = \sum_{i=1}^{s} x_i \pi^*_{\mathfrak{m}_i} Q \tag{7.6}$$

for some $x_1, \ldots, x_s > 0$, then

$$S(g) = \frac{1}{2} \sum_{i=1}^{s} \frac{d_i b_i}{x_i} - \frac{1}{4} \sum_{i,j,k=1}^{s} [ijk] \frac{x_k}{x_i x_j}. \tag{7.7}$$

For the derivation of formulas (7.5) and (7.7), see, e.g., [10, Chap. 7] and [48, Lem. 3.2]. Given $g \in \mathcal{M}$, it is always possible to choose (7.4) in such a way that (7.6) holds. For the proof of this fact, see [52, p. 180].

Suppose $\mathfrak{k}$ is a Lie subalgebra of $\mathfrak{g}$ containing $\mathfrak{h}$ as a proper subset. It will be convenient for us to denote

$$\mathfrak{n} = \mathfrak{k} \ominus \mathfrak{h}, \qquad \mathfrak{l} = \mathfrak{g} \ominus \mathfrak{k}.$$

Let $\mathcal{M}(\mathfrak{k})$ be the space of $\mathrm{Ad}(H)$-invariant scalar products on $\mathfrak{n}$ equipped with the topology it inherits from the second tensor power of $\mathfrak{n}^*$. Our conditions for the solvability of (7.2) will involve an extension of the functional S to $\mathcal{M}(\mathfrak{k})$. More precisely, define

$$\hat{S}(h) = -\frac{1}{2}\mathrm{tr}_h B|_{\mathfrak{n}} - \frac{1}{2}|\Delta(\mathfrak{l}, \mathfrak{n}, \mathfrak{l})|^2_{\mathrm{mix}} - \frac{1}{4}|\Delta(\mathfrak{n}, \mathfrak{n}, \mathfrak{n})|^2_h, \qquad h \in \mathcal{M}(\mathfrak{k}), \tag{7.8}$$

where $|\cdot|_{\mathrm{mix}}$ is the norm induced by $Q|_{\mathfrak{l}}$ and h on $\mathfrak{l} \otimes \mathfrak{n}^* \otimes \mathfrak{l}^*$ and $|\cdot|_h$ is the norm induced by h on $\mathfrak{n} \otimes \mathfrak{n}^* \otimes \mathfrak{n}^*$. Intuitively, the first term and the third term on the right-hand side are the scalar curvature of h viewed as a degenerate metric on G/H. The remaining portion of $\hat{S}(h)$ quantifies the extent to which h 'notices' the interaction between $\mathfrak{n}$ and $\mathfrak{l}$. If $\mathfrak{k} = \mathfrak{g}$, we identify $\mathcal{M}(\mathfrak{k})$ with $\mathcal{M}$. In this case, the second term on the right-hand side of (7.8) vanishes, and $\hat{S}(h)$ equals $S(h)$. If the decomposition (7.4) is such that

$$\mathfrak{k} = \left(\bigoplus_{i \in J_{\mathfrak{k}}} \mathfrak{m}_i \right) \oplus \mathfrak{h}, \qquad h = \sum_{i \in J_{\mathfrak{k}}} y_i \pi^*_{\mathfrak{m}_i} Q, \qquad y_i > 0, \tag{7.9}$$

for some $J_{\mathfrak{k}} \subset \{1, \ldots, s\}$, then

$$\hat{S}(h) = \frac{1}{2} \sum_{i \in J_{\mathfrak{k}}} \frac{d_i b_i}{y_i} - \frac{1}{2} \sum_{i \in J_{\mathfrak{k}}} \sum_{j,k \in J_{\mathfrak{k}}^c} \frac{[ijk]}{y_i} - \frac{1}{4} \sum_{i,j,k \in J_{\mathfrak{k}}} [ijk] \frac{y_k}{y_i y_j},$$

where $J_{\mathfrak{k}}^c$ is the complement of $J_{\mathfrak{k}}$ in $\{1, \ldots, s\}$; see [32, Lem. 2.19].

7.3.5 Non-Maximal Isotropy: The First Existence Theorem

In order to formulate our first existence theorem for (7.2) in the case where H is not maximal in G, we need a lemma and a definition. As before, let $\mathfrak{k}$ be a Lie subalgebra of $\mathfrak{g}$ containing $\mathfrak{h}$ as a proper subset. Denote

$$\mathcal{M}_T(\mathfrak{k}) = \{h \in \mathcal{M}(\mathfrak{k}) \mid \mathrm{tr}_h T|_{\mathfrak{n}} = 1\}.$$

Throughout Sections 7.3.5 and 7.3.6, we assume T is positive-definite.

Lemma 7.3.3 *The quantity $\sigma(\mathfrak{k}, T)$ defined by the formula*

$$\sigma(\mathfrak{k}, T) = \sup\{\hat{S}(h) \mid h \in \mathcal{M}_T(\mathfrak{k})\}$$

satisfies

$$0 \le \sigma(\mathfrak{k}, T) < \infty.$$

For the proof, see [47, Prop. 2.2].

Definition 7.3.4 We call $\mathfrak{k}$ a *T-apical* subalgebra of $\mathfrak{g}$ if $\mathfrak{k}$ meets the following requirements:

1 The inequality $\mathfrak{k} \ne \mathfrak{g}$ holds.
2 There exists a scalar product $h \in \mathcal{M}_T(\mathfrak{k})$ such that

$$\hat{S}(h) = \sigma(\mathfrak{k}, T).$$

3 If $\mathfrak{s}$ is a maximal Lie subalgebra of $\mathfrak{g}$ containing $\mathfrak{h}$, then

$$\sigma(\mathfrak{s}, T) \le \sigma(\mathfrak{k}, T).$$

Intuitively, $\mathfrak{k}$ is T-apical if the metrics supported on $\mathfrak{k}$ have the largest (modified) scalar curvature. The definition also requires that the supremum of $\hat{S}$ on the set of such metrics be attained. The following result was proven by Pulemotov in [47]. It provides a sufficient condition for the solvability of (7.2). As before, $\mathfrak{l}$ stands for $\mathfrak{g} \ominus \mathfrak{k}$.

Theorem 7.3.5 *Suppose that every maximal Lie subalgebra $\mathfrak{s}$ of $\mathfrak{g}$, such that $\mathfrak{h} \subset \mathfrak{s}$, satisfies the following requirement: if $\mathfrak{u} \subset \mathfrak{s} \ominus \mathfrak{h}$ and $\mathfrak{v} \subset \mathfrak{g} \ominus \mathfrak{s}$ are non-zero $\mathrm{Ad}(H)$-invariant spaces, then the representations $\mathrm{Ad}(H)|_{\mathfrak{u}}$ and $\mathrm{Ad}(H)|_{\mathfrak{v}}$ are inequivalent. Let $\mathfrak{k}$ be a T-apical subalgebra of $\mathfrak{g}$ for some positive-definite $T \in \mathcal{M}$. If*

$$4\sigma(\mathfrak{k}, T)\mathrm{tr}_Q T|_{\mathfrak{l}} < -2\mathrm{tr}_Q B|_{\mathfrak{l}} - \langle \mathfrak{l}\mathfrak{l}\mathfrak{l} \rangle, \tag{7.10}$$

then there exists $g \in \mathcal{M}_T$ such that $S(g) \geq S(g')$ for all $g' \in \mathcal{M}_T$. The Ricci curvature of g equals cT for some $c > 0$.

As we mentioned in Section 7.3.1, the functional S is bounded above on $\mathcal{M}_T$. It is, therefore, enough to show that its supremum is attained. Estimates imply that, roughly speaking, S drops down to $-\infty$ in all but a finite number of directions. Each such direction is associated with a maximal Lie subalgebra $\mathfrak{k}$ of $\mathfrak{g}$, and the corresponding asymptotic value of S is $\sigma(\mathfrak{k}, T)$. If $\mathfrak{k}$ is T-apical, this value is the highest. Formula (7.10) ensures that the slope of S is negative as it approaches $\sigma(\mathfrak{k}, T)$. This ensures the existence of a global maximum.

The class of homogeneous spaces satisfying the conditions of Theorem 7.3.5 is extensive. We discuss examples in Section 7.3.9.

The above result is moot if $\mathfrak{g}$ does not have any T-apical subalgebras. However, as Pulemotov showed in [47], at least one such subalgebra exists in most situations.

The quantity on the right-hand side of (7.10) is necessarily non-negative; see [32, Lem. 2.15]. It is useful to rewrite this quantity in terms of the structure constants of M. Suppose the decomposition (7.4) is such that the first formula in (7.9) holds for some $J_{\mathfrak{k}} \subset \{1, \dots, s\}$. Then

$$-2\mathrm{tr}_Q B|_{\mathfrak{l}} - \langle \mathfrak{lll} \rangle = 2 \sum_{i \in J_{\mathfrak{k}}^c} d_i b_i - \sum_{i,j,k \in J_{\mathfrak{k}}^c} [ijk].$$

Perhaps the biggest challenge in applying Theorem 7.3.5 is to compute the quantity $\sigma(\mathfrak{k}, T)$ for a given Lie subalgebra $\mathfrak{k}$. Our next result helps overcome this challenge in a number of important cases. For the proof, see [47, Prop. 2.5]. As above, $\mathfrak{n}$ stands for $\mathfrak{k} \ominus \mathfrak{h}$.

Proposition 7.3.6 *If* $\mathrm{Ad}(H)|_{\mathfrak{n}}$ *is irreducible, then* $\mathcal{M}_T(\mathfrak{k})$ *consists of a single point. In this case,*

$$\sigma(\mathfrak{k}, T) = -\frac{2\mathrm{tr}_Q B|_{\mathfrak{n}} + \langle \mathfrak{nnn} \rangle + 2\langle \mathfrak{nll} \rangle}{4\mathrm{tr}_Q T|_{\mathfrak{n}}}. \tag{7.11}$$

Remark 7.3.7   Suppose K is the connected Lie subgroup of G whose Lie algebra equals $\mathfrak{k}$. The irreducibility assumption on the representation $\mathrm{Ad}(H)|_{\mathfrak{n}}$ in Proposition 7.3.6 means that the homogeneous space K/H is isotropy irreducible.

It is useful to restate (7.11) in terms of the structure constants of M. If $\mathrm{Ad}(H)|_{\mathfrak{n}}$ is irreducible, we can choose the decomposition (7.4) so that

$\mathfrak{n} = \mathfrak{m}_1$. In this case,

$$T|_{\mathfrak{n}} = z_1 Q|_{\mathfrak{m}_1}, \qquad z_1 > 0.$$

Equation (7.11) becomes

$$\sigma(\mathfrak{k}, T) = \frac{1}{d_1 z_1} \left(\frac{1}{2} d_1 b_1 - \frac{1}{4}[111] - \frac{1}{2} \sum_{j,k \in \{2,\ldots,s\}} [1jk] \right).$$

7.3.6 Non-Maximal Isotropy: The Second Existence Theorem

Our next result, Theorem 7.3.9, provides one more sufficient condition for the solvability of (7.2). In some situations, the quantity $\sigma(\mathfrak{k}, T)$ is difficult to compute for all $\mathfrak{k}$ even with Proposition 7.3.6 at hand. Accordingly, it is problematic to determine which $\mathfrak{k}$ are T-apical and to verify condition (7.10). In such situations, Theorem 7.3.9 may be more effective than Theorem 7.3.5.

Let us introduce some additional notation and state a definition. Given a bilinear form R on an $\mathrm{Ad}(H)$-invariant non-zero subspace $\mathfrak{u} \subset \mathfrak{m}$, define

$$\lambda_-(R) = \inf\{R(X, X) \mid X \in \mathfrak{u} \text{ and } Q(X, X) = 1\},$$

$$\omega(\mathfrak{u}) = \min\{\dim \mathfrak{v} \mid \mathfrak{v} \text{ is a non-zero } \mathrm{Ad}(H)\text{-invariant subspace of } \mathfrak{u}\}.$$

Clearly, $\lambda_-(R)$ is the smallest eigenvalue of the matrix of R in a $Q|_{\mathfrak{u}}$-orthonormal basis of $\mathfrak{u}$. The number $\omega(\mathfrak{u})$ always lies between 1 and $\dim \mathfrak{u}$. In fact, $\omega(\mathfrak{u})$ is equal to $\dim \mathfrak{u}$ if $\mathrm{Ad}(H)|_{\mathfrak{u}}$ is irreducible.

Let $\mathfrak{k}$ and $\mathfrak{k}'$ be Lie subalgebras of $\mathfrak{g}$ such that

$$\mathfrak{g} \supset \mathfrak{k} \supset \mathfrak{k}' \supset \mathfrak{h}. \tag{7.12}$$

Denote

$$\mathfrak{l} = \mathfrak{g} \ominus \mathfrak{k}, \qquad \mathfrak{l}' = \mathfrak{g} \ominus \mathfrak{k}', \qquad \mathfrak{j} = \mathfrak{k} \ominus \mathfrak{k}', \qquad \mathfrak{n}' = \mathfrak{k}' \ominus \mathfrak{h}.$$

Definition 7.3.8 We call (7.12) a *simple chain* if $\mathfrak{k}'$ is a maximal Lie subalgebra of $\mathfrak{k}$ and $\mathfrak{h} \neq \mathfrak{k}'$.

We emphasise that this definition allows the equality $\mathfrak{k} = \mathfrak{g}$ but not $\mathfrak{k}' = \mathfrak{k}$. In [32], Gould and Pulemotov proved the following sufficient condition for the solvability of (7.2).

Theorem 7.3.9 *Suppose that every Lie subalgebra $\mathfrak{s} \subset \mathfrak{g}$ such that $\mathfrak{h} \subset \mathfrak{s}$ and $\mathfrak{h} \neq \mathfrak{s}$ meets the following requirements:*

1 *The representations* $\mathrm{Ad}(H)|_{\mathfrak{u}}$ *and* $\mathrm{Ad}(H)|_{\mathfrak{v}}$ *are inequivalent for every pair of non-zero* $\mathrm{Ad}(H)$-*invariant spaces* $\mathfrak{u} \subset \mathfrak{s} \ominus \mathfrak{h}$ *and* $\mathfrak{v} \subset \mathfrak{g} \ominus \mathfrak{s}$.
2 *The commutator* $[\mathfrak{r}, \mathfrak{s}]$ *is non-zero for every* $\mathrm{Ad}(H)$-*invariant 1-dimensional subspace* $\mathfrak{r}$ *of* $\mathfrak{g} \ominus \mathfrak{s}$.

Given a positive-definite $T \in \mathcal{M}$, *if the inequality*

$$\frac{\lambda_-(T|_{\mathfrak{n}'})}{\mathrm{tr}_Q T|_{\mathfrak{j}}} > \frac{2\mathrm{tr}_Q B|_{\mathfrak{n}'} + 2\langle \mathfrak{n}'\mathfrak{l}'\mathfrak{l}'\rangle + \langle \mathfrak{n}'\mathfrak{n}'\mathfrak{n}'\rangle}{\omega(\mathfrak{n}')(2\mathrm{tr}_Q B|_{\mathfrak{j}} + \langle \mathfrak{j}\mathfrak{j}\mathfrak{j}\rangle + 2\langle \mathfrak{j}\mathfrak{l}\mathfrak{l}\rangle)} \tag{7.13}$$

holds for every simple chain of the form (7.12), then there exists a Riemannian metric $g \in \mathcal{M}_T$ *such that* $S(g) \geq S(g')$ *for all* $g' \in \mathcal{M}_T$. *The Ricci curvature of* g *coincides with* cT *for some* $c > 0$.

As we mentioned in Section 7.3.1, the functional S is bounded above on $\mathcal{M}_T$. It is, therefore, enough to show that its supremum is attained.

We will discuss examples of homogeneous spaces satisfying the conditions of Theorem 7.3.9 in Section 7.3.9. Under these conditions, the denominator of the fraction on the right-hand side of (7.13) cannot equal 0, and the fraction itself is necessarily non-negative; see [32, Lem. 2.15]. It is possible to rewrite (7.13) in terms of the structure constants of M, as we rewrote several expressions above; see [32, Sect. 2.2–2.3]. However, the formulas turn out to be quite bulky, and we will not present them here.

As explained above, both Theorem 7.3.5 and Theorem 7.3.9 provides sufficient conditions for the solvability of (7.2). One advantage of the former result over the latter is that it imposes lighter assumptions on the homogeneous space M. In the case where both can be used, Theorem 7.3.5 seems to yield better conclusions; see [47, Rem. 5.2 and 5.5].

7.3.7 The Case of Two Isotropy Summands

Let us discuss the prescribed Ricci curvature problem on M assuming that the isotropy representation of M splits into two inequivalent irreducible summands. The conditions for the solvability of (7.2) given by Theorems 7.3.5 and 7.3.9 are not only sufficient but also necessary when this assumption holds. Moreover, the pair $(g, c) \in \mathcal{M} \times (0, \infty)$ satisfying (7.2) is unique up to scaling of g. Homogeneous spaces with two irreducible isotropy summands were classified in [30, 35]. Several authors have studied their geometry in detail; see, e.g., [6, 16, 48].

Let T be positive-semidefinite and non-zero. Suppose $s = 2$ in every decomposition of the form (7.4), i.e.,

$$\mathfrak{m} = \mathfrak{m}_1 \oplus \mathfrak{m}_2. \tag{7.14}$$

According to Theorem 7.3.2, equation (7.2) necessarily has a solution if $\mathfrak{h}$ is maximal in $\mathfrak{g}$. By [48, Lem. 4.6], the pair $(g, c) \in \mathcal{M} \times (0, \infty)$ is unique up to scaling of g. In this section, we assume that there exists a Lie subalgebra $\mathfrak{s} \subset \mathfrak{g}$ such that

$$\mathfrak{g} \supset \mathfrak{s} \supset \mathfrak{h}, \qquad \mathfrak{h} \neq \mathfrak{s}, \qquad \mathfrak{s} \neq \mathfrak{g}.$$

Let $\mathrm{Ad}(H)|_{\mathfrak{m}_1}$ and $\mathrm{Ad}(H)|_{\mathfrak{m}_2}$ be inequivalent. Without loss of generality, suppose

$$\mathfrak{s} = \mathfrak{m}_1 \oplus \mathfrak{h}.$$

If $\mathfrak{m}_2 \oplus \mathfrak{h}$ is also a Lie subalgebra of $\mathfrak{g}$, then the structure constants [112] and [221] vanish. In this case, all the metrics in $\mathcal{M}$ have the same Ricci curvature, and the analysis of (7.2) is easy; see, e.g., [48, Sect. 4.2]. Therefore, we assume $\mathfrak{m}_2 \oplus \mathfrak{h}$ is not closed under the Lie bracket. This implies $[221] > 0$. There exist $z_1, z_2 \geq 0$ such that $z_1 z_2 \neq 0$ and

$$T = z_1 \pi_{\mathfrak{m}_1}^* Q + z_2 \pi_{\mathfrak{m}_2}^* Q. \tag{7.15}$$

Clearly, if $\mathfrak{u} \subset \mathfrak{s} \ominus \mathfrak{h}$ and $\mathfrak{v} \subset \mathfrak{g} \ominus \mathfrak{s}$ are non-zero $\mathrm{Ad}(H)$-invariant spaces, then $\mathfrak{u} = \mathfrak{m}_1$ and $\mathfrak{v} = \mathfrak{m}_2$. Since $\mathrm{Ad}(H)|_{\mathfrak{m}_1}$ and $\mathrm{Ad}(H)|_{\mathfrak{m}_2}$ are inequivalent, the hypotheses of Theorems 7.3.5 and 7.3.9 are satisfied. Our next result, proven by Pulemotov in [46], settles the prescribed Ricci curvature problem for homogeneous metrics on M under the conditions of this section. Both inequalities (7.10) and (7.13) reduce to formula (7.16) below; see [32, Sect. 3]. Thus, Theorems 7.3.5 and 7.3.9 are optimal in the current setting.

Proposition 7.3.10 *The following statements are equivalent:*

1 *There exist a metric $g \in \mathcal{M}$ and a number $c > 0$ such that the Ricci curvature of g coincides with cT.*
2 *The inequality*

$$(2b_2 d_1 d_2 - d_1[222])z_1 > (2b_1 d_1 d_2 - 2d_2[122] - d_2[111])z_2 \tag{7.16}$$

is satisfied.

When these statements hold, the pair $(g, c) \in \mathcal{M} \times (0, \infty)$ is unique up to scaling of g.

The method behind the proof of Proposition 7.3.10 in [46] seems to be effective even if T has mixed signature. However, we will not consider this case here.

7.3.8 Homogeneous Spheres

Exploiting the results in Sections 7.3.5 and 7.3.7, we can obtain a wealth of information about the solvability of (7.2) for homogeneous metrics on the sphere $\mathbb{S}^d$. The nature of this information depends strongly on the dimension d. Transitive actions of Lie groups on $\mathbb{S}^d$ were classified in the 1940s by Borel, Montgomery and Samelson; see, e.g., [2, Exam. 6.16]. This classification shows that the homogeneous space M is a sphere if and only if G, H and d are as in Table 7.1. It will be convenient for us to assume that Q is induced by the round metric of curvature 1.

Table 7.1 *Homogeneous structures on spheres;*
$n \geq 2$ in cases (1)–(3) and $n \geq 1$ in cases (4)–(6)

	G	H	d
(1)	$SO(n+1)$	$SO(n)$	n
(2)	$SU(n+1)$	$SU(n)$	$2n+1$
(3)	$U(n+1)$	$U(n)$	$2n+1$
(4)	$Sp(n+1)$	$Sp(n)$	$4n+3$
(5)	$Sp(n+1)Sp(1)$	$Sp(n)Sp(1)$	$4n+3$
(6)	$Sp(n+1)U(1)$	$Sp(n)U(1)$	$4n+3$
(7)	$SU(2)$	$\{e\}$	3
(8)	$Spin(9)$	$Spin(7)$	15
(9)	$Spin(7)$	G_2	7
(10)	G_2	$SU(3)$	6

In cases (1), (9) and (10), the isotropy representation of M is irreducible. Consequently, all G-invariant (0,2)-tensor fields on M are the same up to a scalar multiple, and the analysis of (7.2) is easy. In cases (2), (3), (5) and (8), the isotropy representation of M splits into two inequivalent irreducible summands. Given a positive-semidefinite, non-zero, G-invariant T, formulas (7.14) and (7.15) hold with $z_1, z_2 \geq 0$. The spheres

$\mathbb{S}^{2n+1}$, $\mathbb{S}^{4n+3}$ and $\mathbb{S}^{15}$ admit the generalised Hopf fibrations

$$\mathbb{S}^1 \hookrightarrow \mathbb{S}^{2n+1} \to \mathbb{CP}^n, \qquad \mathbb{S}^3 \hookrightarrow \mathbb{S}^{4n+3} \to \mathbb{HP}^n, \qquad \mathbb{S}^7 \hookrightarrow \mathbb{S}^{15} \to \mathbb{S}^8.$$

We can assume that $\mathfrak{m}_1$ is vertical and $\mathfrak{m}_2$ is horizontal. This assumption does not involve any loss of generality. In cases (5) and (8), Proposition 7.3.10 leads to the following result; see [15, Sect. 3].

Proposition 7.3.11   *Suppose M is as in case (5) (resp., case (8)). Let T be a positive-semidefinite, non-zero and $\mathrm{Sp}(n+1)\mathrm{Sp}(1)$-invariant (respectively, $\mathrm{Spin}(9)$-invariant). A metric $g \in \mathcal{M}$ solving (7.2) for some $c > 0$ exists if and only if T satisfies (7.15) with $(2n+4)z_1 > z_2$ (respectively, $14z_1 > 3z_2$). When it exists, this metric is unique up to scaling.*

In cases (2) and (3), the situation is somewhat different. It is easy to see that every $\mathrm{SU}(n+1)$-invariant metric on $\mathbb{S}^{2n+1}$ is necessarily $\mathrm{U}(n+1)$-invariant. Proposition 7.3.10 leads to the first assertion of Proposition 7.3.12 below. In case (7), the space M is a Lie group, and the isotropy representation of M splits into three equivalent 1-dimensional summands. We have the usual Hopf fibration

$$\mathbb{S}^1 \hookrightarrow \mathbb{S}^3 \to \mathbb{S}^2.$$

The question of solvability of (7.2) was settled, in part, by Hamilton in [34], and later completed by Buttsworth in [13]. Specifically, the following result holds.

Proposition 7.3.12   *Let M be a sphere of odd dimension $d \geq 3$.*

1. *Assume M is as in case (2) or (3). Given a positive-semidefinite, non-zero, $\mathrm{SU}(n+1)$-invariant tensor field T, there exists a metric $g \in \mathcal{M}$, unique up to scaling, satisfying (7.2) for some $c > 0$.*

2. *Assume M is as in case (7) and T is a positive-semidefinite, non-zero, left-invariant tensor field. If T is non-degenerate, then there is a metric $g \in \mathcal{M}$, unique up to scaling, satisfying (7.2) for some $c > 0$. If $\mathrm{rank}\, T = 1$, then there is a 2-parameter family of such metrics. If $\mathrm{rank}\, T = 2$, none exists.*

Remark 7.3.13   It is possible to state and prove Proposition 7.3.11 and the first assertion of Proposition 7.3.12 avoiding the homogeneous space formalism. Instead, one may use the language of Hopf fibrations as in [15, Rem. 3.2].

Finally, we turn to cases (4) and (6). If M equals $\mathrm{Sp}(n+1)/\mathrm{Sp}(n)$, then the isotropy representation of M splits into 4 irreducible summands, i.e.,

$$\mathfrak{m} = \mathfrak{m}_1 \oplus \mathfrak{m}_2 \oplus \mathfrak{m}_3 \oplus \mathfrak{m}_4. \tag{7.17}$$

Three of these summands, say, $\mathfrak{m}_1$, $\mathfrak{m}_2$ and $\mathfrak{m}_3$, are 1-dimensional and equivalent. Given a positive-definite, $\mathrm{Sp}(n+1)$-invariant tensor field T, we can choose the decomposition (7.17) so that

$$T = z_1 \pi_{\mathfrak{m}_1}^* Q + z_2 \pi_{\mathfrak{m}_2}^* Q + z_3 \pi_{\mathfrak{m}_3}^* Q + q \pi_{\mathfrak{m}_4}^* Q \tag{7.18}$$

for some $z_1, z_2, z_3, q > 0$. The following sufficient condition for the solvability of (7.2) was obtained by Buttsworth *et al.* in [15].

Theorem 7.3.14 *Let M be as in case (4). Suppose T is a positive-definite, $\mathrm{Sp}(n+1)$-invariant tensor field on M satisfying (7.18). If*

$$(2n+4)\min\{z_1, z_2, z_3\} > q, \tag{7.19}$$

then there exists a metric $g \in \mathcal{M}$ such that (7.2) holds for some $c > 0$.

This result was proven in [15] by means of topological degree theory. It can also be derived from Theorem 7.3.5 above. The homogeneous sphere $\mathrm{Sp}(n+1)/\mathrm{Sp}(n)$ is, thus, an example of a space with equivalent isotropy summands on which this theorem applies.

It is possible to replace (7.19) with a less restrictive formula that is more difficult to verify. We will not discuss this here; see [15, Thm. 5.4].

If M is as in case (6), then a positive-definite $\mathrm{Sp}(n+1)\mathrm{U}(1)$-tensor field T on M can be written in the form (7.18) with two of the z_is equal to each other. In this situation, again, Theorem 7.3.14 provides a sufficient condition for the solvability of (7.2).

The results stated above cover all homogeneous structures on spheres. By analogy with Proposition 7.3.11 and Remark 7.3.13, we can use Proposition 7.3.10 and generalised Hopf fibrations to address the prescribed Ricci curvature problem on projective spaces. We will not discuss the details here; see [15].

7.3.9 Further Examples

Aside from homogeneous spheres and projective spaces, Theorems 7.3.5 and 7.3.9 provide easy-to-verify sufficient conditions for the solvability of (7.2) on large classes of generalised Wallach spaces and generalised flag manifolds. Previous literature contains little information concerning the prescribed Ricci curvature problem on such spaces. However, several other aspects of their geometry have been investigated thoroughly; see the survey [4].

Theorem 7.3.5 is an effective tool for solving (7.2) on generalised Wallach spaces with inequivalent isotropy summands. The reader will find classifications of such spaces in [19, 41]. There are several infinite families and 10 isolated instances, excluding products. We will illustrate the usage of Theorem 7.3.5 by considering an example. For a more detailed discussion, see [47].

Example 7.3.15   Let M be the generalised Wallach space $E_6/\mathrm{Sp}(3)\mathrm{Sp}(1)$ with the decomposition (7.4) chosen as in [41]. Suppose that $Q = -B$ and that T is an element of $\mathcal{M}$. The formula

$$T = -z_1 \pi_{\mathfrak{m}_1}^* B - z_2 \pi_{\mathfrak{m}_2}^* B - z_3 \pi_{\mathfrak{m}_3}^* B \tag{7.20}$$

holds for some $z_1, z_2, z_3 > 0$. Theorem 7.3.5 implies that a Riemannian metric $g \in \mathcal{M}$ satisfying (7.2) for some $c > 0$ exists if the triple (z_1, z_2, z_3) lies in the set

$$\{(x, y, z) \in (0, \infty)^3 \,|\, 3x \le 2y, \ 5x \le 6z, \ 7y + 3z < 20x\}$$
$$\cup \{(x, y, z) \in (0, \infty)^3 \,|\, 3x \ge 2y, \ 5y \le 9z, \ 21x + 18z < 52y\}$$
$$\cup \{(x, y, z) \in (0, \infty)^3 \,|\, 5x \ge 6z, \ 5y \ge 9z, \ 5x + 10y < 36z\}.$$

Generalised flag manifolds form an important class of homogeneous spaces with applications across a range of fields. The reader will find classifications in [5, 3, 7, 20]. There are numerous infinite families and isolated instances. As shown by Gould and Pulemotov in [32], if M is a generalised flag manifold, then M satisfies the hypotheses of Theorem 7.3.9. The simple chains are easy to identify, and inequality (7.13) is straightforward to check, as long as the structure constants of M are known. For the computation of these constants for a variety of spaces, see [37, 5, 6, 3, 7, 20].

Evidently, generalised flag manifolds satisfy the hypotheses of Theorem 7.3.5 as well. The T-apical subalgebras are relatively easy to find, and the verification of (7.10) is straightforward, if the isotropy representation of M splits into 5 or fewer irreducible summands. Let us provide an example. We refer to [47, Sect. 5] for further examples.

Example 7.3.16   Suppose M is the generalised flag manifold $G_2/\mathrm{U}(2)$ in which $\mathrm{U}(2)$ corresponds to the long root of G_2. Let the decomposition (7.4) be as in [3]. Given $T \in \mathcal{M}$, formula (7.20) holds for some $z_1, z_2, z_3 > 0$. Theorem 7.3.5 implies that a Riemannian metric $g \in \mathcal{M}$

satisfying (7.2) for some $c > 0$ exists if (z_1, z_2, z_3) is in the set

$$\{(x, y, z) \in (0, \infty)^3 \,|\, 9y \leq 2z, \ x + z < 12y\}$$
$$\cup \{(x, y, z) \in (0, \infty)^3 \,|\, 9y \geq 2z, \ 6x + 3y < 10z\}.$$

7.3.10 Ricci Iterations

The results discussed above lead to new existence theorems for Ricci iterations. More precisely, let $(g_i)_{i=1}^{\infty}$ be a sequence of Riemannian metrics on a smooth manifold. One calls $(g_i)_{i=1}^{\infty}$ a *Ricci iteration* if

$$\mathrm{Ric}\, g_{i+1} = g_i \tag{7.21}$$

for $i \in \mathbb{N}$. First considered by Rubinstein in [49], sequences satisfying (7.21) have been investigated intensively in the framework of Kähler geometry; see the survey [50] and also [21] for the most recent and comprehensive results. One may interpret (7.21) as a discretisation of the Ricci flow, as explained in, e.g., [50, Sect. 6]. This observation establishes an intriguing link between prescribed curvature problems and the theory of geometric evolutions.

In [48], Pulemotov and Rubinstein initiated the study of Ricci iterations in the homogeneous framework. Relying on Theorem 7.3.2 and Proposition 7.3.10, they achieved a deep understanding of (7.21) on spaces with two inequivalent irreducible isotropy summands, and proved existence and compactness results for (7.21) on spaces with maximal isotropy. They also observed an interesting connection between Ricci iterations and ancient solutions to the Ricci flow. On the basis of this work, Buttsworth *et al.* investigated (7.21) on homogeneous spheres and projective spaces in [15]. One of the results of [15] establishes the existence and the convergence of Ricci iterations by means of the inverse function theorem. This approach turns out to be effective even on spaces without a homogeneous structure, as Buttsworth and Hallgren demonstrate in [14].

7.4 Open Questions and Non-Compact Homogeneous Spaces

While the results surveyed in Section 7.3 provide a large amount of information about the prescribed curvature problem for homogeneous metrics, and even settle the problem in some cases, a number of questions remain open. For instance, what necessary conditions for the solvability of (7.2) can one state on compact spaces with more than two isotropy summands? In what circumstances is it possible to establish the uniqueness of g on such spaces (up to scaling)? What approaches can one use to study the solvability of (7.2) when M is compact and T has mixed signature? These questions are, in fact, still open on spheres of dimension $4n + 3$ acted upon by $\mathrm{Sp}(n + 1)$.

7.4.1 The Non-Compact Case

The investigation of the prescribed Ricci curvature problem for homogeneous metrics on *non-compact* spaces is a promising area of research. So far, progress has been scarce. On the other hand, while equations (7.1) and (7.2) appear to be difficult to solve in the homogeneous non-compact setting, a number of related questions are more tractable. For instance, what are the possible signatures of the Ricci curvature of a left-invariant metric on a connected Lie group $\mathcal{G}$? In [40], Milnor proposed an answer to this question assuming $\mathcal{G}$ was 3-dimensional and unimodular. His arguments relied on the observation that any left-invariant metric can be diagonalised in a basis satisfying 'nice' commutation relations. Subsequently, in [33], Ha and Lee obtained counterparts of Milnor's results for non-unimodular $\mathcal{G}$. Possible signatures of the Ricci curvature were further studied by Kremlev and Nikonorov on 4-dimensional Lie groups and by Boucetta, Djiadeu Ngaha and Wouafo Kamga on nilpotent groups of all dimensions; see [39, 12]. Several other researchers addressed questions related to the solvability of (7.1) and (7.2) in indirect ways. Namely, in [38], Kowalski and Nikcevic extended the results of [40, 33] by finding all possible triples of numbers that could be eigenvalues of the Ricci curvature of a left-invariant metric in 3 dimensions. In [31], Eberlein investigated the space of metrics with prescribed Ricci operator on 2-step nilpotent Lie groups.

Let us make a few observations concerning the solvability of (7.1) on non-compact homogeneous spaces. These observations mainly deal with the non-existence of metrics with non-negative Ricci curvature. Equality (7.1) cannot hold for the metric on a non-compact Riemannian homogeneous space if the tensor field T is positive-definite. Indeed, the Bonnet–Myers theorem implies the following result.

Proposition 7.4.1 *If a connected Riemannian homogeneous space has positive-definite Ricci curvature, then it must be compact.*

Using the Cheeger–Gromoll splitting theorem, one can prove the following extension of this result (see [10, Thm. 6.65 and Rem. 6.66(f)]).

Proposition 7.4.2 *If a non-compact, connected Riemannian homogeneous space has positive-semidefinite Ricci curvature, then it is isometric to the Riemannian product of a compact Riemannian homogeneous space with positive-semidefinite Ricci curvature and a Euclidean space with standard metric.*

A natural question arises: can the metric on a non-compact Riemannian homogeneous space satisfy (7.1) with $T = 0$? The following result, known as the Alekseevskii–Kimel'fel'd theorem, provides the answer.

Theorem 7.4.3 *A Ricci-flat Riemannian homogeneous space is isometric to the Riemannian product of a flat torus and a Euclidean space with standard metric.*

One can derive this result from Proposition 7.4.2 and the Bochner theorem (see [10, Thm. 1.84]). It was originally proven in [1] by other methods.

Theorem 7.4.3 would be false, of course, without the homogeneity assumption. One can see this, for example, by considering Calabi–Yau manifolds.

7.4.2 Unimodular Lie Groups of Dimension 3

In [13], Buttsworth settled the question of the solvability of equation (7.2) for left-invariant Riemannian metrics on unimodular Lie groups in 3 dimensions. These results provide insight into the prescribed Ricci curvature problem on non-compact homogeneous spaces. We summarise them in Theorem 7.4.4 below.

Consider a 3-dimensional connected unimodular Lie group $\mathcal{G}$ with Lie algebra $\mathfrak{r}$. It is well known that $\mathcal{G}$ necessarily possesses a Milnor frame, i.e., it has a basis $\{V_1, V_2, V_3\}$ of $\mathfrak{r}$ such that

$$[V_i, V_j] = \varepsilon_{ijk}\lambda_k V_k, \qquad i, j \in \{1, 2, 3\},$$

where ε_{ijk} is the Levi-Civita symbol, $k \in \{1, 2, 3\}$ is distinct from i and j, and $\lambda_k \in \mathbb{R}$. By scaling and reordering, we can assume the following:

(a) Each λ_k lies in $\{-2, 0, 2\}$.
(b) There are more non-negative λ_ks than negative ones.
(c) The equality $\lambda_3 = 0$ holds if any of the λ_ks is 0.

(d) The equality $\lambda_1 = 2$ holds unless $\lambda_k = 0$ for all k.

With these additional constraints, the triple $(\lambda_1, \lambda_2, \lambda_3)$ is the same for all Milnor frames on $\mathcal{G}$. Moreover, this triple determines the Lie algebra $\mathfrak{r}$ uniquely; see, e.g., [40] and the first column in Table 7.2.

Let T be a left-invariant tensor field on $\mathcal{G}$. In [13], Buttsworth proved the following result.

Theorem 7.4.4 *A left-invariant Riemannian metric g satisfying (7.2) for some $c > 0$ exists if and only if there is a Milnor frame $\{V_1, V_2, V_3\}$ such that the above constraints (a)–(d) and the following three statements hold:*

1 *The tensor field T is diagonalisable in $\{V_1, V_2, V_3\}$.*
2 *The signs of the eigenvalues (T_1, T_2, T_3) of the matrix of T in the basis $\{V_1, V_2, V_3\}$ appear in the second column of Table 7.2.*
3 *The eigenvalues (T_1, T_2, T_3) satisfy the conditions in the third column of Table 7.2.*

It is natural to ask whether the constant c in (7.2) is uniquely determined by T and whether the metric g satisfying (7.2) for given T and $c > 0$ is in any sense unique. The answers to these questions are provided in the fourth and fifth columns of Table 7.2. The symbol $\sim$ denotes equality up to scaling.

Results of [40, 33] on the possible signatures of the Ricci curvature yield the information contained in the first two columns. Thus, Theorem 7.4.4 extends these results.

Table 7.2 Existence and uniqueness of solutions to (7.2) on $\mathcal{G}$.

Lie group $(\lambda_1, \lambda_2, \lambda_3)$	Signs of (T_1, T_2, T_3)	Conditions on (T_1, T_2, T_3)	Is c the same for all solutions?	$g_1 \sim g_2$ for any solutions (g_1, c) and (g_2, c)
SO(3) or SU(2) $(2, 2, 2)$	$(+, +, +)$	—	yes	yes
	$(+, 0, 0)$	—	yes	no
	$(0, +, 0)$	—	yes	no
	$(0, 0, +)$	—	yes	no
	$(+, -, -)$	$(T_1 + T_2 + T_3)^3 \geq 27 T_1 T_2 T_3$	no	yes
	$(-, +, -)$	$(T_1 + T_2 + T_3)^3 \geq 27 T_1 T_2 T_3$	no	yes
	$(-, -, +)$	$(T_1 + T_2 + T_3)^3 \geq 27 T_1 T_2 T_3$	no	yes
SL$(2, \mathbb{R})$ $(2, 2, -2)$	$(+, -, -)$	$T_3 + T_1 > 0$	yes	yes
	$(-, +, -)$	$T_3 + T_2 > 0$	yes	yes
	$(-, -, +)$	$\max\{-T_1, -T_2\} < T_3$ or $\min\{-T_1, -T_2\} > T_3$ or $T_3 = -T_1 = -T_2$	yes	yes
			yes	yes
	$(-, 0, 0)$	—	yes	no
	$(0, -, 0)$	—	yes	no
E(2) $(2, 2, 0)$	$(0, 0, 0)$	—	no	no
	$(+, -, -)$	$T_1 + T_2 > 0$	yes	yes
	$(-, +, -)$	$T_1 + T_2 > 0$	yes	yes
E(1, 1) $(2, -2, 0)$	$(0, 0, -)$	—	yes	no
	$(+, -, -)$	$T_1 + T_2 > 0$	yes	yes
	$(-, +, -)$	$T_1 + T_2 > 0$	yes	yes
H$_3$ $(2, 0, 0)$	$(+, -, -)$	—	yes	yes
$\mathbb{R}^3$ $(0, 0, 0)$	$(0, 0, 0)$	—	no	no

References

[1] D. Alekseevskii, B. Kimel'fel'd, Structure of homogeneous Riemann spaces with zero Ricci curvature, Funct. Anal. Appl. 9 (1975) 97–102.

[2] M.M. Alexandrino, R.G. Bettiol, Lie groups and geometric aspects of isometric actions, Springer-Verlag, Cham, 2015. .

[3] S. Anastassiou, I. Chrysikos, The Ricci flow approach to homogeneous Einstein metrics on flag manifolds, J. Geom. Phys. 61 (2011) 1587–1600.

[4] A. Arvanitoyeorgos, Progress on homogeneous Einstein manifolds and some open problems, Bull. Greek Math. Soc. 58 (2015) 75–97.

[5] A. Arvanitoyeorgos, I. Chrysikos, Invariant Einstein metrics on flag manifolds with four isotropy summands, Ann. Glob. Anal. Geom. 37 (2010) 185–219.

[6] A. Arvanitoyeorgos, I. Chrysikos, Invariant Einstein metrics on generalized flag manifolds with two isotropy summands, J. Aust. Math. Soc. 90 (2011) 237–251.

[7] A. Arvanitoyeorgos, I. Chrysikos, Y. Sakane, Homogeneous Einstein metrics on generalized flag manifolds with five isotropy summands, Intern. J. Math. 24 (2013) 1350077.

[8] T. Aubin, Some nonlinear problems in Riemannian geometry, Springer-Verlag, Berlin, 1998.

[9] A. Baldes, Non-existence of Riemannian metrics with prescribed Ricci tensor, in: Nonlinear problems in geometry (Mobile, AL, 1985), Contemp. Math. 51, Amer. Math. Soc., Providence, RI, 1986, pp. 1–8.

[10] A. Besse, Einstein manifolds, Springer-Verlag, Berlin, 1987.

[11] C. Böhm, Homogeneous Einstein metrics and simplicial complexes, J. Differ. Geom. 67 (2004) 79–165.

[12] M. Boucetta, M.B. Djiadeu Ngaha, J. Wouafo Kamga, The signature of the Ricci curvature of left-invariant Riemannian metrics on nilpotent Lie groups, Differ. Geom. Appl. 47 (2016) 26–42.

[13] T. Buttsworth, The prescribed Ricci curvature problem on three-dimensional unimodular Lie groups, Math. Nachr. 292 (2019) 747–759.

[14] T. Buttsworth, M. Hallgren, Local stability of Einstein metrics under the Ricci iteration, preprint, arXiv:1907.10222 [math.DG].

[15] T. Buttsworth, A. Pulemotov, Y.A. Rubinstein, W. Ziller, On the Ricci iteration for homogeneous metrics on spheres and projective spaces, to appear in Transf. Groups, preprint, arXiv:1811.01724 [math.DG].

[16] M. Buzano, Ricci flow on homogeneous spaces with two isotropy summands, Ann. Global Anal. Geom. 45 (2014) 25–45.

[17] G. Calvaruso, Three-dimensional homogeneous Lorentzian metrics with prescribed Ricci tensor, J. Math. Phys. 48 (2007) 123518.

[18] J. Cao, D.M. DeTurck, The Ricci curvature equation with rotational symmetry, Amer. J. Math. 116 (1994) 219–241.

[19] Z. Chen, Y. Kang, K. Liang, Invariant Einstein metrics on three-locally-symmetric spaces, Commun. Anal. Geom. 24 (2016) 769–792.

[20] I. Chrysikos, Y. Sakane, The classification of homogeneous Einstein metrics on flag manifolds with $b_2(M) = 1$, Bull. Sci. Math. 138 (2014) 665–692.

[21] T. Darvas, Y.A. Rubinstein, Convergence of the Kähler-Ricci iteration, Anal. PDE 12 (2019) 721–735.

[22] Ph. Delanoë, Local solvability of elliptic, and curvature, equations on compact manifolds, J. reine angew. Math. 558 (2003) 23–45.

[23] E. Delay, Inversion of some curvature operators near a parallel Ricci metric II: non-compact manifold with bounded geometry, Ark. Mat. 56 (2018) 285–297.

[24] D.M. DeTurck, Existence of metrics with prescribed Ricci curvature: local theory, Invent. Math. 65 (1981/82) 179–207.

[25] D.M. DeTurck, Metrics with prescribed Ricci curvature, in: Seminar on differential geometry (S.-T. Yau, ed.), Princeton University Press, Princeton, NJ, 1982, pp. 525–537.

[26] D.M. DeTurck, The Cauchy problem for Lorentz metrics with prescribed Ricci curvature, Comp. Math. 48 (1983) 327–349.

[27] D.M. DeTurck, Prescribing positive Ricci curvature on compact manifolds, Rend. Sem. Mat. Univ. Politec. Torino 43 (1985) 357–369.

[28] D.M. DeTurck, N. Koiso, Uniqueness and nonexistence of metrics with prescribed Ricci curvature, Ann. Inst. H. Poincaré Anal. Non Linéaire 1 (1984) 351–359.

[29] D. DeTurck, H. Goldschmidt, Metrics with prescribed Ricci curvature of constant rank. I. The integrable case, Adv. Math. 145 (1999) 1–97.

[30] W. Dickinson, M. Kerr, The geometry of compact homogeneous spaces with two isotropy summands, Ann. Global Anal. Geom. 34 (2008) 329–350.

[31] P. Eberlein, Riemannian 2-step nilmanifolds with prescribed Ricci tensor, in: Geometric and probabilistic structures in dynamics (K. Burns et al., eds.), Contemp. Math. 630, Amer. Mathe. Soc., Providence, RI, 2008, pp. 167–195.

[32] M.D. Gould, A. Pulemotov, The prescribed Ricci curvature problem on homogeneous spaces with intermediate subgroups, to appear in Commun. Anal. Geom., preprint, arXiv:1710.03024 [math.DG].

[33] K. Ha, J. B Lee, Left invariant metrics and curvatures on simply connected three-dimensional Lie groups, Math. Nachr. 282 (2009) 868–898.

[34] R.S. Hamilton, The Ricci curvature equation, in: Seminar on nonlinear partial differential equations (S.-S. Chern, ed.), Math. Sci. Res. Inst. Publ. 2, Springer-Verlag, New York, 1984, pp. 47–72.

[35] C. He, Cohomogeneity one manifolds with a small family of invariant metrics, Geom. Dedicata 157 (2012) 41–90.

[36] J.L. Kazdan, Applications of partial differential equations to some problems in differential geometry, lecture notes, 2006, available at http://hans.math.upenn.edu/~kazdan/japan/japan.pdf (accessed on 20/04/2019).

[37] M. Kimura, Homogeneous Einstein metrics on certain Kähler *C*-spaces, in: Recent topics in differential and analytic geometry (T. Ochiai, ed.), Adv. Stud. Pure Math. 18-I, Academic Press, Boston, MA, 1990, pp. 303–320.

[38] O. Kowalski, S. Nikcevic, On Ricci eigenvalues of locally homogeneous Riemannian 3-manifolds, Geom. Dedicata 62 (1996) 65–72.

[39] A. Kremlev, Y. Nikonorov, The signature of the Ricci curvature of left-invariant Riemannian metrics on four-dimensional Lie groups. The unimodular case, Siberian Adv. Math. 19 (2008) 245–267.

[40] J. Milnor, Curvatures of left invariant metrics on Lie groups, Adv. Math. 21 (1976) 293–329.

[41] Y.G. Nikonorov, Classification of generalized Wallach spaces, Geom. Dedicata 181 (2016) 193–212.

[42] Y.G. Nikonorov, E.D. Rodionov, V.V. Slavskii, Geometry of homogeneous Riemannian manifolds, J. Math. Sci. (N.Y.) 146 (2007) 6313–6390.

[43] R. Pina, J.P. dos Santos, Group-invariant solutions for the Ricci curvature equation and the Einstein equation, J. Differ. Equ. 266 (2019) 2214–2231.

[44] A. Pulemotov, Metrics with prescribed Ricci curvature near the boundary of a manifold, Math. Ann. 357 (2013) 969–986.

[45] A. Pulemotov, The Dirichlet problem for the prescribed Ricci curvature equation on cohomogeneity one manifolds, Ann. Mat. Pura Appl. 195 (2016) 1269–1286.

[46] A. Pulemotov, Metrics with prescribed Ricci curvature on homogeneous spaces, J. Geom. Phys. 106 (2016) 275–283.

[47] A. Pulemotov, Maxima of curvature functionals and the prescribed Ricci curvature problem on homogeneous spaces, J. Geom. Anal. 30 (2020) 987–1010.

[48] A. Pulemotov, Y.A. Rubinstein, Ricci iteration on homogeneous spaces, Trans. AMS 371 (2019) 6257–6287.

[49] Y.A. Rubinstein, The Ricci iteration and its applications, C. R. Acad. Sci. Paris 345 (2007) 445–448.

[50] Y.A. Rubinstein, Smooth and singular Kähler–Einstein metrics, in: Geometric and spectral analysis (P. Albin et al., eds.), Contemp. Math. 630 and Centre de Recherches Mathématiques Proceedings, Amer. Math. Soc., Providence, RI, and Centre de Recherches Mathématiques, Montreal, QC, 2014, pp. 45–138.

[51] G. Smith, The Bianchi identity and the Ricci curvature equation, PhD thesis, The University of Queensland, 2016.

[52] M.Y. Wang, W. Ziller, Existence and nonexistence of homogeneous Einstein metrics, Invent. Math. 84 (1986) 177–194.

8
Singular Yamabe and Obata Problems
A. Rod Gover and Andrew K. Waldron

Abstract

A conformal geometry determines a distinguished, potentially singular, variant of the usual Yamabe problem, where the conformal factor can change sign. When a smooth solution does change sign, its zero locus is a smoothly embedded separating hypersurface that, in dimension three, is necessarily a Willmore energy minimiser or, in higher dimensions, satisfies a conformally invariant analog of the Willmore equation. In all cases the zero locus is critical for a conformal functional that generalises the total Q-curvature by including extrinsic data. These observations lead to some interesting global problems that include natural singular variants of a classical problem solved by Obata.

8.1 Introduction

On a closed Riemannian manifold $(M, \bar{g})$ the Yamabe problem concerns finding a conformally related metric $g = e^{2\omega}\bar{g}$, for some $\omega \in C^\infty(M)$, that has constant scalar curvature. The statement and ultimate solution of this problem by Yamabe, Trudinger, Aubin, and Schoen [38, 35, 2, 34] were milestones in conformal geometry and geometric analysis. Another – apparently unrelated – problem of great interest has been that of finding critical points of the Willmore energy [36, 29]. The Willmore energy $\mathcal{W}$ for a closed surface Σ in Euclidean 3-space is given by

$$\mathcal{W} = \int_\Sigma (H^2 - K),$$

where H is the surface mean curvature, and K its Gauss curvature. A key property of this energy is that it is invariant under conformal transformations [37]. We shall call the Euler–Lagrange equation for this energy the *Willmore equation*; this equation determines a conformally invariant quantity termed the *Willmore invariant*.

Recently it has been observed that there is a rather interesting link between natural generalisations of these problems. For that one replaces the usual Yamabe problem with a singular variant; namely, on a Riemannian manifold $(M, \bar{g})$ of dimension d one seeks a smooth (here and throughout smooth means C^∞) function σ such that

$$|d\sigma|^2_{\bar{g}} - \frac{2\sigma}{d}\left(\Delta^{\bar{g}}\sigma + \frac{\sigma}{2(d-1)}Sc^{\bar{g}}\right) = 1. \qquad (8.1)$$

The Laplacian $\Delta^g = \nabla^a\nabla_a$ used above is the "negative energy" convention, and ∇ is the Levi-Civita connection of g; an obvious index notation has been employed. This is the usual Yamabe-type equation requiring that the scalar curvature satisfies $Sc^g = -d(d-1)$, where $g = \sigma^{-2}\bar{g}$, except that we allow the possibility that σ changes sign. In the latter case the metric g is singular along the zero locus $\mathcal{Z}(\sigma)$ of σ, and it is evident from the equation and setup here that this zero locus is an embedded hypersurface (i.e., codimension-1 submanifold) with an induced conformal structure. See Proposition 8.3.1 below for more detail. More striking is that Equation (8.1) puts an interesting restriction on the conformal embedding of $\mathcal{Z}(\sigma)$; if $d = 3$ it is necessarily Willmore (i.e., a Willmore energy minimiser), and in higher dimensions this condition *defines* a conformally invariant generalisation of the Willmore equation; see Theorem 8.2.4 which follows [1, 19, 20]. It turns out that this higher Willmore equation is the Euler–Lagrange equation (with respect to variation of embedding) of an action [23], which in fact can be expressed as an integral of a quantity that generalises, by the addition of extrinsic curvature terms [18, 21, 15], the Branson Q-curvature of [5, 6] (see the review [8]).

The picture just described captures some of the important local aspects of the link between (higher) Willmore minimisers and solutions to the singular Yamabe equation, Equation (8.1). A key purpose of the current chapter is to point out some very interesting features and questions linked to the global version of this problem on closed manifolds. This culminates in the main questions introduced in Section 8.4. In Section 8.2 we review briefly some of the background and mention informally a singular variant of one of the Obata problems. Section 8.3 describes some key tools from conformal tractor calculus required to handle hypersurfaces embedded in conformal manifolds.

One might also consider setting the left hand side of (8.1) to other constants. In Riemannian signature, the choice -1 forbids σ from having a zero locus, and thus Equation (8.1) recovers the (positive scalar curvature) Yamabe problem that seeks g in a given conformal class of metrics such that $\mathrm{Sc}^g = d(d-1)$. Hence there is then no singular case. Requiring the left hand side of (8.1) to vanish corresponds to seeking a scalar flat metric from the conformal class. In this case there is a singular version. For example, the pullback to sphere of the Euclidean metric via stereographic projection extends to a global solution on the sphere. Here we are primarily interested in sign changing solutions of (8.1) where σ is a defining function for a hypersurface.

Acknowledgements A.W. was also supported by a Simons Foundation Collaboration Grant for Mathematicians ID 317562. A.W. and A.R.G. gratefully acknowledge support from the Royal Society of New Zealand via Marsden Grants 16-UOA-051 and 19-UOA-008.

8.2 Background and a Singular Obata Problem

Let (M^d, c) be a closed, orientable, Riemannian signature, conformal manifold, where $c \ni [g] = [\Omega^2 g]$ denotes an equivalence class of smooth, conformally related metrics with $0 < \Omega \in C^\infty M$, and $d \geq 3$. The trace-free Schouten tensor of $g \in c$, defined by

$$\mathring{P}^g = \frac{1}{d-2}\left(Ric^g - \frac{1}{d}\, g\, Sc^g\right)$$

obeys the conformal transformation law

$$\mathring{P}^{\Omega^2 g}_{ab} = \mathring{P}^g_{ab} - \nabla^g_{(a}\Upsilon_{b)\circ} + \Upsilon_{(a}\Upsilon_{b)\circ}\,.$$

Here ∇ is the Levi-Civita connection, $\Upsilon := \mathrm{d}\log\Omega$, and we have employed an abstract index notation to denote sections of tensor bundles over M as well as the notation $X_{(ab)\circ} := X_{ab} - \frac{1}{d}\, g_{ab}\langle g, X\rangle_g$ for projection to the trace free part of $X \in \Gamma(\odot^2 T^* M)$; this projection is independent of the choice of $g \in c$. Also note that $\langle X, Y\rangle_g := g^{ac} g^{bd} X_{ab} Y_{cd}$ and $|X|^2_g := \langle X, X\rangle$. Metrics whose trace-free Schouten tensor vanishes are called *Einstein*.

Now, supposing that g has constant scalar curvature, the Bianchi identity gives (suppressing the g dependence)

$$\nabla^a \mathring{P}_{ab} = 0\,.$$

Thus, in that case it follows that $\int_{(M,g)} \mathring{P}^{ab}\nabla_a\nabla_b\Omega^{-1} = 0$, or, equivalently,

$$\int_{(M,g)} \Omega^{-1}\mathring{P}^{ab}\left(\nabla_a\Upsilon_b - \Upsilon_a\Upsilon_b\right) = 0.$$

Calling $\Omega^2 g = \bar{g}$, we see that $\int_{(M,g)} \Omega^{-1}\langle\mathring{P}^g, \mathring{P}^{\bar{g}}\rangle = \int_{(M,g)} \Omega^{-1}|\mathring{P}^g|^2$. The Schwarz inequality then implies that

$$\int_{(M,g)} \Omega^{-1}|\mathring{P}^g|_g^2 \leq \int_{(M,g)} \Omega^{-1}|\mathring{P}^{\bar{g}}|_g^2.$$

The above simple argument due to Schoen [33] implies that if two metrics are conformally related, and one is Einstein while the other has constant scalar curvature, then they must in fact both be Einstein. This proves the following classical theorem due to Obata [32].

Theorem 8.2.1 *Let $(S^d, \mathbf{c}_{\text{round}})$ be the sphere equipped with its standard, conformally flat, class of metrics. Then if $g \in \mathbf{c}_{\text{round}}$ has constant scalar curvature, g must be Einstein.*

It is easy to construct an example of this phenomenon.

Example 8.2.2 Let $h : S^d \to [-1,1]$ be the height function on the standard sphere in $\mathbb{R}^{d+1}$,

$$S^d := \{x^2 + y^2 + \cdots + z^2 = 1\},$$

given by $h(x, y, \ldots, z) = z$. Then if $\bar{g}$ is the standard sphere metric, an elementary computation shows that, where it is defined, the conformally related metric

$$g = \frac{\bar{g}}{(h - k)^2}$$

has constant scalar curvature

$$Sc^g = d(d - 1)(k^2 - 1).$$

We give a simple argument in Section 8.3.1 that, for $k > 1$, the metric g is indeed Einstein and, since its Weyl tensor necessarily vanishes, is isometric to the standard sphere metric.

An intriguing feature of this example is that when $k = 1$ the metric g becomes singular at the north pole $z = 1$. However, in that case, the scalar curvature vanishes and $(S^d \setminus \{z = 1\}, g)$ is isometric to Euclidean space. This can be checked explicitly by writing $\bar{g} = d\theta^2 + \sin^2\theta\, ds^2_{S^{d-1}}$, where $h = z = \cos\theta$, and then noting that the change of coordinates $r = \sin\theta/(1 - \cos\theta)$ gives the required isometry between g and the flat metric $dr^2 + r^2 ds^2_{S^{d-1}}$. Geometrically this is the stereographic projection of S^d to the hyperplane $z = 1$.

When $|k| < 1$, g becomes singular along a hypersurface of height k. The Riemannian manifold (S_+, g_+), given by the above data restricted to $z > k$, is isometric to the Poincaré ball. The same applies to (S_-, g_-) with $z < k$. Observe that the conformal rescaling function $\Omega = \frac{1}{h-k}$ is singular along the hypersurface $\Sigma = S^{d-1}$ at $z = k$, and changes sign in S_- because the function $h - k$ is a defining function for the hypersurface embedding $\Sigma \hookrightarrow S^d$. These facts are recovered in Section 8.3.1 from a different perspective that unifies the three cases discussed above and provides information about the geometry of the embedding $\Sigma \hookrightarrow M$.

Given a Riemannian manifold $(M^d, \bar{g})$ and a smooth function σ we define

$$S(\bar{g}, \sigma) := |d\sigma|^2_{\bar{g}} - \frac{2\sigma}{d}\left(\Delta^{\bar{g}}\sigma + \frac{\sigma}{2(d-1)}\, Sc^{\bar{g}}\right), \qquad (8.2)$$

which obeys

$$S(\Omega^2\bar{g}, \Omega\sigma) = S(\bar{g}, \sigma),$$

for any $0 < \Omega \in C^\infty M$. When the part of the jet of σ given by $\left(\sigma, d\sigma, -\frac{1}{d}(\Delta^{\bar{g}}\sigma + \frac{\sigma}{2(d-1)}Sc^{\bar{g}})\right)$ is nowhere vanishing, $S(\bar{g}, \sigma)$ has a natural interpretation as a curvature that we term the *S-curvature*. Indeed, away from the zero locus Σ of σ,

$$S(g, 1) = -\frac{Sc^g}{d(d-1)}, \text{ where } g = \sigma^{-2}\bar{g}, \qquad (8.3)$$

so the S-curvature smoothly extends the scalar curvature of the singular metric $g = \sigma^{-2}\bar{g}$ to all of M.

Given its nature as a curvature that extends (and generalises) the scalar curvature, it is interesting to consider the problem of finding functions σ that yield $S(\bar{g}, \sigma)$ constant. If σ has fixed sign then this boils down to the usual Yamabe problem on closed manifolds. However, in general other solutions are possible, and hence it is interesting to consider the case that σ has a zero locus. Note that, in view of (8.2), a non-trivial zero locus is impossible if $S(\bar{g}, \sigma)$ is strictly negative. If $S(\bar{g}, \sigma) = 0$ then only isolated zeros are possible [14]. But observe that when $S(\bar{g}, \sigma)$ is strictly positive, the zero locus of σ, if non-empty, is a hypersurface with defining function σ (see [14]). Here we shall be most interested in this last case, and, if σ satisfies

$$S(\bar{g}, \sigma) = 1, \qquad (8.4)$$

then the singular metric $g = \sigma^{-2}\bar{g}$ has (by virtue of Equation (8.3)) constant negative scalar curvature $-d(d-1)$ on $M\backslash\Sigma$. Hence the above display is called the *singular Yamabe equation* [30]. For Σ closed and orientable, it is known that one-sided solutions for σ exist [28, 3, 30, 1].

This brings us to a distinct but closely related local problem. Namely, given an embedded hypersurface Σ, can we find a local defining function for Σ that is distinguished in the sense that σ solves $S(\bar{g},\sigma) = 1$? It turns out that, for smooth solutions σ, this is obstructed by an interesting conformal invariant of the extrinsic geometry. For example, for surfaces Σ we have the following result.

Proposition 8.2.3 *Let $(M^3, \bar{g})$ be a Riemannian three-manifold. Then a surface Σ admits a defining function satisfying*

$$S(\bar{g}, \sigma) = 1 + \sigma^4 T,$$

for some $T \in C^\infty M$, if and only if it is Willmore.

This follows from the work of [1], where a formula is given for T, and observations made in, for example, [19, 20] that in fact T can be identified with the functional gradient of the Willmore energy, which we term the *Willmore invariant*.

Proposition 8.2.3 is a special case of a rather uniform picture:

Theorem 8.2.4 *Given a hypersurface $\Sigma \hookrightarrow (M^d, \bar{g})$ embedded in a Riemannian d-manifold, there exists a smooth defining function such that*

$$S(\bar{g}, \sigma) = 1 + \sigma^d B,$$

where the obstruction $B_\Sigma := B|_\Sigma$ is a hypersurface conformal invariant and so, in particular, depends only on the conformal embedding $\Sigma \hookrightarrow (M^d, [\bar{g}])$.

The first and last parts of this theorem are in [1]. The link to the Willmore invariant led to the program, initiated in [19], to use this construction to define and investigate an analogous notion of a "Willmore invariant" in higher dimensions. Conformal hypersurface invariants are defined in [19], and in detail in [20] where Theorem 8.2.4 is established in a new way that is coordinate free and which reveals the nature of the obstruction function B_Σ. The quantity B_Σ obstructs smooth solutions σ to the equation $S(\bar{g}, \sigma) = 1$ subject to the boundary condition that Σ is the zero locus of σ. As mentioned, when $d = 3$ the obstruction B_Σ is the functional gradient of the Willmore energy functional, and thus defines the Willmore invariant discussed above. It has leading term $-\frac{1}{3}\Delta_\Sigma H$, where H is the hypersurface mean curvature. For hypersurfaces of higher dimension the obstruction B_Σ is a conformal hypersurface invariant that generalises this and, in particular, for d odd has the leading term

$$\Delta_\Sigma^{\frac{d-1}{2}} H, \quad \text{up to a non-zero constant.}$$

Thus we term B_Σ a *higher Willmore invariant*.

It happens that the obstructions B_Σ are variational: namely in d dimensions, generalised Willmore energy functionals exist with functional gradient B_Σ [23, 21] (and thus their critical points are hypersurfaces that admit smooth defining functions subject to $S(\bar{g}, \sigma) = 1 + \sigma^{d+1} T$ [19, 20, 22]).

With these preliminaries established we can now state a *singular Obata problem*, which is an obvious singular analog of the problem answered by the Obata theorem, Theorem 8.2.1.

Question 8.2.5 Let $(S^d, \bar{g})$ be the standard round sphere. Does there exist a smooth function $\sigma \in C^\infty S^d$ such that

$$S(\bar{g}, \sigma) = 1,$$

and such that $g = \sigma^{-2}\bar{g}$ is not Einstein where it is non-singular?

Recall that a hypersurface embedded in a conformal manifold $(M, [\bar{g}])$ is totally umbilic if the second fundamental form of $\bar{g}$ is pure trace, and this condition is independent of the choice of metric representative $\bar{g}$. A necessary condition for the singular metric $g = \sigma^{-2}\bar{g}$ to be Einstein with negative scalar curvature, away from the hypersurface Σ defined by σ, is that Σ is totally umbilic [14, 27]. Thus *if we fix the hypersurface Σ*, the trace-free second fundamental form obstructs the existence of Einstein solutions to (8.1) with $\mathcal{Z}(\sigma) = \Sigma$. The Clifford torus of the next example is important to us because it is a simple, and rather symmetric, Willmore surface that is not umbilic. (In particular, it provides a nice candidate for the zero locus of a non-Einstein solution of the singular Yamabe problem, as in Question 8.4.3 and Question 8.4.4.)

Example 8.2.6 To construct the Clifford torus, note that S^3 equipped with its standard conformally flat class of metrics may be realised as the space of lightlike lines $\mathcal{N}_+$ in $\mathbb{R}^5 \ni (x, y, X, Y, z)$ defined by

$$x^2 + y^2 + X^2 + Y^2 = z^2.$$

The pullback of the Minkowski metric

$$\tilde{g} = dx^2 + dy^2 + dX^2 + dY^2 - dz^2$$

to the $z = 1$ section of $\mathcal{N}_+$ is the standard round sphere metric $\bar{g}$ on S^3, while conformally related metrics are obtained by changing this choice of section. Calling

$$r^2 = x^2 + y^2 \text{ and } R^2 = X^2 + Y^2,$$

the surface $\Sigma = \{r = \frac{1}{\sqrt{2}} = R, z = 1\} \hookrightarrow S^3$ is Willmore; indeed Σ is the Clifford torus. That Σ is Willmore may be easily verified using Theorem 8.3.4. For that, note that the defining function

$$\sigma = \left[\frac{r - R}{\sqrt{2}} \left(1 - \frac{2}{3} \left(\frac{r - R}{\sqrt{2z}} \right)^2 \right) \right]^\star \in C^\infty S^3 \qquad (8.5)$$

obeys $S(\bar{g}, \sigma) = 1 + \sigma^4 T$ with T smooth; in the above $\star$ denotes the pullback from $C^\infty \mathbb{R}^5 \to C^\infty S^3$. To check this computation, note that if $\sigma = \tilde{\sigma}^\star$ for some $\tilde{\sigma} \in C^\infty \mathbb{R}^5$ of homogeneity one, then, using the ambient formula for the Thomas D-operator, the S-curvature enjoys an ambient formula

$$S(\bar{g}, \sigma) = \left[|d\tilde{\sigma}|_{\tilde{g}}^2 - \frac{2}{3} \tilde{\sigma} \Delta^{\tilde{g}} \tilde{\sigma} \right]^\star ;$$

see for example [7]. This model for the Clifford torus is shown in Figure 8.1.

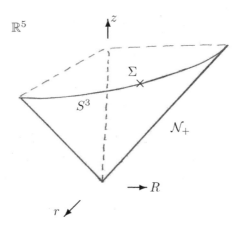

Figure 8.1 Model for the Clifford torus.

The angle coordinates are suppressed in Figure 8.1, so each point (save for when a radius r or R vanishes) represents a torus T^2.

By using polar coordinates (r, θ) and (R, Θ) for the xy- and XY-planes of the $z = 1$ hyperplane, calling

$$\tan \tau = \left[\frac{r - R}{r + R}\right]^{\star} \in [-1, 1], \text{ where } -\pi/4 \leq \tau \leq \pi/4,$$

the round S^3 metric in coordinates (τ, θ, Θ) becomes

$$ds^2 = d\tau^2 + \frac{1}{2}(d\theta^2 + d\Theta^2) + \frac{1}{2}\sin 2\tau(d\theta^2 - d\Theta^2).$$

Thus, away from circles at $\tau = \pm\pi/4$ (so $r^{\star} = 0$ or $R^{\star} = 0$), the sphere S^3 is foliated by tori of constant τ. The two circles correspond to a radius of a torus cycle degenerating to zero. The Clifford torus is the zero mean curvature torus at $\tau = 0$ and has flat torus metric

$$ds_\Sigma^2 = \frac{1}{2}(d\theta^2 + d\Theta^2).$$

The second fundamental form for the embedding $\Sigma \hookrightarrow S^3$ in these coordinates is

$$I\!I = \frac{1}{2}(d\theta^2 - d\Theta^2).$$

This is trace free, i.e. $I\!I = \mathring{I\!I}$, so Σ is a minimal surface, but the embedding $\Sigma \hookrightarrow (S^3, [\bar{g}])$ is not umbilic.

Question 8.2.5 leads to other natural questions, all of which have natural extensions to generally curved manifolds. To develop these, we first present key elements of the theory of conformal hypersurface embeddings.

8.3 Tractor Calculus for Hypersurface Embeddings

We say that a conformal manifold (M, c) is *almost Einstein* (AE) if there exists a metric $\bar{g} \in c$ and smooth function σ such that, away from the zero locus $\mathcal{Z}(\sigma)$, the metric

$$g = \frac{\bar{g}}{\sigma^2} \tag{8.6}$$

has vanishing trace-free Schouten tensor

$$\mathring{P}^g = 0. \tag{8.7}$$

Note that, in the special case when M is a d-manifold with dimension $(d - 1)$ boundary given by $\mathcal{Z}(\sigma) = \partial M$, and in addition σ is a defining function for ∂M (so $\mathrm{d}\sigma \neq 0$ at all points of ∂M), then metrics g obeying (8.6) are said to be *conformally compact*. When the Einstein condition (8.7) also holds, then such metrics g are called *Poincaré–Einstein*, or *asymptotically Poincaré–Einstein* when g solves (8.7) asymptotically to the highest order uniquely determined by the conformal class of metrics on ∂M determined by $\bar{g}$; see [11] for details.

Returning to the more general AE setting, notice that because g depends only on the ratio $\bar{g}/\sigma^2$, and thus equivalently only on equivalence classes of metric function pairs $[\bar{g}, \sigma] = [\Omega^2 g, \Omega\sigma]$ with $0 < \Omega \in C^\infty M$, the AE condition is better stated in terms of conformal densities defined as follows: a *weight* $w \in \mathbb{C}$ *conformal density* is a section of the line bundle

$$\mathcal{E}M[w] := \left((\wedge^d M)^2\right)^{-\frac{w}{2d}}.$$

The pair $[\bar{g}; \sigma]$ define a section $\boldsymbol{\sigma}$ of $\mathcal{E}M[1]$ via $\boldsymbol{\sigma}\,(\omega_{\bar{g}})^{-\frac{1}{d}}$. Moreover, tautologically, the metric $\bar{g}$ defines a section $\boldsymbol{g} \in \Gamma(\odot^2 T^* M \otimes \mathcal{E}M[2])$ by $\boldsymbol{g} := \bar{g}\,(\omega_{\bar{g}})^{-\frac{2}{d}}$; we call $\boldsymbol{g}$ the *conformal metric*. In addition, where this is defined, $g = \boldsymbol{\sigma}^{-2}\boldsymbol{g}$. In general, a *true scale* $0 < \boldsymbol{\tau} \in \Gamma(\mathcal{E}M[1])$ canonically defines a metric $g_{\boldsymbol{\tau}} := \boldsymbol{\tau}^{-2}\boldsymbol{g}$.

The AE equation, Equation (8.7), is overdetermined, so it is propitious to study its prolongation to a triple of sections

$$(\boldsymbol{\sigma}, \boldsymbol{n}, \boldsymbol{\rho}) \in \Gamma(\mathcal{E}M[1] \oplus T^* M[1] \oplus \mathcal{E}M[-1]),$$

where, for any vector bundle B over M, we define $B[w] := B \otimes \mathcal{E}M[w]$. Then, given any true scale $\boldsymbol{\tau}$, the conformal manifold $(M, \boldsymbol{c})$ is AE if and only if [4, 14]

$$\nabla^{\boldsymbol{\tau}}\boldsymbol{\sigma} - \boldsymbol{n} = 0,$$

$$\nabla^{\boldsymbol{\tau}}\boldsymbol{n} + \boldsymbol{\sigma}P^{g_{\boldsymbol{\tau}}} + \boldsymbol{\rho}\boldsymbol{g} = 0,$$

$$\nabla^{\boldsymbol{\tau}}\boldsymbol{\rho} - P^{g_{\boldsymbol{\tau}}}(\boldsymbol{n}, \cdot) = 0.$$

In the above, for example, $\nabla^{\boldsymbol{\tau}}\boldsymbol{\sigma} = [g_{\boldsymbol{\tau}}, \nabla^{g_{\boldsymbol{\tau}}}\sigma_{\boldsymbol{\tau}}] \in \Gamma(T^* M[1])$ and the inverse of the conformal metric $\boldsymbol{g}$ is used to contract $\boldsymbol{n}$ with the Schouten tensor; the remaining new notations should then be self explanatory. The choice of $\boldsymbol{\tau}$ in the above condition is irrelevant, since a different choice gives a linear (pointwise) combination of the three stated equations. In particular, away from $\mathcal{Z}(\boldsymbol{\sigma})$, the choice $\boldsymbol{\tau} = \boldsymbol{\sigma}$ gives that P^g is proportional to the metric and hence that the trace-free Schouten tensor $\mathring{P}^g = 0$.

In fact, the above system can be re-expressed as a linear connection acting on a section of a suitable, conformally invariant bundle. This bundle is the *tractor bundle*

$$\mathcal{T}M = \mathcal{E}M[1] \dotplus T^*M[1] \dotplus \mathcal{E}M[-1],$$

and its sections are called (standard) tractors. The semi-direct sum notation $\dotplus$ indicates that the tractor bundle is a disjoint union of Whitney sum vector bundles $\mathcal{E}M[1] \oplus T^*M[1] \oplus \mathcal{E}M[-1]$ indexed by $\bar{g} \in c$ (so equivalently true scales), quotiented by the equivalence on sections with indices labeled by true scales τ and τ', given by

$$(v^+, v, v^-)_\tau \sim \left(v^+, v + \Upsilon v^+, v^- - g^{-1}(\Upsilon, v + \tfrac{1}{2}\tau^2 \Upsilon v^-)\right)_{\tau'}, \quad (8.8)$$

where $\Upsilon := d\log(\tau'/\tau)$. Observe that the above formula implies that the first non-zero entry of a tractor is conformally invariant; this is called the *projecting part*. The appropriate linear connection is the *tractor connection* $\nabla^\mathcal{T} : \Gamma(\mathcal{T}M) \to \Gamma(\mathcal{T}M \otimes T^*M)$, defined for a metric labeled by the true scale τ, according to

$$\nabla^\mathcal{T}(v^+, v, v^-)_\tau := \left(\nabla^\tau v^+ - v, \nabla^\tau v + v^+ P^{g_\tau} + v^- g, \nabla^\tau v^- - P^{g_\tau}(v, \cdot)\right)_\tau. \tag{8.9}$$

The tractor bundle $\mathcal{T}M[1]$ enjoys a distinguished section X termed the *canonical tractor*, defined in any choice of scale τ by

$$X_\tau := (0, 0, 1).$$

Note that $\tau^{-1}X$ is a standard tractor and, from Equation (8.9), we see that the tractor connection obeys a non-degeneracy condition

$$\tau \nabla^\mathcal{T}(\tau^{-1}X)_\tau = (0, g, 0).$$

In the above terms, the AE condition (8.7) becomes the parallel condition on sections $I \in \Gamma(\mathcal{T}M)$ [4]:

$$\nabla^\mathcal{T} I = 0.$$

A necessary condition for I to be parallel is that, for any choice of true scale τ, this tractor is determined in terms of σ according to

$$I = \left(\sigma, \nabla^\tau \sigma, -\tfrac{1}{d}\tau^{-2}(\Delta^{g_\tau} + J^{g_\tau})\sigma\right)_\tau. \tag{8.10}$$

Tractors determined in terms of a section $\sigma \in \Gamma(\mathcal{E}M[1])$ in this way are termed *scale tractors* (the notation I_σ will sometimes be employed to make the dependence on σ clear).

The tractor bundle enjoys a conformally invariant tractor metric $h \in \Gamma(\odot^2 \mathcal{T} M)$ that is preserved by the tractor connection $\nabla^{\mathcal{T}}$ and given by (suppressing the choice of τ and recycling the notations for sections U, V of $\mathcal{T} M$ used above)

$$h(U, V) = u^+ v^- + u^- v^+ + g^{-1}(\boldsymbol{u}, \boldsymbol{v}) \in C^\infty M.$$

Observe that, given a scale tractor I, the corresponding scale σ is given in terms of the tractor metric and canonical tractor by

$$\sigma = h(I, X).$$

Now, if I is the scale tractor determined by $\sigma \in \Gamma(\mathcal{E}M[1])$, then the function

$$S(\boldsymbol{\sigma}) = h(I, I) =: I^2$$

equals the S-curvature defined in Equation (8.2) upon identifying σ with the metric–function pair $(\bar{g}, \sigma)$. Thus, the singular Yamabe equation, Equation (8.4), becomes

$$I^2 = 1.$$

Writing this condition out using Equation (8.10) with $\sigma = [\bar{g}, \sigma]$ gives

$$I^2 = |d\sigma|_{\bar{g}}^2 - \frac{2\sigma}{d} \left(\Delta^{\bar{g}} \sigma + J^{\bar{g}} \sigma \right) = 1.$$

Hence, along $\mathcal{Z}(\sigma)$, it follows that $d\sigma \neq 0$, so by the implicit function theorem $\Sigma = \mathcal{Z}(\sigma)$ is a smoothly embedded hypersurface. The same applies when $I^2 > 0$. Thus we have

Proposition 8.3.1 (Gover [14]) *Let (M, c) be a conformal manifold and I the scale tractor of $\sigma \in \Gamma(\mathcal{E}M[1])$ such that*

$$I^2 > 0.$$

Then, if $\Sigma \neq \emptyset$,

$$M = M_- \sqcup \Sigma \sqcup M_+,$$

where Σ is a smoothly embedded separating hypersurface. Moreover, the complements $M \backslash M_{\mp}$ are conformal compactifications of

$$M_\pm = \{ P \in M | \pm \sigma(P) > 0 \}.$$

This situation is depicted in Figure 8.2.

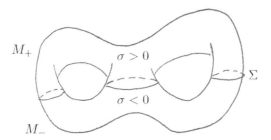

Figure 8.2 Stratification by scale.

The geometric setup depicted in Figure 8.2 is ideally suited to the study of conformal hypersurface embeddings $\Sigma \hookrightarrow (M, \boldsymbol{c})$. When considering Riemannian hypersurface embeddings $\Sigma \hookrightarrow (M, \bar{g})$, a key ingredient is the unit conormal $\hat{n}$ to Σ, which may be computed from any defining function σ by the formula $\hat{n} = d\sigma'\big|_\Sigma$, where the defining function $\sigma' := \sigma/|\nabla\sigma|_{\bar{g}}$ is a normalised improvement of σ. For conformal embeddings, an analogous, conformally invariant tractor object is the *normal tractor* $N \in \Gamma(\mathcal{T}M|_\Sigma)$, which, for a choice of metric $g_\tau \in \boldsymbol{c}$, is given by

$$N = (0, \hat{n}, -H^{g_\tau})_\tau.$$

Here we have used that the unit conormal rescales as a weight one-density under conformal changes of metric and so defines $\hat{n} \in \Gamma(\mathcal{T}M|_\Sigma)$. Also H^{g_τ} is the mean curvature of Σ computed in the g_τ metric. It is not difficult to check that upon replacing the true scale τ with a new true scale τ', this changes in concordance with Equation (8.8). Next we define the conformal analog of a defining function.

Definition 8.3.2 A density $\sigma \in \Gamma(\mathcal{E}M[1])$ satisfying

$$I^2\big|_\Sigma \neq 0, \quad \text{where } \Sigma \text{ is the zero locus of } \sigma,$$

is called a *defining density*.

Note that the zero locus of a defining density subject to $I^2\big|_\Sigma > 0$ is a hypersurface.

The next result relates the normal tractor to a defining density for Σ.

Proposition 8.3.3 (Gover [14, 22])   *Let σ be a defining density for Σ and*

$$\sigma' := \hat{\sigma}\left(1 - \frac{d}{4(d-1)}\left(I_{\hat{\sigma}}^2 - 1\right)\right),$$

with $\hat{\sigma} := \sigma/\sqrt{I_\sigma^2}$. Then

$$I_{\sigma'}^2 = 1 + \sigma^2 b,$$

for some $b \in \Gamma(\mathcal{E}M[-2])$, and

$$I_{\sigma'}\big|_\Sigma = N.$$

Observe that, not only does the above proposition give a formula for the normal tractor, but also it is the first step in constructing an asymptotic solution to the singular Yamabe problem for Σ. As we discussed earlier, an important result of Andersson, Chruściel, and Friedrich is that this can be solved to order d in σ [1]. In the language of densities, Theorem 8.2.4 is stated as follows (see [20]).

Theorem 8.3.4   *For any conformally embedded hypersurface $\Sigma \hookrightarrow (M^d, c)$, there exists a defining density σ, unique up to the addition of smooth terms proportional to σ^d, such that its scale tractor obeys*

$$I^2 = 1 + \sigma^d B \qquad\qquad (8.11)$$

for some $B \in \Gamma(\mathcal{E}M[-d])$. Moreover,

$$B_\Sigma = B|_\Sigma$$

is an invariant of the conformal embedding.

The density B_Σ is called the *obstruction density* because it obstructs smooth solutions to the singular Yamabe problem, where here we mean that the defining density is smooth across Σ (when the obstruction density is non-vanishing, one-sided smooth solutions are known to still be possible, but with logarithmic behavior around Σ [28, 3, 30, 1]). For conformally embedded surfaces, the obstruction density may be obtained as the functional gradient of the Willmore energy functional. Of course, B_Σ is the density-valued analogue of the higher Willmore invariant introduced earlier.

Another important invariant of conformal hypersurface embeddings is the trace free part of the second fundamental form $\mathring{I\!I} \in \Gamma(\odot^2 T^* M \otimes \mathcal{E}M[1]|_\Sigma)$. Let us call defining densities that obey Equation (8.11) *unit*. Then, from Proposition 8.3.3 and [4] (see also [26, 10]) we have the following result.

Proposition 8.3.5 *Let σ be a unit defining density for $\Sigma \hookrightarrow (M, c)$. Then, the projecting part of the restriction of $\nabla^T I_\sigma$ to Σ equals the trace free second fundamental form $\mathring{I\!I}$.*

Note that total umbilicity of the boundary of a Poincaré–Einstein manifold is a direct corollary of Proposition 8.3.5.

Just as the Gauss formula relates the projection of the ambient Levi-Civita connection to its counterpart intrinsic to the hypersurface Σ through the second fundamental form, a similar result holds for the tractor connection. For that, along Σ the normal tractor N and tractor metric h give a canonical Whitney direct sum bundle decomposition

$$\mathcal{T}M|_\Sigma = \mathcal{T}M^\perp \oplus \mathcal{T}M^\|.$$

Moreover, the rank $d + 1$ bundle $\mathcal{T}M^\perp$ (whose sections V obey $h(V, N) = 0$) is isomorphic to the tractor bundle intrinsic to (Σ, c_Σ), where c_Σ is the conformal class of metrics on Σ induced from c on M. From Equation (8.9), we see that the difference between the projection of the ambient tractor connection to $\mathcal{T}M^\perp$ and the tractor connection depends on the difference between their respective Schouten tensors; this is measured by the *Fialkow tensor*

$$\boldsymbol{F} \stackrel{\tau}{=} (P^{g_\tau})^\top - P^{g_\tau^\Sigma} + H^{g_\tau}\mathring{I\!I} + \frac{1}{2}(H^{g_\tau})^2 g_\tau^\Sigma$$

(an equals sign adorned by τ is used to indicate a formula computed in a choice of scale τ), where the ambient dimension $d \geq 4$. In the above formula, g_τ^Σ denotes the metric induced by g_τ along Σ, the subscript $\top$ projects the tangent bundle of M along Σ to directions orthogonal to the unit conormal, and we have used that this bundle projection gives a rank $(d - 1)$ bundle isomorphic to $T\Sigma$.

The conformally invariant, extrinsically coupled tensors $\hat{n}, \mathring{I\!I}, \boldsymbol{F}$, and B_Σ are key ingredients of a conformal tensor calculus of $\Sigma \hookrightarrow (M, c)$. The computation of more interesting quantities such as higher Willmore energies requires the introduction of a pair of new tractor operators. Firstly, the *Thomas D-operator* gives a mapping

$$\Gamma(\mathcal{T}^\Phi M[w]) \to \Gamma(\mathcal{T}^\Phi M[w - 1] \otimes \mathcal{T}M),$$

where Φ indicates any tensor product of tractor and conformal density bundles. In a self-explanatory matrix notation (see [10]),

$$D \stackrel{\tau}{=} \begin{pmatrix} w(d + 2w - 2) \\ (d + 2w - 2)\nabla^\top \\ -\tau^{-2}(\Delta^\top + wJ^{g_\tau}) \end{pmatrix},$$

where $\Delta^\mathcal{T}$ is the rough Laplace-type operator built in the standard way from the metric g_τ and the tractor connection $\nabla^\mathcal{T}$. Note that the scale tractor I of $\sigma \in \Gamma(\mathcal{E}M[1])$ is given by $I = \frac{1}{d}D\sigma$. Contracting the Thomas D-operator with the scale tractor I, using the tractor metric, gives its *Laplace–Robin operator* denoted $I \cdot D$. Upon restriction to the hyper-surface zero locus Σ of σ, this gives a conformally invariant Robin-type combination of normal and Dirichlet operators due to Cherrier [9]. Away from Σ, computed in the metric g_σ, the Laplace–Robin operator gives the Laplacian modified by scalar curvature, which, when acting on conformal densities at the critical weight $w = 1 - \frac{d}{2}$, recovers the Yamabe operator $-\Delta^{g_\sigma} - (1 - \frac{d}{2})J^{g_\sigma}$.

Suitable critical powers of the Laplace–Robin operator, upon restriction to Σ, give extrinsically coupled Laplacian powers [18, 22]. These may be viewed as extrinsic generalisations of the conformal Laplacian powers of Graham *et al.* [24].

Recall that what we term the higher Willmore invariant is the scalar conformal invariant B_Σ of Theorem 8.2.4. This is the obstruction to smooth solutions of the singular Yamabe problem. It was shown by Graham that there is an action (or "energy") that has B_Σ as its functional gradient with respect to variations of embedding [23]. This energy arises as an anomaly term in the volume asymptotics associated with a metric satisfying the approximate singular Yamabe problem, as in Theorem 8.2.4. Subsequently it was proved in [21] that this energy arises from the integral of a local Riemannian invariant that nicely generalises (by including extrinsic data) Branson's celebrated Q-curvature. To obtain this requires the notion of a log-density. A log-density is a section of an associated bundle induced by the additive representation $y \mapsto y - \log t$ of $\mathbb{R}_+ \ni t$ to the bundle of metrics c over M, which itself is an $\mathbb{R}_+$ principal bundle. Given a true scale $\tau = [g, \tau]$, the pair $(g, \log \tau)$ determines a log-density that we call $\log \tau$. The Laplace–Robin operator can be extended to act on log-densities and maps these to weight $w = -1$ conformal densities. With this all in place we then have (in summary of some results in [21, 22]) the following theorem.

Theorem 8.3.6 *Let $\Sigma \hookrightarrow (M, c)$ be a conformally embedded hyper-surface, let $\bar{g} = g/\tau^2$ be a metric determined by a true scale τ, and call*

$$Q^\tau := (I \cdot D)^{d-1} \log \tau\big|_\Sigma \in \Gamma(\mathcal{E}\Sigma[1 - d]),$$

where I is the scale tractor of a unit defining density σ for Σ. Then the *integral*

$$\mathcal{A} = \int_\Sigma Q^\tau$$

is independent of the choice of $\bar{g} \in \mathbf{c}$. Moreover, if the singular metric $g = \mathbf{g}/\sigma^2$ is asymptotically Poincaré–Einstein and d is odd, then $(-1)^{d-1}((d-2)!!)^{-2}Q^\tau$ is the Branson Q-curvature of $(\Sigma, \mathbf{c}_\Sigma)$ determined by $\bar{g}$.

In the above we have used that the bundles $\mathcal{E}M[w]|_\Sigma$ and $\mathcal{E}\Sigma[w]$ are isomorphic. Results of the above type, where a complicated-to-compute object along a hypersurface is the restriction of a simple bulk quantity, are called *holographic formulae*. The conformal density Q^τ gives an extrinsic analogue of the Branson Q-curvature, while its conformally invariant integral $\mathcal{A}$ is a generalised Willmore energy. In fact, the functional $\frac{(-1)^{d-1}}{(d-1)!\,(d-2)!}\mathcal{A}$ is the anomaly coefficient of the log divergence of the renormalised volume expansion for the singular metric g [21]. Moreover, the failure of the *extrinsic Q-curvature* Q^τ to be conformally invariant is controlled by the extrinsic conformal Laplacian powers described above. Explicit low dimensional formulae for the extrinsic Q-curvature are given in the following example.

Example 8.3.7   When $\Sigma \hookrightarrow (M^3, \mathbf{c})$ is a conformally embedded surface,

$$\int_\Sigma Q^\tau = \pi \chi_\Sigma - \frac{1}{4}\int_\Sigma \mathrm{tr}_{g_\Sigma}\mathring{I\!I}^2.$$

Here χ denotes the Euler characteristic while the second term is a manifestly conformally invariant bending energy. For $\Sigma \hookrightarrow (M^4, \mathbf{c})$ a closed hypersurface embedded in a conformal four-manifold [22, 13],

$$\int_\Sigma Q^\tau = \frac{2}{3}\int_\Sigma \mathrm{tr}_{g_\Sigma}(\mathring{I\!I} \circ F).$$

In the above examples, we have used the conformal metric along Σ to construct an endomorphism of $T\Sigma$ from the trace free second fundamental form and Fialkow tensor; this is the meaning of the symbol tr_{g_Σ}. The analogue of the extrinsic Q-curvature for four manifold embeddings has been computed in [25].

8.3.1 The Sphere

Much of the above is nicely illustrated on the sphere equipped with its standard round conformal structure. In this case the conformal structure is conformally flat and the tractor connection ∇^T has trivial holonomy. Thus any standard tractor at a point can be extended to a parallel tractor I on the sphere.

In fact this is easily established explicitly, as the conformal d-sphere arises as the ray projectivisation $\mathbb{P}_+$ of the future null cone $\mathcal{N}_+$ of the (Lorentzian signature) Minkowski metric $\tilde{g} := \text{diagonal}(-1, 1, \ldots, 1)$ on $\mathbb{R}^{d+2}$ (see Example 8.2.6). Then the tractor connection arises from parallel transport in $\mathbb{R}^{d+2}$ viewed as an affine space (see [7, 10] for more detail). It follows that each (constant) vector $\tilde{I}$ in $\mathbb{R}^{d+2}$ determines a corresponding parallel tractor I on the sphere and vice versa. In this picture the canonical tractor X corresponds to the restriction to $\mathcal{N}_+$ of the Euler vector field $\tilde{X}$ of $\mathbb{R}^{d+2}$. Thus if a parallel tractor I is timelike, and normalised to satisfy $I^2 = h(I, I) = -1$ say, then $\boldsymbol{\sigma} := h(I, X)$ has no zero locus, and the corresponding Einstein metric $g = \boldsymbol{\sigma}^{-2}\mathbf{g}$ has scalar curvature $d(d - 1)$. Thus it is the usual round sphere metric. If on the other hand I is a null parallel tractor then $\boldsymbol{\sigma} := h(I, X)$ has an isolated zero corresponding to where $\tilde{X}$ is parallel to $\tilde{I}$. Since $I^2 = 0$, the scalar curvature vanishes and the Einstein metric $g = \boldsymbol{\sigma}^{-2}\mathbf{g}$ on the complement is thus Ricci flat. Hence it is isometric to Euclidean space. Similarly if a parallel tractor I is spacelike, satisfying $I^2 = 1$, then the zero locus Σ of $\boldsymbol{\sigma} := h(I, X)$ is a hypersurface corresponding to the intersection of the hyperplane $\tilde{h}(\tilde{I}, \tilde{X}) = 0$ with $\mathcal{N}_+$, and the corresponding Einstein metric $g = \boldsymbol{\sigma}^{-2}\mathbf{g}$ on $S^d \setminus \Sigma$ has scalar curvature $-d(d-1)$. As it is a conformally flat Einstein metric it is isometric to the hyperbolic metric. Moreover, since I is in particular parallel along Σ, it follows from Proposition 8.3.5 that Σ is totally umbilic. This, with each of the connected parts of $S^d \setminus \Sigma$, provides a conformal compactification of hyperbolic space which (by stereographic projection) is conformally equivalent to the usual Poincaré ball, see Figure 8.3.

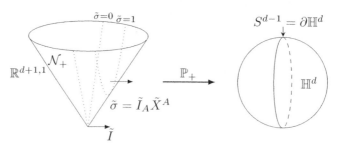

Figure 8.3 Almost Einstein structures on the sphere.

8.4 Singular Yamabe and Obata Problems

Let (M, c) be a conformal manifold equipped with a smooth and sign changing solution σ of the singular Yamabe problem $S(c, \sigma) = 1$. It follows that $\Sigma = \mathcal{Z}(\sigma) \neq \emptyset$, and then, from Proposition 8.3.1 and Theorem 8.3.4, that Σ is a smoothly embedded separating hypersurface that satisfies the higher Willmore equation $B_\Sigma = 0$. Thus, locally, such solutions potentially provide an interesting route to accessing and studying higher Willmore hypersurfaces [19, 20, 22].

The global problem is potentially even more intriguing. The existence of sign changing solutions σ to the singular Yamabe problem is itself interesting, and then, when such exist, the zero locus Σ is a closed embedded higher Willmore hypersurface. There are examples with Σ totally umbilic and the metric $g = \sigma^{-2}g$ Einstein on the complement of Σ. We have seen this above on the sphere in Section 8.3.1 (following [14]), and there are also examples on suitable products of the sphere with Einstein manifolds [16, 17]. This leads to our main questions.

Question 8.4.1 Do there exist closed conformal manifolds (M, c) that admit a smooth, sign changing singular Yamabe scale $\sigma \in \Gamma(\mathcal{E}[1])$ such that the S-curvature $S = I^2$ obeys $S = 1$ but with $g = \sigma^{-2}g$ not Einstein on $M \setminus \mathcal{Z}(\sigma)$?

In fact the situation would be most interesting if the zero locus were not totally umbilic. Thus the following is an interesting problem.

Question 8.4.2 Do there exist closed conformal manifolds (M, c) that admit a sign changing singular Yamabe scale $\sigma \in \Gamma(\mathcal{E}[1])$ such that the S-curvature $S = I^2$ obeys $S = 1$ and such that $\Sigma = \mathcal{Z}(\sigma)$ is not totally umbilic?

Note that a positive answer to this implies a positive answer to Question 8.4.1 because if $g = \sigma^{-2}g$ is Einstein then the scale tractor I is parallel everywhere, but along Σ, agrees with the normal tractor, which is thus parallel and so, as discussed earlier $\Sigma = \mathcal{Z}(\sigma)$ is totally umbilic.

There are refinements of these questions where we assume the initial conformal manifold includes an Einstein metric. The first here is a variant of what we called earlier the singular Obata problem.

Question 8.4.3 Do there exist closed Einstein manifolds $(M, \bar{g})$ that admit a sign changing singular Yamabe scale $\sigma \in \Gamma(\mathcal{E}[1])$ such that the S-curvature $S = I^2$ obeys $S = 1$ but with $g = \sigma^{-2}g$ not Einstein on $M \setminus \mathcal{Z}(\sigma)$?

Once again one can ask for a stronger result, as follows.

Question 8.4.4   Do there exist closed Einstein manifolds $(M, \bar{g})$ that admit a sign changing singular Yamabe scale $\boldsymbol{\sigma} \in \Gamma(\mathcal{E}[1])$ such that the S-curvature $S = I^2$ obeys $S = 1$ and such that $\Sigma = \mathcal{Z}(\boldsymbol{\sigma})$ is not totally umbilic?

The former of course generalises Question 8.2.5 for the sphere. Questions 8.4.2 and 8.4.4 are particularly interesting because the obstruction density $\boldsymbol{B}_\Sigma$ is a higher Willmore invariant [19, 20, 22].

To the best of our knowledge, the above problems, as stated, are completely open. There are other versions of singular Yamabe problems that have been studied by a number of authors, see e.g. [12, 30, 31] and references therein, but these do not appear to have a direct link to the Willmore equation and its higher dimensional analogs.

References

[1] L. Andersson, P. T. Chruściel and H. Friedrich, *On the regularity of solutions to the Yamabe equation and the existence of smooth hyperboloidal initial data for Einstein's field equations*, Commun. Math. Phys. **149** (1992), 587–612.

[2] T. Aubin, *Équations différentielles non linéaires et problème de Yamabe concernant la courbure scalaire*, J. Math. Pures Appl., **55** (9) (1976), 269–296.

[3] P. Aviles and R. C. McOwen, *Complete conformal metrics with negative scalar curvature in compact Riemannian manifolds*, Duke. Math. J. **56** (1988), 395–398.

[4] T. N. Bailey, M. G. Eastwood and A. R. Gover, *Thomas's structure bundle for conformal, projective and related structures*, Rocky Mountain J. Math. **24** (4) (1994), 1191–1217.

[5] T. P. Branson, *Sharp inequalities, the functional determinant, and the complementary series*, Trans. Amer. Math. Soc. **347**(10) (1995), 3671–3742.

[6] T. P. Branson and B. Ørsted, *Explicit functional determinants in four dimensions*, Proc. Amer. Math. Soc. **113** (1991), 669–682.

[7] A. Čap and A.R. Gover, *Standard tractors and the conformal ambient metric construction*, Ann. Global Anal. Geom. **24** (2003), 231–295.

[8] S.-Y. Chang, M. Eastwood, B. Ørsted and Paul C. Yang, *What is Q-curvature?*, Acta Appl. Math. **102**(2) (2008), 119–125.

[9] P. Cherrier, *Problèmes de Neumann non linéaires sur les variétés riemanniennes*, J. Funct. Anal. **57** (2) (1984), 154–206.

[10] S. Curry and A. R. Gover, *An introduction to conformal geometry and tractor calculus, with a view to applications in general relativity*, in

T. Daudé (ed.), Asymptotic Analysis in General Relativity, pp. 86–170, Cambridge: Cambridge University Press, 2018, arXiv:1412.7559.

[11] C. Fefferman and C. R. Graham, *Conformal invariants*, in The Mathematical Heritage of Élie Cartan (Lyon, 1984). Astérisque 1985, Numero Hors Serie, pp. 95–116.

[12] D. L. Finn, *On the negative case of the singular Yamabe problem*, J. Geom. Anal. **9** (1999), 73–92.

[13] M. Glaros, A. R. Gover, M. Halbasch and A. Waldron, *Singular Yamabe problem Willmore energies*, J. Geom. Phys. **138** (2019), 168–193, arXiv:1508.01838.

[14] A. R. Gover, *Almost Einstein and Poincaré-Einstein manifolds in Riemannian signature*, J. Geom. Phys. **60**(2) (2010), 182–204, arXiv:0803.3510.

[15] A. R. Gover, C. Arias and A. Waldron, *Conformal geometry of embedded manifolds with boundary from universal holographic formulae* (2019) arXiv:1906.01731.

[16] A. R. Gover and F. Leitner, *A class of compact Poincaré-Einstein manifolds: properties and construction*, Commun. Contemp. Math. **12** (2010), 629–659.

[17] A. R. Gover and F. Leitner, *A sub-product construction of Poincaré-Einstein metrics*, Internat. J. Math. **20** (2009), 1263–1287.

[18] A. R. Gover and A. Waldron, *Boundary calculus for conformally compact manifolds*, Indiana Univ. Math. J. **63**(1) (2014), 119–163, arXiv:1104.2991.

[19] A. R. Gover and A. Waldron, *Submanifold conformal invariants and a boundary Yamabe problem*, in M. del mar Gonzalez *et al* (eds.), Extended Abstracts Fall 2013: Geometrical Analysis, Trends in Mathematics Research Perspectives CRM Barcelona, vol. 3. Cham: Springer, 2015 pp. 21–26. arXiv:1407.6742.

[20] A. R. Gover and A. Waldron, *Conformal hypersurface geometry via a boundary Loewner-Nirenberg-Yamabe problem*, Commun. Anal. Geom. to appear, arXiv:1506.02723.

[21] A. R. Gover and A. Waldron, *Renormalized volume*, Commun. Math. Phys. **354** (2017), 1205–1244, arXiv:1603.07367.

[22] A. R. Gover and A. Waldron, *A calculus for conformal hypersurfaces and new higher Willmore energy functionals*, Adv. Geom. **20**(1) (2020), 29–60, arXiv:1611.04055.

[23] C. R. Graham, *Volume renormalization for singular Yamabe metrics*, Proc. Amer. Math. Soc. **145** (2017), 1781–1792, arxiv1606.00069.

[24] C. R. Graham, R. Jenne, R. L. Mason and G. Sparling, *Conformally invariant powers of the Laplacian. I. Existence*, J. London Math. Soc. (2) **46** (1992), 557–565.

[25] C. R. Graham and N. Riechert, *Higher-dimensional Willmore energies via minimal submanifold asymptotics* (2017), preprint arXiv:1704.03852.

[26] D. Grant, *A conformally invariant third order Neumann-type operator for hypersurfaces*, Master's thesis, University of Auckland, New Zealand, 2003.

[27] C. R. LeBrun, *H-space with a cosmological constant*, Proc. R. Soc. Lond. A **380** (1982), 171–185.

[28] C. Loewner and L. Nirenberg, *Partial differential equations invariant under conformal or projective transformations*, in L. V. Ahlfors, I. Kra and B. Maskit (eds.), Contributions to Analysis (A Collection of Papers dedicated to Lipman Bers), pp. 245–272. New York, Academic Press, 1974.

[29] F. Coda-Marques and A. Neves, *Min-max theory and the Willmore conjecture*, Ann. of Math. (2) **179** (2014), 683–782.

[30] R. Mazzeo, *Regularity for the singular Yamabe problem*, Indiana Univ. Math. J. **40** (1991), 1277–1299.

[31] M. Medina, M. Musso and J. Wei, *Desingularization of Clifford torus and nonradial solutions to the Yamabe problem with maximal rank*, J. Funct. Anal. **276** (2019), 2470–2523.

[32] M. Obata, *The conjectures on conformal transformations of Riemannian manifolds*, J. Differ. Geom. **6** (1972), 247–258.

[33] R. Schoen, *Variational theory for the total scalar curvature functional for Riemannian metrics and related topics*, in Topics in Variational Calculus, Lecture Notes in Mathematics **1365**, pp. 120–154. Cham: Springer, 1980.

[34] R. Schoen, *Conformal deformation of a Riemannian metric to constant scalar curvature*, J. Differ. Geom. **20** (1984), 479–495.

[35] N. S. Trudinger, *Remarks concerning the conformal deformation of Riemannian structures on compact manifolds*, Ann. Scuola Norm. Sup. Pisa (3), **22** (1968), 265–274.

[36] T. J. Willmore, *Note on embedded surfaces*, An. Sti. Univ. "Al. I. Cuza" Iaşi Sect. I a Mat. (N.S.) **11B** (1965), 493–496.

[37] T. J. Willmore, *A survey on Willmore immersions*, in F. Dilen (ed.), Geometry and Topology of Submanifolds IV, pp. 11–16. Singapore: World Scientific, 1991.

[38] H. Yamabe, *On a deformation of Riemannian structures on compact manifolds*, Osaka J. Math. **12** (1960), 21–37.

9

Einstein Metrics, Harmonic Forms and Conformally Kähler Geometry

Claude LeBrun

Abstract

The author has elsewhere given a complete classification of the compact oriented Einstein 4-manifolds that satisfy $W^+(\omega,\omega) > 0$ for some self-dual harmonic 2-form ω, where W^+ denotes the self-dual Weyl curvature. In this chapter, similar results are obtained when $W^+(\omega,\omega) \geq 0$, provided the self-dual harmonic 2-form ω is transverse to the zero section of $\Lambda^+ \to M$. However, this transversality condition plays an essential role in the story; dropping it leads one into wildly different territory where entirely different phenomena predominate.

9.1 Introduction

Recall that a Riemannian metric h is said to be *Einstein* [3] if it has constant Ricci curvature, or in other words if it solves the Einstein equation

$$r = \lambda h \qquad (9.1)$$

for some real number λ, where r is the Ricci tensor of h. When this happens, λ is called the *Einstein constant* of h, and of course has the same sign as the Einstein metric's scalar curvature.

Dimension four seems to represent a sort of "Goldilocks zone" for the Einstein equation. In lower dimensions, Einstein metrics are extremely rigid, in the sense that they necessarily have constant sectional curvature, and so do not really exhibit any interesting local differential geometry. In higher dimensions, on the other hand, they are extremely flexible, existing in such profusion on familiar manifolds [6, 7, 32] that their local geometry seems to offer little clue as to the identity of the manifold where they reside. By contrast, dimension four seems "just right" for (9.1), as four-dimensional Einstein metrics exhibit a well-tempered combination of local flexibility and global rigidity that often makes their geometry perfectly reflect the manifold on which they live. For example, if M is a compact real or complex-hyperbolic 4-manifold, a 4-torus, or $K3$, the moduli space of Einstein metrics on M is known explicitly, and moreover turns out to be connected [3, 4, 16].

Unfortunately, however, we do not have a similarly complete understanding of the moduli space of Einstein metrics on most of the 4-manifolds where this moduli space is non-empty. An important family of test-cases is provided by the *del Pezzo surfaces*, here understood to mean the smooth compact oriented 4-manifolds that support complex structures with ample anti-canonical line bundle. Up to diffeomorphism, there are exactly ten such manifolds, namely $S^2 \times S^2$ and the nine connected sums $\mathbb{CP}_2 \# m\overline{\mathbb{CP}}_2$, $m = 0, 1, \ldots, 8$. These 4-manifolds are completely characterized [8] by two properties: they admit Einstein metrics with $\lambda > 0$, and they also admit symplectic structures. However, it is currently unclear whether the known Einstein metrics on these spaces sweep out the entire Einstein moduli space. One of our main objectives here will be to generalize and strengthen a characterization of the known Einstein metrics on del Pezzo surfaces previously proved by the author in [20].

In order to formulate our results, first recall that the bundle of 2-forms $\Lambda^2 \to M$ over an oriented Riemannian 4-manifold (M, h) decomposes invariantly as the Whitney sum

$$\Lambda^2 = \Lambda^+ \oplus \Lambda^- \tag{9.2}$$

of the eigenspaces of the Hodge star operator $\star : \Lambda^2 \to \Lambda^2$. Sections of the $(+1)$-eigenbundle Λ^+ are called self-dual 2-forms, while the sections of the (-1)-eigenbundle Λ^- are called anti-self-dual 2-forms. The decomposition (9.2) is moreover *conformally invariant*, meaning that it is unchanged by multiplying the metric by an arbitrary positive function.

One important consequence of the decomposition (9.2) is that it induces an invariant decomposition of the Riemann curvature tensor $\mathcal{R}$ into simpler pieces. Indeed, if we identify the Riemann curvature tensor with the self-adjoint endomorphism $\mathcal{R} : \Lambda^2 \to \Lambda^2$ of the 2-forms defined by

$$\varphi_{ab} \longmapsto \frac{1}{2} \mathcal{R}^{cd}{}_{ab} \varphi_{cd}$$

and known as the *curvature operator*, then (9.2) allows us to decompose

$\mathcal{R}$ into irreducible pieces:

$$\mathcal{R} = \left(\begin{array}{c|c} W^+ + \frac{s}{12} & \overset{\circ}{r} \\ \hline \overset{\circ}{r} & W^- + \frac{s}{12} \end{array} \right), \qquad (9.3)$$

where s denotes the scalar curvature, $\overset{\circ}{r} = r - \frac{s}{4}g$ is the trace-free Ricci curvature, and where the remaining pieces $W^\pm$, known as the *self-dual* and *anti-self-dual* Weyl tensors, are the trace-free parts of the endomorphisms of $\Lambda^\pm$ induced by $\mathcal{R}$. Remarkably, the corresponding pieces $(W^\pm)^a{}_{bcd}$ of the Riemann curvature tensor are both *conformally invariant*, in the sense that they remain unaltered if the metric is multiplied by an arbitrary smooth positive function.

Let us now assume that (M, h) is a *compact* oriented Riemannian 4-manifold. The Hodge theorem then tells us that every de-Rham class on M has a unique harmonic representative. In particular, there is a canonical isomorphism

$$H^2(M, \mathbb{R}) \cong \{\varphi \in \Gamma(\Lambda^2) \mid d\varphi = 0, \ d \star \varphi = 0\}.$$

However, since the Hodge star operator $\star$ defines an involution of the right-hand side, we obtain a direct-sum decomposition

$$H^2(M, \mathbb{R}) = \mathcal{H}_h^+ \oplus \mathcal{H}_h^-, \qquad (9.4)$$

where

$$\mathcal{H}_h^\pm = \{\varphi \in \Gamma(\Lambda^\pm) \mid d\varphi = 0\}$$

are the spaces of self-dual and anti-self-dual harmonic forms. Since the conditions of being closed and belonging to $\Lambda^\pm$ are both conformally invariant, it follows that the spaces $\mathcal{H}^\pm$ are both conformally invariant, too. Moreover, the dimensions $b_+ = \dim \mathcal{H}^\pm$ of these spaces are completely metric independent, and can easily be shown to be oriented homotopy invariants of the 4-manifold M.

Now, if (M, h) is a compact oriented Riemannian 4-manifold, and if $\omega \in \mathcal{H}^+$ is a fixed self-dual harmonic 2-form, the quantity

$$W^+(\omega, \omega) := \langle W^+(\omega), \omega \rangle = \frac{1}{4}(W^+)^{abcd}\omega_{ab}\omega_{cd}$$

transforms in an extremely simple manner under conformal rescaling;

namely, if we change our metric by

$$h \rightsquigarrow u^2 h$$

for some positive function u, then the quantity in question changes by

$$W^+(\omega, \omega) \rightsquigarrow u^{-6} W^+(\omega, \omega).$$

In particular, the *sign* of this quantity at a given point is unchanged by conformal rescalings. This makes this hybrid measure of curvature particularly compelling when $b_+(M) = 1$, because in this case there is, up to a non-zero constant factor, only one non-trivial choice of ω, and the sign of $W^+(\omega, \omega)$ at each point then becomes a natural global conformal invariant of (M, h).

The main result of [20] was that if a compact 4-dimensional Einstein manifold satisfies

$$W^+(\omega, \omega) > 0 \qquad\qquad (9.5)$$

for some self-dual harmonic 2-form ω, then (M, h) is one of the known Einstein metrics on some del Pezzo surface. Conversely, the known Einstein metrics on del Pezzo surfaces all have this property. Combining these two observations then shows, as a corollary, that the known Einstein metrics on these spaces exactly sweep out one connected component of the Einstein moduli space. Here it is worth noting that every del Pezzo surface has $b_+ = 1$, so that condition (9.5) represents a rather natural characterization of the known Einstein metrics on these 4-manifolds.

On the other hand, since condition (9.5) trivially implies that both W^+ and ω are nowhere zero, it might seem desirable to relax this overly stringent condition by merely requiring that $W^+(\omega, \omega)$ be non-negative. What we will show here is that this can indeed be done, provided one imposes an interesting and natural condition on the 2-form. Namely, if ω is a harmonic self-dual 2-form on a compact oriented Riemannian 4-manifold (M, h), one says that ω is *near-symplectic* if its graph is transverse to the zero section of the rank-3 vector bundle $\Lambda^+ \to M$. This is a *generic* condition, as has come to be understood through the work of Taubes [27, 29] and others [14, 18, 24]; indeed, on any smooth compact oriented 4-manifold with $b_+ \neq 0$, the set of metrics admitting a near-symplectic self-dual harmonic 2-form is open and dense. Of course, a dimension count immediately reveals that the zero locus of a near-symplectic self-dual harmonic 2-form ω on (M, h) is automatically a (possibly empty) finite disjoint union Z of circles:

$$Z \approx \sqcup_{j=1}^n S^1. \qquad\qquad (9.6)$$

Imposing this reasonable assumption on the behavior of ω will actually allow us to prove some natural generalizations of the main result of [20]. Namely, the main results of the present chapter are the following:

Theorem 9.1.1 *Let (M, h) be a compact oriented Einstein 4-manifold that carries a near-symplectic self-dual harmonic 2-form ω such that*

$$W^+(\omega, \omega) \geq 0, \qquad W^+(\omega, \omega) \not\equiv 0. \tag{9.7}$$

Then $W^+(\omega, \omega) > 0$ everywhere, M is diffeomorphic to a del Pezzo surface, and h is conformally related to a positive-scalar-curvature extremal Kähler metric g on M with Kähler form ω. Conversely, every del Pezzo surface admits an Einstein metric h satisfying (9.7) for a self-dual harmonic 2-form ω that is nowhere zero (and hence near-symplectic).

Theorem 9.1.2 *Let (M, h) be a compact oriented $\lambda \geq 0$ Einstein 4-manifold that carries a near-symplectic self-dual harmonic 2-form ω such that*

$$W^+(\omega, \omega) \geq 0 \tag{9.8}$$

everywhere. Then ω is nowhere zero, and h is conformally related to an extremal Kähler metric g on M with Kähler form ω. Moreover, M is diffeomorphic to a del Pezzo surface, a K3 surface, an Enriques surface, an Abelian surface, or a hyper-elliptic surface. Conversely, each of these complex surfaces admits a $\lambda \geq 0$ Einstein metric h satisfying (9.8) for a self-dual harmonic 2-form ω that is nowhere zero (and hence near-symplectic).

Theorem 9.1.3 *The near-symplectic hypothesis in Theorem 9.1.1 is essential: counter-examples show that the result fails without this assumption.*

The proofs of these main results can be found Section 9.4 below, following the proofs, in Sections 9.2–9.3, of the technical results that underpin these theorems.

Acknowledgements This project was originally conceived during the Australian–German workshop on Differential Geometry in the Large held in Creswick, Australia, in February, 2019, while the author was on sabbatical leave as a Simons Fellow in Mathematics. The author would thus like to offer his profound thanks to both the Simons Foundation

220   *C. LeBrun*

and the MATRIX Institute. He would also like to heartily thank Cliff
Taubes for his helpful, encouraging, and amusing comments on the first
version of the manuscript.

The author was supported in part by a Simons Fellowship and by NSF
grant DMS-1906267.

9.2 An Integral Weitzenböck Formula

Let (M, h) be a compact oriented Riemannian 4-manifold with *harmonic
self-dual Weyl curvature*, in the sense that $\delta W^+ := -\nabla \cdot W^+ = 0$. When
h is Einstein, this property automatically holds, by virtue of the second
Bianchi identity. We will further assume throughout that h is at least
C^4. The latter assumption is of course innocuous in the Einstein case,
as elliptic regularity for (9.1) implies that Einstein metrics are always
[10] real-analytic in harmonic coordinates.

We will henceforth also assume that $b_+(M) \neq 0$. This is equivalent to
saying that (M, h) admits a self-dual harmonic 2-form $\omega \not\equiv 0$. We now
choose some such form, and regard it as fixed for the remainder of the
discussion. Let $Z \subset M$ denote the zero set of ω. Since ω is self-dual by
assumption,

$$\omega \wedge \omega = \omega \wedge \star\omega = |\omega|_h^2 d\mu_h,$$

and it therefore follows that ω is actually a symplectic form on the open
set $X := M - Z$, where ω is non-zero. Moreover, the Riemannian metric
g on X defined by $g = 2^{-1/2}|\omega|_h h$ is then an *almost-Kähler metric*, in
the sense that g is related to the symplectic form ω by $g = \omega(\cdot, J\cdot)$ for a
unique almost-complex structure J on X.

Let us now re-express the conformal relationship between our two
metrics as

$$h = f^2 g,$$

where $f = 2^{1/4}|\omega|_h^{-1/2}$. The fact that h satisfies $\delta W^+ = 0$ then implies
[23] that g satisfies $\delta(fW^+) = 0$. Since our assumptions imply that g
is also at least C^4, we therefore have [9, 12, 20, 23] the Weitzenböck
formula

$$0 = \nabla^*\nabla(fW^+) + \frac{s}{2}fW^+ - 6fW^+ \circ W^+ + 2f|W^+|^2 I \qquad (9.9)$$

for fW^+, which for notational simplicity has been represented here as
a trace-free section of $\mathrm{End}(\Lambda^+)$, while s and ∇ respectively denote the

scalar curvature and Levi-Civita connection of our almost-Kähler metric g on X.

Our strategy is now to contract (9.9) with $\omega \otimes \omega$, integrate on $X = M - Z$, and then try to integrate by parts in order to throw the Bochner Laplacian $\nabla^* \nabla$ onto $\omega \otimes \omega$. In order to accomplish this, we first exhaust X by domains X_ε with smooth boundary, where X_ε is the region where $|\omega|_h \geq \varepsilon$, where $\varepsilon > 0$ is any regular value of the smooth non-negative function $|\omega|_h : X \to \mathbb{R}$. Integrating by parts on X_ε then has the following effect:

Lemma 9.2.1   *There is a constant C, independent of $\varepsilon \in (0,1)$, but depending on (M, h, ω), such that*

$$\left| \int_{X_\varepsilon} \left[\langle \nabla^* \nabla (fW^+), \omega \otimes \omega \rangle - \langle fW^+, \nabla^* \nabla (\omega \otimes \omega) \rangle \right] d\mu_g \right| \leq C \varepsilon^{-3/2} \mathrm{Vol}(\partial X_\varepsilon, h),$$

where all terms in the integral on the left are computed with respect to g, but where the 3-dimensional boundary volume on the right is computed with respect to h.

Proof   By the divergence version of Stokes' theorem, we have

$$\int_{X_\varepsilon} \langle \nabla^* \nabla (fW^+), \omega \otimes \omega \rangle d\mu_g = \int_{X_\varepsilon} \langle -\nabla \cdot \nabla fW^+, \omega \otimes \omega \rangle d\mu_g$$

$$= -\int_{X_\varepsilon} \nabla \cdot \langle \nabla fW^+, \omega \otimes \omega \rangle d\mu_g$$

$$+ \int_{X_\varepsilon} \langle \nabla fW^+, \nabla(\omega \otimes \omega) \rangle d\mu_g$$

$$= -\int_{\partial X_\varepsilon} \langle \nabla_\nu fW^+, \omega \otimes \omega \rangle da_g$$

$$+ \int_{X_\varepsilon} \langle \nabla fW^+, \nabla(\omega \otimes \omega) \rangle d\mu_g$$

$$= -\int_{\partial X_\varepsilon} \nabla_\nu \langle fW^+, \omega \otimes \omega \rangle da_g$$

$$+ \int_{\partial X_\varepsilon} \langle fW^+, \nabla_\nu(\omega \otimes \omega) \rangle da_g$$

$$+ \int_{X_\varepsilon} \nabla \cdot \langle fW^+, \nabla(\omega \otimes \omega) \rangle d\mu_g$$

$$+ \int_{X_\varepsilon} \langle fW^+, -\nabla \cdot \nabla(\omega \otimes \omega) \rangle d\mu_g$$

$$= -\int_{\partial X_\varepsilon} \nabla_\nu \langle fW^+, \omega \otimes \omega \rangle da_g$$

$$+ 2\int_{\partial X_\varepsilon} \langle fW^+, \nabla_\nu(\omega \otimes \omega) \rangle da_g$$

$$+ \int_{X_\varepsilon} \langle fW^+, \nabla^*\nabla(\omega \otimes \omega) \rangle d\mu_g$$

$$= -\int_{\partial X_\varepsilon} \nabla_\nu [fW^+(\omega, \omega)] da_g$$

$$+ 4\int_{\partial X_\varepsilon} fW^+(\omega, \nabla_\nu \omega) da_g$$

$$+ \int_{X_\varepsilon} \langle fW^+, \nabla^*\nabla(\omega \otimes \omega) \rangle d\mu_g,$$

where ν is the outward-pointing unit normal of ∂X_ε with respect to g, and where da_g is the g-induced volume 3-form on the boundary. Here, every term is thus understood to be computed with respect to g.

We now estimate the boundary integrals by first re-expressing them in terms of the original metric $h = f^2 g$. For emphasis and clarity, we will temporarily use $\hat{\nu} = f^{-1}\nu$ to denote the unit normal of ∂X_ε with respect to h, and $\hat{\nabla}$ to denote the Levi-Civita connection of h, which differs from the Levi-Civita connection of ∇ of g by

$$\delta^a_b \beta_c + \delta^a_c \beta_b - \beta_d h^{da} h_{bc},$$

where $\beta = d\log f = -\frac{1}{2} d\log |\omega|_h$. In other cases where the meaning of a term depends on a choice of metric, we will indicate the metric used by means of a subscript; for example, since index-raising is needed to define $W^+(\omega, \omega)$, one has

$$[W^+(\omega, \omega)]_g = f^6 [W^+(\omega, \omega)]_h.$$

With these conventions in hand, we thus have

$$\left| \int_{\partial X_\varepsilon} \nabla_\nu [fW^+(\omega, \omega)]_g da_g \right| = \left| \int_{\partial X_\varepsilon} f\nabla_{\hat{\nu}} [f^7 W^+(\omega, \omega)]_h f^{-3} da_h \right|$$

$$\leq 7 \left| \int_{\partial X_\varepsilon} f^4 (\nabla_{\hat{\nu}} f)[W^+(\omega, \omega)]_h da_h \right|$$

$$+ \left| \int_{\partial X_\varepsilon} f^5 \nabla_{\hat{\nu}} [W^+(\omega, \omega)]_h da_h \right|$$

$$\leq 7\left|\int_{\partial X_\varepsilon} f^5 |f^{-1}df|_h |W^+|_h |\omega|_h^2 \, da_h\right|$$

$$+\left|\int_{\partial X_\varepsilon} f^5 |\hat{\nabla} W^+|_h |\omega|_h^2 \, da_h\right|$$

$$+2\left|\int_{\partial X_\varepsilon} f^5 |W^+|_h |\omega|_h |\hat{\nabla}\omega|_h \, da_h\right|$$

$$= 7\left|\int_{\partial X_\varepsilon} 2^{1/4}|\omega|_h^{-3/2} |d|\omega|_h|_h |W^+|_h \, da_h\right|$$

$$+\left|\int_{\partial X_\varepsilon} 2^{5/4}|\omega|_h^{-1/2} |\hat{\nabla} W^+|_h \, da_h\right|$$

$$+2\left|\int_{\partial X_\varepsilon} 2^{5/4}|\omega|_h^{-3/2} |W^+|_h |\hat{\nabla}\omega|_h \, da_h\right|$$

$$\leq C_1 \varepsilon^{-3/2} \mathrm{Vol}^{(3)}(\partial X_\varepsilon, h),$$

where $C_1 = \sqrt[4]{2}\left[11(\max_M |W^+|_h)(\max_M |\hat{\nabla}\omega|_h) + 2\max_M |\hat{\nabla} W^+|_h\right]$.
(In the last step, we have used the Kato inequality $|d|\omega|| \leq |\hat{\nabla}\omega|$, and
have remembered that $\varepsilon < 1$ by hypothesis.) Similarly,

$$\left|\int_{\partial X_\varepsilon} f W^+(\omega, \nabla_\nu \omega)_g \, da_g\right| = \left|\int_{\partial X_\varepsilon} f \cdot f^6 W^+(\omega, \nabla_{f\hat{\nu}}\omega)_h f^{-3} \, da_h\right|$$

$$= \left|\int_{\partial X_\varepsilon} f^5 W^+(\omega, \nabla_{\hat{\nu}}\omega)_h \, da_h\right|$$

$$\leq 2\left|\int_{\partial X_\varepsilon} f^5 |W^+|_h |\omega|_h |\nabla\omega|_h \, da_h\right|$$

$$\leq 2\left|\int_{\partial X_\varepsilon} f^5 |W^+|_h |\omega|_h |\hat{\nabla}\omega|_h \, da_h\right|$$

$$+6\left|\int_{\partial X_\varepsilon} f^5 |W^+|_h |\omega|_h^2 |\beta|_h \, da_h\right|$$

$$= 2\left|\int_{\partial X_\varepsilon} f^5 |W^+|_h |\omega|_h |\hat{\nabla}\omega|_h \, da_h\right|$$

$$+3\left|\int_{\partial X_\varepsilon} f^5 |W^+|_h |\omega|_h |d|\omega|_h|_h \, da_h\right|$$

$$\leq 5\left|\int_{\partial X_\varepsilon} 2^{5/4}|\omega|_h^{-3/2} |W^+|_h |\hat{\nabla}\omega|_h \, da_h\right|$$

$$\leq C_2 \varepsilon^{-3/2} \mathrm{Vol}^{(3)}(\partial X_\varepsilon, h),$$

where $C_2 = 10\sqrt[4]{2}(\max_M |W^+|_h)(\max_M |\hat{\nabla}\omega|_h)$. Setting $C = C_1 + 4C_2$, and referring back to our integration-by-parts calculation, we thus see that the claim now follows immediately from the triangle inequality. $\quad\square$

So far, we have only assumed that ω is a non-trivial self-dual harmonic form on (M, h). However, the information we have just gleaned becomes much more useful when ω happens to be near-symplectic:

Lemma 9.2.2 *Let ω be a near-symplectic self-dual harmonic 2-form on a compact oriented Riemannian 4-manifold. Let $X = M - Z$ be the complement of the zero set Z of ω, set $f = 2^{1/4}|\omega|_h^{-1/2}$ on X, and let $g = f^{-2}h$ be the almost-Kähler metric on (X, ω) obtained by conformally rescaling h to make $|\omega|_g \equiv \sqrt{2}$. Then*

$$\int_X \langle \nabla^*\nabla(fW^+), \omega \otimes \omega \rangle \, d\mu_g = \int_X \langle fW^+, \nabla^*\nabla(\omega \otimes \omega) \rangle \, d\mu_g, \quad (9.10)$$

where the integrands on both sides are defined with respect to g, and where both moreover belong to L^1. In particular, both integrals are finite, and may be treated either as improper Riemann integrals or as Lebesgue integrals.

Proof To say that ω is near-symplectic means, by definition, that the section ω of $\Lambda^+ \to M$ is transverse to the zero section along its zero locus $Z \approx \sqcup_{j=1}^n S^1$. In particular, the derivative of ω along Z induces an isomorphism between the normal bundle of $Z \subset M$ and the vector bundle $\Lambda^+|_Z \to Z$. This moreover allows us to construct a diffeomorphism between a sufficiently small tubular neighborhood $\mathcal{U}$ of Z and $Z \times B_\varepsilon^3$, where $B_\varepsilon^3 \subset \mathbb{R}^3$ is the standard 3-ball of some small radius ε, by combining the nearest-point projection $\mathcal{U} \to Z$ with the components of ω relative to some orthornormal framing of the vector bundle $\Lambda^+ \to \mathcal{U}$. (Here, we are using the fact that $\Lambda^+|_{\mathcal{U}}$ is necessarily trivial because Λ^+ is oriented, $\mathbf{SO}(3)$ is connected, and $\mathcal{U}$ deform retracts to a union of circles.) Via this diffeomorphism, the function $|\omega|_h$ on $\mathcal{U}$ then just becomes the standard radius function on B_ε^3. Moreover, after reducing the size of ε if necessary, the Riemannian metric h on $\mathcal{U}$ becomes quasi-isometric to the standard flat product metric h_0 on $Z \times B_\varepsilon^3$, in the sense that $h_0/\kappa < h < \kappa h_0$ for some constant $\kappa > 1$, and where we have $|\omega|_h \geq \varepsilon$ on the complement $M - \mathcal{U}$ of $\mathcal{U}$. It then follows that the hypersurfaces $(\partial X_\varepsilon, h)$ are uniformly quasi-isometric to $(Z \times S_\varepsilon^2, h_0)$, so there consequently exists a positive constant $L = 4\pi|Z|\kappa^{3/2}$ such that

$$\mathrm{Vol}^{(3)}(\partial X_\varepsilon, h) < L\varepsilon^2$$

for all $\varepsilon \in (0, \varepsilon)$. Combining this with Lemma 9.2.1 then tells us that

$$\left| \int_{X_\varepsilon} \left[\langle \nabla^* \nabla(fW^+), \omega \otimes \omega \rangle - \langle fW^+, \nabla^* \nabla(\omega \otimes \omega) \rangle \right] d\mu_g \right| \leq CL\sqrt{\varepsilon}$$

for all $\varepsilon \in (0, \varepsilon)$. In particular, this implies that

$$\lim_{\varepsilon \searrow 0} \int_{X_\varepsilon} \left[\langle \nabla^* \nabla(fW^+), \omega \otimes \omega \rangle - \langle fW^+, \nabla^* \nabla(\omega \otimes \omega) \rangle \right] d\mu_g = 0. \quad (9.11)$$

To prove the claim, it therefore suffices to show that both integrands in (9.10) are absolutely integrable, and so belong to L^1. To see this, first notice that

$$\int_X \left| \langle \nabla^* \nabla(fW^+), \omega \otimes \omega \rangle_g \right| d\mu_g \leq 2 \int_X \left| \nabla^* \nabla(fW^+) \right|_g d\mu_g$$

$$= 2 \int_X f^2 \left| [\nabla \cdot \nabla(fW^+)]_g \right|_h f^{-4} d\mu_h$$

$$\leq 2 \int_X \left| \nabla^* \nabla(fW^+) \right|_h d\mu_h$$

$$+ 8 \int_X \left| \nabla(\beta \otimes fW^+) \right|_h d\mu_h$$

$$+ 10 \int_X \left| \beta \otimes \nabla(fW^+) \right|_h d\mu_h$$

$$+ 40 \int_X \left| \beta \otimes \beta \otimes fW^+ \right|_h d\mu_h$$

$$\leq 2 \int_X f \left| \nabla^* \nabla W^+ \right|_h d\mu_h$$

$$+ 22 \int_X |\nabla f|_h |\nabla W^+|_h d\mu_h$$

$$+ 10 \int_X |\nabla \nabla f|_h |W^+|_h d\mu_h$$

$$+ 50 \int_X f^{-1} |\nabla f|_h^2 |W^+|_h d\mu_h$$

$$\leq C_3 \int_M \left[|\omega|_h^{-1/2} + |\nabla|\omega|_h^{-1/2}|_h \right.$$

$$+ |\omega|_h^{1/2} |\nabla|\omega|_h^{-1/2}|_h^2$$

$$\left. + |\nabla \nabla|\omega|_h^{-1/2}|_h \right] d\mu_h$$

$$\leq C_3 \int_M \left[|\omega|_h^{-1/2} + \frac{1}{2}|\omega|_h^{-3/2}|\nabla\omega|_h \right.$$
$$+ \frac{23}{4}|\omega|_h^{-5/2}|\nabla\omega|_h^2 + 2|\omega|_h^{-3/2}|\nabla\nabla\omega|_h \Big] d\mu_h$$
$$\leq C_4 \int_M |\omega|_h^{-5/2} d\mu_h$$
$$< \infty,$$

where C_3 is a positive constant depending on (M, h), C_4 is a positive constant depending on (M, h, ω), and where, as in the remainder of the chapter, ∇ denotes the Levi-Civita connection $\hat{\nabla}$ of h when its relation to h is clearly indicated by a subscript. Here, in the last step, we have used the fact that $|\omega|^{-5/2}$ is comparable, near $Z = M - X$, to $r^{-5/2}$ on $B_\varepsilon^3 \times S^1$, where $r = |\vec{x}|$ is the distance from the origin in the ε-ball $B_\varepsilon^3 \subset \mathbb{R}^3$, and therefore has finite integral because

$$\int_{B_\varepsilon^3} |\vec{x}|^{-5/2} dx^1 \wedge dx^2 \wedge dx^3 = 4\pi \int_0^\varepsilon r^{-5/2} r^2 dr = 8\pi\sqrt{\varepsilon} < \infty.$$

In much the same way,

$$\int_X \left| \langle fW^+, \nabla^*\nabla(\omega\otimes\omega)\rangle_g \right| d\mu_g$$

$$\leq 2\sqrt{2}\int_X f|W^+|_g|\nabla^*\nabla\omega|_g d\mu_g + 2\int_X f|W^+|_g|\nabla\omega|_g^2 d\mu_g$$

$$\leq 2^{3/2}\int_X f^3|W^+|_h f^4 \Big[|\nabla^*\nabla\omega|_h + 2|\nabla\beta|_h|\omega|_h$$
$$+ 4|\beta|_h|\nabla\omega|_h + |\beta|^2|\omega|_h \Big] f^{-4} d\mu_h$$

$$+ 2\int_X f^3|W^+|_h f^6 \Big[|\nabla\omega|_h^2 + 4|\beta|_h|\omega|_h|\nabla\omega|_h$$
$$+ 4|\beta|_h^2|\omega|_h^2 \Big] f^{-4} d\mu_h$$

$$\leq 2^{3/2}\int_X |W^+|_h \Big[f^3|\nabla^*\nabla\omega|_h + 2f^2|\nabla(f^{-1}\nabla f)|_h|\omega|_h$$
$$+ 4f^2|\nabla f|_h|\nabla\omega|_h + f|\nabla f|^2|\omega|_h \Big] d\mu_h$$

$$+ 2\int_X |W^+|_h \Big[f^5|\nabla\omega|_h^2 + 4f^4|\nabla f|_h|\omega|_h|\nabla\omega|_h$$
$$+ 4f^3|\nabla f|_h^2|\omega|_h^2 \Big] d\mu_h$$

$$\leq C_5 \int_X \Big[|\omega|_h^{-3/2} |\nabla^* \nabla \omega|_h + |\omega|_h^{-2} |\nabla \omega|_h^2$$

$$+ |\omega|_h^{-2} |\nabla \nabla \omega|_h^2 + |\omega|_h^{-5/2} |\nabla \omega|_h^2 \Big] d\mu_h$$

$$\leq C_6 \int_M |\omega|_h^{-5/2} d\mu_h < \infty,$$

where C_5 and C_6 are positive constants depending, respectively, on (M, h) and (M, h, ω). Thus, the integrands in (9.10) both belong to L^1, and (9.11) therefore implies that the two integrals in (9.10) are equal. □

Since we are thus entitled to carry out the desired integration by parts in the near-symplectic case, (9.9) therefore implies an interesting integral Weitzenböck formula when h also satisfies $\delta W^+ = 0$.

Proposition 9.2.3 *Let ω be a near-symplectic self-dual harmonic 2-form on a compact oriented Riemannian 4-manifold (M, h) with $\delta W^+ = 0$. Let $X = M - Z$ be the complement of the zero set Z of ω, set $f = 2^{1/4} |\omega|_h^{-1/2}$ on X, and let $g = f^{-2} h$ be the almost-Kähler metric on (X, ω) obtained by conformally rescaling h to make $|\omega|_g \equiv \sqrt{2}$. Then g satisfies*

$$\int_X \Big[\langle W^+, \nabla^* \nabla (\omega \otimes \omega) \rangle + \frac{s}{2} W^+ (\omega, \omega) - 6|W^+(\omega)|^2 + 2|W^+|^2 |\omega|^2 \Big] f \, d\mu_g = 0,$$

both as a Lebesgue integral and as an improper Riemann integral.

Proof Contraction of (9.9) with $\omega \otimes \omega$ tells us that

$$\langle \nabla^* \nabla (f W^+), \omega \otimes \omega \rangle + \frac{s}{2} f W^+ (\omega, \omega) - 6f |W^+(\omega)|^2 + 2f |W^+|^2 |\omega|^2 = 0$$

on (X, g), so integration certainly tells us that

$$\int_X \Big[\langle \nabla^* \nabla (f W^+), \omega \otimes \omega \rangle + \frac{s}{2} f W^+ (\omega, \omega) - 6f |W^+(\omega)|^2 + 2f |W^+|^2 |\omega|^2 \Big] d\mu_g = 0.$$

However, because the first term is L^1, equation (9.9) tells us that the same is also true of the sum of the remaining terms, and Lemma 9.2.2 therefore allows us to rewrite the above expression as

$$\int_X \Big[\langle f W^+, \nabla^* \nabla (\omega \otimes \omega) \rangle + f \frac{s}{2} W^+ (\omega, \omega) - 6f |W^+(\omega)|^2 + 2f |W^+|^2 |\omega|^2 \Big] d\mu_g = 0.$$

Collecting the common of factor of f now yields the desired result. □

9.3 Some Almost-Kähler Geometry

When an oriented Riemannian manifold (M, h) with $\delta W^+ = 0$ carries a near-symplectic self-dual harmonic 2-form ω, we saw in Proposition 9.2.3 that, if we set $f = 2^{1/4}|\omega|_h^{-1/2}$ on the open set X where $\omega \neq 0$, the conformally related almost-Kähler metric $g = f^{-2}h$ then satisfies an integral Weitzenböck formula on X. In order to exploit this effectively, we will need a universal identity previously pointed out in [20]:

Lemma 9.3.1 *Any 4-dimensional almost-Kähler manifold satisfies*

$$\langle W^+, \nabla^*\nabla(\omega \otimes \omega)\rangle = [W^+(\omega,\omega)]^2 + 4|W^+(\omega)|^2 - sW^+(\omega,\omega)$$

at every point.

Proof First notice that the oriented Riemannian 4-manifold (X, g) satisfies

$$\Lambda^+ \otimes \mathbb{C} = \mathbb{C}\omega \oplus K \oplus \overline{K},$$

where $K = \Lambda_J^{2,0}$ is the canonical line bundle of the almost-complex structure J defined by $\omega = g(J\cdot, \cdot)$. Locally choosing a unit section φ of K, we thus have

$$\nabla\omega = \alpha \otimes \varphi + \bar{\alpha} \otimes \bar{\varphi}$$

for a unique 1-form $\alpha \in \Lambda_J^{1,0}$, since $\nabla_{[a}\omega_{bc]} = 0$ and $\omega^{bc}\nabla_a\omega_{bc} = 0$. If

$$\circledast : \Lambda^+ \times \Lambda^+ \to \odot_0^2\Lambda^+$$

denotes the symmetric trace-free product, we therefore have

$$(\nabla_e\omega) \circledast (\nabla^e\omega) = 2|\alpha|^2\varphi \circledast \bar{\varphi} = -\frac{1}{4}|\nabla\omega|^2\omega \circledast \omega$$

and we thus deduce that

$$\begin{aligned}
\langle W^+, \nabla^*\nabla(\omega \otimes \omega)\rangle &= 2W^+(\omega, \nabla^*\nabla\omega) - 2W^+(\nabla_e\omega, \nabla^e\omega) \\
&= 2W^+(\omega, \nabla^*\nabla\omega) + \frac{1}{2}|\nabla\omega|^2 W^+(\omega,\omega) \\
&= 2W^+\left(\omega, 2W^+(\omega) - \frac{s}{3}\omega\right) \\
&\quad + \left[W^+(\omega,\omega) - \frac{s}{3}\right]W^+(\omega,\omega) \\
&= -\frac{2}{3}sW^+(\omega,\omega) + 4|W^+(\omega)|^2 \\
&\quad + \left[W^+(\omega,\omega) - \frac{s}{3}\right]W^+(\omega,\omega) \\
&= [W^+(\omega,\omega)]^2 + 4|W^+(\omega)|^2 - sW^+(\omega,\omega),
\end{aligned}$$

where we have used the Weitzenböck formula

$$0 = \nabla^*\nabla\omega - 2W^+(\omega) + \frac{s}{3}\omega$$

for the harmonic self-dual 2-form ω, as well as the associated key identity

$$\frac{1}{2}|\nabla\omega|^2 = W^+(\omega,\omega) - \frac{s}{3} \tag{9.12}$$

resulting from the fact that $|\omega|^2 \equiv 2$.   □

In conjunction with Proposition 9.2.3, this now yields the following:

Theorem 9.3.2 *Let ω be a near-symplectic self-dual harmonic 2-form on a compact oriented Riemannian 4-manifold (M, h) with $\delta W^+ = 0$. Let $X = M - Z$ be the complement of the zero set Z of ω, set $f = 2^{1/4}|\omega|_h^{-1/2}$ on X, and let $g = f^{-2}h$ be the almost-Kähler metric on (X, ω) obtained by conformally rescaling h to make $|\omega|_g \equiv \sqrt{2}$. Then the almost-Kähler metric g satisfies*

$$\int_X \left[8\left(|W^+|^2 - \frac{1}{2}|W^+(\omega)^\perp|^2 \right) - sW^+(\omega,\omega) \right] f \, d\mu_g = 0, \tag{9.13}$$

where s is the scalar curvature of g, and where $W^+(\omega)^\perp$ denotes the orthogonal projection of $W^+(\omega)$ to the orthogonal complement of $\omega \in \Lambda^+$. Moreover, the integrand belongs to L^1, so the statement holds whether the left-hand-side is construed as a Lebesgue integral or as an improper Riemann integral.

Proof Combining Proposition 9.2.3 with Lemma 9.3.1, we have

$$0 = \int_X \left[\langle W^+, \nabla^*\nabla(\omega \otimes \omega) \rangle + \frac{s}{2}W^+(\omega,\omega) \right.$$
$$\left. - 6|W^+(\omega)|^2 + 2|W^+|^2|\omega|^2 \right] f \, d\mu$$

$$= \int_X \left[\left([W^+(\omega,\omega)]^2 + 4|W^+(\omega)|^2 - sW^+(\omega,\omega) \right) \right.$$
$$\left. + \frac{s}{2}W^+(\omega,\omega) - 6|W^+(\omega)|^2 + 4|W^+|^2 \right] f \, d\mu$$

$$= \int_X \left[[W^+(\omega,\omega)]^2 - \frac{s}{2}W^+(\omega,\omega) - 2|W^+(\omega)|^2 + 4|W^+|^2 \right] f \, d\mu .$$

Since $|W^+(\omega)^\perp|^2 = |W^+(\omega)|^2 - \frac{1}{2}[W^+(\omega,\omega)]^2$, multiplication by 2 thus yields the desired formula (9.13). Moreover, this calculation shows that the integrand is the sum of two L^1 functions, and is therefore itself L^1 by the triangle inequality.   □

Next, we prove a refinement of a point-wise inequality used in [20]:

Lemma 9.3.3   *Any 4-dimensional almost-Kähler manifold satisfies*

$$|W^+|^2 - \frac{1}{2}|W^+(\omega)^\perp|^2 \geq \frac{3}{8}\left[W^+(\omega,\omega)\right]^2 + \frac{1}{2}|W^+(\omega)^\perp|^2$$

at every point.

Proof   If $A = [A_{jk}]$ is any symmetric trace-free 3×3 matrix, the fact that $A_{33} = -(A_{11} + A_{22})$ implies that

$$\sum_{jk} A_{jk}^2 \geq 2A_{21}^2 + A_{11}^2 + A_{22}^2 + (A_{11} + A_{22})^2 = 2A_{21}^2 + \frac{3}{2}A_{11}^2 + 2\left(\frac{A_{11}}{2} + A_{22}\right)^2$$

and we therefore conclude that

$$|A|^2 \geq 2A_{21}^2 + \frac{3}{2}A_{11}^2.$$

If we now let A represent $W^+ : \Lambda^+ \to \Lambda^+$ with respect to an orthogonal basis $\mathfrak{e}_1, \mathfrak{e}_2, \mathfrak{e}_3$ for Λ^+ such that $\omega = \sqrt{2}\mathfrak{e}_1$ and $W^+(\omega)^\perp \propto \mathfrak{e}_2$, this inequality becomes

$$|W^+|^2 \geq |W^+(\omega)^\perp|^2 + \frac{3}{8}\left[W^+(\omega,\omega)\right]^2,$$

and subtracting $\frac{1}{2}|W^+(\omega)^\perp|^2$ from both sides thus proves the claim.   □

This now yields a key inequality:

Lemma 9.3.4   *Let (M, h), ω, X, g, and f be as in Theorem 9.3.2. Then the almost-Kähler metric $g = f^{-2}h$ satisfies*

$$0 \geq \int_X \left[W^+(\omega,\omega)|\nabla\omega|^2 + \frac{8}{3}|W^+(\omega)^\perp|^2\right] f \, d\mu_g, \qquad (9.14)$$

in the sense that the Lebesgue integral on the right is well defined and belongs to $[-\infty, 0]$.

Proof   Theorem 9.3.2 tells us that

$$0 = \int_X \left[8\left(|W^+|^2 - \frac{1}{2}|W^+(\omega)^\perp|^2\right) - sW^+(\omega,\omega)\right] f \, d\mu_g$$

and that the positive and negative parts of the integrand are both L^1 functions. The point-wise inequality of integrands provided by Lemma 9.3.3 therefore implies that

$$0 \geq \int_X \left[3\left[W^+(\omega,\omega)\right]^2 - sW^+(\omega,\omega) + 4|W^+(\omega)^\perp|^2\right] f \, d\mu_g$$

in the Lebesgue sense. After dividing by 3, we can then re-express this as

$$0 \geq \int_X \left[W^+(\omega,\omega)\left(W^+(\omega,\omega) - \frac{s}{3}\right) + \frac{4}{3}|W^+(\omega)^\perp|^2 \right] f \, d\mu_g. \quad (9.15)$$

However, (9.12) tells us that $W^+(\omega,\omega) - \frac{s}{3} = \frac{1}{2}|\nabla\omega|^2$ for any almost-Kähler 4-manifold. Making this substitution in (9.15) and then multiplying by 2 thus yields the desired inequality (9.14). $\square$

In the special case where (M, h, ω) satisfies the conformally invariant condition $W^+(\omega, \omega) \geq 0$, we thus obtain the following:

Proposition 9.3.5 *Let (M, h) be a compact oriented Riemannian 4-manifold that satisfies $\delta W^+ = 0$, and suppose that ω is a near-symplectic self-dual harmonic 2-form on (M, h) that satisfies $W^+(\omega, \omega) \geq 0$. Let X, g, and f be as in Theorem 9.3.2. Then the almost-Kähler manifold (X, g, ω) satisfies*

$$\int_X \left[W^+(\omega,\omega)|\nabla\omega|^2 + \frac{8}{3}|W^+(\omega)^\perp|^2 \right] f \, d\mu_g = 0, \quad (9.16)$$

both as a Lebesgue and as an improper Riemann integral.

Proof The added assumption that $W^+(\omega, \omega) \geq 0$ obviously implies

$$\int_X \left[W^+(\omega,\omega)|\nabla\omega|^2 + \frac{8}{3}|W^+(\omega)^\perp|^2 \right] f \, d\mu_g \geq 0$$

as an extended real number, because the integrand is now non-negative. But, in conjunction with (9.14), this immediately implies that

$$\int_X \left[W^+(\omega,\omega)|\nabla\omega|^2 + \frac{8}{3}|W^+(\omega)^\perp|^2 \right] f \, d\mu_g = 0$$

as a Lebesgue integral. Moreover, since the integrand is also L^1, the integral also necessarily vanishes as an improper Riemann integral. $\square$

This very strong statement now has even stronger consequences:

Proposition 9.3.6 *Let M, h, ω, X, g, and f be as in Proposition 9.3.5. Then either g is a Kähler metric on X whose scalar curvature is given by $s = \mathbf{c}/f$ for some constant $\mathbf{c} > 0$, or else g satisfies $W^+ \equiv 0$, and so is an anti-self-dual metric.*

Proof Since $f > 0$ by construction, and since $W^+(\omega, \omega) \geq 0$ by assumption, both terms in the integrand of (9.16) must vanish identically.

We thus have

$$W^+(\omega,\omega)|\nabla\omega|^2 = 0 \qquad \text{and} \qquad W^+(\omega)^\perp = 0 \qquad (9.17)$$

at every point of X. In particular, $\nabla\omega = 0$ wherever $W^+(\omega,\omega) \neq 0$. If $\mathscr{V} \subset X$ is the open subset where $W^+(\omega,\omega) \neq 0$, the restriction of g to $\mathscr{V}$ is therefore Kähler. On the other hand, since $h = f^2 g$ satisfies $\delta W^+ = 0$, conformal invariance of this equation tells us that g satisfies $\delta(f W^+) = 0$, as previously noted. On $(\mathscr{V}, g)$ we therefore have

$$0 = \omega^{ab}\omega^{cd}\nabla^e(fW^+_{ebcd}) = \nabla^e(fW^+_{ebcd}\omega^{ab}\omega^{cd})$$
$$= \nabla^e(f\frac{s}{3}\omega_{eb}\omega^{ab}) = \frac{1}{3}\nabla^e(fs\,\delta^a_e) = \frac{1}{3}\nabla^a(fs) = \nabla^a[fW^+(\omega,\omega)],$$

since at each point of any Kähler manifold of real dimension 4, the Kähler form ω is an eigenvector of $W^+ : \Lambda^+ \to \Lambda^+$, with eigenvalue one-sixth of the scalar curvature s. This shows that $d[fW^+(\omega,\omega)] = 0$ on $\mathscr{V}$, and therefore, by continuity, on the closure $\overline{\mathscr{V}}$ of $\mathscr{V}$, too. On the other hand, since our definition of $\mathscr{V}$ guarantees that $fW^+(\omega,\omega) \equiv 0$ on the open set $X - \overline{\mathscr{V}}$, we also have $d[fW^+(\omega,\omega)] = 0$ on $X - \overline{\mathscr{V}}$. It follows that $d[fW^+(\omega,\omega)] = 0$ on all of X. Since X is connected, and since $fW^+(\omega,\omega) \geq 0$, we therefore conclude that $fW^+(\omega,\omega) = \mathbf{c}/3$ for some non-negative constant $\mathbf{c} \geq 0$.

If $\mathbf{c} > 0$, $\mathscr{V} = X$, and it follows that (X, g, ω) is a Kähler manifold, with

$$s = 3W^+(\omega,\omega) = \frac{\mathbf{c}}{f}.$$

Otherwise, $\mathbf{c} = 0$, and we have $W^+(\omega,\omega) \equiv 0$. On the other hand, (9.17) also tells us that $W^+(\omega)^\perp \equiv 0$ on X. Substituting these two facts into (9.13) then yields

$$\int_X |W^+|^2 f \, d\mu_g = 0.$$

Thus, when $\mathbf{c} = 0$, we conclude that $W^+ \equiv 0$, and g is therefore anti-self-dual in this remaining case, exactly as claimed. $\qquad \square$

Sharpening these conclusions now supplies our mainspring result:

Theorem 9.3.7 *Let (M, h) be a compact oriented Riemannian 4-manifold with $\delta W^+ = 0$ that admits a near-symplectic self-dual harmonic 2-form ω such that*

$$W^+(\omega,\omega) \geq 0.$$

Then either h satisfies $W^+ \equiv 0$, and so is anti-self-dual, or else $W^+(\omega,\omega)$

*is everywhere positive, in which case M admits a global Kähler metric g
with scalar curvature s > 0 such that h = s⁻²g.*

Proof If (X, g) satisfies $W^+ \equiv 0$, the conformal invariance of this condition implies that (X, h) satisfies $W^+ \equiv 0$, too. But since $X \subset M$ is
dense, it then follows by continuity that h satisfies $W^+ \equiv 0$ on all of M.
Thus, (M, h) must be a compact anti-self-dual manifold in this case.

Otherwise, $W^+ \not\equiv 0$, and Proposition 9.3.6 then guarantees that $g = f^{-2}h$ must be a Kähler metric on $X = M - Z$, with Kähler form ω and

$$3W^+(\omega, \omega) = s = \mathbf{c}f^{-1}$$

for some positive constant $\mathbf{c}$. However, since $h = f^2 g$, we also have

$$[W^+(\omega, \omega)]_h = f^{-6}[W^+(\omega, \omega)]_g,$$

and it therefore follows that

$$[W^+(\omega, \omega)]_h = \frac{\mathbf{c}}{3} \, f^{-7}.$$

But since $f = 2^{1/4}|\omega|_h^{-1/2}$ by construction, this means that

$$[W^+(\omega, \omega)]_h = \mathbf{b}\,|\omega|_h^{7/2} \tag{9.18}$$

on X, where $\mathbf{b} = \sqrt[4]{2}\mathbf{c}/12$ is a positive constant. However, since g is
Kähler, with positive scalar curvature and Kähler form ω, it follows
that W^+ has a repeated negative eigenvalue at every point of X, and
that ω everywhere belongs to the positive eigenspace. This implies that

$$W^+(\omega, \omega) = \sqrt{\frac{2}{3}}|W^+||\omega|^2$$

at every point of X, both for g and for h. Thus (9.18) implies that

$$|W^+|_h = \mathbf{a}|\omega|_h^{3/2} \tag{9.19}$$

everywhere on X, where $\mathbf{a} = \sqrt{\frac{3}{2}}\mathbf{b}$ is another positive constant. However,
since $X \subset M$ is dense, and because the two sides are both continuous
functions, it then follows that (9.19) actually holds on all of M. Now
notice that this implies that $|W^+|$ is everywhere differentiable, and that
W^+ must vanish to first order along Z; thus, $\nabla W^+ = 0$ at every point
of Z, where ∇ denotes the Levi-Civita connection of h. Next, notice that
(9.19) also implies that

$$|d|W^+|_h|_h = \frac{3}{2}\mathbf{a}|\omega|_h^{1/2}|d|\omega|_h|_h$$

on $X = M - Z$. Since the near-symplectic nature of ω moreover guarantees that $|d|\omega|_h|$ is bounded away from zero near Z, we therefore have

$$|d|W^+|_h|_h \geq \mathbf{A}|\omega|_h^{1/2}$$

on some neighborhood $\mathcal{U}$ of Z, where $\mathbf{A} := \frac{3}{2}\mathbf{a} \inf_{\mathcal{U}-Z} |d|\omega|_h|_h$ is another positive constant. By the Kato inequality, we therefore have

$$|\nabla W^+|_h \geq \mathbf{A}|\omega|_h^{1/2}$$

on $\mathcal{U}$. But since h has been assumed throughout to be a C^4 metric, ∇W^+ is a differentiable tensor field, and we have moreover previously observed that this field vanishes along Z. It thus follows that $|\nabla W^+|_h$ is a Lipschitz function that vanishes along Z. But since ω is near-symplectic, $|\omega|_h$ is commensurate with the distance from Z in a small enough neighborhood $\mathcal{U} \supset Z$, and we must therefore have $\mathbf{B}|\omega|_h > |\nabla W^+|_h$ on a sufficiently small neighborhood $\mathcal{U}$ of Z, for some positive constant $\mathbf{B}$. But this then says that

$$\mathbf{B}|\omega|_h > \mathbf{A}|\omega|_h^{1/2}$$

on $\mathcal{U}$, and so implies that

$$|\omega|_h > \frac{\mathbf{A}^2}{\mathbf{B}^2} > 0$$

on $\mathcal{U} - Z$. But since $X - (\mathcal{U} - Z) = M - \mathcal{U}$ is compact, and since $\omega \neq 0$ on X, this implies that $|\omega_h|$ is uniformly bounded away from zero on all of X. But since X is dense in M, it therefore follows by continuity that $|\omega|_h$ is bounded away from zero on all of M. Since Z is by definition the zero set of ω, we are therefore forced to conclude that $Z = \varnothing$.

Thus, g is a globally defined Kähler metric with scalar curvature $s > 0$ such that $h = f^2g = c^2s^{-2}g$ on all of M. By now replacing ω with $\mathbf{c}^{-2/3}\omega$ and thus replacing g with $\mathbf{c}^{-2/3}g$, we can now arrange for h to simply be given by $s^{-2}g$, as promised. □

This tells us quite a bit about the 4-manifolds that carry metrics h of the type covered by Theorem 9.3.7. Indeed [3, 9], if (M, J, g) is a compact Kähler surface of scalar curvature $s > 0$, then $h = s^{-2}g$ is a metric on M with $\delta W^+ = 0$, and with $W^+(\omega, \omega) > 0$ for the Kähler form ω of g. On the other hand, if a compact complex surface (M, J) admits Kähler metrics g with $s > 0$, it is necessarily rational or ruled [33]. Conversely, any rational or ruled surface has arbitrarily small deformations that admit such metrics [13, 26]. Up to oriented diffeomorphism, we can therefore give a complete list of the 4-manifolds that admit solutions of this first type: they are $\mathbb{CP}_2$, $(\Sigma^2 \times S^2)\#k\overline{\mathbb{CP}}_2$, and $\Sigma^2 \rtimes S^2$, where Σ is any compact orientable surface, k is any non-negative integer, and $\Sigma^2 \rtimes S^2$ is the non-trivial oriented 2-sphere bundle over Σ. The moduli space of solutions on any of these manifolds is, moreover, infinite dimensional.

The other class of solutions allowed by Theorem 9.3.7 is rather different, both because the moduli spaces of solutions are always finite dimensional, and because the near-symplectic self-dual harmonic 2-form ω is allowed to have non-empty zero set. Of course, a vast menagerie of smooth compact oriented 4-manifolds with $b_+ \neq 0$ is known to admit anti-self-dual metrics [21, 28], but little is known about when their self-dual harmonic 2-forms ω are near-symplectic. There certainly are many examples with nowhere-zero ω that are not conformally Kähler [15], but there are also related explicit families [5] with $b_+ = 1$ where the self-dual harmonic 2-form ω transmutes from being nowhere zero to having non-empty zero locus. For the latter explicit anti-self-dual manifolds, it seems likely that the self-dual harmonic 2-form ω is usually near-symplectic, but this is equivalent to the non-degeneracy of all critical points for a preferred harmonic function on a quasi-Fuchsian hyperbolic 3-manifold associated with the solution. Perhaps some interested reader will decide that this tractable-looking open problem merits careful investigation!

9.4 The Main Theorems

With the results of Section 9.3 in hand, we are now ready to prove our main theorems, starting with Theorem 9.1.1.

Proof of Theorem 9.1.1 If (M, h) is an oriented 4-dimensional Einstein manifold, the second Bianchi identity implies that $\delta W^+ = 0$. If (M, h) is moreover compact, connected, and admits a near-symplectic self-dual harmonic 2-form ω such that $W^+(\omega, \omega) \geq 0$, the conclusions of Theorem 9.3.7 then apply. Thus, if $W^+(\omega, \omega) > 0$ at some point, we know that $W^+ \not\equiv 0$, and Theorem 9.3.7 then tells us that $W^+(\omega, \omega) > 0$ everywhere, and $h = s^{-2}g$ for some globally defined Kähler metric g on M with scalar curvature $s > 0$. However, any 4-dimensional Einstein metric is Bach-flat, and, because this is a conformally invariant condition, the Kähler metric g must therefore be Bach-flat, too. In particular, this implies [8, 9] that g is an extremal Kähler metric. Moreover, one can also show [17] that the complex structure associated with any such g has $c_1 > 0$, and it therefore follows that M is necessarily diffeomorphic to a del Pezzo surface. Conversely, each del Pezzo diffeotype carries [8, 22, 25, 30, 31] an Einstein metric h which can be written as $s^{-2}g$ for a suitable extremal Kähler metric g with scalar curvature $s > 0$. In fact, h is actually Kähler–Einstein in most cases, the only exceptions being when M is diffeomorphic to $\mathbb{CP}_2 \# \overline{\mathbb{CP}}_2$ or $\mathbb{CP}_2 \# 2\overline{\mathbb{CP}}_2$. □

For each del Pezzo diffeotype, the moduli space of all Einstein metrics h with $W^+(\omega, \omega) > 0$ is actually connected [20]. Moreover, it follows from [19, Thm. A] and a modicum of elementary Seiberg–Witten theory [16, Thm. 3] that, for each del Pezzo M, this moduli space exactly coincides with the moduli space of all conformally Kähler, Einstein metrics. We now prove Theorem 9.1.2.

Proof of Theorem 9.1.2 If (M^4, h) is a compact oriented $\lambda \geq 0$ Einstein manifold that carries a near-symplectic self-dual harmonic ω with $W^+(\omega, \omega) \geq 0$, then Theorem 9.3.7 tells us that either $W^+(\omega, \omega) > 0$ everywhere, or else $W^+ \equiv 0$. Since the former case is covered by Theorem 9.1.1, we may therefore assume that $W^+ \equiv 0$. However, by the Weitzenböck formula for the Hodge Laplacian, the non-trivial self-dual harmonic 2-form ω satisfies

$$0 = \nabla^*\nabla\omega - 2W^+(\omega) + \frac{s}{3}\omega,$$

and, since $W^+ = 0$ and $s = 4\lambda \geq 0$ in our case, taking the inner product with ω and integrating yields

$$0 = \int_M \left[|\nabla\omega|^2 + \frac{4\lambda}{3}|\omega|^2 \right] d\mu_h.$$

We therefore conclude that $\nabla\omega = 0$ and $\lambda = 0$, so that (M^4, h) is necessarily Ricci-flat and Kähler. Thus, after multiplying ω by a positive constant if necessary in order to give it constant length $|\omega|_h \equiv \sqrt{2}$, we see that (M, h) carries an integrable, metric-compatible almost-complex structure J such that $\omega = h(J\cdot, \cdot)$. Moreover, since the Kähler metric h is Ricci-flat, the canonical line bundle K of (M, J) is flat, and $c_1(M, J)$ must therefore be a torsion class. The Kodaira classification of complex surfaces [2, 11] therefore tells us that (M, J) must be a $K3$ surface, an Enriques surface, an Abelian surface, or a hyper-elliptic surface. Conversely, Yau's solution of the Calabi conjecture [35] tells us that each complex surface of one of these types carries a unique Ricci-flat Kähler metric in each Kähler class, and every such Calabi–Yau metric satisfies $W^+ \equiv 0$. $\qquad\qquad\qquad\qquad\qquad\qquad\qquad\qquad\qquad\qquad\qquad\quad$ $\square$

It is worth pointing out that the moduli space of Ricci-flat Kähler metrics is connected. Indeed, since the Kähler cone is contractible for each complex structure, Yau's theorem reduces this statement to the known fact [2] that all the $c_1^{\mathbb{R}} = 0$ complex structures on these 4-manifolds are swept out by a single connected family.

Finally we prove Theorem 9.1.3, by showing that the near-symplectic hypothesis is absolutely essential for Theorem 9.1.1:

Proof of Theorem 9.1.3 Let (M, J, h) be a Kähler–Einstein metric with $\lambda < 0$ on a compact complex surface (M, J) with $p_g(M) := h^{2,0}(M) \neq 0$. (For example, we could take (M, J) to be a smooth quintic hypersurface in $\mathbb{CP}_3$, so that $c_1(M) < 0$ and $p_g(M) = 4$, and let h be the Kähler–Einstein metric whose existence is guaranteed by the Aubin–Yau theorem [1, 34].) Now recall that the self-dual Weyl curvature $W^+ : \Lambda^+ \to \Lambda^+$ of any Kähler surface (M^4, J, g) takes the form

$$\begin{bmatrix} -\frac{s}{12} & & \\ & -\frac{s}{12} & \\ & & \frac{s}{6} \end{bmatrix}$$

in any orthonormal basis $\mathfrak{e}_1, \mathfrak{e}_2, \mathfrak{e}_3$ for Λ^+ in which $\mathfrak{e}_3$ is a multiple of the Kähler form, where s is the scalar curvature. Rather than taking ω to be the Kähler form, we now instead take $\omega = \mathrm{Re}(\varphi)$ for some holomorphic 2-form $\varphi \neq 0$, on (M, J). Of course, the existence of such a φ is guaranteed by our assumption that $h^{2,0} \neq 0$. Notice that φ is automatically self-dual and harmonic as a consequence of standard Kähler identities, and that the same is therefore automatically true of its real part ω.

However, since $\omega \in \mathrm{Re}\, \Lambda^{2,0}$ is everywhere point-wise orthogonal to the Kähler form, we now see that

$$W^+(\omega, \omega) = -\frac{s}{12}|\omega|^2 = \frac{|\lambda|}{3}|\omega|^2 \geq 0,$$

since the Einstein constant λ of h is assumed to be negative. Moreover, since $\omega \not\equiv 0$, this non-negative expression is somewhere positive. On the other hand, the canonical line bundle of (M, J) is non-trivial, because $c_1(K) = -c_1 > 0$, so φ, and therefore ω, must vanish along some non-empty holomorphic curve $\Sigma \subset M$. Thus, $W^+(\omega, \omega)$ vanishes somewhere, and the conclusion of Theorem 9.1.1 therefore fails for this class of examples. □

Of course, in light of counter-examples like those detailed in the proof of Theorem 9.1.3, it is important to explain exactly where the proof of Theorem 9.1.1 breaks down when ω is not near-symplectic. In fact, the key failure occurs at the very beginning of our chain of reasoning, when Lemma 9.2.2 is deduced from Lemma 9.2.1. Recall that Lemma 9.2.1 tells us that the boundary terms arising from integration by parts have size $\sim \varepsilon^{-3/2} \mathrm{Vol}^{(3)}(\partial X_\varepsilon, h)$, where ∂X_ε is the hypersurface where $|\omega|_h = \varepsilon$. In the near-symplectic case, $\mathrm{Vol}^{(3)}(\partial X_\varepsilon, h) \sim \varepsilon^2$, so the boundary terms are no worse than $\varepsilon^{1/2}$, and so vanish in the limit as $\varepsilon \to 0$. By contrast, in the above examples, the zero locus $Z = \Sigma$ of ω has real codimension 2, and we instead have $\mathrm{Vol}^{(3)}(\partial X_\varepsilon, h) \sim \varepsilon$. This means that the boundary terms could in principle blow up as fast as $\varepsilon^{-1/2}$, and so, in particular, can then no longer be expected to become negligible as ε tends to zero.

References

[1] Aubin, Thierry. 1976. Équations du type Monge-Ampère sur les variétés kähleriennes compactes. *C. R. Acad. Sci. Paris Sér. A-B*, **283**(3), Aiii, A119–A121.

[2] Barth, Wolf, Peters, Chris, and Van de Ven, Antonius 1984. *Compact Complex Surfaces*. Ergebnisse der Mathematik und ihrer Grenzgebiete (3), vol. 4. Berlin: Springer-Verlag.

[3] Besse, Arthur L. 1987. *Einstein Manifolds*. Ergebnisse der Mathematik und ihrer Grenzgebiete (3), vol. 10. Berlin: Springer-Verlag.

[4] Besson, Gérard, Courtois, Gilles, and Gallot, Sylvestre 1995. Entropies et rigidités des espaces localement symétriques de courbure strictement négative. *Geom. Func. Anal.*, **5**, 731–799.

[5] Bishop, Christopher, and LeBrun, Claude. 2020. Anti-self-dual 4-manifolds, quasi-Fuchsian groups, and almost-Kähler geometry. Preprint arXiv:1708.03824 [math.DG]; to appear in *Commun. Anal. Geom.*

[6] Böhm, Christoph. 1998. Inhomogeneous Einstein metrics on low-dimensional spheres and other low-dimensional spaces. *Invent. Math.*, **134**(1), 145–176.

[7] Boyer, Charles P., Galicki, Krzysztof, and Kollár, János. 2005. Einstein metrics on spheres. *Ann. of Math. (2)*, **162**(1), 557–580.

[8] Chen, Xiu Xiong, LeBrun, Claude, and Weber, Brian. 2008. On conformally Kähler, Einstein manifolds. *J. Amer. Math. Soc.*, **21**(4), 1137–1168.

[9] Derdziński, Andrzej. 1983. Self-dual Kähler manifolds and Einstein manifolds of dimension four. *Compositio Math.*, **49**(3), 405–433.

[10] DeTurck, Dennis M., and Kazdan, Jerry L. 1981. Some regularity theorems in Riemannian geometry. *Ann. Sci. École Norm. Sup. (4)*, **14**(3), 249–260.

[11] Griffiths, Phillip, and Harris, Joseph. 1978. *Principles of Algebraic Geometry.* New York: Wiley-Interscience.

[12] Gursky, Matthew J. 2000. Four-manifolds with $\delta W^+ = 0$ and Einstein constants of the sphere. *Math. Ann.*, **318**(3), 417–431.

[13] Hitchin, Nigel J. 1975. On the curvature of rational surfaces. Pages 65–80 of: *Differential geometry (Proc. Symp. Pure Math., vol. XXVII, Part 2, Stanford Univ., Stanford, CA, 1973).* Providence, RI: Amer. Math. Soc.

[14] Honda, Ko. 2004. Transversality theorems for harmonic forms. *Rocky Mountain J. Math.*, **34**(2), 629–664.

[15] Kim, Inyoung. 2016. Almost-Kähler anti-self-dual metrics. *Ann. Global Anal. Geom.*, **49**(4), 369–391.

[16] LeBrun, Claude. 1995. Einstein metrics and Mostow rigidity. *Math. Res. Lett.*, **2**(1), 1–8.

[17] LeBrun, Claude. 1997a. Einstein metrics on complex surfaces. Pages 167–176 of: *Geometry and Physics (Aarhus, 1995).* Lecture Notes in Pure and Applied Mathematics, vol. 184. New York: Dekker.

[18] LeBrun, Claude. 1997b. Yamabe constants and the perturbed Seiberg-Witten equations. *Commun. Anal. Geom.*, **5**, 535–553.

[19] LeBrun, Claude. 2012. On Einstein, Hermitian 4-manifolds. *J. Differ. Geom.*, **90**(2), 277–302.

[20] LeBrun, Claude. 2015. Einstein metrics, harmonic forms, and symplectic four-manifolds. *Ann. Global Anal. Geom.*, **48**(1), 75–85.

[21] LeBrun, Claude, and Singer, Michael. 1994. A Kummer-type construction of self-dual 4-manifolds. *Math. Ann.*, **300**(1), 165–180.

[22] Odaka, Yuji, Spotti, Cristiano, and Sun, Song. 2016. Compact moduli spaces of del Pezzo surfaces and Kähler-Einstein metrics. *J. Differ. Geom.*, **102**(1), 127–172.

[23] Penrose, Roger, and Rindler, Wolfgang. 1986. *Spinors and Space-Time. Vol. 2: Spinors and Twistor Methods in Space-Time Geometry.* Cambridge: Cambridge University Press.

[24] Perutz, Tim. 2006. Zero-sets of near-symplectic forms. *J. Symplectic Geom.*, **4**(3), 237–257.

[25] Siu, Yum Tong. 1988. The existence of Kähler-Einstein metrics on manifolds with positive anticanonical line bundle and a suitable finite symmetry group. *Ann. of Math. (2)*, **127**(3), 585–627.

[26] Sung, Myong-Hee. 1997. Kähler surfaces of positive scalar curvature. *Ann. Global Anal. Geom.*, **15**(6), 509–518.

[27] Taubes, Clifford H. 1998. The geometry of the Seiberg-Witten invariants. Pages 299–339 of: *Surveys in Differential Geometry, vol. III (Cambridge, MA, 1996).* Boston, MA: International Press.

[28] Taubes, Clifford Henry. 1992. The existence of anti-self-dual conformal structures. *J. Differ. Geom.*, **36**(1), 163–253.

[29] Taubes, Clifford Henry. 2006. A proof of a theorem of Luttinger and Simpson about the number of vanishing circles of a near-symplectic form on a 4-dimensional manifold. *Math. Res. Lett.*, **13**(4), 557–570.

[30] Tian, Gang. 1990. On Calabi's conjecture for complex surfaces with positive first Chern class. *Invent. Math.*, **101**(1), 101–172.

[31] Tian, Gang, and Yau, Shing-Tung. 1987. Kähler–Einstein metrics on complex surfaces with $C_1 > 0$. *Commun. Math. Phys.*, **112**(1), 175–203.

[32] van Coevering, Craig. 2012. Sasaki-Einstein 5-manifolds associated to toric 3-Sasaki manifolds. *New York J. Math.*, **18**, 555–608.

[33] Yau, Shing Tung. 1974. On the curvature of compact Hermitian manifolds. *Invent. Math.*, **25**, 213–239.

[34] Yau, Shing Tung. 1977. Calabi's conjecture and some new results in algebraic geometry. *Proc. Nat. Acad. Sci. U.S.A.*, **74**(5), 1798–1799.

[35] Yau, Shing Tung. 1978. On the Ricci curvature of a compact Kähler manifold and the complex Monge-Ampère equation. I. *Commun. Pure Appl. Math.*, **31**(3), 339–411.

10

Construction of the Supersymmetric Path Integral: A Survey

Matthias Ludewig

Abstract

This is a survey based on the joint work [17, 19] with Florian Hanisch and Batu Güneysu reporting on a rigorous construction of the supersymmetric path integral associated to compact spin manifolds.

10.1 Introduction

A way to understand the geometry of the loop space[1] LX of a manifold X is by studying its differential forms. On finite-dimensional manifolds, a key feature of differential forms is that they can be *integrated*, which gives a linear functional on the space of differential forms.[2] Of course, one of the fundamental properties of this integration functional is that it is only non-zero on forms of highest degree; at first glance, this seems to make the task of defining such an integration functional in the infinite-dimensional context of the loop space impossible, as there is no top degree.

However, if we fix a Riemannian metric on X and define the *canonical two-form* on LX by setting

$$\omega[v, w] := \int_{\mathbb{T}} \langle v(t), \nabla_{\dot\gamma} w(t) \rangle \, dt \tag{10.1}$$

for $\gamma \in LX$, $v, w \in T_\gamma LX = C^\infty(\mathbb{T}, \gamma^* TX)$, it turns out that there

---

[1] Throughout, we denote $\mathbb{T} = S^1 = \mathbb{R}/\mathbb{Z}$; the smooth loop space is then defined by $LX = C^\infty(\mathbb{T}, X)$.

[2] We remark that even in the finite-dimensional case, if the manifold is non-compact, then the integration functional will only be defined on a suitable subset of *integrable forms*. The same will be true in the infinite-dimensional case.

is a natural way to make sense of the top degree component of the wedge product $e^\omega \wedge \theta$ for suitable forms θ, by using simple analogies to the finite-dimensional situation (here e^ω denotes the exponential of ω in the algebra $\Omega(LX)$ of differential forms). This *top degree component* $[e^\omega \wedge \theta]_{\text{top}}$ should be seen as the pairing of $e^\omega \wedge \theta$ with the volume form corresponding to the L^2-metric on LX (which must remain heuristic as there is no such volume form). Using this notion, an integration functional can then be defined using the Wiener measure.

Relation to the Atiyah–Singer index theorem There is another side to this story of integrating differential forms on the loop space, which is our main motivation. More than 30 years ago, it was observed by Atiyah [3] and Witten that there is a very short and conceptual, but formal, i.e. non-rigorous, proof of the Atiyah–Singer index theorem using a supersymmetric version of the Feynman path integral. In physics terms, this is the path integral of the $\mathcal{N} = 1/2$ supersymmetric σ-model [1]. Reformulating the supergeometry appearing in the work of Alvarez-Gaumé in the language of differential forms, Atiyah was led to consider the differential form integral

$$I[\theta] \stackrel{\text{formally}}{=} \int_{LX} e^{-S+\omega} \wedge \theta \tag{10.2}$$

over the loop space of a Riemannian (spin) manifold X, for suitable differential forms $\theta \in \Omega(LX)$, where

$$S(\gamma) = \frac{1}{2} \int_{\mathbb{T}} |\dot{\gamma}(t)|^2 dt \tag{10.3}$$

is the usual energy functional, and ω is the canonical two-form defined in (10.1). Atiyah proceeds with a series of formal manipulations allowing him to rewrite (10.2) as a Wiener integral. Then, using the Feynman–Kac formula, he identifies this Wiener integral with the supertrace of the heat semigroup associated to the Dirac operator and thus (via the McKean–Singer formula) with the index of the Dirac operator.

On the other hand, the loop space has a natural $\mathbb{T}$-action by rotation of loops, and the differential form $S - \omega$ is closed with respect to the *equivariant differential*

$$d_K := d + \iota_K, \tag{10.4}$$

where ι_K denotes insertion of the generating vector field $K(\gamma) = \dot{\gamma}$ of the rotation action. Hence, if the given θ also satisfies $d_K \theta = 0$, then the composite differential form $e^{-S+\omega} \wedge \theta$ considered above is equivariantly

closed as well. Motivated by this observation, Atiyah formally[3] applies
a Duistermaat–Heckmann type formula [7, 11], in order to localize the
integral to the fixed point set with respect to the rotation action, which
is precisely the set of constant loops. Now there is an obvious inclusion
map $i : X \to \mathsf{L}X$ identifying X with this fixed point set, and one has
the *localization formula*

$$I[\theta] \overset{\text{formally}}{=} \int_X \widehat{A}(X) \wedge i^*\theta. \tag{10.5}$$

It was later observed by Bismut [8] that this can be used to (formally)
prove the twisted Atiyah–Singer index theorem, by considering special
differential forms on $\mathsf{L}X$ defined from the data of a vector bundle with
connection on X, which today are called *Bismut–Chern characters*.

Our work In this chapter, we give an account of a recent project [19,
17] that carries out a rigorous construction of the supersymmetric path
integral map I described above. The map should have the following
properties.

(i) The map I is defined on some large subset of $\Omega(\mathsf{L}X)$ of *integrable
forms*, which at least includes the Bismut–Chern characters defined
by Bismut [8].

(ii) For any integrable differential form θ with $d_K\theta = 0$, the map I
satisfies the localization formula (10.5).

We remark that, in particular, (ii) implies that I is coclosed with
respect to d_K; in physics language, this means that the path integral is
supersymmetric, where the idea is that the functional is invariant under
the odd symmetry generated by d_K. Of course, the properties (i) and
(ii) do not fix I uniquely, since e.g. the functional $I_0(\theta)$, just defined as
the right hand side of (10.5), satisfies both requirements tautologically.
To obtain a reasonable problem, we therefore add the following rather
heuristic requirement.

(iii) The map I is given by formula (10.2) in a suitable sense.

In our work, we construct such a map I. Notice that property (ii)
follows if I is homologous to the map I_0 defined by the right hand side
of (10.5); however, we emphasize that our construction is *geometric*. In
other words, we construct I as a *cochain* rather than an equivalence class
in cohomology.

---

[3] Meaning that one pretends that $\mathsf{L}X$ is finite dimensional.

In fact, we provide two different constructions of the map I: In [19], a stochastic approach is taken to construct I starting from property (iii); it is not necessarily apparent from this approach, however, that the map constructed that way has property (ii). This is fixed in [17], where we use methods from non-commutative geometry to define a map which – using a fancy version of Getzler rescaling – can be shown to satisfy (ii). The equivalence of these constructions is then established in [19].

In this chapter, we proceed by highlighting the first construction, as described in [19]; afterwards, we discuss the second construction, as given in [17]. In the final section, we connect the two approaches and discuss the localization formula (10.5) and its application to the Bismut–Chern characters.

10.2 First Construction: The Top Degree Functional

In this section, we give a quick overview of the construction of the path integral map I portrayed in the introduction, following [19].

The Wiener measure The construction is essentially based on the Wiener measure $\mathbb{W}$, a certain measure on the continuous loop space[4] $\mathsf{L}_c X = C(\mathbb{T}, X)$ of a Riemannian manifold X. The Wiener integral of so-called *cylinder functions* is easy to describe. These are functions $F : \mathsf{L}_c X \to \mathbb{C}$ of the form

$$F(\gamma) = f\big(\gamma(\tau_1), \ldots, \gamma(\tau_N)\big) \tag{10.6}$$

for some $f \in C(M^N)$ and $0 \leq \tau_1 < \cdots < \tau_N < 1$; the formula for their Wiener integral $\mathbb{W}[F]$ is[5]

$$\mathbb{W}[F] \stackrel{\text{def}}{=} \int_X \cdots \int_X f(x_1, \ldots, x_N) \left(\prod_{j=1}^{N} p_{\tau_{j+1}-\tau_j}(x_j, x_{j+1}) \right) \mathrm{d}x_1 \cdots \mathrm{d}x_N, \tag{10.7}$$

where $p_t(x, y)$ is the *heat kernel* of X, i.e. the fundamental solution to the heat equation. By the extension theorem and the continuity theorem of Kolmogorov, this determines $\mathbb{W}[F]$ uniquely for all bounded functions F on $\mathsf{L}_c X$.

---

[4] In fact, the Wiener measure is defined on any space of paths, but here we restrict to the loop space.

[5] In the formula, we adopt the notation $\tau_{N+1} := 1 + \tau_1$ and $x_{N+1} := x_1$.

For $X = \mathbb{R}^n$, one has the explicit formula

$$p_t(x, y) = (2\pi t)^{-n/2} \exp\left(-\frac{|x - y|^2}{2t}\right)$$

for the heat kernel. After inserting this into (10.7) for F a cylinder function, some elementary manipulations give the result

$$\mathbb{W}[F] = \left(\prod_{j=1}^{N} \left(2\pi(\tau_j - \tau_{j-1})\right)^{-n/2}\right) \int_{\mathbb{R}^n} \cdots \int_{\mathbb{R}^n} F(\gamma)e^{-S(\gamma)} dx, \quad (10.8)$$

where $\gamma = \gamma_x$ is the piecewise linear loop with $\gamma(\tau_j) = x_j$ and, as usual, S is the energy functional (10.3). This formula has an extension to manifolds [2, 4, 22]. In fact, formulas like (10.8) go all the way back to Feynman [13], constituting the starting point for his path integral approach to quantum mechanics. Taking the limit over N, (10.8) leads to the heuristic formula

$$\mathbb{W}[F] \stackrel{\text{formally}}{=} \frac{1}{C} \int_{\mathsf{L}_c X} F(\gamma)e^{-S(\gamma)} d\gamma \quad (10.9)$$

for a suitable constant C; in other words, the slogan is that the Wiener measure has the density function e^{-S} with respect to the "Riemannian volume measure" $d\gamma$ on the loop space $\mathsf{L}_c X$. Of course, there are several well-known problems with this formula that make it remain heuristic, first and foremost the non-existence of the measure $d\gamma$ and the infinitude of the constant C.

Formal definition of the path integral map Ignoring the difference between the smooth and the continuous loop space for the moment, we record that we do not know yet how to integrate differential forms, but at least the Wiener measure enables us to integrate *functions* over the loop space. However, if M is an *oriented* (for now finite dimensional) Riemannian manifold, integrating differential forms and functions is essentially the same thing: The two are related by the formula

$$\int_M \theta = \int_M [\theta_y]_{\text{top}} dy \quad (10.10)$$

for differential forms θ, where the left hand side is to be understood as a differential form integral (determined by the orientation) and the right hand side is the integration map for functions determined by the Riemannian structure. While the latter integration map does not depend on the choice of orientation, the integrand

$$[\theta_y]_{\text{top}} \stackrel{\text{def}}{=} \langle \theta_y, \text{vol}_y \rangle,$$

a function on M called the *top degree component* of θ, does, as the sign of the volume form vol depends on the orientation. In supergeometry, this functional is often called the *Grassmann* or *Berezin integral* [5].

The idea now is to apply the observation above to the heuristic formula (10.2) for the path integral map. Starting from this formula, we obtain the chain of identifications

$$\int_{LX} e^{-S+\omega} \wedge \theta \overset{\text{formally}}{=} \int_{LX} [e^{\omega} \wedge \theta_{\gamma}]_{\text{top}} \, e^{-S(\gamma)} \mathrm{d}\gamma \overset{\text{formally}}{=} \mathbb{W}\big[[e^{\omega} \wedge \theta]_{\text{top}}\big];$$

(10.11)

as the first step, we formally applied (10.10) to this infinite-dimensional example, while in the second step, we recognized the right hand side of the heuristic formula (10.9), for the integrand $F(\gamma) = [e^{\omega} \wedge \theta_{\gamma}]_{\text{top}}$.

With a view on the right hand side of (10.11), the non-trivial task that remains is to provide meaning for the top degree component $[e^{\omega} \wedge \theta]_{\text{top}}$ of the differential form $e^{\omega} \wedge \theta$ as a $\mathbb{W}$-integrable function on $L_c X$; this is the main achievement of the paper [19]; we outline the construction below.

Remark 10.2.1   The formal manipulations conducted in (10.11) are more or less well known. However, in the literature, the differential form θ is either constant equal to one (see [3]) or taken to be a Bismut–Chern character (see [8]). In both cases, the top degree component $[e^{\omega} \wedge \theta]_{\text{top}}$ can be defined (and computed) using ad hoc methods. The novelty of the approach taken in our paper [19] is that we allow a very general class of differential forms θ to be plugged into our top degree functional, in order to obtain a general definition of the path integral.

Definition of the top degree functional To explain the definition of our top degree functional, notice that the canonical two-form ω has the form $\omega[v, w] = \langle v, Aw \rangle_{L^2}$ in terms of the L^2 scalar product, where $A = \nabla_{\dot{\gamma}}$, a skew-adjoint operator on $C^{\infty}(\mathbb{T}, \gamma^* TX)$. Now, if V is an arbitrary finite-dimensional, oriented Euclidean vector space and $\omega \in \Lambda^2 V'$ has the form $\omega[v, w] = \langle v, Aw \rangle_V$ for an invertible skew-adjoint operator A on V, one has the result

$$[e^{\omega} \wedge \vartheta_1 \wedge \cdots \wedge \vartheta_N]_{\text{top}} = \mathrm{pf}(A) \, \mathrm{pf}\Big(\langle \vartheta_a, A^{-1} \vartheta_b \rangle_V\Big)_{1 \le a, b \le N}, \quad (10.12)$$

for $\vartheta_1, \ldots, \vartheta_N \in V'$, where pf stands for the *Pfaffian* of a skew-symmetric matrix, see [21, Prop. 1]. In the case that A is not invertible, there is a similar, slightly more complicated formula; for details, see [19]. This allows us to define the top degree functional on the infinite-dimensional

Euclidean vector space $V = C^\infty(\mathbb{T}, \gamma^*TX)$ by analogy. In the case that $A = \nabla_{\dot\gamma}$ is invertible, we can set

$$q(\theta_1 \wedge \cdots \wedge \theta_N) \stackrel{\text{def}}{=} \mathrm{pf}_\zeta(\nabla_{\dot\gamma})\mathrm{pf}\left(\langle\theta_a, \nabla_{\dot\gamma}^{-1}\theta_b\rangle_V\right)_{1\le a,b\le N} \tag{10.13}$$

for $\theta_1, \ldots, \theta_N \in C^\infty(\mathbb{T}, \gamma^*T'X)$, and if $\nabla_{\dot\gamma}$ is not invertible, it is invertible on the orthogonal complement of its (always finite dimensional) kernel, which allows us to employ the generalization of the formula (10.12) mentioned above. Hence, heuristically, $q(\theta)$ is the "top degree component" of the differential form $e^\omega \wedge \theta$.

In (10.13), $\mathrm{pf}_\zeta(\nabla_{\dot\gamma})$ denotes the zeta-regularized Pfaffian of $\nabla_{\dot\gamma}$, a square root of its zeta-regularized determinant. This quantity is not a number but rather an element of the *Pfaffian line* Pf_γ, a certain one-dimensional real vector space canonically associated to γ; this reflects the fact that there is no naïve concept of orientation on the infinite-dimensional vector space V. These Pfaffian lines glue together to the so-called *Pfaffian line bundle* Pf on LX, which is related to the spin condition: By the work of Stolz–Teichner and Waldorf [23, 24], a spin structure on X gives an orientation of the loop space LX, in the sense that it provides a canonical trivialization of the Pfaffian line bundle and turns the top degree component (10.13) into an honest number. This is the reason why the spin condition is important to define our path integral.

Remark 10.2.2   This is analogous to the fact that on a finite-dimensional *non-oriented* manifold, the top degree component is also a section of a real line bundle, the *orientation bundle*, which is trivialized by an orientation.

The main result of [19] is the following formula, which provides a way actually to compute its value and the value of the integral map.

Theorem 10.2.3   *Suppose that X is a spin manifold with spinor bundle Σ. Then the top degree component $q(\theta_N \wedge \cdots \wedge \theta_1)$ defined above is canonically a number, and it is given by the formula*

$$2^{-N/2} \sum_{\sigma\in S_N} \mathrm{sgn}(\sigma) \int_{\Delta_N} \mathrm{str}\left([\gamma\|_{\tau_N}^1]^\Sigma \prod_{a=1}^N \mathbf{c}\big(\theta_{\sigma_a}(\tau_a)\big)[\gamma\|_{\tau_{a-1}}^{\tau_a}]^\Sigma\right) d\tau. \tag{10.14}$$

Here $\Delta_N = \{0 \le \tau_1 \le \cdots \le \tau_N \le 1\}$ is the standard simplex, $[\gamma\|_{\tau_{a-1}}^{\tau_a}]^\Sigma$ denotes parallel translation in the spinor bundle along the loop γ and

c *denotes Clifford multiplication. Moreover,* str *is the supertrace of the spinor bundle.*[6]

Rigorous definition of the path integral map At this point, the top degree map assigns to a certain class of differential forms θ on the loop space $\mathsf{L}X$ of a spin manifold X the smooth function $q(\theta)$ on $\mathsf{L}X$, to be interpreted as the top degree component of $e^\omega \wedge \theta$. The problem now is that we need $q(\theta)$ to be a function on the *continuous* loop space L_cX, in order to be able to integrate with respect to the Wiener measure, as in (10.11).

One problem here when looking at formula (10.14) is that a loop has to be absolutely continuous in order to define the parallel transport along it. A solution to this problem is provided by the notion of *stochastic parallel transport*: As, ultimately, I is defined by Wiener integration, it suffices to have $q(\theta)$ defined as a measurable function only (with respect to the Wiener measure $\mathbb{W}$). This is achieved by interpreting the occurrences of the parallel transport in (10.14) in the stochastic sense, which provides a stochastic extension $\widetilde{q}$ of the function q; for details on the stochastic parallel transport, see e.g. [12, 16, 18].

To discuss the possible integrands θ, notice that since $\mathsf{L}_cX \subset \mathsf{L}X$ is dense, we can consider $\Omega(\mathsf{L}_cX)$ as a subspace of $\Omega(\mathsf{L}X)$.

Notation 10.2.4   Denote by $\mathscr{D} \subseteq \Omega(\mathsf{L}_cX) \subset \Omega(\mathsf{L}X)$ the space of differential forms θ that are wedge products of one-forms that are uniformly bounded.

For elements $\theta \in \mathscr{D}$, the function $\widetilde{q}(\theta)$ is a well-defined measurable function on L_cX. Since it is also bounded by the boundedness of θ, it is moreover integrable, and we define $I : \mathscr{D} \to \mathbb{R}$ by the formula

$$I[\theta] \stackrel{\text{def}}{=} \mathbb{W}\left[\widetilde{q}(\theta) \exp\left(-\frac{1}{8} \int_{\mathbb{T}} \text{scal}\big(\gamma(\tau) \big) \mathrm{d}\tau \right) \right]. \tag{10.15}$$

The main difference between this definition and the formal version (10.11) is the appearance of the exponential including the scalar curvature. While this may seem strange at first glance, this is an important "quantum correction" to the definition; see Remark 10.2.5 below.

Examples We now give some examples of differential forms that are contained in $\mathscr{D}$, together with their I-integrals. We assume X to be a

---

[6] Throughout, we take the *real spinor bundle*, a bundle of irreducible $\text{Cl}(TX)$-Cl_n-bimodules. In any dimension, the space $\text{End}_{\text{Cl}_n}(\Sigma)$ of endomorphisms of Σ commuting with the right Cl_n-action carries a canonical supertrace.

compact spin manifold in this section. Given a differential form $\vartheta \in \Omega^\ell(X)$ and $\tau \in \mathbb{T}$, we can produce a differential ℓ-form $\vartheta(\tau) \in \Omega^\ell(LX)$ by setting

$$\vartheta(\tau)_\gamma[v_1, \ldots, v_\ell] \overset{\text{def}}{=} \vartheta_{\gamma(\tau)}\big[v_1(\tau), \ldots, v_\ell(\tau)\big] \qquad (10.16)$$

for $v_1, \ldots, v_\ell \in T_\gamma LX$. Moreover, for any function $\varphi \in C^\infty(\mathbb{T})$, we can construct another differential form $\overline{\vartheta} \in \Omega^\ell(LX)$ by setting

$$\overline{\vartheta} = \int_{\mathbb{T}} \vartheta(\tau) \mathrm{d}\tau \qquad \text{or} \qquad \overline{\vartheta}_\gamma[v_1, \ldots, v_\ell] = \int_{\mathbb{T}} \vartheta_{\gamma(\tau)}\big[v_1(\tau), \ldots, v_\ell(\tau)\big] \mathrm{d}\tau.$$

$$(10.17)$$

If $\vartheta \in \Omega^1(X)$, then $\overline{\vartheta}$ is uniformly bounded on the loop space by compactness of X, hence $\overline{\vartheta} \in \mathscr{D} \subseteq \Omega^1(LX)$.

On these forms, the integral map is given as follows: Given $\vartheta_a \in \Omega^1(X)$, $a = 1, \ldots, N$, and correspondingly $\overline{\vartheta}_a \in \Omega^1(LX)$ defined by (10.17), then $\overline{\vartheta}_1 \wedge \cdots \wedge \overline{\vartheta}_N \in \mathscr{D}$, and the corresponding integral $I\big[\overline{\vartheta}_1 \wedge \cdots \wedge \overline{\vartheta}_N\big]$ is given by the combinatorial formula

$$2^{-N/2} \sum_{\sigma \in S_N} \mathrm{sgn}(\sigma) \int_{\Delta_N} \mathrm{Str}\Big(e^{-\tau_1 H} \prod_{a=1}^N \mathbf{c}(\vartheta_{\sigma_a}) e^{-(\tau_a - \tau_{a-1})H}\Big) \mathrm{d}\tau, \quad (10.18)$$

where $H = \mathsf{D}^2/2$, with D the Dirac operator. This follows from the explicit formula (10.14); the Wiener integral in (10.15) is then evaluated using a vector-valued Feynman–Kac formula, see e.g. [16].

Remark 10.2.5   The scalar curvature factor of (10.15) is needed because of the Lichnerowicz formula; without it, formula (10.18) would feature the operator $H - \mathrm{scal}/8$ instead of H, which has no good cohomological properties: It turns out that the scalar curvature term is necessary in order to make the functional coclosed (or, in physics lingo, to make the path integral supersymmetric).

10.3 Second Construction: The Chern Character

The preceding construction of the integral map was achieved by a naïve reformulation (10.11) of the heuristic path integral formula (10.2). Its disadvantage is that the domain $\mathscr{D} \subset \Omega(LX)$ where it is defined is quite small; for example, it does *not* contain the Bismut–Chern characters considered below, which are the most interesting integrands due to their rôle played in relation to the index theorem. We therefore now give a different construction of a path integral map, which has a much larger domain of definition and turns out to extend the previous one. A complete account can be found in the paper [17].

The bar construction To set things up, we have to introduce the following algebraic machinery: The *bar complex* associated to a differential graded algebra Ω is the graded vector space[7]

$$\mathsf{B}(\Omega) = \bigoplus_{N=0}^{\infty} \Omega[1]^{\otimes N}. \tag{10.19}$$

The elements of $\mathsf{B}(\Omega)$ are called bar chains and denoted by $(\vartheta_1, \ldots, \vartheta_N)$ for $\vartheta_a \in \Omega$, suppressing the tensor product sign in notation for convenience. $\mathsf{B}(\Omega)$ has a distinguished subspace $\mathsf{B}^{\natural}(\Omega)$, which consists of those elements of $\mathsf{B}(\Omega)$ that are invariant under graded cyclic permutation of the tensor factors. $\mathsf{B}(\Omega)$ has two differentials, a differential d coming from the differential of Ω and the *bar differential* b'; they are given by

$$d(\vartheta_1, \ldots, \vartheta_N) = \sum_{k=1}^{N} (-1)^{n_{k-1}} (\vartheta_1, \ldots, \vartheta_{k-1}, d\vartheta_k, \ldots, \vartheta_N),$$

$$b'(\vartheta_1, \ldots, \vartheta_N) = -\sum_{k=1}^{N-1} (-1)^{n_k} (\vartheta_1, \ldots, \vartheta_{k-1}, \vartheta_k \vartheta_{k+1}, \vartheta_{k+2}, \ldots, \vartheta_N),$$

where $n_k = |\vartheta_1| + \cdots + |\vartheta_k| - k$. The above differentials satisfy $db' + b'd = 0$; hence they turn $\mathsf{B}(\Omega)$ and $\mathsf{B}^{\natural}(\Omega)$ into bicomplexes with total differential $d + b'$. Dually, we have the *codifferential*

$$(\delta\ell)[\vartheta_1, \ldots, \vartheta_N] \stackrel{\text{def}}{=} -\ell\big[(d + b')(\vartheta_1, \ldots, \vartheta_N)\big] \tag{10.20}$$

on the space of linear maps $\ell : \mathsf{B}(\Omega) \to \mathbb{C}$.

The iterated integral map Remember the definition (10.16) of cylinder forms above. *Chen's iterated integral map* [10] also constructs differential forms on the loop space from differential forms on X, this time taking as input elements of the bar complex $\mathsf{B}(\Omega(X))$. For our purposes, we need an extension of this, introduced by Getzler, Jones and Petrack [15]. We consider the differential graded algebra

$$\Omega_{\mathbb{T}}(X) \stackrel{\text{def}}{=} \Omega(X \times \mathbb{T})^{\mathbb{T}},$$

the space of differential forms on $X \times \mathbb{T}$ which are constant in the $\mathbb{T}$-direction. Elements $\vartheta \in \Omega_{\mathbb{T}}(X)$ will always be decomposed into $\vartheta = \vartheta' +$

---

[7] Here $\Omega(X)[1]$ equals $\Omega(X)$ as a vector space, but with degrees shifted by one. Namely, $\vartheta \in \Omega^{k+1}$ if and only if $\vartheta \in \Omega[1]^k$.

$dt \wedge \vartheta''$, where $\vartheta', \vartheta'' \in \Omega(X)$. The differential of $\Omega_{\mathbb{T}}(X)$ is $d_{\mathbb{T}} := d - \iota_{\partial_t}$, where ι_{∂_t} denotes insertion of the canonical vector field ∂_t on the $\mathbb{T}$ factor and d denotes the de-Rham differential on $X \times \mathbb{T}$. In other words, we have

$$d_{\mathbb{T}}\vartheta = d_{\mathbb{T}}(\vartheta' + dt \wedge \vartheta'') = d\vartheta' - dt \wedge d\vartheta'' - \vartheta'',$$

where now, on the right hand side, d denotes the de-Rham differential on X. The version of the *extended iterated integral map* used in this chapter is a map taking $\mathsf{B}(\Omega_{\mathbb{T}}(X))$ to $\Omega(\mathsf{L}X)$; it is defined by the formula

$$\rho(\vartheta_1, \ldots, \vartheta_N) = \int_{\Delta_N} \big(\iota_K \vartheta_1'(\tau_1) + \vartheta_1''(\tau_1)\big) \wedge \cdots \wedge \big(\iota_K \vartheta_N'(\tau_N) + \vartheta_N''(\tau_N)\big) d\tau$$

$$(10.21)$$

for $\vartheta_1, \ldots, \vartheta_N \in \Omega_{\mathbb{T}}(\mathsf{L}X)$, where we recall that $K(\gamma) = \dot{\gamma}$ is the canonical velocity vector field. This allows us to produce many examples of differential forms on the loop space.

The crucial fact about ρ is that its restriction $\rho^\natural$ to cyclic chains,

$$\rho^\natural : \mathsf{B}\big(\Omega_{\mathbb{T}}(X)\big) \supset \mathsf{B}^\natural\big(\Omega_{\mathbb{T}}(X)\big) \longrightarrow \Omega(\mathsf{L}X)^{\mathbb{T}} \subset \Omega(\mathsf{L}X),$$

is a *chain map* in the sense that $\rho^\natural$ sends the total differential $d_{\mathbb{T}} + b'$ to the equivariant differential d_K (defined in (10.4)). Moreover, notice that the degree shift in the definition of $\mathsf{B}(\Omega_{\mathbb{T}}(X))$ ensures that ρ is in fact degree preserving.

The Chern character Now let X be a compact spin manifold. Our second construction of the path integral map I is based on the construction of a closed cochain

$$\mathrm{Ch}_{\mathsf{D}} : \mathsf{B}^\natural\big(\Omega_{\mathbb{T}}(X)\big) \longrightarrow \mathbb{R},$$

called the *Chern character* in [17]. It has the property that it vanishes on the kernel $\ker(\rho)$ of the iterated integral map (10.21), hence Ch_{D} can be pushed forward to a functional on the image of the iterated integral map inside $\Omega(\mathsf{L}X)$.

To define Ch_{D}, we define a cochain F on $\mathsf{B}(\Omega_{\mathbb{T}}(X))$ with values in the algebra of linear operators on $L^2(X, \Sigma)$. Explicitly, F is given on homogeneous elements by the formula

$$F[\vartheta] \stackrel{\text{def}}{=} \mathbf{c}(\vartheta'') + [\mathsf{D}, \mathbf{c}(\vartheta')] - \mathbf{c}(d\vartheta'),$$

$$F[\vartheta_1, \vartheta_2] \stackrel{\text{def}}{=} (-1)^{|\vartheta_1'|}\big(\mathbf{c}(\vartheta_1')\mathbf{c}(\vartheta_2') - \mathbf{c}(\vartheta_1' \wedge \vartheta_2')\big),$$

where D is the Dirac operator; moreover, we set $F[\vartheta_1, \ldots, \vartheta_k] = 0$ whenever $k \geq 3$. The formula for the Chern character $\mathrm{Ch}_{\mathsf{D}}[\vartheta_1, \ldots, \vartheta_N]$ is now

$$2^{-n_N/2} \sum_{s \in \mathscr{P}_N} \int_{\Delta_M} \mathrm{Str}\Big(e^{-\tau_1 H} \prod_{a=1}^{M} F[\vartheta_{s_{a-1}+1}, \ldots \vartheta_{s_a}] e^{-(\tau_a - \tau_{a-1})H}\Big) \mathrm{d}\tau.$$

$$(10.22)$$

Here $\mathscr{P}_N$ denotes the set of all partitions of $\{1, \ldots, N\}$, given by a sequence of numbers $s = \{0 = s_0 < s_1 < \cdots < s_M = N\}$. In particular, as F vanishes when one inputs more than two elements, a summand corresponding to a partition s is zero as soon as there exists an index a with $s_a - s_{a-1} \geq 3$.

Remark 10.3.1   The name "Chern character" stems from the fact that Ch_{D} can be interpreted as the version of a Chern character in non-commutative geometry, namely that of a Fredholm module given by the Dirac operator on X. For details, see [17].

Properties of the Chern character One of the advantages of the second approach to the supersymmetric path integral map is that due to the algebraic character of the construction, it is easier to investigate its properties. As mentioned above, one of the results is that Ch_{D} is *Chen normalized* [17, Thm. 5.5], meaning that it vanishes on the kernel $\ker(\rho^\natural)$ of the iterated integral map, restricted to cyclic chains. This means that we can define its push-forward

$$I' : \Omega(\mathsf{L}X) \supset \mathrm{im}(\rho^\natural) \longrightarrow \mathbb{R}, \qquad I'[\theta] = \mathrm{Ch}_{\mathsf{D}}\big[\rho^\natural(\vartheta_1, \ldots, \vartheta_N)\big]$$

if $\theta = \rho^\natural(\vartheta_1, \ldots, \vartheta_N)$; notice that this is well defined as Ch_{D} is Chen normalized. This gives a second functional on the space of differential forms on the loop space, with domain $\mathrm{im}(\rho^\natural)$. One of the main features of the construction is the fact that Ch_{D} is coclosed, meaning that

$$\delta\mathrm{Ch}_{\mathsf{D}} = 0, \qquad\qquad (10.23)$$

where δ is the codifferential (10.20); see [17, Thm. 4.2, 5.3]. By the compatibility of $\rho^\natural$ with respect to the differentials, this implies that I' is coclosed with respect to the equivariant differential d_K. In other words, for any differential form $\theta \in \mathrm{im}(\rho^\natural)$, we have the following version of *Stokes' theorem*:

$$I'[d_K\theta] = 0,$$

stating that exact forms have vanishing integral. In physics slang, this is the *supersymmetry* of the path integral.

However, much more is true. The operator-theoretic formula (10.22) for Ch_D makes it accessible to Getzler's rescaling technique; a souped up version of this machinery enables us to show the following [17, Thm. 9.1].

Theorem 10.3.2 Ch_D *is cohomologous, as a Chen normalized cochain on* $B^\natural(\Omega_\mathbb{T}(X))$, *to the Chen normalized cochain* μ_0, *defined by*

$$\mu_0[\vartheta_1, \ldots, \vartheta_N] \overset{\text{def}}{=} \frac{1}{(2\pi)^{n/2} N!} \int_X \hat{A}(X) \wedge \vartheta_1'' \wedge \cdots \wedge \vartheta_N'', \qquad (10.24)$$

where $\hat{A}(X)$ *is the Chern–Weil representative of the* $\hat{A}$-*genus of* X.

Remember here that we say that a cochain is Chen normalized if it vanishes on the kernel of ρ. Since $\hat{A}(X)$ is a closed differential form, this implies (10.23), by the usual Stokes theorem. As we discuss below, (10.24) essentially implies the localization formula (10.5) for suitable differential forms.

Comparison to the previous definition Inspecting formula (10.21), we see that

$$\rho(\vartheta_1, \ldots, \vartheta_N) = \int_{\Delta_N} \vartheta_1''(\tau_1) \wedge \cdots \wedge \vartheta_N''(\tau_N) d\tau$$

whenever $\vartheta_a' = 0$ for each a. If each ϑ_a'' has degree one, we have

$$\sum_{\sigma \in S_N} \text{sgn}(\sigma) \rho(\vartheta_{\sigma_1}, \ldots, \vartheta_{\sigma_N}) = \overline{\vartheta}_1'' \wedge \cdots \wedge \overline{\vartheta}_N'',$$

which is contained in $\mathscr{D}$ and hence has a well-defined path integral, as defined in (10.15). On the other hand, an inspection of (10.22) yields

$$Ch_D[\vartheta_1, \ldots, \vartheta_N] = 2^{-N/2} \int_{\Delta_N} \text{Str}\left(e^{-\tau_1 H} \prod_{a=1}^{N} \mathbf{c}(\vartheta_a'') e^{-(\tau_a - \tau_{a-1})H} \right) d\tau.$$

After anti-symmetrization, this coincides with the formula for $I[\overline{\vartheta}_1'' \wedge \cdots \wedge \overline{\vartheta}_N'']$, as calculated in (10.18). In this sense, the two versions of the integral map agree, and from now on we will use the notation I instead of I'.

10.4 Bismut–Chern Characters, Entire Chains and the Localization Formula

As usual, throughout this section X denotes a Riemannian manifold, which is assumed to be compact and spin for all statements related to the path integral map.

Periodic cyclic cohomology Throughout, a general differential form on the loop space is the direct *sum* of its homogeneous components; in other words, we denote

$$\Omega(LX) \overset{\text{def}}{=} \bigoplus_{\ell=0}^{\infty} \Omega^\ell(LX).$$

It is well known, however [20], that in the equivariant cohomology of the loop space, it is important to allow differential forms that are an infinite sum of its homogeneous components; in other words, elements of the direct *product* of the $\Omega^\ell(LX)$. This gives the *periodic equivariant cohomology* $h_\mathbb{T}(LX)$ of the loop space, which is the cohomology of the $\mathbb{Z}_2$-graded complex[8] $\widehat{\Omega}(LX)^\mathbb{T} = \widehat{\Omega}^+(LX)^\mathbb{T} \oplus \widehat{\Omega}^-(LX)^\mathbb{T}$, where

$$\widehat{\Omega}^+(LX) \overset{\text{def}}{=} \prod_{\ell=0}^{\infty} \Omega^{2\ell}(LX), \qquad \widehat{\Omega}^-(LX) \overset{\text{def}}{=} \prod_{\ell=0}^{\infty} \Omega^{2\ell+1}(LX).$$

The corresponding differential is the equivariant differential d_K, see (10.4), which exchanges the even and odd parts.

Bismut–Chern characters Maybe the most prominent example of such a differential form are the Bismut–Chern characters, defined as in Definition 10.4.1.

Definition 10.4.1 Let E be a Hermitian vector bundle with connection ∇ over the manifold X. The *Bismut–Chern character* associated with this data is the equivariantly closed differential form $\mathrm{Ch}(E, \nabla) \in \widehat{\Omega}^+(LX)$ given by the formula

$$\mathrm{Ch}(E, \nabla)_\gamma = \sum_{N=0}^{\infty} (-1)^N \int_{\Delta_N} \mathrm{tr}_E \left([\gamma\|_{\tau_N}^1]^E \prod_{a=1}^{N} R(\tau_a)_\gamma [\gamma\|_{\tau_{a-1}}^{\tau_a}]^E \right) d\tau$$

at $\gamma \in LX$, where R is the curvature of the connection ∇.

Explicitly, the degree $2N$-component $\mathrm{Ch}_N[v_{2N}, \ldots, v_1]$ of $\mathrm{Ch}(E, \nabla)$ is given by

$$2^{-N} \sum_{\sigma \in S_{2N}} \int_{\Delta_N} \mathrm{tr}_E \left([\gamma\|_{\tau_N}^1]^E \prod_{a=1}^{N} R\big(v_{\sigma_{2a}}(\tau_a), v_{\sigma_{2a-1}}(\tau_a)\big) [\gamma\|_{\tau_{a-1}}^{\tau_a}]^E \right) d\tau.$$

---

[8] It is customary in this context to introduce a formal variable of degree 2 and its inverse in order to define the periodic cyclic cohomology. The effect is that the complex and its cohomology are $\mathbb{Z}$-graded, but 2-periodic. Here we reduce modulo 2 right away.

The main properties of the Bismut–Chern character is that it is equivariantly closed, $d_K \mathrm{Ch}(E, \nabla) = 0$ (in other words, $d\mathrm{Ch}_N = \iota_K \mathrm{Ch}_{N+1}$) and that its pullback along the inclusion $i : X \to LX$ is the ordinary Chern character of (E, ∇), defined using Chern–Weil theory:

$$i^* \mathrm{Ch}(E, \nabla) = \mathrm{ch}(E, \nabla). \qquad (10.25)$$

Formally, the following theorem has been observed by Bismut [8] and was his original motivation for the definition of these differential forms. Of course, by the usual argument of McKean–Singer [6, Thm. 3.50], the right hand side of (10.26) below equals $\mathrm{ind}(\mathsf{D}_E)$, the graded index of the twisted Dirac operator D_E.

Theorem 10.4.2 *We have the formula*

$$I\big[\mathrm{Ch}(E, \nabla)\big] = \mathrm{Str}(e^{-\mathsf{D}_E^2/2}). \qquad (10.26)$$

In (10.26), we take I to be the path integral map constructed in Section 10.3. This makes sense as, by the results of [15, Sect. 6], $\mathrm{Ch}(E, \nabla)$ can be written as an iterated integral, i.e. there exist elements $c_N \in \mathsf{B}^{2N}(\Omega_\mathbb{T}(X))$, $N = 0, 1, 2, \ldots$, such that $\rho(c_N) = \mathrm{Ch}_N$. In particular, each Ch_N is contained in the domain of the integral map $\rho_*^\flat \mathrm{Ch}_\mathsf{D}$ constructed in Section 10.3. The identity (10.26) is then proven using [17, Prop. 8.2].

We remark that Ch does *not* directly lie in the domain of the integral map I defined in Section 10.2; in fact, Ch is not even a smooth differential form on $\mathsf{L}_c X$, due to the presence of the parallel transport in its definition. However, interpreting the parallel transport in the stochastic sense, one obtains a differential form on $\mathsf{L}_c X$ with measurable coefficients. The top degree map can be applied to this measurable differential form, which yields a measurable function on $\mathsf{L}_c X$. One can then compute $I[\mathrm{Ch}(E, \nabla)]$ by employing a suitable version of the Feynman–Kac formula, which gives the same result.

Entire cohomology In the discussion of Theroem 10.4.2, we have so far omitted the fact that the Bismut–Chern characters are not contained in $\Omega(LX)$, but only in the extension $\widehat{\Omega}(LX)$ that allows infinite sums of homogeneous forms. In particular, it is not at all clear a priori that $I[\mathrm{Ch}(E, \nabla)]$, defined as the sum of the individual integrals $I[\mathrm{Ch}_N]$, makes any sense. This issue is best discussed under the framework of our second approach to the integral map, where it is related to the entire cohomology of Connes.

For a differential graded algebra Ω, we denote by $\widehat{\mathsf{B}}(\Omega)$ the complex defined by the same formula (10.19) as $\mathsf{B}(\Omega)$, but with a direct product replacing the direct sum. In other words, its elements are arbitrary sums $\sum_{N=0}^{\infty} \theta^{(N)}$, with $\theta^{(N)} \in \Omega[1]^{\otimes N}$, without any convergence requirement. The *entire bar complex* $\mathsf{B}_\varepsilon(\Omega)$ is then a certain subcomplex of $\widehat{\mathsf{B}}(\Omega)$, containing chains that satisfy a certain growth condition; for details, we refer to [17]. One can then show that, for any Bismut–Chern character $\mathrm{Ch}(E, \nabla)$, the chain $c = \sum_{N=0}^{\infty} c_N \in \widehat{B}(\Omega_{\mathbb{T}}(X))$ such that $\mathrm{Ch}(E, \nabla) = \rho(c)$, constructed by Getzler–Jones–Petrack, is entire. Dually, the following result is shown in [17, Thm. 4.1, 5.2]:

Theorem 10.4.3 *The Chern character* Ch_{D} *has a continuous extension to* $\mathsf{B}_\varepsilon(\Omega)$.

Together with the discussion, this gives an a priori reason why the left hand side of (10.26) is well defined.

The localization formula and the index theorem We now explain how to rigorously conduct the proof of the Atiyah–Singer index theorem envisioned by Atiyah [3] and Bismut [8] using our results. The first result is the following localization formula. We say that $\theta \in \widehat{\Omega}(\mathsf{L}X)$ is an entire iterated integral if there exists $c \in \mathsf{B}_\varepsilon^\natural(\Omega_{\mathbb{T}}(X))$ such that $\theta = \rho^\natural(c)$. Here $\mathsf{B}_\varepsilon^\natural(\Omega_{\mathbb{T}}(X))$ denotes the subcomplex of $\mathsf{B}_\varepsilon(\Omega_{\mathbb{T}}(X))$ consisting of chains that are invariant under graded cyclic permutation.

Theorem 10.4.4 *Let* $\theta \in \widehat{\Omega}(\mathsf{L}X)$ *be equivariantly closed, i.e.* $d_K \theta = 0$, *and assume that it is an entire iterated integral. Then*

$$I[\theta] = (2\pi)^{-n/2} \int_X \hat{A}(X) \wedge i^* \theta. \tag{10.27}$$

Proof By the assumption on θ, there exists $c \in \mathsf{B}_\varepsilon^\natural(\Omega_{\mathbb{T}}(X))$ with $\rho^\natural(c) = \theta$. Define $I_0 : \widehat{\Omega}(\mathsf{L}X) \to \mathbb{R}$ by setting $I_0[\theta]$ to be the right hand side of (10.27) and notice that, by the definition (10.21) of the iterated integral map, we have

$$I_0[\theta] = \mu_0[c],$$

where μ_0 is defined in (10.24). Now, by Theorem 10.3.2, there exists a Chen normalized cochain μ' such that $\mathrm{Ch}_{\mathsf{D}} - \mu_0 = \delta\mu'$. Therefore,

$$\begin{aligned} I[\theta] - I_0[\theta] = (I - I_0)[\rho^\natural(c)] &= \left(\mathrm{Ch}_{\mathsf{D}} - \mu_0\right)[c] \\ &= \delta\mu'[c] = -\mu'\left[(d_{\mathbb{T}} + b')c\right]. \end{aligned}$$

Since $\rho^\natural$ is a chain map and θ is equivariantly closed, the calculation

$$\rho^\natural\big((d_\mathbb{T} + b')c\big) = d_K \rho^\natural(c) = d_K \theta = 0$$

shows that $(d_\mathbb{T} + b')c \in \ker(\rho^\natural)$, hence $\mu[(d_\mathbb{T} + b')c] = 0$, as μ is Chen normalized. $\qquad\square$

The localization formula (10.27) is an infinite-dimensional version of the localization formula of equivariant cohomology in finite dimensions, see [7, 11]. Applying it to a Bismut–Chern character $I\big[\mathrm{Ch}(E, \nabla)\big]$ (which is both equivariantly closed and can be represented as an entire iterated integral, as discussed above), we get

$$(2\pi)^{-n/2} \int_X \hat{A}(X) \wedge i^* \mathrm{Ch}(E, \nabla) = (2\pi)^{-n/2} \int_X \hat{A}(X) \wedge \mathrm{ch}(E, \nabla),$$
$$(10.28)$$

where in the last step we used (10.25). Together with our previous formula (10.26) and the McKean–Singer formula, this proves the Atiyah–Singer index theorem.

Odd dimensions We remark that nowhere in the above was it necessary to restrict to even-dimensional manifolds. Of course, in odd dimensions, both (10.28) and (10.26) are zero; for $I[\mathrm{Ch}(E, \nabla)]$, this is true because the path integral is an odd functional in this case; in other words, it evaluates as zero on even-dimensional forms. To obtain a non-trivial result in this case, one uses the *odd Bismut–Chern character* $\mathrm{Ch}(g)$ of Wilson [25], an equivariantly closed, odd element of $\widehat{\Omega}(\mathsf{L}X)$ associated to a map $g : X \to \mathrm{U}(k)$, the unitary group of order k, for some k. This can be represented by an entire iterated integral following the work of Cacciatori–Güneysu [9]. A result similar to Theorem 10.4.2 connects this to the spectral flow of the family $\mathsf{D}_s = \mathsf{D} + s\mathbf{c}(g^{-1}dg)$ of Dirac operators on $\Sigma \otimes \mathbb{C}^k$. This recovers the odd index theorem of Getzler [14].

References

[1] L. Alvarez-Gaumé. Supersymmetry and the Atiyah-Singer index theorem. *Commun. Math. Phys.*, 90(2):161–173, 1983.

[2] L. Andersson and B. K. Driver. Finite-dimensional approximations to Wiener measure and path integral formulas on manifolds. *J. Funct. Anal.*, 165(2):430–498, 1999.

[3] M. F. Atiyah. Circular symmetry and stationary-phase approximation. *Astérisque*, (131):43–59, 1985. Colloquium in honor of Laurent Schwartz, Vol. 1 (Palaiseau, 1983).

[4] C. Bär and F. Pfäffle. Path integrals on manifolds by finite dimensional approximation. *J. Reine Angew. Math.*, 625:29–57, 2008.

[5] F. A. Berezin. *Introduction to superanalysis*, vol. 9 of Mathematical Physics and Applied Mathematics. D. Reidel Publishing Co., Dordrecht, 1987. Edited and with a foreword by A. A. Kirillov; with an appendix by V. I. Ogievetsky; translated from the Russian by J. Niederle and R. Kotecký, translation edited by Dimitri Leites.

[6] N. Berline, E. Getzler, and M. Vergne. *Heat kernels and Dirac operators*. Grundlehren Text Editions. Springer-Verlag, Berlin, 2004. Corrected reprint of the 1992 original.

[7] N. Berline and M. Vergne. Classes caractéristiques équivariantes. Formule de localisation en cohomologie équivariante. *C. R. Acad. Sci. Paris Sér. I Math.*, 295(9):539–541, 1982.

[8] J.-M. Bismut. Index theorem and equivariant cohomology on the loop space. *Commun. Math. Phys.*, 98(2):213–237, 1985.

[9] S. Cacciatori and B. Güneysu. Odd characteristic classes in entire cyclic homology and equivariant loop space homology. *arXiv e-prints*, page arXiv:1805.07449, May 2018.

[10] K.-T. Chen. Iterated integrals of differential forms and loop space homology. *Ann. of Math. (2)*, 97:217–246, 1973.

[11] J. J. Duistermaat and G. J. Heckman. On the variation in the cohomology of the symplectic form of the reduced phase space. *Invent. Math.*, 69(2):259–268, 1982.

[12] M. Émery. *Stochastic calculus in manifolds*. Springer-Verlag, Berlin, 1989. With an appendix by P.-A. Meyer.

[13] R. P. Feynman and A. R. Hibbs. *Quantum mechanics and path integrals*. Dover Publications, Inc., Mineola, NY, 2010.

[14] E. Getzler. The odd Chern character in cyclic homology and spectral flow. *Topology*, 32(3):489–507, 1993.

[15] E. Getzler, J. D. S. Jones, and S. Petrack. Differential forms on loop spaces and the cyclic bar complex. *Topology*, 30(3):339–371, 1991.

[16] B. Güneysu. The Feynman-Kac formula for Schrödinger operators on vector bundles over complete manifolds. *J. Geom. Phys.*, 60(12):1997–2010, 2010.

[17] B. Güneysu and M. Ludewig. The Chern character of θ-summable Fredholm modules over dg algebras and the supersymmetric path integral. *arXiv e-prints*, page arXiv:1901.04721, Jan 2019.

[18] W. Hackenbroch and A. Thalmaier. *Stochastische Analysis*. Eine Einführung in die Theorie der stetigen Semimartingale. Mathematische Leitfäden. [Mathematical Textbooks.] B. G. Teubner, Stuttgart, 1994.

[19] F. Hanisch and M. Ludewig. A rigorous construction of the supersymmetric path integral associated to compact spin manifolds. *arXiv e-prints*, arXiv:1709.10027, 2017.

[20] J. D. S. Jones and S. B. Petrack. The fixed point theorem in equivariant cohomology. *Trans. Amer. Math. Soc.*, 322(1):35–49, 1990.

[21] J. Lott. Supersymmetric path integrals. *Commun. Math. Phys.*, 108(4):605–629, 1987.

[22] M. Ludewig. Path integrals on manifolds with boundary. *Commun. Math. Phys.*, 354(2):621–640, 2017.

[23] S. Stolz and P. Teichner. The spinor bundle on the loop space. people.mpim-bonn.mpg.de/teichner/Math/ewExternalFiles/MPI.pdf, 2005.

[24] K. Waldorf. Spin structures on loop spaces that characterize string manifolds. *Algebr. Geom. Topol.*, 16(2):675–709, 2016.

[25] S. O. Wilson. A loop group extension of the odd Chern character. *J. Geom. Phys.*, 102:32–43, 2016.

11

Tight Models of de-Rham Algebras of Highly Connected Manifolds

Lorenz Schwachhöfer

Abstract

The rational homotopy type of a closed oriented manifold M is determined by the weak equivalence class of its de-Rham algebra $\Omega^*(M)$. In [3] Crowley and Nordström invented the Bianchi–Massey tensor of a DGCA which is invariant under quasi-isomorphisms. In fact, for $(r-1)$-connected $(r > 1)$ manifolds of dimension $n \leq 5r - 3$, this tensor, together with the cohomology ring, completely determines the rational homotopy type. In this chapter we show that each weak equivalence class of Poincaré DGCAs contains a tight graded differential algebra, by which we mean a finite dimensional algebra with a non-degenerate Poincaré pairing which does not contain any properly enclosed quasi-isomorphic subalgebra. This tight differential graded algebra can be described explicitly in terms of the Bianchi–Massey tensor.

11.1 Introduction

By the seminal work of Sullivan [9], it is known that two simply connected CW-complexes X_1, X_2 are rationally equivalent if and only if their rational homotopy algebras $\pi_*(X, \mathbb{Q})$ are weakly equivalent differential graded commutative algebras (DGCAs). In the case of closed simply connected manifolds M_1, M_2 this is equivalent to the weak equivalence of the de-Rham algebras $\Omega^*(M_i)$.

There are numerous known invariants of DGCAs which are preserved under quasi-isomorphisms and hence may help to distinguish weak equivalence classes of DGCAs. One such invariant, called the *Bianchi–Massey tensor*, was introduced by Crowley and Nordström [3] for DGCAs of

Poincaré type, see Definition 11.3.1 below. This is a class which comprises the de-Rham algebras of closed oriented manifolds. In fact, they showed that, for $(r-1)$-connected $(r > 1)$ Poincaré algebras of degree $n \leq 5r - 3$, the Bianchi–Massey tensor is the only invariant (apart from the cohomology algebra), meaning that two such DGCAs are weakly equivalent if and only if their cohomology algebra and their Bianchi–Massey tensors coincide. The degree of the de-Rham algebra $\Omega^*(M)$ equals the dimension of the closed oriented manifold M.

In [5], the Bianchi–Massey tensor was shown to be equivalent to a class in Harrison cohomology and hence to determine an A_3 algebra. Furthermore, it was shown there that each weak equivalence class of a simply connected Poincaré DGCA of Hodge type (see Definition 11.3.4) contains a finite dimensional representative; moreover, any $(r-1)$-connected $(r > 1)$ Poincaré DGCA of Hodge type is almost formal in the sense of [2] if its degree n satisfies $n \leq 4r - 1$, so that e.g. any closed simply connected 7-manifold is almost formal; see Corollary 11.4.5 below.

In this chapter, we aim at identifying a "canonical" finite dimensional representative in each weak equivalence class, similar to the approach in [6]. For this, we introduce the notion of a *tight Poincaré DGCA* as a finite dimensional Poincaré DGCA with a non-degenerate Poincaré pairing that does not admit any proper quasi-isomorphically embedded subalgebra (Definition 11.5.1).

For a graded commutative algebra (GCA) H^*, we follow [3] in setting $\mathcal{K}^* \subset S^2(H^*)$, the kernel of the multiplication map $\cdot : S^2(H^*) \to H^*$.[1]

Theorem 11.1.1 *Let H^* be an $(r-1)$-connected $(r > 1)$ Poincaré GCA of degree $n \leq 5r - 3$. Then there is a bijective correspondence between symmetric bilinear forms*[2] *$\beta \in (S^2(\mathcal{K}^*))^\vee$ on $\mathcal{K}^*$ of degree $n+1$ and isomorphism classes of tight DGCAs Q_β^* with cohomology H^*.*

The construction of the finite dimensional models in [5] implies that each weak equivalence class of DGCAs with the restrictions in Theorem 11.1.1 contains a tight DGCA. However, different β may result in weakly equivalent DGCAs Q_β^*. A symmetric bilinear form $\hat{\beta}$ on $S^2(H^*)$ is said to be of *Riemannian type* if it satisfies

$$\hat{\beta}(h_1 h_2, h_3 h_4) = -(-1)^{|h_2||h_3|}\hat{\beta}(h_1 h_3, h_2 h_4).$$

---

[1] Here, $S^k(V^*)$ for a graded vector space V^* denotes the graded symmetric k-tensors on V^*.

[2] Here $(S^2(\mathcal{K}^*))^\vee$ denotes the dual space of $S^2(\mathcal{K}^*)$.

This terminology is chosen as such a tensor satisfies all (graded) symmetries of a Riemannian curvature tensor. Furthermore, a symmetric bilinear form β on $\mathcal{K}^*$ is said to be of Riemannian type if $\beta = \hat{\beta}_{|\mathcal{K}^*}$ for a symmetric bilinear form $\hat{\beta}$ of Riemannian type on $S^2(H^*)$. With this, we can show the following:

Theorem 11.1.2 *Let H^* be an $(r-1)$-connected $(r > 1)$ Poincaré GCA of degree $n \leq 5r - 3$.*

1. *Each weak equivalence class of DGCAs with cohomology H^* contains a tight DGCA Q^*_β, $\beta \in (S^2(\mathcal{K}^*))^\vee$. In fact, β may be chosen to be of Riemannian type.*

2. *Tight DGCAs $Q^*_{\beta_i}$, $i = 1, 2$, are weakly equivalent if and only if $(\beta_1 - \beta_2)_{|\mathcal{E}^*} = 0$, where $\mathcal{E}^* \subset S^2(\mathcal{K}^*)$ is the kernel of $S^2(\mathcal{K}^*) \hookrightarrow S^2(S^2(H^*)) \xrightarrow{mult} S^4(H^*)$.*

This chapter is organized as follows. In Section 11.2, we recall the relation between the rational homotopy equivalence of closed oriented manifolds or, more generally, of CW-complexes on the one hand and the weak equivalence of DGCAs on the other. In Section 11.3 we recall the definitions of Poincaré GCAs from [3] and Poincaré DGCAs of Hodge type from [5]. In Section 11.4 we recall from [5] that each simply connected Poincaré DGCA $\mathcal{A}^*$ of Hodge type is weakly equivalent to a finite dimensional DGCA $\mathcal{Q}_{small}$, called the *small quotient algebra of $\mathcal{A}^*$*, with a non-degenerate Poincaré pairing. In Section 11.5 we introduce the notion of *tight Poincaré DGCAs* and show Theorem 11.1.1. Once this is established, we recall in Section 11.6 the *Bianchi–Massey tensor* of [3] and compute it for the tight Poincaré DGCAs from Section 11.5 and show Theorem 11.1.2. We also give an explicit description of these models in the case of closed simply connected 7-manifolds.

Acknowledgements   It is a pleasure to thank Diarmuid Crowley for many inspiring discussions, as well as Domenico Fiorenza, Kotaro Kawai and Hông Vân Lê, who are coauthors of the closely related article [5]. Also, the author thanks the Max Planck Institute for Mathematics in the Sciences in Leipzig (Germany) and the Mathematical Research Institute MATRIX in Creswick (Australia) for their hospitality and providing excellent working conditions. The author was supported by the Deutsche Forschungsgemeinschaft by grant SCHW 893/5-1.

11.2 Rational and Weak Equivalence

A continuous map $f : X \to Y$ between CW-complexes is called a *rational homotopy equivalence* if it induces an isomorphism on rational homotopy, i.e., if $f_* : \pi_*(X) \otimes \mathbb{Q} \to \pi_*(Y) \otimes \mathbb{Q}$ is an isomorphism. If X, Y are simply connected, then this is equivalent to requiring that f induces an isomorphism on rational (co-)homology $f_* : H_*(X, \mathbb{Q}) \to H_*(Y, \mathbb{Q})$ or $f^* : H^*(Y, \mathbb{Q}) \to H^*(X, \mathbb{Q})$, respectively [4, Thm. 8.6]. Observe that in general, a rational homotopy equivalence does not admit an inverse, but it generates an equivalence relation by saying that X and Y are rationally equivalent if there are spaces and maps

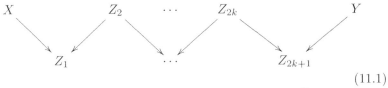

$$(11.1)$$

where all arrows denote rational homotopy equivalences.[3]

As a special case of Sullivan's construction of the localization of topological spaces, it follows that, for each such space X, there is a CW-complex $X_{\mathbb{Q}}$, called the *rationalization of X*, together with a rational homotopy equivalence $X \to X_{\mathbb{Q}}$, and such that $X_{\mathbb{Q}}$ is a *rational space*, meaning that all homotopy groups $\pi_k(X_{\mathbb{Q}})$ are vector spaces over $\mathbb{Q}$. Moreover, $X_{\mathbb{Q}}$ is uniquely defined up to homotopy equivalence [4, Thm. 9.7], so that X and Y are rationally equivalent if and only if the diagram (11.1) may be simplified to

$$(11.2)$$

There is an algebraic analogue to this construction. Namely, recall that a differential commutative graded algebra (DGCA) is a graded vector space $\mathcal{A}^* = \bigoplus_{k=0}^{\infty} \mathcal{A}^k$ with an associative graded commutative product $\cdot$ and a differential $d : \mathcal{A}^* \to \mathcal{A}^*[-1]$, i.e., such that $\mathcal{A}^k \cdot \mathcal{A}^l \subset \mathcal{A}^{k+l}$ and $d\mathcal{A}^k \subset \mathcal{A}^{k+1}$, and with

---

[3] This is called the *localization of the category of simply connected spaces with respect to rational homotopy equivalences*.

$$\alpha \cdot \beta = (-1)^{|\alpha||\beta|}\beta \cdot \alpha,$$
$$d(\alpha \cdot \beta) = (d\alpha) \cdot \beta + (-1)^{|\alpha|}\alpha \cdot (d\beta), \qquad (11.3)$$
$$d^2 = 0,$$

where $\alpha \in \mathcal{A}^{|\alpha|}$ and $\beta \in \mathcal{A}^{|\beta|}$ are homogeneous elements. Thus, the cohomology of $\mathcal{A}^*$ defined by

$$H^*(\mathcal{A}^*) := (\ker d)/(\operatorname{Im} d) \stackrel{\text{not.}}{\equiv} \mathcal{A}_d^*/d\mathcal{A}^*$$

has a well defined product $[\alpha] \cdot [\beta] := [\alpha \cdot \beta]$ which turns $H^*(\mathcal{A}^*)$ into a commutative graded algebra (GCA).

Given two DGCAs $\mathcal{A}^*$ and $\mathcal{B}^*$, a *DGCA-morphism* is a graded morphism of algebras $\varphi : \mathcal{A}^* \to \mathcal{B}^*$ which commutes with the differentials. Then φ induces a homomorphism on cohomologies

$$\varphi_* : H^*(\mathcal{A}^*) \longrightarrow H^*(\mathcal{B}^*),$$

and we call φ a *quasi-isomorphism* if φ_* is an isomorphism. Just as in the case of rational homotopy equivalences, a quasi-isomorphism does not posses an inverse in general, but it generates an equivalence relation by saying that DGCAs $\mathcal{A}^*$ and $\mathcal{B}^*$ are *weakly equivalent* if there are DGCAs $\mathcal{C}_i^*$ and quasi-isomorphisms

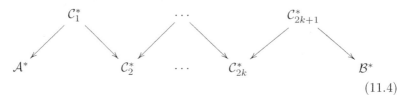

$$(11.4)$$

If $H^*(\mathcal{A}^*)$ is of *finite type*, i.e., all cohomologies $H^k(\mathcal{A}^*)$ are finite dimensional, then there is a DGCA $\mathcal{S}^*$, called the *Sullivan minimal model of* $\mathcal{A}^*$, and a quasi-isomorphism $\mathcal{S}^* \to \mathcal{A}^*$ such that a DGCA $\mathcal{B}^*$ is weakly equivalent to $\mathcal{A}^*$ if and only if $\mathcal{B}^*$ has the same minimal model $\mathcal{S}^*$, so that, in analogy to (11.2), any two weakly equivalent DGCAs $\mathcal{A}^*$ and $\mathcal{B}^*$ can be connected by quasi-isomorphisms

$$(11.5)$$

In fact, $\mathcal{S}^*$ is a free DGCA, generated by (possibly countably many) generators $\{x_n \mid n \in \mathbb{N}\}$ of positive degree, and where each dx_k is a decomposable polynomial in the generators [9, 8].

For instance, for a CW-complex X, the (singular) co-chains with coefficients in a field $\mathbb{F}$, together with the cup product, form a DGCA $C^*(X; \mathbb{F})$ with cohomology $H^*(X, \mathbb{F})$. In particular, a continuous map $f : X \to Y$ induces the DGCA-morphism $\varphi := f^* : C^*(Y, \mathbb{F}) \to C^*(X, \mathbb{F})$, whence in the case $\mathbb{F} = \mathbb{Q}$, φ is a quasi-isomorphim if and only if f is a rational homotopy equivalence. In particular, if X and Y are rationally equivalent, then $C^*(X, \mathbb{Q})$ and $C^*(Y, \mathbb{Q})$ are weakly equivalent, and $S^* := C^*(X_{\mathbb{Q}}, \mathbb{Q})$ is the Sullivan minimal model of $C^*(X, \mathbb{Q})$.

The converse of this statement is also true provided that X and Y are simply connected; in this case X and Y are rationally equivalent if and only if $C^*(X, \mathbb{Q})$ and $C^*(Y, \mathbb{Q})$ are weakly equivalent.

Definition 11.2.1 A DGCA $\mathcal{A}^*$ is called *formal* if it is weakly equivalent to $(H^*(\mathcal{A}^*), d = 0)$. A topological space X is called formal if its rational singular co-chain algebra $C^*(X, \mathbb{Q})$ is formal.

That is, the rational homotopy type of a formal topological space X is determined by its cohomology ring $H^*(X, \mathbb{Q})$ only.

A DGCA over the field $\mathbb{F}$ is called *connected* if $H^0(\mathcal{A}^*) = \mathbb{F}$ and $(r-1)$-*connected* if it is connected and $H^k(\mathcal{A}^*) = 0$ for $k = 1, \ldots, r-1$. A 1-connected DGA is also called *simply connected*.

11.3 Poincaré DGCAs and DGCAs of Hodge Type

In this section we recall the terminology introduced in [5]. In general, for a graded vector space $V^* = \bigoplus_k V^k$ we say that a bilinear pairing $\langle -, - \rangle : V^* \times V^* \to \mathbb{F}$ is *of degree* n if $\langle V^k, V^l \rangle = 0$ whenever $k + l \neq n$. We begin with the following definition.

Definition 11.3.1 (see [3, Def. 2.7]) Let $H^* = \bigoplus_{k \geq 0} H^k$ be a GCA, and let $\int \in (H^n)^{\vee}$, where the latter denotes the dual of H^n. Then the bilinear pairing of degree n on H^* given by

$$\langle \alpha^k, \beta^l \rangle := \int \alpha^k \cdot \beta^l, \tag{11.6}$$

if $k + l = n$, is called the *Poincaré pairing of degree* n *induced by* $\int$.

We call H^* a *Poincaré algebra of degree* n if it is finite dimensional and admits a non-degenerate Poincaré pairing of degree n, i.e., such that $\langle \alpha, H^* \rangle = 0$ if and only if $\alpha = 0$.

Clearly, a Poincaré algebra of degree n is of the form $H^* = \bigoplus_{k=0}^{n} H^k$ whose Betti numbers $b^k = b^k(H^*) := \dim H^k$ satisfy $b^k = b^{n-k}$. If in addition $b^0 = b^n = 1$, then the pairing (11.6) is unique up to multiples.

Note that in [3, Def. 2.7], the degree of a Poincaré algebra is called the *dimension*, but as later we wish to consider the dimension of $\mathcal{A}^*$ as a graded vector space, the notion of degree seems more appropriate.

Definition 11.3.2 (see [3, Def. 2.7]) A *Poincaré DGCA of degree* n is a DGCA $\mathcal{A}^*$ whose cohomology algebra $H^* := H^*(\mathcal{A}^*)$ is a Poincaré algebra of degree n.

Proposition 11.3.3 *Let $\mathcal{A}^*$ be a Poincaré DGCA of degree n. Then there is a Poincaré pairing $\langle -, - \rangle$ of degree n on A^* such that*

$$\langle \alpha, \beta \rangle = \langle [\alpha], [\beta] \rangle_{H^*} \qquad (11.7)$$

for all $\alpha, \beta \in \mathcal{A}_d^*$, where $[-] : \mathcal{A}_d^* \to H^*$ denotes the canonical projection and $\langle -, - \rangle_{H^*}$ is the Poincaré pairing on H^*. In particular,

$$
\begin{aligned}
\langle \alpha^k, \beta^l \rangle &= (-1)^{kl} \langle \beta^l, \alpha^k \rangle, \\
\langle \alpha^k \cdot \beta^l, \gamma^r \rangle &= \langle \alpha^k, \beta^l \cdot \gamma^r \rangle, \\
\langle d\alpha^k, \beta^l \rangle &= (-1)^{k+1} \langle \alpha^k, d\beta^l \rangle.
\end{aligned}
\qquad (11.8)
$$

Proof Pick $\int_{\mathcal{A}^*} \in (\mathcal{A}^n)^\vee$ such that $\int_{\mathcal{A}^*} \alpha_n = \int [\alpha_n]$ for all $\alpha_n \in \mathcal{A}^n$. Then it is straightforward to verify that the Poincaré pairing of degree n induced by $\int_{\mathcal{A}^*}$ has all the asserted properties. $\qquad \square$

Strictly speaking, in order to extend the functional $\int_{\mathcal{A}^*}$ defined on $\mathcal{A}_d^n$ to all of $\mathcal{A}^n$ in the proof of Proposition 11.3.3 we need to make use of the axiom of choice. However, we shall later apply this to the case where either $\mathcal{A}^n$ is finite dimensional, or $\mathcal{A}^{n+1} = 0$, so that $\mathcal{A}_d^n = \mathcal{A}_n$; in these cases, the existence of this extension does not require the axiom of choice.

Observe that this pairing induces another pairing of degree $n+1$ on $d\mathcal{A}^*$, given by

$$\langle\!\langle \alpha, \beta \rangle\!\rangle := \langle d^- \alpha, \beta \rangle, \qquad \alpha, \beta \in d\mathcal{A}^*, \qquad (11.9)$$

where $d^- \alpha$ is an element such that $dd^- \alpha = \alpha$. Indeed, since $d^- \alpha$ is well defined up to adding an element of $\mathcal{A}_d^*$, and $\langle \mathcal{A}_d^*, d\mathcal{A}^* \rangle = 0$ by (11.8), it follows that the pairing (11.9) is well defined. From (11.6) and (11.8) we easily deduce that this pairing is also graded symmetric, i.e.,

$$\langle\langle \alpha^k, \beta^l \rangle\rangle = (-1)^{kl} \langle\langle \beta^l, \alpha^k \rangle\rangle. \tag{11.10}$$

Definition 11.3.4 ([5, Def. 2.2]) Let $\mathcal{A}^*$ be a connected Poincaré DGCA of degree n with a Poincaré pairing $\langle -, - \rangle$.

1 A *harmonic subspace* of $\mathcal{A}^*$ is a graded subspace $\mathcal{H}^* \subset \mathcal{A}_d^*$ complementary to $d\mathcal{A}^*$ and such that $1 \in \mathcal{H}^0$.

2 A *Hodge type decomposition* of $\mathcal{A}^*$ is a direct sum decomposition of the form

$$\mathcal{A}^* = d\mathcal{A}^* \oplus \mathcal{H}^* \oplus \mathcal{B}^*, \tag{11.11}$$

where $\mathcal{H}^* \subset \mathcal{A}_d^*$ is harmonic, such that

$$\langle \mathcal{H}^* \oplus \mathcal{B}^*, \mathcal{B}^* \rangle = 0. \tag{11.12}$$

3 A Poincaré DGCA admitting a Hodge type decomposition is called a *Hodge type DGCA*.

By (11.7) the restriction of the projection $\mathcal{A}_d^* \to H^*(\mathcal{A}^*)$ yields an isometric isomorphism

$$(\mathcal{H}^*, \langle -, - \rangle) \longrightarrow (H^*(\mathcal{A}^*), \langle -, - \rangle_{H^*}). \tag{11.13}$$

In particular, $\mathcal{H}^*$ is finite dimensional and the restriction of the Poincaré pairing to $\mathcal{H}^*$ is non-degenerate.

If $\mathcal{H}^* \subset \mathcal{A}_d^*$ is a harmonic subspace, then we have the decomposition $\mathcal{A}_d^* = d\mathcal{A}^* \oplus \mathcal{H}^*$, and we define the *cocycle of* $\mathcal{H}^*$ to be the linear map

$$\xi_{\mathcal{H}^*} : S^2(H^*) \longrightarrow d\mathcal{A}^*, \qquad \xi_{\mathcal{H}^*}([h_1], [h_2]) := pr_{d\mathcal{A}^*}(h_1 \cdot h_2), h_i \in \mathcal{H}^*. \tag{11.14}$$

Proposition 11.3.5 *Let* $\mathcal{A}^* = d\mathcal{A}^* \oplus \mathcal{H}^* \oplus \mathcal{B}^*$ *be a Poincaré DGCA with a Hodge type decomposition. Then every harmonic subspace* $\hat{\mathcal{H}}^* \subset \mathcal{A}_d^*$ *is of the form*

$$\hat{\mathcal{H}}^* = \{v + \beta(v) \mid v \in \mathcal{H}^*\} \tag{11.15}$$

for some linear map $\beta : \mathcal{H}^* \to d\mathcal{A}^*$ *with* $\beta(1) = 0$. *Defining*

$$\hat{\mathcal{B}}^* := \{x - \beta^\dagger(x) - \frac{1}{2}\beta\beta^\dagger(x) \mid x \in \mathcal{B}^*\}, \tag{11.16}$$

where $\beta^\dagger : \mathcal{B}^* \to \mathcal{H}^*$ *is the unique map satisfying* $\langle \beta^\dagger(x), h \rangle = \langle x, \beta(h) \rangle$ *for all* $x \in \mathcal{B}^*, h \in \mathcal{H}^*$, *the decomposition* $\mathcal{A}^* = d\mathcal{A}^* \oplus \hat{\mathcal{H}}^* \oplus \hat{\mathcal{B}}^*$ *is a Hodge* type decomposition. Moreover, the cocycles of $\mathcal{H}^*$ and $\hat{\mathcal{H}}^*$ are related by the formula

$$\begin{aligned}
\xi_{\hat{\mathcal{H}}^*}([h_1],[h_2]) &= \xi_{\mathcal{H}^*}([h_1],[h_2]) \\
&\quad + h_1 \cdot \beta([h_2]) - \beta([h_1] \cdot [h_2]) + \beta(]h_1]) \cdot h_2 \\
&\quad + \beta([h_1]) \cdot \beta([h_2]). \tag{11.17}
\end{aligned}$$

Proof Since $\mathcal{A}_d^* = d\mathcal{A}^* \oplus \mathcal{H}^* = d\mathcal{A}^* \oplus \hat{\mathcal{H}}^*$, it follows that $\hat{\mathcal{H}}^*$ is of the form (11.16), and it is straightforward to verify that $\mathcal{A}^* = d\mathcal{A}^* \oplus \hat{\mathcal{H}}^* \oplus \hat{\mathcal{B}}^*$ is a Hodge type decomposition and that the cocycle $\xi_{\hat{\mathcal{H}}^*}$ is of the asserted form. □

We define $\mathcal{K}^* \subset S^2(H^*)$ as the kernel of the multiplication map, i.e., via the short exact sequence

$$0 \longrightarrow \mathcal{K}^* \longrightarrow S^2(H^*) \overset{\cdot}{\longrightarrow} H^* \longrightarrow 0. \tag{11.18}$$

Given a Hodge type decomposition (11.11), the restriction $d : \mathcal{B}^* \to d\mathcal{A}^*[-1]$ is a linear isomorphism, whence there is an inverse $d^- : d\mathcal{A}^* \to \mathcal{B}^*[1]$. We may extend d^- to all of $\mathcal{A}^*$ by defining $d^-_{|\mathcal{H}^* \oplus \mathcal{B}^*} = 0$. Thus, $(d^-)^2 = 0$, and

$$dd^-d = d, \qquad d^-dd^- = d^-. \tag{11.19}$$

It follows that the projections in (11.11) are given by

$$pr_{\mathcal{H}^*} = 1 - [d, d^-], \qquad pr_{d\mathcal{A}^*} = dd^-, \qquad pr_{\mathcal{B}^* = d^-\mathcal{A}^*} = d^-d, \tag{11.20}$$

where $[d, d^-] = dd^- + d^-d$ is the super-commutator. Therefore, (11.11) may be written as

$$\mathcal{A}^* = d\mathcal{A}^* \oplus \mathcal{H}^* \oplus d^-\mathcal{A}^* = dd^-\mathcal{A}^* \oplus \mathcal{H}^* \oplus d^-d\mathcal{A}^*, \tag{11.21}$$

and, setting $\mathcal{A}_{d^-}^* := \ker d^- = \mathcal{H}^* \oplus d^-\mathcal{A}^*$, (11.12) implies

$$\langle \mathcal{A}_d^*, d\mathcal{A}^* \rangle = \langle \mathcal{A}_{d^-}^*, d^-\mathcal{A}^* \rangle = 0. \tag{11.22}$$

Example 11.3.6 The quintessential example of a Poincaré algebra of degree n of Hodge type (which motivates our terminology) is the de-Rham algebra $(\Omega^*(M), d)$ of a closed smooth oriented manifold M, with $\int$ given by the integration of n-forms. The Hodge decomposition w.r.t. some Riemannian metric g on M is then a Hodge type decomposition in the sense of Definition 11.3.4 whose harmonic subspace is the space $\mathcal{H}^*(M)$ of $\triangle_g$-harmonic forms. Note that the maps d^* and d^- are related by the formula $d^* = \triangle_g d^-$.

Given a Poincaré pairing on $\mathcal{A}^*$, as a consequence of (11.8), the null-space

$$\mathcal{A}_\perp^* := \{\alpha \in \mathcal{A}^* \mid \langle \alpha, \mathcal{A}^* \rangle = 0\}$$

is a differential ideal of $\mathcal{A}^*$, whence the quotient $\mathcal{Q}^* := \mathcal{A}^*/\mathcal{A}_\perp^*$ is again a DGCA, fitting into the short exact sequence

$$0 \longrightarrow \mathcal{A}_\perp^* \longrightarrow \mathcal{A}^* \longrightarrow \mathcal{Q}^* \longrightarrow 0. \tag{11.23}$$

Then there is an induced non-degenerate pairing on $\mathcal{Q}^*$, satisfying

$$\langle [\alpha], [\beta] \rangle_{\mathcal{Q}^*} = \langle \alpha, \beta \rangle_{\mathcal{A}^*}, \tag{11.24}$$

where $[\cdot] : \mathcal{A}^* \to \mathcal{Q}^*$ is the canonical projection.

Lemma 11.3.7 *If the Poincaré DGCA $\mathcal{A}^*$ admits a Hodge type decomposition, then the differential ideal $\mathcal{A}_\perp^*$ is invariant under d^- and has the decomposition*

$$\mathcal{A}_\perp^* = dd^- \mathcal{A}_\perp^* \oplus d^- d \mathcal{A}_\perp^*. \tag{11.25}$$

In particular, $(\mathcal{A}_\perp^*, d)$ is acyclic, i.e., has trivial cohomology.

Proof The orthogonality relations (11.21) and (11.22) imply that $\mathcal{A}_\perp^* \subset \mathcal{H}_\perp^* = dd^- \mathcal{A}^* \oplus d^- d \mathcal{A}^*$.

If $\alpha^k \in \mathcal{A}_\perp^k$ and $\beta^{n-k+1} \in \mathcal{A}^{n-k+1}$, then by (11.8)

$$\begin{aligned} \langle d^- d\alpha, d\beta^{n-k+1} \rangle &= (-1)^{k+1} \langle dd^- d\alpha, \beta^{n-k+1} \rangle \\ &= (-1)^{k+1} \langle d\alpha, \beta^{n-k+1} \rangle \\ &= -\langle \alpha, d\beta^{n-k+1} \rangle = 0, \end{aligned}$$

where we used $dd^- d = d$. That is, $\langle d^- d\mathcal{A}_\perp^*, d\mathcal{A}^* \rangle = 0$, and then the orthogonality relations (11.22) imply that $\langle d^- d\mathcal{A}_\perp^*, \mathcal{A}^* \rangle = 0$, i.e., $\mathcal{A}_\perp^*$ is invariant under $d^- d$. Thus, because $\mathcal{A}_\perp^* \subset dd^- \mathcal{A}^* \oplus d^- d\mathcal{A}^*$, (11.25) follows. In particular, $\mathcal{A}_\perp^*$ is also invariant under dd^-.

To see that $\mathcal{A}_\perp^*$ is also invariant under d^-, let $\alpha^k \in \mathcal{A}_\perp^k$ and $\beta^{n-k} \in \mathcal{A}^{n-k}$. Then

$$\langle d^- \alpha^k, d\beta^{n-k} \rangle = (-1)^k \langle dd^- \alpha^k, \beta^{n-k} \rangle = 0,$$

as $dd^- \alpha^k \in \mathcal{A}_\perp^k$ by the dd^--invariance of $\mathcal{A}_\perp^k$. Therefore, $\langle d^- \mathcal{A}_\perp^*, d\mathcal{A}^* \rangle = 0$, and then (11.22) implies $\langle d^- \mathcal{A}^*, \mathcal{A}^* \rangle = 0$, i.e., $\mathcal{A}_\perp^*$ is invariant under d^-. This together with (11.25) now implies that $\mathcal{A}_\perp^*$ is acyclic. □

Corollary 11.3.8 *If $\mathcal{A}^*$ is of Hodge type, then the projection $\pi : \mathcal{A}^* \to \mathcal{Q}^*$ is a quasi-isomorphism, $\mathcal{Q}^*$ is of Hodge type and $\langle -, - \rangle_{\mathcal{Q}^*}$ from (11.24) is a non-degenerate Poincaré pairing.*

Proof   The first statement follows from the long exact sequence associated to (11.23) and $H^*(\mathcal{A}_\perp^*) = 0$ by Lemma 11.3.7. Furthermore, (11.21) and (11.25) imply that $d, d^- : \mathcal{Q}^* \to \mathcal{Q}^*$ are well defined, and that there is a decomposition

$$\mathcal{Q}^* \cong dd^- \mathcal{Q}^* \oplus \mathcal{H}^* \oplus d^- d\mathcal{Q}^*,$$

and (11.24) easily implies that this is a Hodge type decomposition and that $\langle -, - \rangle_{\mathcal{Q}^*}$ is the Poincaré pairing.   □

Remark 11.3.9   As the induced pairing on $\mathcal{Q}^*$ is non-degenerate by construction, Corollary 11.3.8 implies that every Hodge type Poincaré DGCA is equivalent to a non-degenerate one. This has been shown in [6, Thm. 1.1] by slightly different means.

11.4 Small Algebras of Hodge Type DGCAs

Given a Poincaré DGCA $\mathcal{A}^*$ with a Hodge type decomposition (11.21), we consider the following class of subalgebras:

Definition 11.4.1   ([5, Def. 3.1]) Let $\mathcal{A}^*$ be a Poincaré DGCA with a Hodge type decomposition (11.21). An $\mathcal{H}^*$-*subalgebra* is a DG-subalgebra of $\mathcal{A}^*$ which is d^--invariant and contains $\mathcal{H}^*$.

If $\mathcal{C}^* \subset \mathcal{A}^*$ is such an $\mathcal{H}^*$-subalgebra, then – as it is closed under both d and d^- – it follows that it admits a Hodge type decomposition analogous to (11.21)

$$\mathcal{C}^* = d\mathcal{C}^* \oplus \mathcal{H}^* \oplus d^- \mathcal{C}^* = dd^- \mathcal{C}^* \oplus \mathcal{H}^* \oplus d^- d\mathcal{C}^*.$$

It is evident from this that $H^*(\mathcal{C}^*) \cong \mathcal{H}^* \cong H^*(\mathcal{A}^*)$, whence the inclusion $\mathcal{C}^* \hookrightarrow \mathcal{A}^*$ is a quasi-isomorphism. As the class of $\mathcal{H}^*$-subalgebras contains $\mathcal{A}^*$ itself and is invariant under arbitrary intersections, it follows that there is a minimal such algebra, namely the intersection of all $\mathcal{H}^*$-subalgebras of $\mathcal{A}^*$.

Definition 11.4.2   ([5, Def. 3.2]) For a Poincaré DGCA $\mathcal{A}^*$ with a Hodge type decomposition (11.21), we define the *small algebra* $\mathcal{A}^*_{\text{small}}$ of $\mathcal{A}^*$ to be the (unique) smallest $\mathcal{H}^*$-subalgebra of $\mathcal{A}^*$. Furthermore, the small quotient of $\mathcal{A}^*$ is defined to be $\mathcal{Q}^*_{\text{small}} := \mathcal{A}^*_{\text{small}}/(\mathcal{A}^*_{\text{small}})_\perp$.

In general, $\mathcal{A}^*_{\text{small}}$ may be of infinite type. However, in the *simply connected* case, it is not, as the following shows.

Proposition 11.4.3 ([5, Prop. 3.3]) *Let $\mathcal{A}^*$ be a simply connected Poincaré DGCA with a Hodge type decomposition (11.21). Then the small algebra $\mathcal{A}^*_{\text{small}}$ is given recursively as*

$$
\begin{cases}
\mathcal{A}^0_{\text{small}} & = \mathbb{F} \cdot 1_{\mathcal{A}^*}, \\
\mathcal{A}^1_{\text{small}} & = 0, \\
\mathcal{A}^k_{\text{small}} & = dd^- \widehat{\mathcal{A}}^k \oplus \mathcal{H}^k \oplus d^- \widehat{\mathcal{A}}^k, \qquad k \geq 2,
\end{cases}
\tag{11.26}
$$

where, for given $l \geq 2$,

$$
\widehat{\mathcal{A}}^l := \operatorname{span}\{\mathcal{A}^{l_1}_{\text{small}} \cdot \mathcal{A}^{l_2}_{\text{small}} \mid l_1, l_2 \geq 2, l_1 + l_2 = l\}
$$

depends on $\mathcal{A}^2_{\text{small}}, \ldots, \mathcal{A}^{l-2}_{\text{small}}$ only. In particular, $\mathcal{A}^*_{\text{small}}$ is of finite type, that is, $\dim \mathcal{A}^k_{\text{small}} < \infty$ for all k, and the small quotient $\mathcal{Q}^*_{\text{small}}$ of $\mathcal{A}^*$ is finite dimensional.

Proof It is straightforward to verify that the space given in (11.26) is an $\mathcal{H}^*$-subalgebra of $\mathcal{A}^*$, and, conversely, it must be contained in any $\mathcal{H}^*$-subalgebra of $\mathcal{A}^*$, showing that (11.26) indeed defines the small algebra. The finite dimensionality of $\mathcal{A}^k_{\text{small}}$ follows from induction on k. As $\mathcal{A}^*_{\text{small}}$ surjects to $\mathcal{Q}^*_{\text{small}}$, it follows that $\mathcal{Q}^*_{\text{small}}$ is of finite type as well; on the other hand, as the Poincaré pairing on $\mathcal{Q}^*_{\text{small}}$ is non-degenerate, it follows that $\mathcal{Q}^k_{\text{small}} = 0$ for $k > n$, whence $\mathcal{Q}^*_{\text{small}}$ is finite dimensional. $\qquad\square$

As an important consequence of this, we obtain the following result.

Theorem 11.4.4 ([5, Cor. 3.5]) *Any simply connected Poincaré DGCA of Hodge type is weakly equivalent to a finite dimensional DGCA with a non-degenerate Poincaré pairing.*

Proof By definition, the restriction of the Poincaré pairing on $\mathcal{A}^*$ to $\mathcal{A}^*_{\text{small}}$ is a Poincaré pairing on $\mathcal{A}^*_{\text{small}}$ with the Hodge type decomposition given in (11.26). In particular, the inclusion $\mathcal{A}^*_{\text{small}} \hookrightarrow \mathcal{A}^*$ is a quasi-isomorphism. Moreover, Corollary 11.3.8 implies that the canonical projection to the quotient $\mathcal{Q}^*_{\text{small}} = \mathcal{A}^*_{\text{small}}/(\mathcal{A}^*_{\text{small}})_\perp$ is a quasi-isomorphism as well, whence the maps

$$
\mathcal{A}^* \longleftarrow \mathcal{A}^*_{\text{small}} \longrightarrow \mathcal{Q}^*_{\text{small}}
$$

are quasi-isomorphisms, showing that $\mathcal{A}^*$ is weakly equivalent to $\mathcal{Q}^*_{\text{small}}$. $\qquad\square$

Let us now use the description of $\mathcal{A}^*_{\text{small}}$ and $\mathcal{Q}^*_{\text{small}}$ in the case where $\mathcal{A}^*$ is $(r-1)$-connected, $r > 1$, i.e., $H^k(\mathcal{A}^*) = 0$ for $k = 1, \ldots, r-1$.

In this case, it follows from the recursion formula in Proposition 11.4.3 that

$$
\mathcal{A}^l_{\text{small}} = \begin{cases} \mathcal{H}^l, & l = 0, \ldots, 2r - 2, \\ \mathcal{H}^{2r-1} \oplus d^-(\mathcal{H}^r \cdot \mathcal{H}^r), & l = 2r - 1. \end{cases} \tag{11.27}
$$

Thus, since $\dim \mathcal{Q}^{n-l}_{\text{small}} = \dim \mathcal{Q}^l_{\text{small}} \leq \dim \mathcal{A}^l_{\text{small}}$ and $\mathcal{Q}^l_{\text{small}} \supset \mathcal{H}^l$, it follows that

$$
\mathcal{Q}^l_{\text{small}} = \mathcal{H}^l, \qquad l = 0, \ldots, 2r - 2, \quad l = n - 2r + 2, \ldots, n,
$$

$$
\tag{11.28}
$$

$$
\mathcal{Q}^{2r-1}_{\text{small}} = \mathcal{H}^{2r-1} \oplus d^-(\mathcal{H}^r \cdot \mathcal{H}^r),
$$

and we obtain the following result.

Corollary 11.4.5 ([5, Cor. 3.11]) *Let $\mathcal{A}^*$ be an $(r - 1)$-connected $(r > 1)$ Poincaré DGCA of Hodge type of degree n. Then $\mathcal{A}^*$ is weakly equivalent to a finite dimensional non-degenerate Poincaré DGCA $\mathcal{Q}^*_{\text{small}}$ for which the differential $d : \mathcal{Q}^{k-1}_{\text{small}} \to \mathcal{Q}^k_{\text{small}}$ is possibly nonzero only for $2r \leq k \leq n - 2r + 1$. In particular,*

1 if $n \leq 4r - 2$, then $\mathcal{A}^*$ is formal;
2 if $n = 4r - 1$, then $d : \mathcal{Q}^{k-1}_{\text{small}} \to \mathcal{Q}^k_{\text{small}}$ vanishes for $k \neq 2r$.

Remark 11.4.6   1. The first statement of Corollary 11.4.5 has been shown by Miller in [7] using the Quillen's functor. In particular, it implies that any closed simply connected manifold of dimension ≤ 6 is formal.

2. A DGCA which is weakly equivalent to a DGCA whose differential vanishes in all but one degree is called *almost formal*. That is, Corollary 11.4.5(2) shows that an $(r - 1)$-connected $(r > 1)$ Poincaré DGCA of Hodge type of degree $n = 4r - 1$ is almost formal.

For instance, the case $r := 2$ implies that any simply connected closed 7-manifold is almost formal; this may be compared with [2, Thm. 4.10], which states that closed G_2-manifolds are almost formal.

3. Let M be a closed n-manifold admitting a Riemannian metric of non-negative Ricci curvature (or, more general, such that all harmonic 1-forms are parallel). Using the Cheeger–Gromoll splitting theorem, one can show that $\Omega^*(M)$ is weakly equivalent to $\mathcal{A}^* \otimes \Lambda^*(H^1(M))$, where $\mathcal{A}^*$ is a simply connected Poincaré DGCA of degree $n - b_1(M)$ [5, Prop. 5.3].

Therefore, if $n - b_1(M) \leq 6$, then, by Corollary 11.4.5(1), $\mathcal{A}^*$ is formal and hence $\mathcal{A}^* \otimes \Lambda^*(H^1(M))$ is formal, whence so is M. In particular, this applies to G_2-manifolds with holonomy properly contained in G_2, as these are Ricci flat and have $b_1(M) > 0$. This generalizes [2, Thm. 4.10] in the non-simply connected case as well.

11.5 Tight DGCAs of Highly Connected DGCAs

We shall assume throughout this section that H^* is an $(r-1)$-connected $(r > 1)$ Poincaré GCA of degree $n \leq 5r - 3$. Motivated by the definition of the small quotient algebra, we introduce the notion of a *tight Poincaré DGCA* (see Definition 11.5.1). We shall give an explicit construction of a tight representative in each weak equivalence class of such algebras.

Definition 11.5.1 A Poincaré DGCA $\mathcal{Q}^*$ is called *tight* if it is of Hodge type with a non-degenerate Poincaré pairing $\langle -, - \rangle$, and if there is no proper quasi-isomorphically embedded sub-DGCA $\hat{\mathcal{Q}}^* \hookrightarrow \mathcal{Q}^*$.

In order to describe the construction, consider a graded vector space

$$\left(\mathcal{V}^* = \bigoplus_{k=2r}^{n+1-2r} \mathcal{V}^k, \langle\!\langle -, - \rangle\!\rangle \right) \tag{11.29}$$

with a non-degenerate graded bilinear form $\langle\!\langle -, - \rangle\!\rangle$ of degree $n + 1$.

Let $\mathcal{B}^* := (\mathcal{V}^*)^{\vee}[n]$, and define $\mathcal{Q}^* := \mathcal{V}^* \oplus H^* \oplus \mathcal{B}^*$ as a vector space. We extend the Poincaré pairing $\langle -, - \rangle$ on $\mathcal{H}^* \cong H^*$ to a non-degenerate pairing of degree n on $\mathcal{Q}^*$ by

$$\begin{aligned}\langle \mathcal{H}^*, \mathcal{V}^* \oplus \mathcal{B}^* \rangle &= \langle \mathcal{V}^*, \mathcal{V}^* \rangle = \langle \mathcal{B}^*, \mathcal{B}^* \rangle = 0, \\ \langle \mathcal{B}^*, \mathcal{V}^* \rangle, &\quad \text{the evaluation map.}\end{aligned} \tag{11.30}$$

Since both $\langle -, - \rangle$ and $\langle\!\langle -, - \rangle\!\rangle$ are non-degenerate of degree n and $n + 1$, respectively, it follows that there is an isomorphism $d^- : \mathcal{V}^* \to \mathcal{B}^*[1]$ satisfying (11.9), and we denote its inverse by $d : \mathcal{B}^* \to \mathcal{V}^*[-1]$.

Extending d and d^- to all of $\mathcal{Q}^*$ by requiring them to vanish on $\mathcal{V}^* \oplus \mathcal{H}^*$ and on $\mathcal{H}^* \oplus \mathcal{B}^*$, respectively, it follows that $d^2 = 0$ and $(d^-)^2 = 0$, and that $\mathcal{V}^* = d\mathcal{Q}^*$, $\mathcal{B}^* = d^- \mathcal{Q}^*$. That is, we have the decomposition

$$\mathcal{Q}^* = \mathcal{V}^* \oplus \mathcal{H}^* \oplus \mathcal{B}^*, \qquad \mathcal{V}^* = d\mathcal{Q}^*, \qquad \mathcal{B}^* = d^- \mathcal{Q}^*. \tag{11.31}$$

We denote the identification $\mathcal{H}^* \leftrightarrow H^*$ by $h \leftrightarrow [h]$; in particular, $[h_1] \cdot [h_2] \in H^* \cong \mathcal{H}^*$ refers to the multiplication in H^*.

Given a graded map $\bar{\xi} : S^2(H^*) \cong S^2(\mathcal{H}^*) \to V^*$, we define a product on Q^* by

$$k_1 \cdot k_2 = \langle k_1, k_2 \rangle \mathrm{vol}_n \qquad \text{for } k_i \in V^* \oplus \mathcal{B}^*, \qquad (11.32)$$

$$V^* \cdot \mathcal{H}^l = 0 \qquad \text{for } l > 0, \qquad (11.33)$$

$$h_1 \cdot h_2 = [h_1] \cdot [h_2] + \bar{\xi}(h_1, h_2) \qquad \text{for } h_i \in \mathcal{H}^*, \qquad (11.34)$$

$$\mathcal{B}^* \cdot \mathcal{H}^l \subset \mathcal{H}^* \qquad \text{for } l > 0, \qquad (11.35)$$

$$\langle d^- k \cdot h_1, h_2 \rangle = \langle\!\langle k, \bar{\xi}(h_1, h_2) \rangle\!\rangle \qquad \text{for } k \in V^*, \ h_i \in \mathcal{H}^*. \quad (11.36)$$

In (11.32), $\mathrm{vol}_n \in \mathcal{H}^n$ denotes the (unique) element for which $\int \mathrm{vol}_n = 1$. Observe that (11.32) implies

$$dQ^* \cdot (Q^*)_d = V^* \cdot (V^* \oplus \mathcal{H}^*) = \mathcal{B}^* \cdot \mathcal{B}^* = 0. \qquad (11.37)$$

It is now straightforward to verify that, with this product, Q^* becomes a Poincaré DGCA with Hodge type decomposition (11.31) and the non-degenerate Poincaré pairing $\langle -, - \rangle$, and $\bar{\xi} = \xi_{\mathcal{H}^*}$ is the cocycle of the harmonic subspace $\mathcal{H}^*$ defined in (11.14).

Lemma 11.5.2 *Let Q^* be a Poincaré DGCA of Hodge type with a non-degenerate Poincaré pairing $\langle -, - \rangle$, and suppose that $Q^k \cong \mathcal{H}^k$ for all $k \leq 2r - 2$. Then $V^* := dQ^*$ is of the form (11.29) with $\langle\!\langle -, - \rangle\!\rangle$ from (11.9), and there is a Hodge type decomposition (11.31) such that the product structure on Q^* is given by (11.32)–(11.36), where $\bar{\xi} := \xi_{\mathcal{H}^*} : S^2(H^*) \to V^*$ is the cocycle map defined in (11.14).*

Proof Pick a Hodge type decomposition (11.31) of Q^*. It is immediate from (11.9) that the pairing $\langle\!\langle -, - \rangle\!\rangle$ on $V^* = dQ^*$ is non-degenerate. Furthermore, $V^k = dQ^{k-1}$, whence $V^k = 0$ for $k \leq 2r - 1$. The non-degenericity of $\langle\!\langle -, - \rangle\!\rangle$ implies that $V^k \cong (V^{n+1-k})^\vee$, so that, in particular, $V^k = 0$ for $k \geq n + 2 - 2r$, i.e., V^* is of the form (11.29).

The non-degenericity of $\langle -, - \rangle$ and the orthogonality relations (11.12) imply that $\mathcal{B}^* \cong (V^*)^\vee[n]$, and the Poincaré pairing $\langle -, - \rangle$ corresponds to the pairing from (11.31) under this identification. In particular, $Q^k \cong (Q^{n-k})^\vee$, so that

$$Q^k \cong \mathcal{H}^k \cong (\mathcal{H}^{n-k})^\vee, \qquad \text{for } k \leq 2r - 2 \text{ and } k \geq n - 2r + 2. \quad (11.38)$$

Since $V^* \oplus \mathcal{B}^*$ has only elements of degree $\geq 2r - 1$, and by assumption $2(2r-1) \geq n - (r-1)$, it follows that

$$(V^* \oplus \mathcal{B}^*) \cdot (V^* \oplus \mathcal{B}^*) \subset \bigoplus_{k=2(2r-1)}^{n} Q^k = Q^n = \mathbb{F}\mathrm{vol}_n,$$

where the latter follows from (11.38) and $\mathcal{H}^k = 0$ for $k = 1, \ldots, r - 1$. Thus, (11.32) follows from (11.6).

By (11.29) it follows that

$$\mathcal{H}^l \cdot (\mathcal{V}^* \oplus \mathcal{B}^*) \subset \bigoplus_{k=2r+l-1}^{n} \mathcal{Q}^k.$$

If $l > 0$ then $\mathcal{H}^l \neq 0$ only if $l \geq r$, in which case $2r + l - 1 \geq 3r - 1 \geq n - (2r - 2)$, so that (11.38) implies $\mathcal{Q}^k = \mathcal{H}^k$ for all $k \geq 2r + l - 1$. That is, $\mathcal{H}^l \cdot (\mathcal{V}^* \oplus \mathcal{B}^*) \subset \mathcal{H}^*$, showing (11.35). Also, $\mathcal{K}^* \subset (\mathcal{Q}^*)_d$ is an ideal, so that $\mathcal{H}^l \cdot \mathcal{K}^* \subset \mathcal{K}^* \cap \mathcal{H}^* = 0$ which shows (11.33).

Equation (11.34) follows immediately from the definition of $\bar\xi = \xi_{\mathcal{H}^*}$ in (11.14), and to show (11.36) let $k \in \mathcal{V}^*$ and $h_1, h_2 \in \mathcal{H}^*$. Then

$$\langle d^- k \cdot h_1, h_2 \rangle \overset{(11.8)}{=} \langle d^- k, h_1 \cdot h_2 \rangle \overset{(11.34)}{=} \langle d^- k, \bar\xi([h_1],[h_2]) \rangle = \langle\!\langle k, \bar\xi([h_1],[h_2]) \rangle\!\rangle.$$

$\square$

Lemma 11.5.3 *There is a one-to-one correspondence of Poincaré DGCAs $\mathcal{Q}^*$ of Hodge type with $\mathcal{Q}^k \cong \mathcal{H}^k$ for all $k \leq 2r - 2$ and a non-degenerate Poincaré pairing $\langle -, - \rangle$, and linear maps $\xi : \mathcal{K}^* \to d\mathcal{Q}^*$. This correspondence is given by the restriction of the cocycle map $\xi_{\mathcal{H}^*} : S^2(H^*) \to d\mathcal{Q}^*$ from (11.14) to $\mathcal{K}^* \subset S^2(H^*)$.*

Proof If $\mathcal{Q}^*$ is as requested, Lemma 11.5.2 implies that $\mathcal{Q}^*$ has a Hodge type decomposition (11.31), and the product is determined by (11.32)–(11.36) for the cocycle map $\bar\xi := \xi_{\mathcal{H}^*}$.

Let $\hat{\mathcal{H}}^* \subset \mathcal{Q}_d^*$ be another harmonic subspace which is hence of the form (11.15) for some linear map $\beta : H^* \to \mathcal{V}^*$. Since $\mathcal{Q}^k \cong \mathcal{H}^k$, and hence $d^- \mathcal{V}^{k+1} = 0$ for $k \leq 2r - 2$ and $d^- : \mathcal{V}^{k+1} \to \mathcal{B}^k$ is an isomorphism, it follows that $\mathcal{V}^k = 0$ for $k < 2r$, whence $|\beta(h)| \geq 2r$. Thus, if $h_1, h_2 \in H^*$ are not 1 and hence of degree $\geq r$, it follows that $|h_1 \cdot \beta(h_2)| \geq 3r > n + 1 - 2r$, whence by (11.29) it follows that $h_1 \cdot \beta(h_2) = 0$ and likewise $\beta(h_1) \cdot h_2 = 0$. Also, $\beta(h_1) \cdot \beta(h_2) \in \mathcal{V}^* \cdot \mathcal{V}^* = 0$. That is,

$$\xi_{\hat{\mathcal{H}}^*}([h_1],[h_2]) = \xi_{\mathcal{H}^*}([h_1],[h_2]) - \beta([h_1] \cdot [h_2]) \tag{11.39}$$

by (11.17). If we now define $\hat{\mathcal{B}}^*$ by (11.16), then we obtain the Hodge type decompositon

$$\mathcal{Q}^* = \mathcal{V}^* \oplus \hat{\mathcal{H}}^* \oplus \hat{\mathcal{B}}^*, \tag{11.40}$$

and the DGCA structure on $\mathcal{Q}^*$ is again determined by (11.32)–(11.36), replacing $\mathcal{H}^*$ by $\hat{\mathcal{H}}^*$ and $\mathcal{B}^*$ by $\hat{\mathcal{B}}^*$ and $\bar{\xi} := \xi_{\mathcal{H}^*}$ by $\bar{\xi}' := \xi_{\hat{\mathcal{H}}^*}$ from (11.40), respectively.

Conversely, $Q_d^* = V^* \oplus \mathcal{H}^*$ is a central extension of H^* by (11.37) whose Hochschild cocycle is given by $\bar{\xi} = \xi_{\mathcal{H}^*}$. Therefore, replacing $\bar{\xi}$ by $\bar{\xi}'$ yields an isomorphic DGCA structure if and only if $\bar{\xi}' - \bar{\xi}$ is a Hochschild coboundary, implying that $\bar{\xi}' := \xi_{\hat{\mathcal{H}}^*}$ is given in (11.39) for some $\beta : H^* \to V^*$. That is, two maps $\bar{\xi}'$ and $\bar{\xi}$ in (11.32)–(11.36) yield isomorphic DGCAs if and only if $(\bar{\xi}' - \bar{\xi})(h_1, h_2) = \beta(h_1 \cdot h_2)$ for some $\beta : H^* \to V^*$, and clearly this is the case if and only if $\xi := \bar{\xi}_{|\mathcal{K}^*} = \bar{\xi}'_{|\mathcal{K}^*}$; that is, Q^* is determined up to isomorphism by $\xi : \mathcal{K}^* \to V^* = dQ^*$, and, since any such ξ is the restriction of some map $\bar{\xi} : S^2(H^*) \to V^*$, the assertion follows. $\square$

Lemma 11.5.4 *Let Q^* be the Poincaré DGCA of Hodge type with $Q^k \cong \mathcal{H}^k$ for all $k \leq 2r - 2$ and a non-degenerate Poincaré pairing $\langle -, - \rangle$ corresponding to the map $\xi : \mathcal{K}^* \to dQ^*$ by Lemma 11.5.3. Then Q^* is tight if and only if ξ is surjective.*

Proof Pick a decomposition $S^2(H^*) = \mathcal{K}^* \oplus \mathcal{N}^*$, and define the map $\bar{\xi} : S^2(H^*) \to V^*$ by $\bar{\xi}_{|\mathcal{K}^*} = \xi$ and $\bar{\xi}_{|\mathcal{N}^*} = 0$.

According to Lemma 11.5.3, there is a Hodge type decomposition (11.40) of Q^* such that the product is given by (11.32)–(11.36) for the map $\bar{\xi}$ from above. It follows that

$$\hat{Q}^* := \bar{\xi}(S^2(H^*)) \oplus \hat{\mathcal{H}}^* \oplus d^-\bar{\xi}(S^2(H^*)) = \xi(\mathcal{K}^*) \oplus \mathcal{H}^* \oplus d^-\xi(\mathcal{K}^*) \subset Q^*$$

is a sub-DGCA of Q^* whose inclusion $\hat{Q}^* \hookrightarrow Q^*$ is a quasi-isomorphism. If ξ is not surjective, then $\hat{Q}^* \subsetneq Q^*$ is a proper quasi-isomorphically embedded sub-DGCA, showing that Q^* is not tight.

Conversely, suppose that ξ is surjective, and let $\hat{Q}^* \subset Q^*$ be a quasi-isomorphically embedded sub-DGCA. Then $\hat{Q}^*$ contains a harmonic subspace $\mathcal{H}^* \subset Q_d^*$. We may choose the Hodge type decomposition (11.40) of Q^* containing $\mathcal{H}^*$ as a factor, whence the product on Q^* is given by (11.32)–(11.36) for some map $\bar{\xi}$ which extends ξ, so that, in particular, $\bar{\xi}$ is surjective as well. Therefore, (11.34) and $\mathcal{H}^* \subset Q^*$ imply that $\bar{\xi}(S^2(H^*)) = V^* \subset \hat{Q}^*$. Since the inclusion map $\hat{Q}^* \hookrightarrow Q^*$ is a quasi-isomorphism, it follows that each $k \in V^*$ is exact in $\hat{Q}^*$, implying that $d^-V^* = \mathcal{B}^* \subset \hat{Q}^*$, whence $\hat{Q}^* = Q^*$. This shows that Q^* is tight. $\square$

Proof of Theorem 11.1.1 Let $\beta = \langle\langle -, - \rangle\rangle_{\mathcal{K}^*}$ on $\mathcal{K}^*$ be a symmetric bilinear form of degree $n+1$, and define the quotient space $V^* := \mathcal{K}^*/\mathcal{K}_\perp^*$ with the canonical projection $\xi : \mathcal{K}^* \to V^*$. Evidently, there is an induced non-degenerate symmetric pairing $\beta_{\mathcal{V}^*}$ on $\mathcal{V}^*$ satisfying

$$\beta_{\mathcal{V}^*}(\xi(k_1), \xi(k_2)) = \beta(k_1, k_2). \tag{11.41}$$

Since H^* is $(r-1)$-connected, it follows that $\mathcal{K}^* \subset S^2(H^*)$ has only elements of degree $\geq 2r$, whence so does $\mathcal{V}^*$. By the non-degenericity of $\beta_{\mathcal{V}^*}$ it follows that $\mathcal{V}^k \cong (\mathcal{V}^{n+1-k})^\vee$ and, in particular, $\mathcal{V}^k = 0$ for $k > n + 1 - 2r$. Therefore, $\mathcal{V}^*$ is of the form (11.29).

Pick an extension $\bar{\xi} : S^2(H^*) \to \mathcal{V}^*$ of ξ and define the algebra $\mathcal{Q}^*_\beta$ by (11.31) with the product given by (11.32)–(11.36). The surjectivity of ξ implies that $\mathcal{Q}^*_\beta$ is a tight DGCA by Lemma 11.5.4, and it is independent of the choice of the extension $\bar{\xi}$ of ξ by Lemma 11.5.3. That is, to a given $\beta \in S^2(\mathcal{K}^*)^\vee$ we have associated a tight DGCA $\mathcal{Q}^*_\beta$ with cohomology H^*.

For the converse, let $\mathcal{Q}^*$ be tight and choose a Hodge type decomposition (11.31). As $\mathcal{Q}^*_{\text{small}} \hookrightarrow \mathcal{Q}^*$ is quasi-isomorphically embedded, it follows that $\mathcal{Q}^*_{\text{small}} = \mathcal{Q}^*$, whence by (11.28), the $(r-1)$-connectedness of $\mathcal{Q}^*$ implies that $\mathcal{Q}^k = \mathcal{H}^k$ for $k \leq 2r - 2$. Thus, Lemma 11.5.2 implies that multiplication in $\mathcal{Q}^*$ is given by (11.32)–(11.36) for some decomposition (11.31) and some cocycle map $\bar{\xi} := \xi_{\mathcal{H}^*} : S^2(H^*) \to \mathcal{V}^*$, and the pairing $\beta_{\mathcal{V}^*} := \langle\!\langle -, - \rangle\!\rangle$ on $\mathcal{V}^* = d\mathcal{A}^*$ defined in (11.9) is non-degenerate.

The tightness of $\mathcal{Q}^*$ and Lemma 11.5.4 imply that $\xi := \bar{\xi}_{|\mathcal{K}^*}$ is surjective, whence the pull-back $\beta := \xi^*(\beta_{\mathcal{V}^*}) \in (S^2(\mathcal{K}^*))^\vee$ satisfies (11.41), so that $\mathcal{V}^* = \mathcal{K}^*/\mathcal{K}^*_\perp$. Therefore, $\mathcal{Q}^* = \mathcal{Q}^*_\beta$. $\square$

11.6 The Bianchi–Massey Tensor

We recall the definition and basic properties of the Bianchi–Massey tensor of a DGCA $\mathcal{A}^*$ introduced by Crowley–Nordström [3], and apply it to the tight DGCAs $\mathcal{Q}^*_\beta$ constructed in Section 11.5.

Let $\mathcal{K}^* \subset S^2(H^*)$ be the kernel of the multiplication map (11.18). We define the space $\mathcal{E}^* \subset S^2(\mathcal{K}^*) \subset S^2(S^2(H^*))$ as the kernel

$$S^2(S^2(H^*)) \tag{11.42}$$

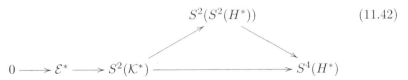

$$0 \longrightarrow \mathcal{E}^* \longrightarrow S^2(\mathcal{K}^*) \longrightarrow S^4(H^*)$$

where the bottom row is exact, and the diagonal maps are induced by the inclusion $\mathcal{K}^* \hookrightarrow S^2(H^*)$ and the multiplication map $S^2(S^2(H^*)) \to S^4(H^*)$, respectively. If we wish to emphasize the dependence of $\mathcal{K}^*$ and

$\mathcal{E}^*$ on the cohomology ring $H^* = H^*(\mathcal{A}^*)$, then we shall denote them by $\mathcal{K}^*_{\mathcal{A}^*}$ and $\mathcal{E}^*_{\mathcal{A}^*}$, respectively.

Given a Poincaré DGCA $\mathcal{A}^*$ with cohomology ring H^*, let $\mathcal{H}^* = \imath(H^*) \subset \mathcal{A}^*_d$ be a harmonic subspace, where $\imath$ is a right inverse of the projection $\mathcal{A}^*_d \to H^*$. Then multiplication induces maps $m_k : S^k(H^*) \to \mathcal{A}^*_d$, $m_k(x_1, \ldots, x_k) := \imath(x_1) \cdots \imath(x_k)$, and, evidently, $\mathcal{K}^* = m_2^{-1}(d\mathcal{A}^*)$. Therefore, we may choose a map $\varepsilon : \mathcal{K}^* \to \mathcal{A}^*[1]$ such that

$$m_2(\varphi) = d\varepsilon(\varphi), \qquad \varphi \in \mathcal{K}^*, \tag{11.43}$$

and furthermore we define the map

$$\hat{\varepsilon} : S^2(\mathcal{K}^*) \longrightarrow \mathcal{A}^*[1], \qquad (\varphi \circ \psi) \longmapsto \varepsilon(\varphi) \cdot m_2(\varphi) = \varepsilon(\varphi) \cdot d\varepsilon(\psi). \tag{11.44}$$

Observe that $d\hat{\varepsilon}(\varphi \circ \psi) = m_2(\varphi) \cdot m_2(\psi) = m_4(\varphi \circ \psi)$. That is, for $e \in \mathcal{E}^* \subset S^2(\mathcal{K}^*)$ we have $d\hat{\varepsilon}(e) = 0$, so that projection onto cohomology yields a map

$$\mathcal{BM}_{\mathcal{A}^*} : \mathcal{E}^* \longrightarrow H^*[1], \qquad \mathcal{BM}_{\mathcal{A}^*}(e) := [\hat{\varepsilon}(e)],$$

which is the *Bianchi–Massey tensor* of $\mathcal{A}^*$ (see [3, Def. 1.1]).

Since ε is uniquely determined up to adding closed elements, it follows that $\hat{\varepsilon}$ is well defined up to adding exact elements, whence $\mathcal{BM}_{\mathcal{A}^*}$ is well defined, independently of the choice of ε. Moreover, it is natural in the sense that for a DGCA-morphism $f : \mathcal{A}^* \to \mathcal{B}^*$ there is an induced commutative diagram

$$
\begin{array}{ccc}
\mathcal{E}^*_{H^*(\mathcal{A}^*)} & \xrightarrow{\ \mathcal{BM}_{\mathcal{A}^*}\ } & H^*(\mathcal{A}^*)[1] \\
\downarrow{\scriptstyle f_*} & & \downarrow{\scriptstyle f_*} \\
\mathcal{E}^*_{H^*(\mathcal{B}^*)} & \xrightarrow{\ \mathcal{BM}_{\mathcal{B}^*}\ } & H^*(\mathcal{B}^*)[1]
\end{array}
$$

Here we denote both the cohomology morphism $H^*(\mathcal{A}^*) \to H^*(\mathcal{B}^*)$ induced by f and its extension to the symmetric tensor algebra

$$S^*(H^*(\mathcal{A}^*)) \to S^*(H^*(\mathcal{B}^*))$$

by the same symbol f_*. In particular, if f is a quasi-isomorphism, then f_* canonically identifies the Bianchi–Massey tensors, so that these are invariants of the weak equivalence class of $\mathcal{A}^*$.

Remarkably, in the case in which $\mathcal{A}^*$ is highly connected, $\mathcal{BM}_{\mathcal{A}^*}$ uniquely determines the weak equivalence class of $\mathcal{A}^*$. Namely, Crowley–Nordström showed the following.

Theorem 11.6.1 ([3, Thm. 1.3]) *Let H^* be an $(r-1)$-connected $(r > 1)$ Poincaré GCA of degree $n \leq 5r - 3$. Then two DGCAs $\mathcal{A}_i^*$, $i = 1, 2$, with cohomology H^* are weakly equivalent if and only if their Bianchi–Massey tensors*

$$\mathcal{BM}_{\mathcal{A}_i^*} : \mathcal{E}^{n+1} \longrightarrow H^n = \mathbb{F}\mathrm{vol}_n$$

coincide.

We say that a bilinear form $\hat{\beta}$ on $S^2(H^*)$ is of *Riemannian type* if it satisfies for all homogeneous $h_i \in H^{|h_i|}$ the symmetry relation

$$\hat{\beta}(h_1 h_2, h_3 h_4) = -(-1)^{|h_2||h_3|}\hat{\beta}(h_1 h_3, h_2 h_4). \tag{11.45}$$

Furthermore we say that a bilinear form β on $\mathcal{K}^*$ is of Riemannian type if $\beta = \hat{\beta}_{|\mathcal{K}^*}$ for $\hat{\beta}$ a bilinear form on $S^2(H^*)$ of Riemannian type. This terminology is due to the fact that tensors of Riemannian type satisfy all (graded) symmetries of a Riemannian curvature tensor.

It follows that there is a decomposition $S^2(S^2(H^*))^\vee = \iota(S4(H^*)) \oplus \mathcal{R}(H^*)$, where $\iota : S^4(H^*)^\vee \to S^2(S^2(H^*))^\vee$ is the dual of the multiplication map and $\mathcal{R}(H^*)$ is the space of bilinear forms of Riemannian type.

Proof of Theorem 11.1.2 We need to compute the Bianchi–Massey tensors $\mathcal{BM}_{\mathcal{Q}_\beta^*}$. By construction, the product structure of $\mathcal{Q}_\beta^*$ is defined by (11.32)–(11.36) for some extension $\bar{\xi} : S^2(H^*) \to \mathcal{V}^*$ of the canonical projection $\xi : \mathcal{K}^* \to \mathcal{V}^*$. Thus, when defining the Bianchi–Massey tensor, the map $\varepsilon : \mathcal{K}^* \to \mathcal{Q}_\beta^*[1]$ from (11.43) may be chosen as

$$\varepsilon(k) := d^-\xi(k).$$

Let $e \in \mathcal{E}^* \subset S^2(\mathcal{K}^*)$ and write it as $e = \sum_i k_1^i \circ k_2^i \in \mathcal{E}^* \subset S^2(\mathcal{K}^*)$ with $k_j^i \in \mathcal{K}^*$. Then, by (11.44),

$$\hat{\varepsilon}(e) = \sum_i d^-k_1^i \circ k_2^i = \sum_i \langle\!\langle k_1^i, k_2^i \rangle\!\rangle \mathrm{vol}_n = \beta(e)\mathrm{vol}_n,$$

that is, the Bianchi–Massey tensor of $\mathcal{Q}_\beta^*$ is determined by the restriction of β to $\mathcal{E}^* \subset S^2(\mathcal{K}^*)$:

$$\mathcal{BM}_{\mathcal{Q}_\beta^*}(e) = \beta(e)\mathrm{vol}_n \qquad \text{for all } e \in \mathcal{E}^*. \tag{11.46}$$

Since any element in $(\mathcal{E}^*)^\vee$ can be realized as the restriction of some $\beta \in (S^2(\mathcal{K}^*))^\vee$ of Riemannian type, and since by Theorem 11.6.1 the Bianchi–Massey tensor determines the weak equivalence type, the statements follow. $\qquad \square$

Example 11.6.2 Let H^* be a simply connected Poincaré algebra of Hodge type of degree 7, so that $r = 2$ and $n = 7 = 5r - 3$. As $\mathcal{K}^*$ has elements of degree $\geq 2r = 4$ only, any bilinear form β of degree $n + 1 = 8$ on $\mathcal{K}^*$ is a bilinear form on

$$\mathcal{K}^4 = \ker(\cdot : H^2 \otimes H^2 \longrightarrow H^4).$$

In fact, we may assume that $\beta \in S^2(\mathcal{K}^4)^\vee$ is the restriction of an element in $\mathcal{R}^8 \subset S^2(S^2(H^2))^\vee$ of Riemannian type to $\mathcal{K}^4$.

Then $\mathcal{V}^* = \mathcal{V}^4 = \mathcal{K}^4/\mathcal{K}^4_\perp$, where $\mathcal{K}^4_\perp$ is the null space of β, so that

$$\mathcal{Q}^k_\beta = \begin{cases} \mathcal{H}^k, & \text{for } k \neq 3, 4, \\ \mathcal{H}^3 \oplus d^- \mathcal{V}^4, & \text{for } k = 3, \\ \mathcal{H}^4 \oplus \mathcal{V}^4, & \text{for } k = 4, \end{cases}$$

and the algebra structure of $\mathcal{Q}^*_\beta$ is given by (11.32)–(11.36) for some extension $\bar{\xi} : S^2(H^2) \to \mathcal{V}^4$ of the canonical projection $\xi : \mathcal{K}^4 \to \mathcal{V}^4$.

In particular, the de-Rham algebra $\Omega^*(M)$ of a closed simply connected 7-manifold M is weakly equivalent to such an DGCA $\mathcal{Q}^*_\beta$.

References

[1] K. Cieliebak, K. Fukaya and J. Latschev, *Homological algebra related to surfaces with boundary*, arXiv:1508.02741.

[2] K.F. Chan, S. Karigiannis and C.C. Tsang, *The $\mathcal{L}_B$-cohomology on compact torsion-free G_2 manifolds and an application to 'almost' formality*, Ann. Global Anal. Geom., **55**(2) (2019), 325–369.

[3] D. Crowley and J. Nordström, *The rational homotopy type of $(n-1)$-connected manifolds of dimension up to $5n - 3$*, J. Topol., **13**(2) (2020), 539–575.

[4] Y. Félix, S. Halperin and J.-C. Thomas, *Rational Homotopy Theory*, Graduate Texts in Mathematics, **205**, Springer-Verlag, Berlin (2001).

[5] D. Fiorenza, K. Kotaro, H.V. Lê and L. Schwachhöfer, *Rational Homotopy Theory*, Graduate Texts in Mathematics, **205**, Springer-Verlag, Berlin (2001).

[6] P. Lambrechts and D. Stanley, *Poincaré duality and commutative differential graded algebras*, Ann. Sci. de l'École Norm. Supérieure, **41**(4) (2008), 497–511.

[7] T. J. Miller, *On the formality of $k - 1$ connected compact manifolds of dimension less than or equal to $4k - 2$*, Illinois J. Math., **23** (1979), 253–258.

[8] D. Sullivan, *Differential forms and the topology of manifolds*, Manifolds-Tokyo (1973) (Proc. Internat. Conf. Manifolds and Related Topics in Topology, Tokyo 1973) (ed. A. Hattori), University of Tokyo Press, 1975, pp. 37–49.

[9] D. Sullivan, *Infinitesimal computations in topology*. Publ. Math. IHES, **47** (1977), 269–331.

Part Three

Recent Developments in Non-Negative
Sectional Curvature

12

Fake Lens Spaces and Non-Negative Sectional Curvature

Sebastian Goette, Martin Kerin and Krishnan Shankar

Abstract

In this short chapter we observe the existence of free, isometric actions of finite cyclic groups on a family of 2-connected 7-manifolds with non-negative sectional curvature. This yields many new examples including fake, and possible exotic, lens spaces.

12.1 Introduction

Riemannian manifolds with positive or non-negative sectional curvature have been of great interest to geometers but there are not many examples nor many obstructions known. For non-negatively curved manifolds the most far-reaching structure theorem is that of Gromov, bounding the total Betti number by a constant depending only on the dimension. Beyond that, given the dearth of examples and theorems, it is of interest to generate new methods and new examples of non-negatively curved manifolds.

In the recent paper [GKS1] we showed that there is a six parameter family $\mathscr{F}$ of 2-connected 7-manifolds, each with the cohomology of an S^3-bundle over S^4 and admitting non-negative sectional curvature. The family $\mathscr{F}$ is a rich source of interesting new examples; it includes all homotopy 7-spheres (each with infinitely many $SO(3)$-invariant metrics) as well as infinitely many examples with non-standard linking form. The latter class represents the first known examples of 2-connected 7-manifolds with non-negative curvature that are not even homotopy equivalent to S^3-bundles over S^4; see [GKS2]. The manifolds $M_{a,b}^7 \in \mathscr{F}$ are each the total space of a Seifert fibration over an orbifold S^4 with generic fiber S^3. The parameters $\underline{a} = (a_1, a_2, a_3)$, $\underline{b} = (b_1, b_2, b_3)$ are each triples of integers satisfying $a_i, b_i \equiv 1 \pmod 4$ for all $i \in \{1, 2, 3\}$, and $\gcd(a_1, a_2 \pm a_3) = 1, \gcd(b_1, b_2 \pm b_3) = 1$. Note that the sub-family corresponding to $a_1 = b_1 = 1$ is precisely the one introduced by K. Grove and W. Ziller in [GZ], and consists of all S^3-bundles over S^4.

Theorem 12.1.1 *There exists a free, isometric action of $\mathbb{Z}_\ell \subset SO(3)$ on $M^7_{\underline{a},\underline{b}} \in \mathscr{F}$ if and only if $\gcd(\ell, a_2 \pm a_3) = 1$ and $\gcd(\ell, b_2 \pm b_3) = 1$ (which implies that ℓ is necessarily odd). In particular, there are infinitely many fake lens spaces in dimension 7 admitting non-negative sectional curvature (see Table 12.1).*

Recall that a fake lens space is a manifold with finite, cyclic fundamental group and universal cover a homotopy sphere, while an exotic lens space is a fake lens space that is homeomorphic, but not diffeomorphic, to a lens space. It would be interesting to obtain a classification of these quotients up to diffeomorphism.

Acknowledgements   It is a pleasure to thank the MATRIX institute in Creswick, Australia, and the organizers of the Australian–German workshop "Differential Geometry in the Large" held there in February 2019, where this work was initiated and discussed. The institute and the workshop provided ideal working conditions for collaborating on this project. S. Goette and M. Kerin received support under the DFG Priority Program 2026 *Geometry at Infinity*. K. Shankar received support from the National Science Foundation.[1]

12.2 $\mathbb{Z}_\ell$ Actions on the Family $\mathscr{F}$

Consider the family of 10-manifolds $P^{10}_{\underline{a},\underline{b}}$ with a cohomogeneity-one action of $S^3 \times S^3 \times S^3$ given by the group diagram

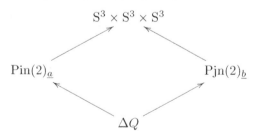

---

[1] The views expressed in this paper are those of the authors and do not necessarily reflect the views of the National Science Foundation.

where ΔQ is the diagonal embedding of $Q = \{\pm 1, \pm i, \pm j, \pm k\} \subseteq S^3$ and

$$\text{Pin}(2)_{\underline{a}} := \{(e^{ia_1\theta}, e^{ia_2\theta}, e^{ia_3\theta})\} \cup \{(e^{ia_1\theta}, e^{ia_2\theta}, e^{ia_3\theta}) \cdot j\},$$
$$\text{Pjn}(2)_{\underline{b}} := \{(e^{jb_1\theta}, e^{jb_2\theta}, e^{jb_3\theta})\} \cup \{i \cdot (e^{jb_1\theta}, e^{jb_2\theta}, e^{jb_3\theta})\}.$$

When $\gcd(a_1, a_2 \pm a_3) = 1$ and $\gcd(b_1, b_2 \pm b_3) = 1$ the subgroup $1 \times \Delta S^3 \subseteq S^3 \times (S^3 \times S^3)$ acts freely and isometrically with quotient $M^7_{\underline{a},\underline{b}}$. The family $\mathscr{F}$ consists of all such spaces, and each $M^7_{\underline{a},\underline{b}} \in \mathscr{F}$ inherits a codimension-one singular Riemannian foliation by biquotients (or double coset manifolds) with regular leaf diffeomorphic to the hypersurface $B_0 := (S^3 \times S^3) /\!/ \Delta Q$, and singular leaves of codimension two diffeomorphic to $B_- := (S^3 \times S^3) /\!/ \text{Pin}(2)_{\underline{a}}$ and $B_+ := (S^3 \times S^3) /\!/ \text{Pjn}(2)_{\underline{b}}$ respectively (both have the integral cohomology of $S^3 \times \mathbb{R}P^2$). In each case, the free action of the group $U \in \{\Delta Q, \text{Pin}(2)_{\underline{a}}, \text{Pjn}(2)_{\underline{b}}\}$ on $S^3 \times S^3$ is described by

$$U \times (S^3 \times S^3) \to S^3 \times S^3,$$

$$\left((u_1, u_2, u_3), \begin{pmatrix} q_1 \\ q_2 \end{pmatrix} \right) \mapsto \begin{pmatrix} q_1\, u_1^{-1} \\ u_2\, q_2\, u_3^{-1} \end{pmatrix}.$$

There is an obvious isometric action by S^3 on the left of the first factor of $S^3 \times S^3$ which commutes with each U action and induces an isometric (leaf-preserving) action of $SO(3)$ on each $M^7_{\underline{a},\underline{b}} \in \mathscr{F}$. In particular, $-1 \in S^3$ always acts trivially on $M^7_{\underline{a},\underline{b}}$.

By the slice theorem, the quotient $M^7_{\underline{a},\underline{b}}$ is the union of disk bundles over the two singular leaves glued along their common boundary, a regular leaf. Furthermore, as noted in Section 12.1, each $M^7_{\underline{a},\underline{b}}$ is a cohomology S^3-bundle over S^4. In particular, $M^7_{\underline{a},\underline{b}}$ is a 2-connected 7-manifold with $H^4(M^7_{\underline{a},\underline{b}}; \mathbb{Z}) = \mathbb{Z}_n$, where $n = \frac{1}{8} \det \begin{pmatrix} a_1^2 & b_1^2 \\ a_2^2 - a_3^2 & b_2^2 - b_3^2 \end{pmatrix}$. Whenever $n \neq 0$, $M^7_{\underline{a},\underline{b}}$ is a rational homology sphere, while $n = \pm 1$ ensures that $M^7_{\underline{a},\underline{b}}$ is a homotopy 7-sphere. In particular, all exotic 7-spheres show up in the family $\mathscr{F}$ [GKS1].

Theorem 12.1.1 is an immediate consequence of the following observation.

Theorem 12.2.1 $\mathbb{Z}_\ell \subseteq SO(3)$ *acts freely and isometrically on* $M^7_{\underline{a},\underline{b}}$ *if and only if* $\gcd(\ell, a_2 \pm a_3) = 1$ *and* $\gcd(\ell, b_2 \pm b_3) = 1$. *In particular,* ℓ *is necessarily odd.*

Proof Consider the isometric (leaf-preserving) action of $\mathbb{Z}_r = \{w \in U(1) \subseteq \mathbb{C} : w^r = 1\} \subseteq S^3$ (the rth roots of unity) on $M^7_{\underline{a},\underline{b}}$ described on each biquotient leaf $(S^3 \times S^3)/U$, $U \in \{\Delta Q, \text{Pin}(2)_{\underline{a}}, \text{Pjn}(2)_{\underline{b}}\}$, by

$$\mathbb{Z}_r \times (S^3 \times S^3)/U \to (S^3 \times S^3)/U,$$

$$\left(w, \begin{bmatrix} q_1 \\ q_2 \end{bmatrix} \right) \mapsto \begin{bmatrix} wq_1 \\ q_2 \end{bmatrix}.$$

As every $\mathbb{Z}_\ell \subseteq SO(3)$ is, up to conjugation, covered by such a $\mathbb{Z}_r \subseteq S^3$, the statement of the theorem is now equivalent to the claim that $\mathbb{Z}_r$ acts freely (resp. effectively freely) on $M^7_{a,b}$ if and only if $r = \ell$ (resp. $r = 2\ell$) for some $\ell \in \mathbb{Z}$ satisfying $\gcd(\ell, a_2 \pm a_3) = 1$ and $\gcd(\ell, b_2 \pm b_3) = 1$. As $a_2 \pm a_3$ and $b_2 \pm b_3$ are all even, it is clear that ℓ must be odd.

Since ΔQ is a subgroup of $\text{Pin}(2)_{\underline{a}}$ and $\text{Pjn}(2)_{\underline{b}}$, it suffices to check freeness of the action on the singular leaves, i.e., for $U \in \{\text{Pin}(2)_{\underline{a}}, \text{Pjn}(2)_{\underline{b}}\}$. Moreover, the argument for freeness of the action on $B_- = (S^3 \times S^3)/\text{Pin}(2)_{\underline{a}}$ is completely analogous to the argument for $B_+ = (S^3 \times S^3)/\text{Pjn}(2)_{\underline{b}}$ by viewing elements of S^3 as being of the form $u + iv$, for $u, v \in \text{span}\{1, j\}$, rather than the usual $u + vj$, for $u, v \in \mathbb{C}$). For this reason, freeness of the action will only be verified on B_-.

To this end, suppose that there is a point in B_- with isotropy, that is, that

$$\begin{bmatrix} wq_1 \\ q_2 \end{bmatrix} = \begin{bmatrix} q_1 \\ q_2 \end{bmatrix}$$

for some $w \in \mathbb{Z}_r$. Then there exists some $z \in U(1) \subseteq \mathbb{C}$ and some $\lambda \in \{0, 1\}$ such that

$$wq_1 = q_1 j^{-\lambda} \bar{z}^{a_1} \quad \text{and} \quad q_2 = z^{a_2} j^\lambda q_2 j^{-\lambda} \bar{z}^{a_3}. \tag{12.1}$$

The goal now is to show that the only possibility at every such point is $w = 1$ (resp. $w \in \{\pm 1\}$) precisely when $\gcd(r, a_2 \pm a_3) = 1$ (resp. $\gcd(r, a_2 \pm a_3) = 2$). Clearly, the analogous conditions involving the parameters $\underline{b}$ arise when considering the singular leaf B_+.

If $\lambda = 0$, then (12.1) implies that $wq_1 = q_1 \bar{z}^{a_1}$ and $q_2 = z^{a_2} q_2 \bar{z}^{a_3}$. Writing $q_2 = u_2 + v_2 j$, with $u_2, v_2 \in \mathbb{C}$, and using the commutation relations for the quaternions yields

$$q_2 = z^{a_2} q_2 \bar{z}^{a_3} \iff u_2 + v_2 j = z^{a_2 - a_3} u_2 + z^{a_2 + a_3} v_2 j.$$

Since $|q_2| = 1$, this forces either $z^{a_2 - a_3} = 1$ or $z^{a_2 + a_3} = 1$.

On the other hand, $wq_1 = q_1 \bar{z}^{a_1} \iff w = q_1 \bar{z}^{a_1} \bar{q}_1$. Therefore,

$$1 = w^r = (q_1 \bar{z}^{a_1} \bar{q}_1)^r = q_1 \bar{z}^{ra_1} \bar{q}_1.$$

Conjugating both sides by $\bar{q}_1$, it follows that $\bar{z}^{ra_1} = 1$.

Setting $d_{\pm} = \gcd(ra_1, a_2 \pm a_3)$, these identities together yield $z^{d_{\pm}} = 1$. Moreover, since $\gcd(a_1, a_2 \pm a_3) = 1$, it follows that $d_{\pm} = \gcd(r, a_2 \pm a_3)$. In particular, it is clear that $d_{\pm} = 1$ implies $z = 1$ and, hence, $w = 1$, since $w = q_1 \bar{z}^{a_1} \bar{q}_1$. Similarly, if $d_{\pm} = 2$, then $z \in \{\pm 1\}$ and, hence, $w \in \{\pm 1\}$.

If, on the other hand, $d_{\pm} > 2$, then the $\mathbb{Z}_r$ action cannot even be effectively free. Indeed, choose $z \in U(1)$ such that $z^{d_{\pm}} = 1$ and $z \neq \pm 1$. Since r is divisible by $d_{\pm}$ by definition, it follows that $z^r = (z^{d_{\pm}})^{r/d_{\pm}} = 1$ and, therefore, $z, z^{a_1} \in \mathbb{Z}_r$. Notice, however, that $\gcd(a_1, a_2 \pm a_3) = 1$ implies $\gcd(a_1, d_{\pm}) = 1$ and, thus, $\gcd(2a_1, d_{\pm}) = \gcd(2, d_{\pm})$. This ensures that $z^{a_1} \neq \pm 1$ since, otherwise, the identities $z^{2a_1} = 1$ and $z^{d_{\pm}} = 1$ would imply that $z^{\gcd(2, d_{\pm})} = 1$ and, hence, that $z \in \{\pm 1\}$, a contradiction. Now, setting $w = z^{a_1} \in \mathbb{Z}_r \backslash \{\pm 1\}$ yields, for example,

$$w \cdot \begin{bmatrix} 1 \\ 1 \end{bmatrix} = \begin{bmatrix} z^{a_1} \\ 1 \end{bmatrix} = \begin{bmatrix} z^{a_1} \bar{z}^{a_1} \\ \bar{z}^{a_2 - a_3} \end{bmatrix} = \begin{bmatrix} 1 \\ \bar{z}^{a_2 - a_3} \end{bmatrix} = \begin{bmatrix} \cdot & 1 \\ (\bar{z}^{d_{\pm}})^{(a_2 - a_3)/d_{\pm}} \end{bmatrix} = \begin{bmatrix} 1 \\ 1 \end{bmatrix}.$$

Suppose, finally, that $\lambda = 1$. In this case, (12.1) yields the equalities $wq_1 = q_1 \bar{j} \bar{z}^{a_1}$ and $q_2 = z^{a_2} j q_2 \bar{j} \bar{z}^{a_3}$. The first equality can be rewritten as $w = -q_1 z^{a_1} j \bar{q}_1$, which implies that $z^{a_1} j = -\bar{q}_1 w q_1$. Notice that $z^{a_1} j \in \text{span}\{j, k\}$, which implies that $\text{Re}(w) = \text{Re}(\bar{q}_1 w q_1) = 0$ and, therefore, that $w \in \{\pm i\} \cap \mathbb{Z}_r$. However, $\pm i \in \mathbb{Z}_r$ if and only if $r \equiv 0 \mod 4$, which is impossible if $\gcd(r, a_2 \pm a_3), \gcd(r, b_2 \pm b_3) \in \{1, 2\}$. $\square$

Example 12.2.2 Actions on homotopy 7-spheres: To illustrate some specific examples we exhibit parameters along with values of ℓ for which there are fake lens spaces with non-negative sectional curvature. Note that in the case when the universal cover is *not* the standard 7-sphere, the manifold cannot be diffeomorphic to a lens space. Therefore, computing differential invariants for these manifolds could yield infinitely many examples of exotic lens spaces with non-negative sectional curvature. We exhibit a few examples of actions on some homotopy spheres (which are determined up to oriented diffeomorphism by the Eells–Kuiper invariant μ, [EK]). Note that the last example is a non-Milnor sphere, i.e., a homotopy 7-sphere that is *not* diffeomorphic to an S^3-bundle over S^4.

Table 12.1 *Free, isometric $\mathbb{Z}_\ell$ actions on homotopy 7-spheres*

$(\mathbf{a_1}, \mathbf{a_2}, \mathbf{a_3})$	$(\mathbf{b_1}, \mathbf{b_2}, \mathbf{b_3})$	$\mu(\underline{\mathbf{a}}, \underline{\mathbf{b}})$	Free, isometric $\mathbb{Z}_\ell$ action
$(5, -3, 1)$	$(-7, 5, -3)$	$27/28$	all odd ℓ
$(985, -3, 1)$	$(1393, 5, -3)$	$6/28$	all odd ℓ
$(29, -3, 1)$	$(41, 5, -3)$	$20/28$	all odd ℓ
$(17, -47, 33)$	$(-15, -219, 217)$	$5/28$	$\gcd(\ell, 2 \cdot 5 \cdot 7 \cdot 109) = 1$

Remark 12.2.3   We recently noticed that the action by $\mathbb{Z}_2$ on the co-homogeneity one manifold $P^{10}_{\underline{a},\underline{b}}$, generated by the element $(1, 1, -1)$, induces a free, isometric $\mathbb{Z}_2$-action on every $M^7_{\underline{a},\underline{b}}$. This action can also be viewed as the antipodal map on the S^3 fibers of the correspoinding Seifert fibration. See [W] for study of such actions on the Milnor spheres. We also note that the product action $\mathbb{Z}_\ell \times \mathbb{Z}_2 \cong \mathbb{Z}_{2\ell}$ is also free under the same conditions on the greatest common divisor as in Theorem 12.1.1.

References

[EK] J. Eells and N. Kuiper, *An invariant for certain smooth manifolds*, Ann. Mat. Pura Appl., **60** (1962), 93–110.

[GKS1] S. Goette, M. Kerin and K. Shankar, *Highly connected 7-manifolds and non-negative sectional curvature*, Ann. of Math. (2) **191**(3) (2020), 829–892.

[GKS2] S. Goette, M. Kerin and K. Shankar, *Highly connected 7-manifolds, the linking form and non-negative sectional curvature*, preprint arXiv:2003.04907, 2020.

[GZ] K. Grove and W. Ziller, *Curvature and symmetry of Milnor spheres*, Ann. of Math. (2) **152** (2000), 331–367.

[W] J. Wermelinger, *Moduli space of nonnegatively curved metrics on Milnor sphere quotients*, preprint arXiv:2006.13690, 2020.

13

Collapsed 3-Dimensional Alexandrov Spaces: A Brief Survey

Fernando Galaz-García, Luis Guijarro and Jesús Núñez-Zimbrón

Abstract

We survey two recent developments in the topic of three-dimensional Alexandrov spaces: the topological classification of closed collapsed three-dimensional Alexandrov spaces and the geometrization of sufficiently collapsed closed three-dimensional Alexandrov spaces.

13.1 Introduction

Alexandrov spaces (with curvature bounded below) are metric generalizations of complete Riemannian manifolds with a uniform lower sectional curvature bound. In addition to its intrinsic interest, Alexandrov geometry plays an important role in the proof of finiteness results for certain families of closed (i.e. compact and without boundary) Riemannian manifolds (see, for example, the survey [12]). Indeed, by Gromov's precompactness theorem, the family $\mathcal{M}_k^D(n)$ of closed Riemannian n-manifolds with sectional curvature $\sec \geq k$ and diameter bounded above by $D > 0$ is precompact in the Gromov–Hausdorff topology. Moreover, limits of sequences in $\mathcal{M}_k^D(n)$ are Alexandrov spaces with curvature bounded below by k. More generally, if X is the Gromov–Hausdorff limit of a sequence $\{X_i^n\}_{i=1}^\infty$ of compact n-dimensional Alexandrov spaces with curvature bounded below by k, then X is an Alexandrov space with curvature bounded below by k and (Hausdorff) dimension at most n. Topologically, the case where the limit X is n-dimensional is well understood. Indeed, by Perelman's stability theorem [21], the elements X_i^n of the sequence are homeomorphic to X for i sufficiently large. The complementary phenomenon, in which the dimension of X is strictly less than n, is known as *collapse*. Note that, by Perelman's stability theorem, if sequences in a precompact family of closed Riemannian n-manifolds with a uniform lower sectional curvature bound do not collapse, then the family must consist of finitely many homeomorphism types. This is the case, for example, for the family $\mathcal{M}_{k,v}^D(n)$ of closed Riemannian n-manifolds with sec $\geq k$, diameter at most $D > 0$ and volume bounded below by $v > 0$.

A simple example of collapse is furnished by rescaling the Riemannian metric of a given flat n-dimensional torus T^n, $n \geq 2$, by $1/k$, $k = 1, 2\ldots$ In this way, one obtains a sequence $\{T_k^n\}$ of flat n-tori whose diameter decreases as $k \to \infty$. In this case, the sequence of flat tori collapses to a point. By rescaling appropriate factors of T^n we may obtain sequences of flat tori which collapse to flat tori of dimension strictly less than n. Further examples of collapse with a uniform lower sectional curvature bound may be obtained by rescaling the orbits of isometric compact Lie group actions on a given closed Riemannian manifold. In this case, the sequence of metrics converges to the orbit space of the action, which is an Alexandrov space and is, in general, not a manifold.

Motivated by the preceding considerations, one may attempt to understand the topological consequences of collapse. In the Riemannian category, a thorough analysis of collapse of Riemannian 3-manifolds was carried out by Shioya and Yamaguchi in [28, 29]. More recently, Mitsuishi and Yamaguchi obtained topological classification and structure results for collapsed Alexandrov spaces of dimension 3 (see [18]), while the authors of the present survey obtained the geometrization of closed, sufficiently collapsed irreducible 3-dimensional Alexandrov spaces (see [9]), thus extending the Riemannian results to the case of Alexandrov spaces. In this note we give a brief account of these results in Alexandrov geometry in the hope of sparking the interest of the reader.

This chapter is organized as follows. Section 13.2 contains a summary of basic results in Alexandrov geometry. In Section 13.3 we recall the basic results on the topological structure of general Alexandrov spaces of dimension three. Finally, in Section 13.4, we present the topological structure and classification results for closed collapsed 3-dimensional Alexandrov spaces, as well as the geometrization of closed, sufficiently collapsed irreducible Alexandrov spaces of dimension 3.

Acknowledgements This survey is partially based on a talk given by F.G-G. in the "Australian–German Workshop on Differential Geometry" in February 2019 at the MATRIX Institute, in Australia. It is a pleasure to thank the organizers of the workshop, as well as the MATRIX Institute, for their hospitality and for their invitation to submit a manuscript to the proceedings of the meeting.

F.G-G. was supported in part by research grants MTM2014–57769–3-P and MTM2017–85934–C3–2–P from the Ministerio de Economía y Competitividad de Espana (MINECO), and by the Deutsche Forschungsgemeinschaft grant GA 2050 2-1 within the Priority Program SPP 2026 "Geometry at Infinity". L.G. was supported by research grants MTM2014-57769-3-P and MTM2017-85934-C3-2-P from the MINECO, and by IC-MAT Severo Ochoa project SEV-2015-0554 (MINECO). J.N-Z. was supported by a DGAPA-UNAM postdoctoral fellowship and CONACyT project CB2016-283988-F.

13.2 Basic Alexandrov Geometry

In this section we will recall the notation and main aspects of the theory of Alexandrov spaces (of curvature bounded below). Standard references in the subject are [4, 5] (see also the recent manuscript [1]), and we refer the reader to these sources for a detailed account of the theory.

In order to introduce the definition of an Alexandrov space, we first recall some concepts. Alexandrov spaces fall within the class of the so-called *length spaces*. A metric space (X,d) is a *length space* whenever, for every $x,y \in X$,

$$d(x,y) = \inf \left\{ L(\gamma) \mid \gamma(a) = x,\ \gamma(b) = y \right\}.$$

Here, the infimum is taken over all continuous curves $\gamma : [a,b] \to X$ (for some $a \leq b$) and $L(\gamma)$ stands for the *length* of γ. The length of such a curve is defined as

$$L(\gamma) = \sup \left\{ \sum_{i=1}^{n-1} d\left(\gamma(t_i), \gamma(t_{i+1})\right) \right\},$$

where the supremum is taken over all finite partitions

$$a = t_0 \leq t_1 \leq \ldots \leq t_n = b$$

of $[a,b]$. We require two technical assumptions on any length space X: completeness and local compactness. This ensures the existence of *geodesics* between each pair of points $x,y \in X$, that is, continuous curves $\gamma : [a,b] \to X$ such that $\gamma(a) = x$, $\gamma(b) = y$ and $L(\gamma) = d(x,y)$. A geodesic joining x and y will be denoted by $[xy]$. Note that such a geodesic might not be unique, as can be readily seen by considering geodesics in a round sphere.

One of the concepts playing a central role in Alexandrov geometry is that of the *model spaces*. Given a real number k, the 2-dimensional *model space* M_k^2 is defined to be the complete, simply-connected 2-dimensional Riemannian manifold of constant sectional curvature k. In other words, depending on the sign of k, M_k^2 is isometric to one of the following spaces:

- $\mathbb{S}_k^2$, the sphere of constant curvature $k > 0$;
- $\mathbb{E}^2$, the Euclidean plane of curvature 0; or
- $\mathbb{H}_k^2$, the hyperbolic plane of constant curvature $k < 0$.

Other prominent objects in the theory are *geodesic triangles*. A geodesic triangle $\triangle\, pqr$ in a length space (X, d) is a collection of three points $p, q, r \in X$ and three geodesics $[pq]$, $[qr]$ and $[rp]$. Once a geodesic triangle $\triangle\, pqr$ in X is given, one says that a geodesic triangle $\triangle\, \widetilde{p}\widetilde{q}\widetilde{r}$ in M_k^2 is a *comparison triangle for* $\triangle\, pqr$ if $d(p, q) = |p, q|$, $d(q, r) = |q, r|$ and $d(r, p) = |r, p|$. Here $|\cdot, \cdot|$ stands for the usual length metric on M_k^2.

With these definitions in hand, we may now recall the definition of an Alexandrov space. We will say that a length space (X, d) has *curvature bounded below by* $k \in \mathbb{R}$, denoted by $\mathrm{curv}(X, d) \geq k$ (or simply $\mathrm{curv}(X) \geq k$), if, for every $x \in X$, there exists an open neighborhood $U \subset X$ of x such that for every geodesic triangle $\triangle\, pqr$ and any comparison triangle $\triangle\, \widetilde{p}\widetilde{q}\widetilde{r}$ in M_k^2 the so called T_k-*property* holds: For every $s \in [pq]$ and $\widetilde{s} \in [\widetilde{p}\widetilde{q}]$ such that $d(p, s) = |\widetilde{p}, \widetilde{s}|$, $d(r, s) \geq |\widetilde{r}, \widetilde{s}|$.

Definition 13.2.1 An *Alexandrov space* is a complete and locally compact length space (X, d) such that $\mathrm{curv}(X) \geq k$ for some $k \in \mathbb{R}$.

It is worth noting that there are several equivalent definitions of Alexandrov spaces (see [4, Thm. 4.3.5]). Here we just mention the following *monotonicity of angles condition*: Let $\gamma_1, \gamma_2 : [a, b] \to X$ be geodesics such that $\gamma_1(0) = \gamma_2(0)$. Then X is an Alexandrov space of curv $\geq k$ if and only if the function

$$\theta_k(s, t) := \angle \widetilde{\gamma_1(s)\gamma_1(0)\gamma_2(t)} \tag{13.1}$$

is monotone non-increasing in $s, t \in [a, b]$. Here, $\triangle\, \widetilde{\gamma_1(s)\gamma_1(0)\gamma_2(t)}$ is a comparison triangle for $\triangle\, \gamma_1(s)\gamma_1(0)\gamma_2(t)$.

One of the most powerful tools available in Alexandrov geometry is the following globalization theorem, essentially asserting that once the T_k-property is satisfied locally, then it holds in the large (see [4, Thm. 10.3.1]).

Theorem 13.2.2 (Globalization theorem) *Let X be an Alexandrov space with* $\mathrm{curv}(X) \geq k$. *Then the T_k-property is satisfied for any geodesic triangle in X.*

The most familiar examples of Alexandrov spaces are smooth, complete Riemannian manifolds of sectional curvature bounded below. This is guaranteed by the Toponogov distance comparison theorem, [23, Thm. 12.2.2]. The same result implies naturally that if $l < k$ and X is an Alexandrov space of curv $\geq k$, then $\mathrm{curv}(X) \geq l$. However, the class of Alexandrov spaces includes non-smooth spaces. For example, the boundary of an open and convex set in a Euclidean space $\mathbb{R}^n$, regarded with the induced metric, is a non-negatively curved space [27]. There are a number of constructions available to produce new Alexandrov spaces from known examples. In this way one can produce Alexandrov spaces which are not homeomorphic to manifolds. Let us mention the most commonly used constructions.

- *Cartesian products.* Let X and Y be Alexandrov spaces with curv $\geq k$ and $k \leq 0$. The Cartesian product $X \times Y$ with the usual product metric is an Alexandrov space of curv $\geq k$. For $k > 0$, the product is a space of curv $\geq k$ only in the case that one of the spaces is a single point.
- *Euclidean cones.* Let (X, d) be a metric space with $\mathrm{diam}(X) \leq \pi$. Recall that the cone over X is the metric space $(K(X), d_K)$ obtained from $X \times [0, \infty)$ by collapsing $X \times \{0\}$ to a point. The metric d_K is given by

$$d_K\left((x_1, t_1), (x_2, t_2)\right) = \sqrt{t_1^2 + t_2^2 - 2t_1 t_2 \cos d(x_1, x_2)}.$$

  The cone $K(X)$ is an Alexandrov space of curv ≥ 0 if and only if X is an Alexandrov space of curv ≥ 1.
- *Spherical suspensions.* Let (X, d) be a metric space with $\mathrm{diam}(X) \leq \pi$. The spherical suspension $(\mathrm{Susp}(X), d_S)$ of X is the metric space obtained from $X \times [0, \pi]$ by collapsing $X \times \{0\}$ and $X \times \{\pi\}$ to single points. A metric d_S is then defined by the equation

$$\cos d_S\left((x_1, t_1), (x_2, t_2)\right) = \cos t_1 \cos t_2 + \sin t_1 \sin t_2 \cos d(x_1, x_2).$$

  If X is an Alexandrov space of curv ≥ 1, then $\mathrm{Susp}(X)$ is an Alexandrov space of curv ≥ 1.

These constructions are in fact special cases of *warped products*. The fact that they indeed produce spaces of curvature bounded below can be easily seen from the so called fiber independence theorem, [1, Thm.

10.1.3]. The problem of obtaining necessary and sufficient conditions for a warped product to be an Alexandrov space was solved in [2, 3]. One of the most important features in Alexandrov geometry is the following:

- *Gromov–Hausdorff limits.* Let $\{X_i\}_{i=1}^{\infty}$ be an infinite sequence of compact Alexandrov spaces with $\mathrm{curv}(X_i) \geq k$ for all i. If X_i converges in the Gromov–Hausdorff sense to a metric space X, then X is an Alexandrov space of curv $\geq k$.

It is in the context of Gromov–Hausdorff limits where the phenomenon of *collapse* occurs. We say that a sequence of compact Alexandrov spaces $\{X_i\}_{i=1}^{\infty}$ of (Hausdorff) dimension n which converges in the Gromov–Hausdorff sense to an Alexandrov space X *collapses* if dim $X < n$.

As the previous examples indicate, the local geometry and topology of an Alexandrov space may be vastly different from that of a manifold. Nevertheless, there are certain similarities with the manifolds case. It is known that the Hausdorff dimension of an Alexandrov space is either a non-negative integer or infinite. In fact, if the Hausdorff dimension is finite, then, as in the manifold setting, the Hausdorff dimension coincides with the topological dimension [5, Cor. 6.5]. For simplicity we focus on finite-dimensional spaces below.

In the smooth category, the local structure of the space is completely determined by the infinitesimal picture, in the sense that, for every point in a smooth manifold there exists a neighborhood diffeomorphic to the tangent space at the said point. A similar relationship is available in Alexandrov geometry via the *space of directions*. In order to recall this concept, we outline some definitions.

Firstly, let us recall that the monotonicity of angles condition implies the well possessedness of angles between two geodesics which share a starting point. Let X be an Alexandrov space of curv $\geq k$, and assume that $\gamma_1, \gamma_2 : [a, b] \to X$ are two geodesics with $p := \gamma_1(0) = \gamma_2(0)$. As the function $\theta_k(s, t)$ of Equation (13.1) is monotone non-increasing and takes values in $[0, \pi]$, one can define the *angle between γ_1 and γ_2* by

$$\angle(\gamma_1, \gamma_2) := \lim_{s,t} \theta_k(s, t).$$

It is worth noting that the angle is, in fact, independent of k. An equivalence relation between geodesics emanating from the same point is then obtained: Two such geodesics are equivalent if they make a null angle. A *geodesic direction at $p \in X$* is an equivalence class of geodesics having p as a starting point. The collection of all geodesic directions at a point p

has the structure of a (possibly incomplete) metric space when equipped with the angle as metric. The completion of the space of geodesic directions at p is the *space of directions of X at p* and is denoted by $\Sigma_p X$ (or simply by Σ_p).

To obtain a tangent space at $p \in X$ there are at least two natural procedures one can use:

(i) Consider the cone over Σ_p.
(ii) Consider a *blow-up of X at p*, i.e. the pointed Gromov–Hausdorff limit of balls $B(p, r_i)$ with the restricted metric rescaled by a factor of $1/r_i$, where $r_i \to 0$. Such a limit exists and is independent of the choice of sequence.

These two methods give rise to isometric metric spaces denoted by $T_p X$ (see [4, Thm. 10.9.3]), directly implying the following structural properties of Σ_p, [4, Cor. 10.9.6].

Theorem 13.2.3   *Let X be an n-dimensional Alexandrov space and $p \in X$. Then the following hold:*

(1) Σ_p *is a compact $(n-1)$-dimensional Alexandrov space.*
(2) *If $n \geq 2$, then $\mathrm{curv}(\Sigma_p) \geq 1$.*
(3) *If $n = 1$, then Σ_p either consists of two points or a single point.*

The local topology of X at a point p is determined by $T_p X$, a fact that is asserted in the following result of Perelman [22].

Theorem 13.2.4 (Conical neighborhood theorem)   *Let X be an Alexandrov space and $p \in X$. Then, any sufficiently small neighborhood of p is pointed-homeomorphic to $T_p X$.*

Being one of the most powerful tools in Alexandrov geometry, the previous result allows one, among several other applications, to define inductively the *boundary* of an Alexandrov space. One-dimensional spaces are topological manifolds. Hence the boundary of such a space is defined in the usual manner. Assuming that the boundary of $(n-1)$-dimensional spaces has been defined, one says that a point p in an n-dimensional space is in the boundary if Σ_p has non-empty boundary. The boundary ∂X of an Alexandrov space X is a closed subset of Hausdorff codimension 1.

Once one has a well-posed concept of boundary, one can construct more examples of Alexandrov spaces from pairs of them by gluing along the boundaries:

- *Gluing along the boundary.* Let X_1 and X_2 be Alexandrov spaces of curv $\geq k$ with non-empty boundaries such that ∂X_1 is isometric to ∂X_2 when considered with the induced metrics. Let $f : \partial X_1 \to \partial X_2$ be an isometry. Then the adjunct space $X_1 \cup_f X_2$ is an Alexandrov space of curv $\geq k$ [24, Thm. 2.1]. It is possible to glue along more general subsets known as *extremal subsets* in some circumstances [17].

The conical neighborhood theorem suggests the following terminology. A point p on an n-dimensional Alexandrov space X is said to be *topologically regular* if Σ_p is homeomorphic to a sphere $\mathbb{S}^{n-1}$. Otherwise, p is said to be *topologically singular*. Furthermore, p is *metrically regular* if Σ_p is isometric to the unit round sphere $\mathbb{S}^{n-1}$ and *metrically singular* otherwise. In contrast to Riemannian manifolds, an Alexandrov space can be topologically regular (that is, each of its points is topologically regular) but have metrically singular points, (see, for example, [25, Exam. 97]). However, the subset of topologically singular points of X is dimensionally not very large. The codimension of the subset of topologically singular points which are not boundary points is at least 3. This is a consequence of the fact that Alexandrov spaces have a canonical stratification by topological manifolds (see [4, Thm. 10.10.1], [22, Thm. III structure theorem]). This fact will play an important role in the following section.

13.3 Three-Dimensional Alexandrov Spaces

As previously seen, Alexandrov spaces are generalizations of Riemannian manifolds, and, as such, it is interesting to study their topology. In this section we will focus on the three-dimensional case, with the intention of providing the necessary background for the main results in this survey.

It should be observed that Alexandrov spaces of dimensions one or two are respectively topological curves or surfaces. This was already proven in the original [5] (see also [4]), but it is also a consequence of Perelman's conical neighborhood theorem (see Theorem 13.2.4) and the classification of Alexandrov spaces of positive curvature in dimensions zero and one (that are, respectively, one or two points, and circles of length less than 2π or an interval of length less than or equal to π). Therefore, the first dimension where interesting new phenomena occur is dimension 3.

To understand the difference between genuine 3-dimensional Alexandrov spaces and 3-manifolds, it is helpful to think again of Perelman's conical neighborhood theorem; the spaces of directions that can appear in a 3-dimensional Alexandrov space will be compact Alexandrov surfaces of positive curvature. The list of these is also found in [5], where it was proven that, ignoring those with non-empty boundary, they are topologically 2-spheres or projective planes. The former will give rise to manifold points, while the latter will correspond to singular points. We collect these observations in the following statement. As is customary, we will say that a compact space without boundary is *closed*.

Lemma 13.3.1 *Let X be a closed 3-dimensional Alexandrov space. If X is not homeomorphic to a closed topological 3-manifold, then there is an even number of points $p_1, \ldots, p_k$ in X such that X is homeomorphic to the union of k disjoint cones over $\mathbb{R}P^2$s, and a non-orientable 3-dimensional compact manifold Y with a boundary formed by k connected components equal to $\mathbb{R}P^2$.*

Proof Observe that if a point $p \in X$ has $\mathbb{R}P^2$ as its space of directions, then, by Perelman's conical neighborhood theorem, there is a neighborhood U of p such that any other point in $U \setminus \{p\}$ has $\mathbb{S}^2$ as its space of directions. Thus, singular points in X are isolated and, since X is compact, there can only be a finite number of such p's.

Denote by $p_1, \ldots, p_k$ the singular points in X. For each $1 \le i \le k$, choose a conical neighborhood of p_i, U_i, such that the collection $\{U_i\}$ consists of pairwise disjoint sets. Then $Y := X \setminus \cup U_i$ is a compact 3-manifold with boundary equal to a disjoint union of k copies of $\mathbb{R}P^2$. Since each boundary component of Y has a collared neighborhood, Y contains two-sided $\mathbb{R}P^2$'s, and is therefore non-orientable. A simple application of Lefschetz duality shows that k is an even number (see, for instance, [11, Exer. 28.25]). $\qquad\square$

We can improve upon the preceding description by considering the orientable double cover of Y, as we illustrate in the following proposition.

Proposition 13.3.2 *Let X be a closed 3-dimensional Alexandrov space with singular points. Then there is an orientable closed 3-dimensional manifold M and an orientation-reversing involution $a : M \to M$ with a finite number of isolated fixed points such that X is homeomorphic to the quotient space M/a.*

Proof Let Y be the compact 3-manifold obtained above by removing disjoint open neighborhoods of the singular points of X. Since Y has some boundary components homeomorphic to the projective plane, Y contains two-sided $\mathbb{R}P^2$'s, and is therefore non-orientable. Denote by $\bar{Y}$ its orientable 2-fold cover, and its covering map by $\bar{a} : \bar{Y} \to \bar{Y}$; since $\bar{Y}$ is orientable but Y is not, $\bar{a}$ is orientation reversing. Observe also that $\bar{Y}$ has the same number of boundary components as Y, but containing only $\mathbb{S}^2$'s. Capping out these boundary components by 3-disks, D^3, we get a closed orientable 3-manifold M. The involution $\bar{a} : \bar{Y} \to \bar{Y}$ can be extended to the whole M by identifying each disk D with the Euclidean 3-ball B, and using the involution mapping each point $x \mapsto -x$ in B. It is clear that the extended involution $a : M \to M$ will have only fixed points corresponding to the center of the disks, and will therefore be isolated. □

The above topological description of 3-dimensional Alexandrov spaces is quite useful, since it allows us to switch from the category of metric spaces to that of 3-manifolds, where a lot of information is available. Note that singular Alexandrov 3-spaces are homeomorphic to non-orientable 3-dimensional orbifolds.

The next natural step is to return to the metric category, and consider whether lifting the metric from X to M produces something of interest. This is contemplated in the following lemma due to Grove and Wilking [13, Sect. 5].

Lemma 13.3.3   *Let X be a closed three-dimensional Alexandrov space with curvature bounded below by k, with $k \geq 0$, and assume that X is not a topological manifold. If M is the orientable double branched cover of X in Lemma 13.3.2, then the following hold:*

(1) *The metric in X can be lifted to M so that M is an Alexandrov space with curvature bounded below by k.*

(2) *The involution $a : M \to M$ is an isometry.*

The proof of this lemma appears in [13], although as it also includes the case of Alexandrov spaces of dimension 4, it is at times brief; for a more detailed version, entirely adapted to dimension three, the reader can refer to [6].

13.3.1 Geometric 3-Alexandrov Spaces

Geometric 3-manifolds can be considered the building blocks of arbitrary 3-dimensional closed manifolds, as Thurston's geometrization conjecture shows. It is then natural to ask about the corresponding notion for Alexandrov spaces. Recall that the eight Thurston geometries are $\mathbb{S}^3$, $\mathbb{E}^3$, $\mathbb{H}^3$, $\mathbb{S}^2 \times \mathbb{R}$, $\mathbb{H} \times \mathbb{R}$, Nil, Sol and $\widetilde{SL_2(\mathbb{R})}$ (see [26]).

Definition 13.3.4 We say that an Alexandrov space X^3 has a given Thurston geometry (see [26]) if X^3 can be written as a quotient of the corresponding geometry by some cocompact lattice. In that case, we will say that such an Alexandrov 3-space is *geometric*.

The main difference with the manifold case is that we allow for fixed points in the lattice action.

13.3.2 Geometrization of 3-Alexandrov Spaces

Recall that the usual geometrization of closed 3-manifolds requires the manifold to be divided into pieces: first one takes the decomposition into prime manifolds using 2-spheres to subdivide, and later one performs a Jaco–Shalen–Johannson decomposition using 2-tori. In our case, since Alexandrov 3-spaces with singular points contain a non-orientable core, we will require more subdividing surfaces.

Definition 13.3.5 We say that a closed three-dimensional Alexandrov space X admits a geometric decomposition if there exists a collection of spheres, projective planes, tori and Klein bottles that decompose X into geometric pieces.

Theorem 13.3.6 (Geometrization of 3-dimensional Alexandrov spaces) *A closed three-dimensional Alexandrov space admits a geometric decomposition into geometric three-dimensional Alexandrov spaces.*

The proof of the above result can be found in [8]. For an overview of further results on three-dimensional Alexandrov spaces, we refer the reader to [7].

13.4 Collapsed Three-Dimensional Alexandrov Spaces

Let $\{X_i\}_{i=1}^{\infty}$ be a sequence of n-dimensional Alexandrov spaces with diameters uniformly bounded above by $D > 0$ and curv $\geq k$ for some $k \in \mathbb{R}$. After passing to a subsequence, Gromov's precompactness theorem implies that there exists an Alexandrov space Y with diam$Y \leq D$ and curv$Y \geq k$ such that $X_i \xrightarrow{\mathrm{GH}} Y$. As in the Riemannian case, the sequence X_i is said to *collapse* to Y if $\dim Y < n$. We will also say that an n-dimensional Alexandrov space X *collapses* (or that it is a *collapsing Alexandrov space*) if there exists a sequence of Alexandrov metrics $\{d_i\}_{i=1}^{\infty}$ on X, such that $\{(X, d_i)\}_{i=1}^{\infty}$ is a collapsing sequence. In this section, the last in this survey, we give an overview of the available structure and classification results for collapsed Alexandrov 3-spaces.

13.4.1 General Structure Results

In Riemannian geometry, collapse imposes strong geometric and topological restrictions on the spaces on which it occurs. Indeed, Shioya and Yamaguchi obtained comprehensive structure results for closed, collapsed three-dimensional Riemannian 3-manifolds [27]. In the Alexandrov category, Mitsuishi and Yamaguchi carried out an exhaustive study of collapsed, closed, three-dimensional Alexandrov spaces, and we summarize their results in this section. The main difference between the collapse of three-dimensional Alexandrov spaces and that of three-dimensional Riemannian manifolds resides in the fact that in the Alexandrov case collapse can occur along the fibers of a "generalized" Seifert fibration. Collapsed Alexandrov 3-spaces can be described as unions of certain pieces. Before stating the general structure results, let us give a brief account of those pieces where topological singularities arise.

The space B(pt) Let $D^2 \times \mathbb{S}^1 \subset \mathbb{R}^2 \times \mathbb{C}$ be equipped with the usual flat product metric. An isometric involution α on $D^2 \times \mathbb{S}^1$ is defined by

$$\alpha((x, y), e^{i\theta}) := ((-x, -y), e^{-i\theta}).$$

The space B(pt) $:= D^2 \times \mathbb{S}^1/\alpha$ is an Alexandrov space of curv ≥ 0 with two topologically singular points corresponding to the image in the quotient of the points $((0, 0), e^{i0})$ and $((0, 0), e^{i\pi})$, which are fixed by α (see [18, Exam. 1.2]). There is a projection $p : \mathrm{B(pt)} \to K_1(\mathbb{S}^1)$ sending an interval joining the topologically singular points to the vertex o of the cone. To describe it, observe that the quotient of $D^2 \subset \mathbb{R}^2$ by the involution $(x, y) \mapsto (-x, -y)$ is homeomorphic to D^2, and metrically is isometric to $K_1(\mathbb{S}^1)$, where the $\mathbb{S}^1$ taken has length π. The projection $p : \mathrm{B(pt)} \to K_1(\mathbb{S}^1)$ is then obtained by mapping

$$[(x, y), e^{i\theta}] \to [(x, y)].$$

This projection is a fibration on $K_1(\mathbb{S}^1) \setminus \{o\}$.

The space $\mathrm{B(pt)}$ can also be described as follows (see lines after [18, Exam. 2.60]): take two cones over $\mathbb{R}P^2$, select a disk D_i^2, $i = 0, 1$, on each $\mathbb{R}P^2$-boundary, and glue both cones using some homeomorphism $\varphi : D_0^2 \to D_1^2$. The resulting space does not depend on the gluing homeomorphism φ, and is homeomorphic to $\mathrm{B(pt)}$. It is clear that its boundary is obtained by taking two Möbius bands glued by their boundaries, i.e. the boundary of $\mathrm{B(pt)}$ is a Klein bottle.

Spaces with 2-dimensional souls We now describe three different closed Alexandrov 3-spaces as quotients of certain involutions:

(i) $\mathrm{B}(S_2) := \mathbb{S}^2 \times [-1, 1]/(\sigma, -\mathrm{id})$, where $\mathbb{S}^2$ is a sphere of non-negative curvature in the Alexandrov sense with an isometric involution σ of $\mathbb{S}^2$ topologically conjugate to the involution on the 2-sphere given by the suspension of the antipodal map on the circle. The resulting space is homeomorphic to $\mathrm{Susp}(\mathbb{R}P^2) \setminus \mathrm{int}(D^3)$, where $D^3 \subset \mathrm{Susp}(\mathbb{R}P^2)$ is a closed 3-ball consisting of topologically regular points (see [18, Rem. 2.62]).

(ii) $\mathrm{B}(S_4) := T^2 \times [-1, 1]/(\sigma, -\mathrm{id})$, where T^2 is a flat torus and the involution $\sigma : T^2 \to T^2$ maps (z_1, z_2) to $(\bar{z}_1, \bar{z}_2)$ (observe that $\mathbb{T}^2/\sigma$ is homeomorphic to $\mathbb{S}^2$). This space has four topologically singular points, corresponding to the four fixed points of the involution; this can be seen by observing that at each such point, the differential of the involution acts as the antipodal map on the unit tangent sphere. Its oriented branched cover is $\mathbb{T}^2 \times [-1, 1]$.

(iii) $\mathrm{B}(\mathbb{R}P^2) := K^2 \times [-1, 1]/(\sigma, -\mathrm{id})$, where K^2 is a flat Klein bottle and $\sigma : K^2 \to K^2$ is an isometric involution topologically conjugate to the unique involution on K^2 whose quotient is $\mathbb{R}P^2$.

Generalized Seifert fiber spaces A *generalized Seifert fibration* of a topological 3-orbifold M over a topological 2-orbifold B (both possibly with boundaries) is a map $f : M \to B$ whose fibers are homeomorphic to circles or bounded closed intervals. It is required that, for every $x \in B$, there is a neighborhood U_x homeomorphic to a 2-disk such that

(i) if $f^{-1}(x)$ is homeomorphic to a circle, then there is a fiber-preserving homeomorphism of $f^{-1}(U_x)$ to a Seifert fibered solid torus in the usual sense, and

(ii) if $f^{-1}(x)$ is homeomorphic to an interval, then there exists a fiber-preserving homeomorphism of $f^{-1}(U_x)$ to the space B(pt), with respect to the fibration $(B(\mathrm{pt}), p^{-1}(o)) \to (K_1(\mathbb{S}^1), o)$.

Furthermore, for any compact component C of ∂B there is a collar neighborhood N of C in B such that $f|_{f^{-1}(N)}$ is a usual circle bundle over N. We say that M is a *generalized Seifert fibered space* and we use the notation $M = \mathrm{Seif}(B)$.

Generalized solid tori and Klein bottles A *generalized solid torus* (respectively, *generalized solid Klein bottle*) is a topological 3-orbifold Y with boundary homeomorphic to a torus (respectively, a Klein bottle). It admits a map $Y \to \mathbb{S}^1$ such that the fibers are homeomorphic to either a 2-disk or a Möbius band, and the fiber type can only change at a finite number of *corner points* in $\mathbb{S}^1$. We refer the reader to [18, Def. 1.4] for the precise definitions.

I-bundles over the Klein bottle These are obtained as disk bundles of certain line bundles over the Klein bottle K^2. They are easily described as quotients of $\mathbb{R}^3$ under certain isometric actions. Except for the trivial bundle $K^2 \times I$, the rest are as follows.

(i) $K^2 \widetilde{\times} I$: this is the disk bundle in the orientable 3-manifold obtained as the quotient of $\mathbb{R}^3$ under the group generated by

$$(x, y, z) \overset{\widetilde{\tau}}{\to} (x + 2, y, z), \qquad (x, y, z) \overset{\widetilde{\sigma}}{\to} (-x, y + 1, -z).$$

Its boundary is given by a 2-torus.

(ii) $K^2 \widehat{\times} I$: this is the disk bundle in the non-orientable 3-manifold obtained as the quotient of $\mathbb{R}^3$ under the group generated by

$$(x, y, z) \overset{\widehat{\tau}}{\to} (x + 1, y, -z), \qquad (x, y, z) \overset{\widehat{\sigma}}{\to} (-x, y + 1, -z).$$

Its boundary is given by a Klein bottle.

The identity map in $\mathbb{R}^3$ induces a two-fold Riemannian covering map $\pi : K^2 \widetilde{\times} I \to K^2 \widehat{\times} I$. At the fundamental group level, π is an injective homomorphism that sends $\widetilde{\tau} \to \widehat{\tau}^2$ and $\widetilde{\sigma} \to \widehat{\sigma}$. Furthermore, since the fundamental group of K^2 is the dihedral group, and this group contains a unique subgroup of index 2, it follows that $K^2 \tilde{\times} I$ is the unique two-fold cover of $K^2 \hat{\times} I$.

With these pieces now in hand, we are ready to state the topological classification and structure theorems for closed, collapsed Alexandrov 3-spaces obtained by Mitsuishi and Yamaguchi in [18]. These results are obtained via a thorough analysis of the local structure of the limit spaces and we refer the reader to [18] for more details. We divide the presentation according to the dimension of the limit space, starting with the case where it is two dimensional. We always assume that our spaces are connected.

13.4.1.1 Collapse to Dimension Two

In this case, the limit space of the collapsing sequence is a compact two-dimensional Alexandrov space, possibly with boundary. Hence, the limit space is topologically a surface.

Theorem 13.4.1 (Collapse to a compact surface without boundary) *Let $\{X_i\}_{i=1}^\infty$ be a sequence of closed, three-dimensional Alexandrov spaces with* $\operatorname{curv} X_i \geq -1$ *and* $\operatorname{diam} X_i \leq D$. *If X_i GH-converges to a two-dimensional Alexandrov space X^* without boundary, then, for sufficiently large i, X_i is homeomorphic to a generalized Seifert space over X^*.*

In the preceding theorem, singular fibers may occur over *essential singular points* in X^*, i.e. over points whose space of directions has radius at most $\pi/2$.

Theorem 13.4.2 (Collapse to a compact surface with boundary) *Let $\{X_i\}_{i=1}^\infty$ be a sequence of closed, three-dimensional Alexandrov spaces with* $\operatorname{curv} X_i \geq -1$ *and* $\operatorname{diam} X_i \leq D$. *If X_i GH-converges to a two-dimensional Alexandrov space X^* with non-empty boundary, then, for sufficiently large i, there exist a generalized Seifert fiber space $\operatorname{Seif}_i(X^*)$ over X^* and generalized solid tori or generalized Klein bottles $\pi_{i,k}:$ $Y_{i,k} \to (\partial X^*)_k$ over each component $(\partial X^*)_k$ of ∂X^* such that X_i is homeomorphic to the union of $\operatorname{Seif}_i(X^*)$ and the $Y_{i,k}$, glued along their boundaries, where the fibers of $\operatorname{Seif}_i(X^*)$ over boundary points $x \in (\partial X^*)$ are identified with $\partial \pi_{i,k}^{-1}(x) \approx \mathbb{S}^1$.*

13.4.1.2 Collapse to Dimension One

In this case, the limit space of the collapsing sequence is a compact one-dimensional Alexandrov space, possibly with boundary. Hence, the limit space is topologically a circle or a compact interval.

Theorem 13.4.3 (Collapse to a circle)   *Let* $\{X_i\}_{i=1}^{\infty}$ *be a sequence of closed, three-dimensional Alexandrov spaces with* $\mathrm{curv} X_i \geq -1$ *and* $\mathrm{diam} X_i \leq D$. *If* X_i *GH-converges to a circle, then, for* i *sufficiently large,* X_i *is homeomorphic to the total space of fiber bundle over* $\mathbb{S}^1$ *with fiber homeomorphic to one of* $\mathbb{S}^2$, $\mathbb{R}P^2$, T^2 *or the Klein bottle* K^2. *In particular,* X_i *is a topological manifold.*

Theorem 13.4.4 (Collapse to a compact interval)   *Let* $\{X_i\}_{i=1}^{\infty}$ *be a sequence of closed, three-dimensional Alexandrov spaces with* $\mathrm{curv} X_i \geq -1$ *and* $\mathrm{diam} X_i \leq D$. *If* X_i *GH-converges to an interval* $I \approx [-1, 1]$, *then, for* i *sufficiently large,* X_i *is homeomorphic to a union* $B_i^- \cup B_i^+$ *of two spaces* $B_i^{\pm}$ *glued along their boundary* $\partial B_i^- = \partial B_i^+$. *The boundary* $\partial B_i^{\pm}$ *is homeomorphic to one of* $\mathbb{S}^2$, $\mathbb{R}P^2$, T^2 *or the Klein bottle* K^2. *The topology of the spaces* $B_i^{\pm}$ *is determined as follows:*

(1) *If* $\partial B_i^{\pm} \approx \mathbb{S}^2$, *then* $B_i^{\pm}$ *is homeomorphic to one of* D^3, $\mathbb{R}P^3 - \mathrm{int} D^3$ *or* $\mathrm{B}(S_2)$ *with* $S_2 \approx \mathbb{S}^2$.
(2) *If* $\partial B_i^{\pm} \approx \mathbb{R}P^2$, *then* $B_i^{\pm}$ *is homeomorphic to* $K_1(P^2)$.
(3) *If* $\partial B_i^{\pm} \approx T^2$, *then* $B_i^{\pm}$ *is homeomorphic to one of* $\mathbb{S}^1 \times D^2$, $\mathbb{S}^1 \times \mathrm{Mb}$, $K^2 \tilde{\times} I$ *or* $\mathrm{B}(S_4)$.
(4) *If* $\partial B_i^{\pm} \approx K^2$, *then* $B_i^{\pm}$ *is homeomorphic to one of* $\mathbb{S}^1 \tilde{\times} D^2$, $K^2 \tilde{\times} I$, $\mathrm{B}(\mathrm{pt})$, *or* $\mathrm{B}(S_2)$ *with* $S_2 \approx \mathbb{R}P^2$.

13.4.1.3 Collapse to a Point

The last case to consider is collapse to a zero-dimensional space, i.e. to a point.

Theorem 13.4.5 (Collapse to a point)   *Let* $\{X_i\}_{i=1}^{\infty}$ *be a sequence of closed, three-dimensional Alexandrov spaces with* $\mathrm{curv} X_i \geq -1$ *and* $\mathrm{diam} X_i \leq D$. *If* X_i *GH-converges to a point, then, for* i *sufficiently large,* X_i *is homeomorphic to some space among the following:*

(1) *generalized Seifert fiber spaces as in the conclusion of Theorem 13.4.1 with base an Alexandrov surface with non-negative curvature;*
(2) *spaces in the conclusion of Theorem 13.4.2 with base an Alexandrov surface with non-negative curvature;*
(3) *spaces in the conclusion of Theorems 13.4.3 and 13.4.4;*
(4) *closed Alexandrov three-dimensional spaces with non-negative curvature having finite fundamental group.*

By the work in [8], a manifold in item (4) in Theorem 13.4.5 is homeomorphic to a three-dimensional spherical space form or to one of $\mathrm{Susp}(\mathbb{R}P^2)$, $\mathrm{Susp}(\mathbb{R}P^2)\#\mathrm{Susp}(\mathbb{R}P^2)$ or $\mathbb{R}P^3\#\mathrm{Susp}(\mathbb{R}P^2)$.

13.4.2 Geometrization of Sufficiently Collapsed Three-Dimensional Alexandrov Spaces

We conclude this chapter with a brief discussion of the geometrization of sufficiently collapsed closed Alexandrov 3-spaces. We refer the reader to [9] for further details.

Recall the eight Thurston geometries: $\mathbb{S}^3$, $\mathbb{R}^3$, $\mathbb{H}^3$, $\mathbb{S}^2 \times \mathbb{R}$, $\mathbb{H}^2 \times \mathbb{R}$, $\widetilde{SL_2(\mathbb{R})}$, Nil and Sol. As stated in Section 13.3, a closed (i.e. compact and without boundary) 3-manifold is *geometric* if it admits a geometric structure modeled on one of these geometries. In this context, Shioya and Yamaguchi [28] obtained a geometrization result for sufficiently collapsed Riemannian 3-manifolds. More precisely, they showed that, for any $D > 0$, there exists a constant $\varepsilon = \varepsilon(D) > 0$ such that if a closed, prime 3-manifold admits a Riemannian metric with diameter at most D and sectional curvature bounded below by -1 with volume $< \varepsilon$, then it admits a geometric structure modeled on one of the seven geometries $\mathbb{S}^3$, $\mathbb{R}^3$, $\mathbb{S}^2 \times \mathbb{R}$, $\mathbb{H}^2 \times \mathbb{R}$, $\widetilde{SL_2(\mathbb{R})}$, Nil and Sol (see [28, Cor. 0.9]).

Recall that a non-trivial closed 3-manifold M is *prime* if it cannot be presented as a connected sum of two non-trivial closed 3-manifolds. A closed 3-manifold is *irreducible* if every embedded 2-sphere bounds a 3-ball. It is known that, with the exception of manifolds homeomorphic to $\mathbb{S}^3$, $\mathbb{S}^1 \times \mathbb{S}^2$ or $\mathbb{S}^1 \widetilde{\times} \mathbb{S}^2$ (the non-trivial 2-sphere bundle over $\mathbb{S}^1$), a closed 3-manifold is prime if and only if it is irreducible (see [14, Lem. 3.13]). Since $\mathbb{S}^1 \times \mathbb{S}^2$ and $\mathbb{S}^1 \widetilde{\times} \mathbb{S}^2$ are geometric, one can think of the geometrization of sufficiently collapsed prime Riemannian 3-manifolds as a result pertaining to irreducible 3-manifolds. Therefore, in seeking a generalization to Alexandrov spaces, one may focus on the irreducible case. This leads to the following definition of *irreducibility* for this more general class of spaces.

Definition 13.4.6 Let X be a closed Alexandrov 3-space. We say that X is *irreducible* if every embedded 2-sphere in X bounds a 3-ball, and, in the case that the set of topologically singular points of X is non-empty, it is further required that every 2-sided $\mathbb{R}P^2$ bounds a $K(\mathbb{R}P^2)$, a cone over $\mathbb{R}P^2$.

With this definition in hand, we may now state the geometrization of sufficiently collapsed Alexandrov spaces.

Theorem 13.4.7 ([9, Thm. A]) *For any $D > 0$ there exists $\varepsilon = \varepsilon(D) > 0$ such that, if X is a closed, irreducible Alexandrov 3-space with* curv ≥ -1, diam$X \leq D$, *and* vol$X < \varepsilon$, *then X admits a geometric structure modeled on one of the seven geometries $\mathbb{R}^3$, $\mathbb{S}^3$, $\mathbb{S}^2 \times \mathbb{R}$, $\mathbb{H}^2 \times \mathbb{R}$, $\widetilde{SL_2(\mathbb{R})}$,* Nil *and* Sol.

Here, the *volume* of an Alexandrov 3-space is its 3-dimensional Hausdorff measure, normalized so that the volume of 3-dimensional Riemannian manifolds agrees with the usual Riemannian volume. As in the Riemannian case, one can rule out hyperbolic geometry by combining the fact that the simplicial volume of a collapsing Alexandrov space is zero (see [19, Cor. 1.7]) with the fact that the simplicial volume of a hyperbolic manifold must be bounded below by the Riemannian volume (see [30, Thm. 6.2]).

Theorem 13.4.7 is proven by carefully studying the metric and topological structure of collapsed irreducible Alexandrov 3-spaces and their orientable double branched covers (in the case where the space is not a manifold). Combining Theorems 13.4.1–13.4.5 with the irreducibility hypothesis, one obtains fairly explicit topological descriptions of closed, collapsed, irreducible Alexandrov 3-spaces, exhibiting them as geometric 3-manifolds or their quotients by orientation-reversing involutions with only isolated fixed points. Classification results of Alexandrov spaces with (local) circle actions (see [10, 20]) and classical results on involutions on 3-manifolds (see [15, 16]) also play an important role in the proof. We refer the reader to [9] for precise details, as the proof is based on a case-by-case analysis and is of a rather technical nature.

References

[1] S. Alexander, V. Kapovitch, and A. Petrunin, *Alexandrov geometry: preliminary version no. 1*, preprint arXiv:1903.08539 [math.DG].

[2] S. Alexander and R. Bishop, *Curvature bounds for warped products of metric spaces*, Geom. Funct. Anal. 14 (2004), no. 6, 1143–1181.

[3] S. Alexander and R. Bishop, *Warped products admitting a curvature bound*, Adv. Math. 303 (2016), 88–122.

[4] D. Burago, Y. Burago, and S. Ivanov, *A course in metric geometry*, Graduate Studies in Mathematics, vol. 33, Amer. Math. Soc., Providence, RI, 2001.

[5] Y. Burago, M. Gromov, and G. Perelman, *A. D. Aleksandrov spaces with curvatures bounded below* (in Russian), Uspekhi Mat. Nauk 47 (1992), no. 2(284), 3–51, 222; translation in Russ. Math. Surv. 47 (1992), no. 2, 1–58.

[6] Q. Deng, F. Galaz-García, L. Guijarro, and M. Munn, *Three-dimensional Alexandrov spaces with positive or nonnegative Ricci curvature*, Pot. Anal. 48 (2018), no. 2, 223–238.

[7] F. Galaz-García, *A glance at three-dimensional Alexandrov spaces*, Front. Math. China 11 (2016), no. 5, 1189–1206.

[8] F. Galaz-García and L. Guijarro, *On three-dimensional Alexandrov spaces*, Int. Math. Res. Not. (2015), no. 14, 5560–5576.

[9] F. Galaz-García, L. Guijarro, and J. Núñez-Zimbrón, *Sufficiently collapsed irreducible Alexandrov 3-spaces are geometric*, Indiana Univ. Math. J. 69 (2020), 977–1005.

[10] F. Galaz-García and J. Núñez-Zimbrón, *Three-dimensional Alexandrov spaces with local isometric circle actions*, Kyoto J. Math. Advance publication (2020), 23 pages. doi:10.1215/21562261-2019-0047.

[11] M. J. Greenberg and J. R. Harper, *Algebraic topology. A first course*, Mathematics Lecture Note Series, vol. 58, Benjamin Cummings Publishing Co., Inc., Advanced Book Program, Reading, MA, 1981.

[12] K. Grove, *Finiteness theorems in Riemannian geometry*, Explorations in complex and Riemannian geometry, Contemporary Mathematics, vol. 332, Amer. Math. Soc., Providence, RI, 2003, pp. 101–120.

[13] K. Grove and B. Wilking, *A knot characterization and 1-connected non-negatively curved 4-manifolds with circle symmetry*, Geom. Topol. 18 (2014), no. 5, 3091–3110.

[14] J. Hempel, *3-manifolds*, AMS Chelsea Publishing, Providence, RI, 2004. Reprint of the 1976 original.

[15] P. K. Kim and J. L. Tollefson, *Splitting the PL involutions of nonprime 3-manifolds*, Michigan Math. J. 27 (1980), no. 3, 259–274.

[16] K. W. Kwun, *Scarcity of orientation-reversing PL involutions of lens spaces*, Michigan Math. J. 17 (1970), 355–358.

[17] A. Mitsuishi, *Self and partial gluing theorems for Alexandrov spaces with a lower curvature bound*, preprint arXiv:1606.02578.

[18] A. Mitsuishi and T. Yamaguchi, *Collapsing three-dimensional closed Alexandrov spaces with a lower curvature bound*, Trans. Amer. Math. Soc. 367 (2015), 2339–2410.

[19] A. Mitsuishi and T. Yamaguchi, *Locally Lipschitz contractibility of Alexandrov spaces and its applications*, Pacific J. Math. 270 (2014), no. 2, 393–421.

[20] J. Núñez-Zimbrón, *Closed three-dimensional Alexandrov spaces with isometric circle actions*, Tohoku Math. J. (2) 70 (2018), no. 2, 267–284.

[21] G. Perelman, *Alexandrov spaces with curvatures bounded from below II*, preprint (1991).

[22] G. Perelman, *Elements of Morse theory on Aleksandrov spaces* (in Russian), translated from Algebra i Analiz 5 (1993), no. 1, 232–241; translation in St. Petersburg Math. J. 5 (1994), no. 1, 205–213.

[23] P. Petersen, *Riemannian geometry*, Third edition. Graduate Texts in Mathematics, vol. 171. Springer, Cham, 2016.

[24] A. Petrunin, *Applications of quasigeodesics and gradient curves*, In K. Grove and P. Petersen (eds.), Comparison geometry (Berkeley, CA, 1993–94), Math. Sci. Res. Inst. Publ. vol. 30, Cambridge University Press, Cambridge, 1997, pp. 203–219.

[25] C. Plaut, *Metric spaces of curvature $\geq k$*. In R. J. Daverman and R. B. Sher (eds.), Handbook of geometric topology, North-Holland, Amsterdam, 2002, pp. 819–898.

[26] P. Scott, *The geometries of 3-manifolds*, Bull. London Math. Soc. 15 (1983), no. 5, 401–487.

[27] K. Shiohama, *An introduction to the geometry of Alexandrov spaces*, Lecture Notes Series, vol. 8. Seoul National University, Research Institute of Mathematics, Global Analysis Research Center, Seoul, 1993.

[28] T. Shioya and T. Yamaguchi. *Collapsing three-manifolds under a lower curvature bound*, J. Differ. Geom. 56 (2000), no. 1, 1–66.

[29] T. Shioya and T. Yamaguchi, *Volume collapsed three-manifolds with a lower curvature bound*, Math. Ann. 333 (2005), no. 1, 131–155.

[30] W. P. Thurston, *The Geometry and topology of 3-manifolds*, Lecture notes, Princeton, 1978.

14

Pseudo-Angle Systems and the Simplicial Gauss–Bonnet–Chern Theorem

Stephan Klaus

Abstract

In a previous article, we proved a simplicial version of the Gauss–Bonnet–Chern theorem which expresses the Euler characteristics of a Euclidean simplicial complex K of any dimension as a sum of its Gauss curvatures over all vertices: $\sum_{x \in K_0} \kappa(x) = \chi(K)$, where the simplicial Gauss curvature $\kappa(x)$ is given as a logarithmic sum over the normed dihedral angle defects of all simplices which are adjacent to the vertex x. Here, we generalize this result to systems of pseudo-angles which are certain functions defined on the moduli space of Euclidean simplices. Then we consider combinatorial Riemannian manifolds and introduce simplicial sectional curvature. In the final section, we consider its relation to the simplicial Gauss pseudo-curvature. In particular, the hypothetical existence of a systems of pseudo-angles with certain properties would imply the Hopf conjecture for manifolds of positive sectional curvature.

14.1 Introduction

This chapter is the second part of our combinatorial version and generalization of the Gauss–Bonnet–Chern theorem in differential geometry:

$$\frac{1}{(2\pi)^n} \int_M Pf^n(\Omega) = \chi(M^{2n}).$$

Here χ denotes the Euler characteristic of a closed smooth manifold M^{2n} of even dimension, and $Pf^n(\Omega)$ denotes the Pfaffian of the curvature 2-form matrix of a Riemannian metric g on M. Note that this famous theorem only makes sense in even dimension, because the Euler characteristic of an odd-dimensional manifold vanishes by Poincaré duality and there is no Pfaffian, for a matrix of odd dimension. Note also that it is not necessary to assume orientability of M because the integral is stable under a local orientation change: the local orientation is also used to define the Pfaffian, and an orientation change flips the signs of the integration *and* of *Pf*.

In [7] we proved a simplicial version of the Gauss–Bonnet–Chern theorem for any Euclidean simplicial complex K (i.e., a simplicial complex with a compatible Euclidean metric on all simplices): $\sum_{x \in K_0} \kappa(x) = \chi(K)$. Here, K does not need to be a combinatorial manifold or of even dimension. We recall the necessary definitions in Section 14.2.

In Section 14.3, we generalize this result to systems of pseudo-angles Θ which are certain sets of universal functions defined on the moduli space $\mathcal{S}^n$ of all Euclidean n-simplices. We consider some examples for Θ and in Theorem 14.3.5 prove

$$\sum_{x \in K_0} \kappa^\Theta(x) = \chi(K),$$

where $\kappa^\Theta(x)$ denotes the simplicial Gauss pseudo-curvature associated to Θ (see Definition 14.3.4).

In Section 14.4 we introduce combinatorial Riemannian manifolds $M = (M, d, e)$. Their local models are flat Euclidean stars forming a set $\mathcal{T}^n$, and the local simplicial isomorphisms to the vertex stars of M have to respect *radial distances* in stars (i.e., distances from the center of the star to outer vertices). In contrast to this, *peripheral distances* (i.e., distances of two adjacent outer vertices) are not respected in general.

In Section 14.5 we define simplicial sectional curvature (see Definition 14.5.1)

$$sec : M_{2,0} \to \mathbb{R},$$

$$sec(x_0; x_i, x_j) := 6 \frac{|v_{0i} - v_{0j}|^2 - d_{ij}^2}{d_{0i}^2 d_{0j}^2 + 2\langle v_{0i}, v_{0j} \rangle^2}$$

as an analogue of sectional curvature in differential geometry, i.e. as a measure of how peripheral distances are changing with respect to a flat model. The formula follows from an approximation of the generalized law of cosines for a surface of constant curvature. Although this definition is quite obvious, it seems to be a new notion.

In Section 14.6 we consider the relation of simplicial sectional curvature sec to simplicial Gauss pseudo-curvature κ^Θ. We show that the hypothetical existence of a systems of pseudo-angles Θ with certain properties would imply the Hopf conjecture for manifolds of positive sectional curvature. Unfortunately, this existence problem is a complicated question concerning the combinatorial geometry of Euclidean stars of positive simplicial sectional curvature which is completely open.

Acknowledgements I would like to thank Ivan Izmestiev, Ruth Kellerhals, Wolfgang Kühnel and Wilderich Tuschmann for helpful discussions and comments on this topic. I would also like to thank the organizers of the Australian–German Workshop on Differential Geometry in the Large at the Matrix Institute on 4–15 February 2019 where I could give a talk on this subject, and the unknown referee for many valuable hints.

14.2 The Simplicial Gauss–Bonnet–Chern Theorem

We recall the necessary definitions and results from [7]. Let K be an abstract simplicial complex (always assumed to be finite), where we denote the set of r-simplices by K_r and its number by c_r. In particular, K_0 is the vertex set of K and K_1 is the set of edges. The Euler characteristic of K is given by $\chi(K) = c_0 - c_1 + c_2 - c_3 \pm \ldots$ and is a homotopy invariant of the geometric realization of K.

Definition 14.2.1 A *Euclidean simplicial complex* K is defined as a simplicial complex with a metric which is a Euclidean (flat) metric on every simplex. Such a metric is fixed by specifying all edge lengths, which form a function

$$d : K_1 \to \mathbb{R}_+.$$

Note that not every set d of given edge lengths can be realized by Euclidean simplices: e.g., in dimension 2, the triangle inequalities have to be satisfied. It is a classical result that the existence of a Euclidean m-simplex σ with prescribed edge lengths d_{ij}, $0 \le i < j \le m$, is equivalent to certain non-linear inequalities for the d_{ij}, which we call the *higher triangle inequalities*:

Definition 14.2.2 The *Cayley–Menger determinant* Γ of a system $d_{ij} > 0$, $0 \le i < j \le m$, is given by

$$\Gamma := \begin{vmatrix} 0 & 1 & 1 & 1 & .. & 1 \\ 1 & 0 & d_{01}^2 & d_{02}^2 & .. & d_{0m}^2 \\ 1 & d_{10}^2 & 0 & d_{12}^2 & .. & d_{1m}^2 \\ 1 & d_{20}^2 & d_{21}^2 & 0 & .. & d_{2m}^2 \\ ... & & & & & \\ 1 & d_{m0}^2 & d_{m1}^2 & d_{m2}^2 & .. & 0 \end{vmatrix},$$

and, for a sub-simplex $\tau \subset \sigma$, Γ_τ is defined as the sub-determinant given by restriction.

Theorem 14.2.3 ([3, Chap. 9.7]) *A system $d_{ij} > 0$, $0 \le i < j \le m$, can be realized by an Euclidean m-simplex σ if and only if*

$$(-1)^{dim(\tau)+1}\Gamma_\tau > 0$$

for every sub-simplex $\tau \subset \sigma$ (including $\tau = \sigma$). The realization is then unique up to isometry.

A pair (K, d) of an abstract simplicial complex with an edge length function statisfying this condition for all sub-simplices of K is called *realizable*. Hence, a Euclidean simplicial complex is the same object as a realizable simplicial complex (K, d).

Now, the flat metric allows us to define higher dihedral angles at a sub-simplex in a Euclidean simplex (see [6]):

Definition 14.2.4 Let $\sigma \subset \mathbb{R}^m$ be a Euclidean m-simplex and τ be a k-dimensional sub-simplex. Then the *normed dihedral angle at τ with respect to σ* is defined as

$$\theta(\tau \subset \sigma) := \frac{vol(D_\varepsilon^m \cap \sigma)}{vol(D_\varepsilon^m)},$$

where D_ε^m denotes a small disc (radius $\varepsilon \approx 0$) around an inner point x of τ and vol denotes the volume.

Note that $\theta(\tau \subset \sigma) \in \,]0, 1]$, $\theta(\sigma \subset \sigma) = 1$ and $\theta(\tau \subset \sigma) = \frac{1}{2}$ for all sub-simplices τ of co-dimension 1. In the formula, it is also possible to replace D_ε^m by its boundary S_ε^{m-1}. Moreover, we can restrict the disc or sphere to the orthogonal complement N of τ at x which is an affine space $N \subset \mathbb{R}^m$ of dimension $m - k$; this number is called the *co-dimension* of the dihedral angle and is a measure of its complexity. Then we also have

$$\theta(\tau \subset \sigma) = \frac{vol(D_\varepsilon^m \cap N \cap \sigma)}{vol(D_\varepsilon^m \cap N)} = \frac{vol(S_\varepsilon^{m-1} \cap N \cap \sigma)}{vol(S_\varepsilon^{m-1} \cap N)}.$$

Note that $S_\varepsilon^{m-1} \cap N$ is a small sphere of dimension $m - k - 1$; this number can be called the *effective (co-)dimension* of the dihedral angle. The first interesting cases of dihedral angles occur in co-dimension 2 and have effective dimension 1 (as a classical angle in plane geometry).

The following theorem of Poincaré can be considered as the *generalization of the angle sum in a triangle* (see [6, Chap. 1]):

Theorem 14.2.5 (Poincaré [6]) *Let σ be a Euclidean m-simplex. Then we have*

$$\sum_\tau (-1)^{dim(\tau)} \theta(\tau \subset \sigma) = 0, ;$$

where the sum is taken over all sub-simplices τ of σ.

The next two definitions are cited from the previous article [7].

Definition 14.2.6 For a Euclidean simplicial complex (K, d) and any simplex τ of K, we define the *normed dihedral angle defect around τ* by

$$\delta(\tau) := 1 - \sum_{\sigma \text{ max}} \theta(\tau \subset \sigma),$$

where the sum is over all *maximal* simplices σ of K which contain τ. Here, maximality is defined with respect to the inclusion relation of simplices. By definition, $\delta(\tau)$ is a real number and $\delta(\tau) = 0$ if τ is maximal.

Definition 14.2.7 We define the *simplicial Gauss curvature* at a vertex $x \in K_0$ as the logarithmic sum over the normed dihedral angle defects of all simplices which are adjacent to the vertex x:

$$\kappa(x) := \delta(x) - \frac{1}{2} \sum_{x \in \sigma^1} \delta(\sigma^1) + \frac{1}{3} \sum_{x \in \sigma^2} \delta(\sigma^2) - \frac{1}{4} \sum_{x \in \sigma^3} \delta(\sigma^3) \pm \ldots$$

$$= \sum_{k \geq 0} (-1)^k \frac{1}{k+1} \sum_{\sigma \in K_k, x \in \sigma} \delta(\sigma).$$

Now we can cite our main result from [7]:

Theorem 14.2.8 ([7], simplicial Gauss–Bonnet–Chern theorem) *For any Euclidean simplicial complex (K, d) we have*

$$\sum_\tau (-1)^{dim(\tau)} \delta(\tau) = \chi(K),$$

where the sum is over all simplices τ of K (simplex version), and

$$\sum_{x \in K_0} \kappa(x) = \chi(K),$$

i.e. the sum over all simplicial Gauss curvatures gives the Euler characteristic (vertex version).

14.3 Systems of Pseudo-Angles

Unfortunately, the functional dependence of the normed dihedral angles θ from the edge lengths d_{ij} of a simplex is very complicated. Their range of definition is a complicated semi-algebraic set given by the Cayley–Menger determinant conditions. In dimensions 2 and 3 it is possible to express θ by trigonometric functions and their inverse functions. As a consequence, co-dimension 2 and 3 dihedral angles can be computed explicitly. But already for the solid angle in dimension 4, there is a formula involving the dilogarithm [9]. To illustrate this, here is the general result of Aomoto (note that by Definition 14.2.4 the computation of θ can be reduced to the computation of the volume of *spherical simplices*, i.e. subsets of the form $S^{m-1} \cap \sigma$ in the unit sphere S^{m-1}):

Theorem 14.3.1   (Aomoto [1]; see also the simplified formula in [10] and [2]) *Let ω be the volume of the d-dimensional spherical simplex which is given by the $\binom{d+1}{2}$ (not normed) dihedral angles $\alpha_{i,j}$ with $0 \leq i < j \leq d$. Then ω is given by the hypergeometric series*

$$\omega = C \sum_{m \in \mathbb{N}_+^D} b_m \prod_{k=0}^{d-1} \Gamma(\frac{1}{2}(m(k) - 1)),$$

where the sum is over all $m = (m_{ij})$ in $\mathbb{N}_+^D$ with $D := \binom{d}{2}$, Γ is the Euler gamma function with $m(k) := m_{0,k} + \cdots + m_{k-1,k} + m_{k,k+1} + \cdots + m_{k,d-1}$, and C and b_m are explicit constants computed from a certain matrix associated to the co-dimension 2 dihedral angles of the simplex (the lengthy details can be found in [10] and [2]).

Thus it is reasonable to ask if the analytically very complicated dihedral angles can be replaced by other functions of edge lengths such that a simplicial Gauss–Bonnet–Chern theorem still holds true. To this end, we consider the moduli space of Euclidean simplices.

Definition 14.3.2 The *moduli space* $\mathcal{S}^n$ of all Euclidean n-simplices is defined as the set of isometry classes of Euclidean n-simplices σ, i.e. convex hulls of $n+1$ ordered points $p_0, p_1, \ldots, p_n$ in $\mathbb{R}^n$ in general position (i.e. affinely independent).

Note that the row vectors $p_1 - p_0$, $p_2 - p_0, \ldots, p_n - p_0$ establish a bijection from the set of simplices up to translation to the general linear group $GL_\mathbb{R}(n)$. Hence $\mathcal{S}^n$ can be identified with the non-compact homogeneous space $GL_\mathbb{R}(n)/O(n)$. In fact, because of polar decomposition this space is diffeomorphic to $\mathbb{R}^N$ with $N := \binom{n+1}{2}$.

Moreover, up to isometry a simplex σ is uniquely given by its edge lengths $d_{ij} \in \mathbb{R}_+$. Because of the realization theorem, Theorem 14.2.3, we get a second canonical identification

$$\mathcal{S}^n = \{(d_{ij}) \in \mathbb{R}_+^N \mid (-1)^{dim(\tau)+1}\Gamma_\tau > 0 \ \forall \tau \subset [n]\}$$

with $[n] := \{0, 1, \ldots, n\}$ and Γ_τ the Cayley–Menger determinant associated to the sub-simplex τ. In particular, $\mathcal{S}^n$ is an open semi-algebraic subset of $\mathbb{R}^N$. The symmetric group Σ_{n+1} acts on $\mathcal{S}^n$ by reordering the vertices.

Normed dihedral angles can be considered as functions,

$$\theta_\tau : \mathcal{S}^n \to]0, 1] \subset \mathbb{R},$$

for every non-empty subset $\tau \subset [n]$. By their definition it is obvious that they satisfy the following properties:

- *Positivity:* The θ_τ are positive functions of the variables d_{ij}.
- *Smoothness:* The θ_τ are smooth functions of the variables d_{ij}.
- *Symmetry:* If π is a permutation of $[n]$ (reordering of vertices), then $\theta_{\pi_*\tau}(\pi_*\sigma) = \theta_\tau(\sigma)$.
- *Balance:* By Poincaré's theorem we have $\sum_\tau (-1)^{dim(\tau)}\theta_\tau = 0$, where the sum is over all non-empty subsets $\tau \subset [n]$.
- *Normality:* It holds that $\theta_{[n]} = 1$ and $\theta_\tau = \frac{1}{2}$ for any co-dimension 1 sub-simplex τ.

Note that symmetry is sufficient in order to define dihedral angles $\theta(\tau \subset \sigma)$ for pairs of Euclidean simplices where the vertices are *not* ordered. Moreover, all functions θ_τ can be reduced to the case $\tau = \{0, 1, \ldots, k\}$ for $k = 0, 1, \ldots, n$ because of symmetry.

There is another important property of dihedral angles which allows us to connect geometry (flatness) and topology (vanishing Euler characteristics) in our combinatorial Gauss–Bonnet–Chern theorem (see [7, remark at bottom of p. 1355]):

- *Flatness:* If K is a Euclidean star in $\mathbb{R}^n$ around a vertex x that is homeomorphic to an n-dimensional disc, then the angle defects $\delta(\tau) := 1 - \sum_\sigma \max \theta(\tau \subset \sigma)$ vanish for all simplices τ adjacent to x.

Definition 14.3.3 Let n be a fixed dimension. A *system of pseudo-angles* Θ is a set of functions,

$$\theta'_\tau : \mathcal{S}^n \to \mathbb{R},$$

for every non-empty subset $\tau \subset [n]$, that are symmetric and balanced in the sense above (but not necessarily positive, smooth, normal or flat).

Here are some examples of systems Θ:

1. Normed dihedral angles (this is the 'classical case').
2. The constant system $\theta'_\tau(\sigma) = c_d$ with constants $c_d \in \mathbb{R}$ depending only on $d := dim(\tau)$ such that $\sum_{d=0}^n (-1)^d \binom{n}{d} c_d = 0$. Note that the choice $c_d = 0$ for all d just gives for κ^Θ below the *local Euler density* λ (see [7], p. 1347).
3. A given system multiplied by a constant $c \in \mathbb{R}$.
4. The sum of two given systems Θ_1 and Θ_2.
5. Here is a more exotic example in dimension $n = 2$: Let $\theta'_{[2]} := 1$ and $\theta'_{ij} := \frac{1}{2}$ (hence this is a normed system) and define

$$\theta'_0 := \frac{d_{12}}{2(d_{01} + d_{02} + d_{12})}$$

   with θ'_1 and θ'_2 defined symmetrically. Balance is obviously satisfied.
6. We can choose arbitrary functions θ'_τ that are symmetric and force balance by changing only the top pseudo-angle $\theta'_{[n]}$.

Thus the set of systems of pseudo-angles is an infinite-dimensional vector space. It seems most interesting to consider small deformations of the classical case.

Definition 14.3.4 For a system of pseudo-angles Θ, the associated *pseudo-angle defects* δ^Θ and *simplicial Gauss pseudo-curvature* κ^Θ are defined by the same formulae as in the classical case.

Theorem 14.3.5 *Let Θ be a system of pseudo-angles, δ^Θ its associated pseudo-angle defects and κ^Θ its associated simplicial Gauss pseudo-curvature. For any Euclidean simplicial complex (K, d) we have*

$$\sum_\tau (-1)^{dim(\tau)} \delta^\Theta(\tau) = \chi(K),$$

where the sum is over all simplices τ of K (simplex version), and

$$\sum_{x \in K_0} \kappa^{\Theta}(x) = \chi(K);$$

i.e. the sum over all simplicial Gauss pseudo-curvatures gives the Euler characteristic (vertex version).

Proof Our proof in [7, Thm. 4.1, p. 1356] uses only symmetry and balance of normed dihedral angles and thus applies verbatim to the more general case of systems of pseudo-angles:

$$\sum_{\tau} (-1)^{dim(\tau)} \delta^{\Theta}(\tau) = \sum_{\tau} (-1)^{dim(\tau)} \left(1 - \sum_{\sigma \text{ max}} \theta^{\Theta}(\tau \subset \sigma) \right)$$

$$= \sum_{\tau} (-1)^{dim(\tau)} - \sum_{\sigma \text{ max}} \sum_{\tau} (-1)^{dim(\tau)} \theta^{\Theta}(\tau \subset \sigma),$$

where the last sum vanishes because of balance. The statement for $\kappa^{\Theta}(x)$ follows by re-summation over all vertices because of the trivial fact that each τ has $1 + dim(\tau)$ vertices x.   □

Question 14.3.6   We conclude this section with a problem in any fixed dimension n. Is the following statement true: If a system Θ of pseudo-angles is positive, continuous, normed and flat, does it then coincide with the classical system of higher dihedral angles? It is not difficult to see that in dimension 2 this is true.

14.4 Combinatorial Riemannian Manifolds

We now introduce combinatorial Riemannian manifolds. This seems to be a new notion which was, to the best knowledge of the author, not considered before in this form. Note that it is not a Riemannian manifold together with a smooth triangulation, but a Euclidean simplicial complex together with additional combinatorial data which guarantee that it is a PL-manifold. Thus it is an object which is defined by finitely many combinatorial data including lengths of edges. First we need a local model for them.

Definition 14.4.1   A *flat Euclidean n-star T* is a simplicial complex in $\mathbb{R}^n$ such that 0 is a vertex, T is the star of 0 and T is homeomorphic to an n-disc. Let $\mathcal{T}^n$ denote the set of flat Euclidean n-stars.

In the following definition, we call distances in a star from the center to outer vertices *radial distances*, and distances of two adjacent outer vertices *peripheral distances*.

Definition 14.4.2   A *combinatorial Riemannian n-manifold* is a triple (M, d, e) that consists of a Euclidean simplicial complex (M, d) and a map e that associates to each vertex $x_0 \in M_0$ an embedding

$$e_{x_0} : |star_{x_0}| \to \mathbb{R}^n,$$

which is linear on each simplex of $star_{x_0}$ (hence PL-linear) such that the image of e_{x_0} is a flat Euclidean n-star and, for all vertices $x_i \in star_{x_0}$ adjacent to x_0, the *radial* edge lengths d_{0i} coincide with the lengths $|v_{0i}|$ of the vectors $v_{0i} := e_{x_0}(x_i)$.

We call the flat star $T_{x_0} := im(e_{x_0}) \subset \mathbb{R}^n$ the *flat tangent disc* at x_0 and e_{x_0} can be considered as the local inverse of a PL-linear exponential map. The map e_{x_0} *respects radial distances* but in general *not peripheral distances* d_{ij}. In particular, a combinatorial Riemannian n-manifold is a PL-manifold of dimension n, but carries more structure (namely, a PL-metric and special 'local flattenings').

Flatness of a combinatorial Riemannian n-manifold is thus equivalent to the property that e also respects all the peripheral distances. In this special case, the e_{x_0} are isometries and thus local diffeomorphisms, inducing a smooth structure on the geometric realization of M. For a general combinatorial Riemannian n-manifold, this is not true.

Note that for a flat combinatorial Riemannian manifold and a flat system of pseudo-angles, the pseudo-angle defects and the simplicial Gauss pseudo-curvature vanish by definition. This refines the fact that the Euler characteristics of such manifolds vanish.

In turn, it is quite obvious that for a closed Riemannian n-manifold, together with a finite geodesic triangulation which is fine enough, one can define an associated combinatorial Riemannian n-manifold such that both spaces are homeomorphic. Here, the triangulation has to be fine enough in order to be sure that the geodesic distances satisfy the higher triangle inequalities and thus form a realizable system of distances.

14.5 Simplicial Sectional Curvature

Let (M, d, e) be a combinatorial Riemannian n-manifold. If all the inverse maps $e_{x_0}^{-1}$ (i.e. the local simplicial exponential maps) reduce all peripheral distances, we call M *positively curved*; if they enlarge all peripheral distances, we call M *negatively curved*.

In order to quantify this and to define the sectional curvature of a combinatorial Riemannian n-manifold, we recall the generalized law of cosines for a surface F of constant curvature κ. On a sphere S^2_r with center $0 \in \mathbb{R}^3$ and radius r, we have $\kappa = \frac{1}{r^2}$. We consider the three points

$$p_0 := (r, 0, 0), \quad p_1 := (r \cos \alpha, r \sin \alpha, 0),$$

$$p_2 := (r \cos \beta, r \sin \beta \cos \varphi, r \sin \beta \sin \varphi).$$

In particular, φ is the angle at p_0 of the geodesic triangle defined by the three points. Their geodesic distances on S^2_r are given by $a = \alpha r$, $b = \beta r$ and $c = \gamma r$, with

$$\cos \gamma = \frac{1}{r^2} \langle p_1, p_2 \rangle = \cos \alpha \cos \beta + \sin \alpha \sin \beta \cos \varphi.$$

Now we expand the cosines and sines up to order 5,

$$\cos \delta = 1 - \frac{1}{2} \delta^2 + \frac{1}{24} \delta^4 \mp \ldots, \qquad \sin \delta = \delta - \frac{1}{6} \delta^3 + \frac{1}{120} \delta^5 \mp \ldots,$$

which yields up to order 5

$$1 - \frac{1}{2} \gamma^2 + \frac{1}{24} \gamma^4 \mp \ldots = 1 - \frac{1}{2}(\alpha^2 + \beta^2 - 2\alpha\beta \cos \varphi)$$

$$+ \frac{1}{24}(\alpha^4 + 6\alpha^2 \beta^2 + \beta^4) - \frac{1}{6}(\alpha^3 \beta + \alpha\beta^3) \cos \varphi \pm \ldots$$

A short computation gives a corresponding expansion of c^2 up to total order 5 in a and b:

$$c^2 = a^2 + b^2 - 2ab \cos \varphi - \frac{\kappa}{6}(1 + 2 \cos^2 \varphi) a^2 b^2 \pm \ldots$$

with higher terms of order ≥ 6. It is easy to see that the expansion of c^2 has no terms of total odd order in a and b. This expansion also remains true in the case of negative curvature κ. For us, this is the starting point in the definition of the combinatorial analogue of sectional curvature.

Here, three points x_0, x_1, x_2 on F are given with geodesic distances $d_{01} = a$, $d_{02} = b$ and $d_{12} = d$, and φ is the angle of the spherical triangle at x_0.

In particular, if v_{01} and v_{02} denote the vectors in the tangent space $T_{x_0} F$ that are mapped by the exponential map to x_1 and x_2, we have $|v_{01} - v_{02}|^2 = a^2 + b^2 - 2ab \cos(\varphi)$ by the Euclidean law of cosines.

As κ represents the sectional curvature, this motivates the following definition:

Definition 14.5.1 The *simplicial sectional curvature* of a combinatorial Riemannian n-manifold (M, d, e) is defined as

$$sec(x_0; x_i, x_j) := 6\frac{|v_{0i} - v_{0j}|^2 - d_{ij}^2}{d_{0i}^2 d_{0j}^2 + 2\langle v_{0i}, v_{0j}\rangle^2}$$

and is a real function defined on the *set of pointed triangles*

$$(x_0; x_i, x_j) \in M_{2,0}.$$

Here we have used the notation (for M a simplicial complex)

$$M_{r,s} = \{(\sigma, \tau) | \tau \subset \sigma, \tau \in M_s, \sigma \in M_r\},$$

which is a relative version of the notation M_r for the set of r-simplices of M. Note that this function *sec*, which is defined on the set of pointed triangles $M_{2,0}$,

$$sec : M_{2,0} \to \mathbb{R},$$

is the simplicial analogue of the sectional curvature in differential geometry

$$sec : G_2(TM) \to \mathbb{R}.$$

Here, the triangles which are adjacent to the vertex x_0 play the role of the tangent planes P through x_0, which are elements of the *Grassmann plane bundle* for a smooth Riemannian manifold M:

$$G_2(TM) := \{(P, x) | x \in M, P \le T_x M, dim(P) = 2\}.$$

14.6 Simplicial Sectional Curvature and the Hopf Conjecture

We recall the following classical conjecture:

Conjecture 14.6.1 (Hopf conjecture) If M is a closed smooth Riemannian manifold of even dimension n and of positive sectional curvature, then its Euler characteristics $\chi(M)$ is positive.

Because of the Gauss–Bonnet–Chern theorem, it is obvious to consider the connection between sectional curvature and Gauss curvature. If it is possible to show from $sec > 0$ that $\kappa > 0$, then it follows that $\chi(M) >$

0 by the positivity of the Gauss–Bonnet–Chern integrand. In fact, as $sec = \kappa$ for $n = 2$, the Hopf conjecture is trivially true in this dimension. Moreover, Milnor has shown (see [4]) that $sec > 0 \Rightarrow \kappa > 0$ for $n = 4$, thus the Hopf conjecture is also true in this dimension.

However, in dimension $n = 6$, Geroch ([5], see also [8]) has given an example of an algebraic curvature tensor R with $sec_R > 0$ but $\kappa = Pf^n(\Omega_R)$ negative. Thus for $n \geq 6$ it seems generally not to be possible to decide the Hopf conjecture by the Gauss–Bonnet–Chern theorem.

We can now also formulate the Hopf conjecture in the simplicial case:

Conjecture 14.6.2 (Hopf conjecture for combinatorial Riemannian manifolds) If (M, d, e) is a (closed) combinatorial Riemannian manifold of even dimension n and of positive simplicial sectional curvature, then its Euler characteristics $\chi(M)$ is positive.

Because of the existence of geodesic triangulations which are fine enough, *the Hopf conjecture for combinatorial Riemannian manifolds would imply the classical Hopf conjecture.*

Again we can consider the connection between simplicial sectional curvature and simplicial Gauss pseudo-curvature. As we have here the freedom to choose a suitable system Θ of pseudo-angles, the chances are better to be able to show that $sec > 0$ implies positivity of κ^Θ.

As an example, the following statement concerning the Hopf conjecture is true. Note that the peripheral distances d_{ij} of a vertex star $star(x_0)$ in a combinatorial Riemannian manifold can be recovered from the distances of the corresponding flat Euclidean star T_{x_0} together with the simplicial sectional curvature function sec by the formula

$$d_{ij}^2 = |v_{0i} - v_{0j}|^2 - \frac{1}{6} sec(x_0; x_i, x_j)(d_{0i}^2 d_{0j}^2 + 2\langle v_{0i}, v_{0j}\rangle^2),$$

which holds for all triangles adjacent to x_0.

Definition 14.6.3 We call the flat Euclidean star T together with the function sec restricted to the pointed triangles $(x_0; x_i, x_j)$ a *curved star* (T, sec), where realizability of the 'non-flat' peripheral distances d_{ij} is understood. If $sec > 0$, then we speak of a *positively curved star*.

Note that, for each curved star, the simplicial Gauss pseudo-curvature κ^Θ can be defined at the vertex x_0. In fact, a small neighbourhood of x_0 is sufficient to define all necessary dihedral angles.

Theorem 14.6.4 *If for every finite set of positively curved stars there exists a system Θ of pseudo-angles such that their simplicial Gauss* *pseudo-curvatures* κ^{Θ} *are all positive, then the Hopf conjecture is generally true.*

We call such a (hypothetical) system Θ of pseudo-angles an *adapted system* for the given finite set of positively curved stars.

Proof Let (M, d, e) be a combinatorial Riemannian n-manifold of even dimension n and of positive simplicial sectional curvature. Hence we get a positively curved star at each vertex and there are only finitely many vertices. According to the assumption, there exists an adapted system Θ of pseudo-angles. As the simplicial Gauss pseudo-curvatures κ^{Θ} are positive for all vertices, the simplex version of Theorem 14.3.5 yields $\chi(M) > 0$. □

Unfortunately, this existence problem of adapted systems is a complicated question concerning the combinatorial geometry of positively curved stars which is completely open at the moment.

Here are some considerations, questions and speculations which could lead to some progress in the future:

- It is crucial to better understand simplicial Gauss pseudo-curvature as a pairing,

$$\kappa : \mathcal{C} \times \{\Theta\} \to \mathbb{R},$$

  where $\mathcal{C} = \{(T, sec)\}$ denotes the set of all curved stars and $\{\Theta\}$ is the set of all systems of pseudo-angles (both in a fixed dimension n).
- For a fixed combinatorial type τ of a flat star T (i.e. if we fix the abstract simplicial complex τ underlying T), the restriction $\mathcal{C}^{\tau} \subset \mathcal{C}$ yields a non-compact manifold $\mathcal{C}^{\tau}$. Let $F \subset \{\Theta\}$ be some finite dimensional family of smooth systems of pseudo-angles, containing the classical dihedral angles. Then $\kappa : \mathcal{C}^{\tau} \times F \to \mathbb{R}$ is a smooth map and we can study its derivatives around flat stars and classical dihedral angles (where κ vanishes).
- This approach could be used to construct an adapted system Θ if *one* positively curved star (T, sec) is given.
- Then one could study the properties of a linear combination $\sum a_i \Theta_i$ where each Θ_i is adapted to one positively curved star T_i. Maybe, for suitable coefficients a_i, this could give an adapted system Θ for all T_i.
- In order to understand the failure of positivity of κ for given Θ, one could also study the inverse set $\mathcal{C}_+^{\Theta} := (\kappa^{\Theta})^{-1}(\mathbb{R}_+)$ of curved stars in $\mathcal{C}$ that yield positive simplicial Gauss pseudo-curvature with respect to Θ.

- If a smooth Riemannian manifold can be geodesically triangulated with all curved stars in $\mathcal{C}_+^{\Theta}$ for some system Θ, then we also have $\chi(M) > 0$.

The author hopes to come back to this subject in the future.

References

[1] Kazuhiko Aomoto, *Analytic structure of Schläfli function*, Nagoya Math. J. 68 (1977), pp. 1–16.

[2] Matthias Beck, Sinai Robins, Steven V. Sam, *Positivity theorems for solid-angle polynomials*, Beitr. Algebra Geom. 51 (2010), pp. 493–507, with corrigendum in Beitr. Algebra Geom. 56 (2015), pp. 775–776.

[3] Marcel Berger, *Geometry I*, Universitext. Springer, Berlin Heidelberg (1987).

[4] R. L. Bishop and S. I. Goldberg, *Some implications of the generalized Gauss-Bonnet theorem*, Trans. Amer. Math. Soc. 112 (1964), pp. 508–535.

[5] R. Geroch, *Positive sectional curvature does not imply positive Gauss-Bonnet integrand*, Proc. Amer. Math. Soc. 54 (1976), pp. 267–270.

[6] Heinz Hopf, *Differential Geometry in the Large*, Lecture Notes in Mathematics, Vol. 1000, Springer, Berlin Heidelberg (1989).

[7] Stephan Klaus, *On the combinatorial Gauss-Bonnet theorem for general Euclidean simplicial complexes*, Front. Math. China 11 (2016), pp. 1345–1362.

[8] Paul F. Klembeck, *On Geroch's counterexample to the algebraic Hopf conjecture*, Proc. Amer. Math. Soc. 59 (1976), pp. 334–336.

[9] Jun Murakami, *Volume formulas for a spherical tetrahedron*, Proc. Amer. Math. Soc. 140 (2012), no. 9, pp. 3289–3295.

[10] Jason M. Ribando, *Measuring solid angles beyond dimension three*, Discrete Comput. Geom. 36 (2006), pp. 479–487.

15

Aspects and Examples on Quantitative Stratification with Lower Curvature Bounds

Nan Li

Abstract

We survey results on quantitative stratifications on Alexandrov spaces and limits of manifolds with lower Ricci curvature bounds. A main goal is to outline the ingredients and compare results between Alexandrov and Ricci cases. This chapter also contains the proof for the following results which were mentioned in [16] without proof. (1) Any compact set Γ with $\operatorname{diam}(\Gamma) \leq 1$ in an n-dimensional Alexandrov space with curv≥ -1 can be covered by at most $N(n, \varepsilon)$ closed sets and each of the sets ε looks like an annulus in a metric cone. (2) For any $1 \leq k \leq n - 2$, there exists a sequence of n-dimensional manifolds M_i with $\sec_{M_i} \geq 0$, which non-collapsed converges to an Alexandrov space X, so that the (k, ε)-singular set on X is a Cantor set with positive k-dimensional Hausdorff measure. The special case $k = n - 2$ was shown in [16].

15.1 Introduction

In this chapter we focus on Alexandrov spaces and non-collapsed Gromov–Hausdorff limits of manifolds with lower Ricci curvature bounds. The reader is assumed to be familiar with most basic concepts of Alexandrov geometry (see [1], [2], [4]) and manifolds with lower Ricci curvature bounds (see [5], [6]).

Let $\mathcal{M}(n, \kappa)$ be the isometric class of n-dimensional Riemannian manifolds M with Ricci curvature $\operatorname{Ric}_M \geq (n - 1)\kappa$. Let $\mathcal{M}(n, \kappa, v) = \{(M, p) : p \in M \in \mathcal{M}(n, \kappa), \operatorname{Vol}(B_1(p)) \geq v > 0\}$. By the Cheeger–Gromov compactness theorem, $\mathcal{M}(n, \kappa)$ is pre-compact in the pointed

Gromov–Hausdorff topology. Let $\mathcal{M}_\infty(n, \kappa)$ and $\mathcal{M}_\infty(n, \kappa, v)$ be the closure of $\mathcal{M}(n, \kappa)$ and $\mathcal{M}(n, \kappa, v)$, respectively.

Let $\text{Alex}^n(\kappa)$ denote the collection of n-dimensional Alexandrov spaces with (sectional) curvature $\geq \kappa$. For $X \in \text{Alex}^n(\kappa)$, the Toponogov comparison holds in the sense that the geodesic triangles are "fatter" than the corresponding triangles in the 2-dimensional space form $\mathbb{S}^2_\kappa$ with constant curvature κ. For example, the Gromov–Hausdorff limits of Riemannian manifolds with $\sec \geq \kappa$ are Alexandrov spaces with curvature $\geq \kappa$. The quotient space $M/G \in \text{Alex}(\kappa)$ if M is a compact Riemannian manifold with $\sec_M \geq \kappa$ and group G acts on M isometrically.

As a comparison between $\mathcal{M}_\infty(n, \kappa)$ and $\text{Alex}^n(\kappa)$, we would like to point out that not every Alexandrov space is isometric to a non-collapsed limit of Riemannian manifolds with uniform lower sectional curvature bound, due to some topological restrictions (see [14], [19]). It is an open question whether every Alexandrov space is a collapsed limit of Riemannian manifolds with uniform lower sectional curvature bound.

In Section 15.2, we begin with a classical stratification of singular sets and discuss the motivation of the quantitative stratification. In Section 15.3, we discuss the notion of quantitative stratifications and the related results for spaces in $\mathcal{M}_\infty(n, \kappa, v)$ and $\text{Alex}^n(\kappa)$. In Section 15.4, we discuss some of the main ingredients and required formulas, for which the complete treatments can be found in [8], [10], [11], [13], [16]. Similar treatments can also be found in many different areas such as [3], [7], [9], [12], [15] and [17]. The aim of our discussion is to give the readers a sense of the general framework and to show how it is applied. In this section, we also prove a new covering theorem for Alexandrov spaces, Theorem 15.3.14. In Section 15.5, we give a proof for Theorem 15.3.13, which shows that the quantitative singular set $\mathcal{S}^k_\varepsilon$ can be a "fat" Cantor set for every $1 \leq k \leq n - 2$.

Convention and notation Let Z be a length metric space.

(1) Let $C(Z) = Z \times [0, \infty]/(a, 0) \sim (b, 0)$ denote the metric cone over Z, whose distance

$$d_{C(Z)}((a, t_1), (b, t_2)) = \left(t_1^2 + t_2^2 - 2t_1 t_2 \cos(\min\{d_Z(a, b), \pi\})\right)^{1/2}$$

is defined by the Euclidean law of cosines. The quotient of the equivalent class $\{(a, 0) : a \in Z\}$ is called the cone point.

(2) Let $\dim_{\mathcal{H}}(Z)$ denote the Hausdorff dimension.

(3) Let $\mathcal{H}^n(Z)$ denote the n-dimensional Hausdorff measure.

(4) Let $N(c_1, c_2, \ldots, c_\ell)$ and $C(c_1, c_2, \ldots, c_\ell)$ denote positive constants depending only on constants $c_1, c_2, \ldots, c_\ell$.

(5) Let $[ab]$ denote a geodesic connecting points a and b.

(6) Let $A_r^R(p) = \{x : r \le d(p, x) \le R\}$ denote the closed annulus centered at p. If $r = 0$ then $A_0^R(p) = \bar{B}_R(p)$ is simply a closed ball.

Acknowledgements   The author was partially supported by the PSC-CUNY Research Award #61533-00 49.

15.2 Stratification of Singular Sets

The metric space X we will discuss in this chapter is either in $\mathcal{M}_\infty(n, \kappa)$ or $\mathrm{Alex}^{\,n}(\kappa)$. Thus the tangent cone $T_p X = \lim_{r_i \to 0} (X, p, r_i^{-1} d)$ exists (but is not necessarily unique for $X \in \mathcal{M}_\infty(n, \kappa, v)$) for every $p \in X$. A classical definition of the singular set $\mathcal{S}(X)$ is the collection of points $p \in X$ for which not every tangent cone $T_p X$ is isometric to $\mathbb{R}^n$. Based on the number of $\mathbb{R}$ factors that a tangent cone splits off isometrically, $\mathcal{S}(X)$ has a natural stratification. Let us begin with the following notion.

Definition 15.2.1 (Symmetric space)   Given a metric space Y and $k \in \mathbb{N}$, we say that Y is k-symmetric if Y is isometric to $\mathbb{R}^k \times C(\Sigma)$ for some metric space Σ.

Now the singular set $\mathcal{S} = \mathcal{S}(X)$ has a stratification

$$\mathcal{S} = \mathcal{S}^{n-1} \supseteq \mathcal{S}^{n-2} \supseteq \cdots \supseteq \mathcal{S}^1 \supseteq \mathcal{S}^0, \tag{15.1}$$

where

$$\mathcal{S}^k \equiv \{p \in X : \text{ no tangent cone at } p \text{ is } (k+1)\text{-symmetric}\}. \tag{15.2}$$

Example 15.2.2   Let X be a unit cube in $\mathbb{R}^3$. The 0-singular $\mathcal{S}^0$ is the set of 8 vertices, because the tangent cones at these points are a metric cone over a $\frac{1}{8}$-sphere, which do not split off any $\mathbb{R}$ factor isometrically. The 1-singular set $\mathcal{S}^1$ is the collection of 12 edges whose tangent cones are isometric to the quarter plane product with $\mathbb{R}^1$. The 2-singular set $\mathcal{S}^2$ is the collection of 6 faces whose tangent cones are half spaces $\mathbb{R}^2 \times \mathbb{R}_+$. Lastly, the interior points have tangent cone $\mathbb{R}^3$. They are considered as the regular points. In fact, they are locally isometric to balls in $\mathbb{R}^3$.

The following theorems are the dimension estimates for $\mathcal{S}^k$.

Theorem 15.2.3 ([6]) *If* $X \in \mathcal{M}_\infty(n, \kappa, v)$, *then* $\dim_{\mathcal{H}}(\mathcal{S}^k) \leq k$ *and* $\mathcal{S}^{n-1} \setminus \mathcal{S}^{n-2} = \varnothing$. *In particular,* $\dim_{\mathcal{H}}(\mathcal{S}) \leq n - 2$.

Theorem 15.2.4 ([4]) *If* $X \in \mathrm{Alex}^n(\kappa)$, *then* $\dim_{\mathcal{H}}(\mathcal{S}^k) \leq k$. *If* X *has no boundary, then* $\mathcal{S}^{n-1} \setminus \mathcal{S}^{n-2} = \varnothing$ *and* $\dim_{\mathcal{H}}(\mathcal{S}) \leq n - 2$.

These results are useful in many cases. However, they are not good enough for some types of analysis. For example, the dimension control doesn't prevent the singular set from being dense (see [18] for an example). This would be an issue for instance in the computation of some integrals. In this case, an upper bound of $\mathcal{H}^n(B_r(\mathcal{S}^k) \cap B_1)$ in the form of cr^α, $\alpha > 0$, would be needed (see [11]). However this is impossible if $\mathcal{S}^k$ is dense. In the following we describe an example from [18] and [16], which shows that $\mathcal{S}^0(X)$, where $X \in \mathrm{Alex}^3(0)$, is dense but a quantitative version of it is finite.

Example 15.2.5   There exist Alexandrov spaces (in fact, non-collapsed Gromov–Hausdorff limits of manifolds with $\sec \geq 0$) whose singular set is dense. Such a space was constructed in [18]. Begin with a regular tetrahedron X_1 in $\mathbb{R}^3$. Suppose that convex polyhedra X_k with triangular faces Δ_i, $i = 1, 2 \ldots, 4 \cdot 3^{k-1}$, have been constructed. Let x_i be the centroid of face Δ_i. Let $y_i \in \mathbb{R}^3$ so that $d(y_i, X_k) = d(y_i, x_i) = d_k^i > 0$. Let Y_i be the tetrahedron formed by y^i and Δ_i. Define $X_{k+1} = X_k \cup (\cup_i Y_i)$. The constants $d_k^i = d_k^i(X_k)$ can be chosen small enough so that X_{k+1} is convex. We have that $\partial X_k \in \mathrm{Alex}^2(0)$ for all k. Thus $Y = \lim_{i \to \infty} \partial X_k \in \mathrm{Alex}^2(0)$. It is easy to see that if all X_k are convex, then $\max_i\{d_k^i\} \to 0$ as $k \to \infty$.

The set of singular points $\mathcal{S}^0(Y) \supseteq \bigcup_{i,k}\{x_k^i\}$ is dense in Y. However, $|\mathcal{S}_\varepsilon^0| < N(\varepsilon)$, asserted by Theorem 15.3.7. For this example, we can get an explicit estimate using the Gauss–Bonnet formula. For each $p \in Y$, we have that the tangent cone $T_p(Y) = C(\mathbb{S}_\beta^1)$, with $0 < \beta \leq 1$. Let $\theta_p = 2\pi\beta$ be the cone angle. Then we have $\mathcal{S}_\varepsilon^0 = \{p \in Y : \theta_p \leq 2\pi - \varepsilon\}$. Note that for any $p \in Y$ the Gaussian curvature $K_p \geq 0$ and $K_p = (2\pi - \theta_p)\delta_p$ if $p \in \mathcal{S}_\varepsilon^0$, where δ_p is the Dirac delta function at p. By the Gauss–Bonnet formula, we have

$$4\pi = \int_Y K \geq \sum_{p \in \mathcal{S}_\varepsilon^0} (2\pi - \theta_p) \geq \sum_{p \in \mathcal{S}_\varepsilon^0} \varepsilon. \tag{15.3}$$

Thus $|\mathcal{S}_\varepsilon^0| \leq \frac{4\pi}{\varepsilon}$. $\qquad\square$

This example shows that there could be more significant estimates in terms of a quantitative stratification of the singular sets. These estimates, together with the study of the structure of $\mathcal{S}_\varepsilon^k(X) \setminus \mathcal{S}_\varepsilon^{k-1}(X)$, can provide us with more information on the structure of X. For example, see Theorem 15.3.12:

15.3 Quantitative Stratification

15.3.1 Definitions

Definition 15.3.1 (Quantitative symmetric space)   Given $x \in X$, we say that $B_r(x)$ is (k, ε)-symmetric if there exists a k-symmetric space Y such that $d_{GH}(B_r(x), B_r(y)) \leq \varepsilon r$, where $y \in Y$ is a cone point.

Definition 15.3.2 (Quantitative singular sets)   Given $k, \varepsilon, r > 0$ and a metric space X:

(1) The r-scale (k, ε)-singular set is defined as

$$\mathcal{S}_{\varepsilon, r}^k \equiv \{x \in X : \text{there is no } r \leq s < 1$$
$$\text{so that } B_s(x) \text{ is } (k+1, \varepsilon)\text{-symmetric}\}. \quad (15.4)$$

(2) The (k, ε)-singular set is defined as

$$\mathcal{S}_\varepsilon^k \equiv \bigcap_{r>0} \mathcal{S}_{\varepsilon, r}^k = \{x \in X : \text{there is no } 0 < s < 1 \text{ so that}$$
$$B_s(x) \text{ is } (k+1, \varepsilon)\text{-symmetric}\}. \quad (15.5)$$

It is easy to see that

$$\mathcal{S}^k = \bigcup_{\varepsilon>0} \mathcal{S}_\varepsilon^k = \bigcup_{\varepsilon>0} \left(\bigcap_{r>0} \mathcal{S}_{\varepsilon, r}^k \right). \quad (15.6)$$

Remark 15.3.3   The singular set $\mathcal{S}_{\varepsilon, r}^k$ also makes sense on smooth manifolds, even though obviously $\mathcal{S}^k = \varnothing$ on these spaces. For example, let X be the space glued from two unit cubes (see Example 15.2.2) in $\mathbb{R}^3$ along an identity map between their boundaries (called the doubling of the unit cube). Then $\mathcal{S}_{\varepsilon, r}^k(X)$, $k = 0, 1$ and 2, are r-tubular neighborhoods around the glued vertices, edges and faces, respectively. By smoothing X we obtain a sequence of 3-manifolds M_i with $\sec_{M_i} \geq 0$ and $M_i \to X$. On every M_i we have $\mathcal{S}^k(M_i) = \varnothing$, $k = 0, 1, 2$, since it is smooth. However, for i large, $\mathcal{S}_{\varepsilon, r}^k(M_i)$ at least contains a cr-neighborhood around the subset which Gromov–Hausdorff converges to $\mathcal{S}_\varepsilon^k(X)$.

The above quantitative notion was initiated by Cheeger and Naber in [10], which works well for $M \in \mathcal{M}_\infty(n, \kappa, v)$. For Alexandrov spaces, one can use the following stronger notion, which was defined in [16].

Definition 15.3.4 (Splitting and quantitative splitting)

(1) Given a metric space Y and $k \in \mathbb{N}$, we say that Y is k-splitting if Y is isometric to $\mathbb{R}^k \times Z$ for some metric space Z.

(2) Given a metric space X we say that a metric ball $B_r(x) \subseteq X$ is (k, ε)-splitting if there exists a k-splitting space Y and $y \in Y$ such that $d_{GH}(B_r(x), B_r(y)) \leq \varepsilon r$.

Definition 15.3.5 (Strong quantitative singular sets)   Given $k, \varepsilon, r > 0$ and metric space X:

(1) The strong r-scale (k, ε)-singular set is defined as

$$\widetilde{\mathcal{S}}^k_{\varepsilon, r} \equiv \left\{ x \in X : B_r(x) \text{ is not } (k+1, \varepsilon)\text{-splitting} \right\}. \tag{15.7}$$

(2) The strong (k, ε)-singular set is defined as

$$\widetilde{\mathcal{S}}^k_\varepsilon \equiv \big\{ x \in X : \text{there is no } 0 < s < 1 \text{ so that}$$
$$B_s(x) \text{ is } (k+1, \varepsilon)\text{-splitting}\big\}. \tag{15.8}$$

By Corollary 15.4.9 and Corollary 15.4.17, the following results hold.

Proposition 15.3.6   *For any $k, \varepsilon, r, v > 0$, there exist $\eta_i(n, \varepsilon, v) > 0$ and $\delta_i(n, \varepsilon) > 0$, $i = 1, 2$, such that the following hold:*

(1) If $X \in \mathcal{M}_\infty(n, -1, v)$, then $\mathcal{S}^k_{\varepsilon, \eta_1 r} \subseteq \widetilde{\mathcal{S}}^k_{\eta_2, r}$.
(2) If $X \in \mathrm{Alex}^n(-1)$, then $\mathcal{S}^k_{\varepsilon, \delta_1 r} \subseteq \widetilde{\mathcal{S}}^k_{\delta_2, r}$ and $\widetilde{\mathcal{S}}^k_{\varepsilon, r} \subseteq \mathcal{S}^k_{\delta_1, r}$.

Proposition 15.3.6 says that if $X \in \mathrm{Alex}^n(-1)$, then notions of strong r-scale and r-scale (k, ε)-singular sets are equivalent, i.e. $\widetilde{\mathcal{S}}^k_{\varepsilon, r} = \mathcal{S}^k_{\varepsilon, r}$. In this chapter, for Alexandrov spaces we may use either $\widetilde{\mathcal{S}}^k_{\varepsilon, r}$ or $\mathcal{S}^k_{\varepsilon, r}$ and the results will always hold for both notions. Note that the statement $\widetilde{\mathcal{S}}^k_{\varepsilon, \eta_1 r} \subseteq \mathcal{S}^k_{\eta_2, r}$ doesn't hold for $X \in \mathcal{M}_\infty(n, -1, v)$. See Remark 15.4.13 for an explanation. Thus the notion of strong r-scale (k, ε)-singular set $\widetilde{\mathcal{S}}^k_{\varepsilon, r}$ is indeed strictly stronger than $\mathcal{S}^k_{\varepsilon, r}$ for the Ricci case. Moreover, the estimates for $\mathcal{S}^k_{\varepsilon, r}$ on Alexandrov spaces in Section 15.3.2 are not true for the Ricci case.

15.3.2 Results

For the cube in Example 15.2.2, we have that $\dim_{\mathcal{H}}(\mathcal{S}_\varepsilon^k) \leq k$, $\mathcal{H}^k(\mathcal{S}_\varepsilon^k) \leq c(k)$ and $\mathcal{H}^k(\mathcal{S}_{\varepsilon,r}^k) \leq c(k)r^{3-k}$. Moreover, we have that every $\mathcal{S}_\varepsilon^k$, away from the zero $\mathcal{H}^k$-measure subset $\mathcal{S}_\varepsilon^{k-1}$, is a k-manifold. These results are all expected to be true in general. Most of them can be done. However, there are exceptions. Now let us describe the known results for $\mathcal{S}_\varepsilon^k$ on $\mathcal{M}_\infty(n, \kappa, v)$ and $\text{Alex}^n(\kappa)$ and make comparisons.

Theorem 15.3.7 (Packing estimate [8], [16])   *The following hold for any $n \in \mathbb{N}$ and $\varepsilon, v > 0$:*

(1) If $(X, p) \in \mathcal{M}_\infty(n, -1, v)$, $x_i \in \mathcal{S}_{\varepsilon,r}^k(X) \cap B_1(p)$ and $\{B_r(x_i)\}$ are disjoint with $r \leq 1$ for all $i \in \mathbb{I}$, then $|\mathbb{I}| < C(n, \varepsilon, v)r^{-k}$.

(2) If $X \in \text{Alex}^n(-1)$, $x_i \in \widetilde{\mathcal{S}}_{\varepsilon, \beta r_i}^k(X) \cap B_1(p)$ and $\{B_{r_i}(x_i)\}$ are disjoint with $r_i \leq 1$ for all $i \in \mathbb{I}$, then

$$\sum_{i \in \mathbb{I}} r_i^k < C(n, \varepsilon). \tag{15.9}$$

In particular, if $x_i \in \widetilde{\mathcal{S}}_{\varepsilon,r}^k(X) \cap B_1(p)$ and $\{B_r(x_i)\}$ are disjoint with $r \leq 1$, then $|\mathbb{I}| < C(n, \varepsilon)r^{-k}$.

A major difference between the Ricci and Alexandrov cases is that the estimate for $X \in \text{Alex}^n(-1)$ doesn't depend on volume. Another difference is Theorem 15.4.19, which holds for Alexandrov spaces but not for the Ricci case. Theorem 15.3.7 reveals a strong geometric control for the distribution of singular sets.

Theorem 15.3.8 (Hausdorff measure estimate [8], [16])   *The following hold for any $n \in \mathbb{N}$ and $\varepsilon, v > 0$:*

(1) If $(X, p) \in \mathcal{M}_\infty(n, -1, v)$, then

$$\mathcal{H}^n\big(B_r(\mathcal{S}_{\varepsilon,r}^k) \cap B_1(p)\big) \leq C(n, \varepsilon, v)\, r^{n-k} \tag{15.10}$$

and

$$\mathcal{H}^k\big(\mathcal{S}_\varepsilon^k \cap B_1(p)\big) < C(n, \varepsilon, v). \tag{15.11}$$

(2) If $(X, p) \in \text{Alex}^n(-1)$, then

$$\mathcal{H}^n\big(B_r(\widetilde{\mathcal{S}}_{\varepsilon,r}^k) \cap B_1(p)\big) \leq C(n, \varepsilon)\, r^{n-k} \tag{15.12}$$

and

$$\mathcal{H}^k\big(\widetilde{\mathcal{S}}_\varepsilon^k \cap B_1(p)\big) < C(n, \varepsilon). \tag{15.13}$$

Based on this theorem and Example 15.2.5, the following results were conjectured in [16].

Conjecture 15.3.9   For any $(X, p) \in \text{Alex}^n(-1)$, we have

$$\mathcal{H}^k(\mathcal{S}^k_\varepsilon \cap B_1(p)) < C(n)\varepsilon^{1-(n-k)}.$$

Conjecture 15.3.10   For any $(X, p) \in \text{Alex}^n(-1)$, we have

$$\sum_{i=0}^{\infty} \varepsilon_{i+1}^{(n-k)-1} \mathcal{H}^k\left(\left(\mathcal{S}^k_{\varepsilon_{i+1}} \setminus \mathcal{S}^k_{\varepsilon_i}\right) \cap B_1(p)\right) < C(n),$$

where $\varepsilon_i = 2^{-i}$.

The following result is concerned with the rectifiability of $\mathcal{S}^k$.

Theorem 15.3.11 (Rectifiability [8], [16]) $\mathcal{S}^k(X)$ *is k-rectifiable for any $X \in \mathcal{M}_\infty(n, -1, v) \cup \text{Alex}^n(-1)$ and $0 \le k \le n$.*

Theorem 15.3.12 (Manifold structure [4], [6], [8])   *The following hold for any $n \in \mathbb{N}$ and $v > 0$:*

(1) *If $(X, p) \in \mathcal{M}_\infty(n, -1, v)$, then there is $\varepsilon(n, v) > 0$ so $X \setminus S^{n-2}_\varepsilon(X)$ is bi-Hölder homeomorphic to a smooth Riemannian manifold. Here $S^{n-2}_\varepsilon(X)$ is $(n-2)$-rectifiable so $\mathcal{H}^{n-2}(S^{n-2}_\varepsilon(X) \cap B_1(p)) \le C(n, \varepsilon, v)$.*

(2) *If $X \in \text{Alex}^n(-1)$, then there exists $\varepsilon(n) > 0$ so that $X \setminus S^{n-1}_\varepsilon(X)$ is locally bi-Lipschitz to a ball in $\mathbb{R}^n$. For any $p \in S^{n-1}_\varepsilon(X) \setminus S^{n-2}_\varepsilon(X)$, there is $r > 0$ so that $B_r(p)$ is bi-Lipschitz to a ball in the half space $\mathbb{R}^{n-1} \times \mathbb{R}_{\ge 0}$, centered at the origin. In particular, if $X \in \text{Alex}^n(-1)$ has no boundary, then $X \setminus S^{n-2}_\varepsilon(X)$ is locally bi-Lipschitz to a ball in $\mathbb{R}^n$. Here $S^{n-2}_\varepsilon(X)$ is $(n-2)$-rectifiable with $\mathcal{H}^{n-2}(S^{n-2}_\varepsilon(X) \cap B_1(x)) \le C(n, \varepsilon)$ for every $x \in X$.*

It was asked in both the Ricci and Alexandrov cases whether $\mathcal{S}^k_\varepsilon$ carries a k-manifold structure, away from a zero $\mathcal{H}^k$-measure subset. Based on Theorem 15.3.12, this is true for $X \in \mathcal{M}_\infty(n, \kappa, v) \cup \text{Alex}^n(\kappa)$, if $k = n$ or $n - 1$. However, for every $1 \le k \le n - 2$, there are counterexamples.

Theorem 15.3.13   *Let $n \ge 3$ and $1 \le k \le n - 2$. Let $\mathfrak{C} = \bar{B}_1(0^k) \setminus \cup_{\ell=1}^\infty U_\ell \subset \mathbb{R}^k$ be a closed set, where U_ℓ are disjoint open convex subsets in $\mathbb{R}^k$, there exist $X \in \text{Alex}^n(0)$ and $\varepsilon > 0$ small, such that the following hold:*

(1) $\mathcal{S}(X) = \mathcal{S}^k_\varepsilon(X)$ *and* $\mathcal{S}^{k-1}(X) = \varnothing$.

(2) *There is a bi-Lipschitz embedding $\varphi : \bar{B}_1(0^k) \to X$, such that $\varphi(\mathfrak{C}) = \mathcal{S}_\varepsilon^k(X)$.*

(3) *There is a sequence of n-dimensional manifolds M_i with $\sec_{M_i} \geq 0$ such that $\lim_{i \to \infty} M_i = X$.*

In particular, for any $0 \leq \alpha \leq k \leq n - 2$, there exists $X \in \mathrm{Alex}^{\,n}(0)$ such that $\mathcal{S}_\varepsilon^k$ is an α-dimensional Cantor set with $\mathcal{H}^\alpha(\mathcal{S}_\varepsilon^k) > 0$. In this case, $\mathcal{S}_\varepsilon^k$ doesn't contain any manifold points. Such an example also shows that the rectifiability theorem, Theorem 15.3.11, is sharp. Theorem 15.3.13 was proved in [16] for the special case $k = n - 2$. As pointed out in [16], the proof for the general case is similar but much more technical. For completeness, we will give the detailed proof in Section 15.5.

Let us end this section by stating a new covering theorem for Alexandrov spaces which is interesting and potentially useful by itself. This theorem will be proved in Section 15.4.4. Given a point $p \in X$ and $\varepsilon > 0$, we define a $\mathbb{Z}_2$-valued function $T_p^\varepsilon(r, R)$ to describe the symmetry of metric balls $B_s(p)$ over scales $0 \leq r \leq s \leq R$. Define $T_p^\varepsilon(r, R) = 0$ if there exists a cone space $C(\Sigma)$, depending on p, r, R, ε but not on $s \in [r, R]$, so that

$$d_{GH}\Big(B_s(p), B_s(p^*)\Big) \leq \varepsilon s \tag{15.14}$$

for every $s \in [r, R]$, where $p^* \in C(\Sigma)$ is the cone point. Otherwise we define $T_p^\varepsilon(r, R) = 1$.

Given a point p, $R > r \geq 0$ and $\varepsilon > 0$, define $T_p^\varepsilon(r, R) \equiv 0$ if there exists a metric space Σ such that

$$d_{GH}\Big(B_s(p), B_s(p^*)\Big) \leq \varepsilon s \tag{15.15}$$

for every $s \in [r, R]$, where $p^* \in C(\Sigma)$ is the cone point. If $T_p^\varepsilon(r, R) = 0$, we say that $B_s(p)$ is uniformly $(0, \varepsilon)$-symmetric for $r \leq s \leq R$. Otherwise we define $T_p^\varepsilon(r, R) \equiv 1$. Note that $T_p^\varepsilon(r, R) = 0$ requires that Σ doesn't depend on the scale $s \in [r, R]$.

Theorem 15.3.14 (Good-scale annuli covering) *Let $n \in \mathbb{N}$, $\frac{1}{10} > \varepsilon, \lambda > 0$, $\Lambda > 2$ and $X \in \mathrm{Alex}^{\,n}(-1)$. Any closed subset $\Gamma \subseteq \bar{B}_1 \subseteq X$ can be covered by at most $N(n, \varepsilon, \lambda, \Lambda)$ closed annuli $\mathcal{C} = \left\{ A_{\lambda_i r_i}^{r_i}(x_i) \right\}$, so that the following hold for every i:*

(A.1) $x_i \in \Gamma$;

(A.2) $0 \leq\leq \lambda$;

(A.3) $T_{x_i}^\varepsilon(\Lambda^{-1}\lambda_i r_i, \Lambda r_i) = 0$.

Remark 15.3.15 Note that the constant N does not depend on the volume, and that each of $B_s(x_i)$ is uniformly $(0, \varepsilon)$-symmetric for $\Lambda^{-1}\lambda_i r_i \leq s \leq \Lambda r_i$.

Remark 15.3.16 For the Ricci case, one can find a similar covering, but

(1) N needs to depend on the lower volume bound;

(2) a uniform symmetry such as (A.3) cannot be true.

15.4 Key Ingredients and Framework

In this section we discuss some of the key ingredients that are needed to prove the estimates of $\mathcal{S}_\varepsilon^k$ in Section 15.3.2. The main idea is to cover $B_1(p)$ by controlled numbers of subsets which are Gromov–Hausdorff close to cones or splitting spaces. In Sections 15.4.1 and 15.4.2, we explain the properties of these subsets. In Section 15.4.3, we explain the technique of dimension reduction, which can be viewed as an interaction between the cone structure and the splitting structure. In Section 15.4.4, we explain the covering technique mentioned in Section 15.4.3 by giving a proof of Theorem 15.3.14. At the end of Section 15.5, we will use Theorem 15.3.14 to prove $|\mathcal{S}_\varepsilon^0(X) \cap B_1(p)| < C(n, \varepsilon)$ with $(X, p) \in \text{Alex}^n(-1)$. This can be viewed as a simple application of the framework explained in the previous sections.

The Ricci case is similar but much more technical. This is mainly due to the non-uniform symmetric structure (see Corollary 15.4.9) and the probability-like sub-splitting structure (see Remark 15.4.13).

15.4.1 Monotonicity Formula and Bad Scales

For the Ricci case, a monotonic formula is built on the following volume comparison theorem and the rigidity theorem.

Theorem 15.4.1 (Bishop–Gromov comparison) *For any $p \in M \in \mathcal{M}(n, \kappa)$, the volume ratio function $\mathcal{V}_r^\kappa(p) = \dfrac{\text{Vol}(B_r(p))}{\text{Vol}(B_r(O))} \leq 1$ is decreasing in $r > 0$, where $O \in \mathbb{S}_k^n$ is a point in the n-dimensional space form of constant curvature κ.*

Theorem 15.4.2 (Almost volume cone implies almost metric cone [5]) *Given $n \in \mathbb{N}$ and $\varepsilon > 0$, there exists $\delta(n, \varepsilon) > 0$ so that the following holds for any $p \in M \in \mathcal{M}(n, -\delta)$. If $\mathcal{V}_2^0(p) \geq (1 - \delta)\mathcal{V}_1^0(p)$, then $B_1(p)$ is $(0, \varepsilon)$-symmetric.*

Note that both of the above theorems hold for Alexandrov spaces. However, using these results will force the further estimates to depend on the volume. To obtain a volume free estimate, the following stronger monotonic formula is needed. Given $p, a, b \in X \in \mathrm{Alex}^n(\kappa)$, let $\widetilde{\measuredangle}_\kappa \left(p \, \substack{a \\ b} \right)$ denote the comparison angle $\widetilde{\measuredangle} \left(\widetilde{p} \, \substack{\widetilde{a} \\ \widetilde{b}} \right)$ at $\widetilde{p}$, where $\widetilde{p}, \widetilde{a}, \widetilde{b} \in \mathbb{S}_k^2$ with $d(\widetilde{p}, \widetilde{a}) = d(p, a)$, $d(\widetilde{p}, \widetilde{b}) = d(p, b)$ and $d(\widetilde{a}, \widetilde{b}) = d(a, b)$. Toponogov comparison follows directly from the definition of Alexandrov space.

Theorem 15.4.3 (Toponogov comparison)   *Given* $X \in \mathrm{Alex}^n(\kappa)$, $p \in X$, $x \in X$ *and* $y \in X$, $x' \in [px]$ *and* $y' \in [py]$, *we have* $\widetilde{\measuredangle}_\kappa \left(p \, \substack{x \\ y} \right) \leq \widetilde{\measuredangle}_\kappa \left(p \, \substack{x' \\ y'} \right) \leq \pi$.

The packing number is used to substitute the volume.

Definition 15.4.4 (Packing)   Let X be a metric space and $S \subseteq X$ with $\mathrm{diam}(S) < \infty$. For $\varepsilon > 0$, we say that a subset $\mathbf{x} \equiv \{x_i\} \subseteq S$ is an ε-*subpacking* if

$$d(x_i, x_j) \geq \varepsilon \cdot \mathrm{diam}(S) \text{ for every } i \neq j. \tag{15.16}$$

An ε-subpacking $\mathbf{x}$ is said to be a packing if it is also $\varepsilon \cdot \mathrm{diam}(S)$-dense in S. Define the ε-*packing number*

$$P_\varepsilon(S) \equiv \sup\{|\mathbf{x}| : \mathbf{x} \text{ is an } \varepsilon\text{-subpacking for } S\}. \tag{15.17}$$

Definition 15.4.5 (Inducing function)   Let $p \in X$, $R > 0$, and for each $x \in \bar{B}_R(p) \setminus \{p\}$ we fix a geodesic $\gamma_{px} = \gamma_{px}^R$ connecting p and x. Given $0 < r < R$, we define the inducing function $\varphi_r^R : \bar{B}_R(p) \to \bar{B}_r(p)$, $x \mapsto \bar{x}$, where $\bar{x} \in \gamma_{px}^R$ is the point with $d(p, \bar{x}) = \frac{r}{R} \cdot d(p, x)$. Now let $\{x_i\}_{i=1}^N$ be an ε-subpacking of $\bar{B}_R(p)$ and $0 < r < R$, then we call the collection of points $\{\varphi_r^R(x_i)\}_{i=1}^N$ the induced subpacking in $\bar{B}_r(p)$ of $\{x_i\}_{i=1}^N$.

Theorem 15.4.6 (Almost packing cone implies almost metric cone [16])   *There is a universal constant* $c > 0$ *such that the following holds for any* $n \in \mathbb{N}$ *and* $\varepsilon \in (0, c)$. *Let* $(X, p) \in \mathrm{Alex}^n(-\varepsilon)$ *and* $0 \leq r \leq \frac{1}{2}R \leq c$. *Let* $\mathbf{x}(p, R) = \{x_i\}_{i=1}^N$ *be an* ε-*packing of* $B_R(p)$. *We have*

$$T_p^{\varepsilon^{0.1}}(r, R) = 0 \tag{15.18}$$

if both of the following are satisfied:

(1) $P_\varepsilon(p, r) = N = P_\varepsilon(p, R)$;
(2) $r^{-1}d(\varphi_r^R(x_i), \varphi_r^R(x_j)) \leq R^{-1}d(x_i, x_j) + \varepsilon$, *for every* $1 \leq i, j \leq N$.

Here $\varphi_r^R : \bar{B}_R(p) \to \bar{B}_r(p)$ *is the inducing function defined as in Definition 15.4.5.*

Given $(M,p) \in \mathcal{M}(n,-\delta,v)$ and $0 < s \leq 1$, Theorem 15.4.2 says that if $B_s(p)$ is not $(0,\varepsilon)$-symmetric, then $V_s^0(p)$ is increased by at least $\delta(n,\varepsilon,v) > 0$ from $V_{2s}^0(p)$. Combine this with the monotonicity formula $c(n,v) \leq V_{s_1}^0(p) \leq V_{s_2}^0(p) \leq 1$, where $0 < s_2 \leq s_1 \leq 1$. One concludes that there are at most $N(n,\varepsilon,v)$ integers $\alpha \geq 0$, for which $B_{2^{-\alpha}}(p)$ is not $(0,\varepsilon)$-symmetric. A similar conclusion can be made for Alexandrov spaces, based on Theorem 15.4.3 and Theorem 15.4.6. In order to make a clean and effective statement, the notion of bad scales is introduced below.

Definition 15.4.7 (Bad scales)   Let $r_\alpha = 2^{-\alpha}$, where $\alpha \in \mathbb{Z}$.

(1) The Ricci case. The (weak) ε-*bad scales* $Bad^\varepsilon(x)$ is the collection of r_α for which $B_s(x)$ is not $(0,\varepsilon)$-symmetric for some $r_{\alpha+2} \leq s \leq r_\alpha$.

(2) The Alexandrov case. The strong ε-*bad scales* $\widetilde{Bad}^\varepsilon(x) = \{r_{\beta_{(j)}}\} \subseteq \{r_\alpha\}$ is defined with respect to the uniform symmetry. Let $r_{\beta_{(0)}} = r_0 = 1$ and

$$r_{\beta_{(k+1)}} = \begin{cases} r_{\beta_{(k)}+1}, & \text{if } T_x^\varepsilon(r_{\beta_{(k)}+1}, r_{\beta_{(k)}}) = 1; \\ & \text{if there exists } \alpha \geq \beta_{(k)}+1 \text{ such that} \\ r_\alpha, & T_x^\varepsilon(r_\alpha, r_{\beta_{(k)}}) = 0 \text{ but } T_x^\varepsilon(r_{\alpha+1}, r_{\beta_{(k)}}) = 1. \end{cases}$$

Theorem 15.4.8 (Finiteness of bad scales)   *The following hold for any* $n, \varepsilon, v > 0$:

(1) If $(X,x) \in \mathcal{M}_\infty(n,-1,v)$, *then* $|Bad^\varepsilon(x)| \leq N(n,\varepsilon,v)$.

(2) If $(X,x) \in \text{Alex}^n(-1)$, *then* $|\widetilde{Bad}^\varepsilon(x)| \leq N(n,\varepsilon)$.

Corollary 15.4.9   *The following hold for any* $n \geq 1$, $1/4 > \lambda, \varepsilon > 0$ *and* $0 < R \leq 1$.

(1) There exists $\eta = \eta(n,\varepsilon,v,\lambda) > 0$ *so that if* $(X,x) \in \mathcal{M}_\infty(n,-1,v)$, *then there exists* $r_x \geq \eta R$, *such that* $Bad^\varepsilon(x) \cap [\lambda r_x, r_x] = \varnothing$, *and thus* $B_s(x)$ *is* $(0,\varepsilon)$-*symmetric for every* $s \in [\lambda r_x, r_x]$.

(2) There exists $\eta = \eta(n,\varepsilon,\lambda) > 0$ *so that if* $(X,x) \in \text{Alex}^n(-1)$, *then there exists* $r_x \geq \eta R$, *such that* $\widetilde{Bad}^\varepsilon(x) \cap [\lambda r_x, r_x] = \varnothing$, *and thus* $B_s(x)$ *is uniformly* $(0,\varepsilon)$-*symmetric for* $s \in [\lambda r_x, r_x]$.

Note that the $(0,\varepsilon)$-symmetric property in (1) cannot be strengthened to be uniformly $(0,\varepsilon)$-symmetric.

15.4.2 Splitting Theory

There are various notions of splitting maps for manifolds with lower Ricci curvature bounds. The following is used in [8] and [11].

Definition 15.4.10 (Harmonic ε-splitting map)   The map $u : B_r(p) \to \mathbb{R}^k$ is said to be a harmonic (k, ε)-splitting map if the following are satisfied:

(1) $\Delta u = 0$;
(2) $\int_{B_r(p)} \left| \langle \nabla u_i, \nabla; u_j \rangle - \delta^{ij} \right| < \varepsilon$;
(3) $\sup_{B_r(p)} |\nabla u| < 1 + \varepsilon$;
(4) $r^2 \int_{B_r(p)} \left| \nabla^2 u \right|^2 < \varepsilon^2$.

Theorem 15.4.11 ([8])   *Given $n \in \mathbb{N}$ and $\varepsilon > 0$, there exists $\delta(n, \varepsilon) > 0$ such that the following hold for any $(M, p) \in \mathcal{M}(n, -\delta)$:*

(1) *If there is a (k, δ)-splitting map $u : B_2(p) \to \mathbb{R}^k$, then $B_1(p)$ is (k, ε)-splitting.*
(2) *If $B_2(p)$ is (k, δ)-splitting, then there exists a (k, ε)-splitting map $u : B_1(p) \to \mathbb{R}^k$.*

By Theorem 15.4.11 and maximal function theory, we have the following result. In words, it says that if some ball almost-splits off a Euclidean factor, then most sub-balls continue to almost-split off this factor.

Corollary 15.4.12   *For any $n, \varepsilon > 0$ and $0 < r \leq 1$, there exists $\delta(n, \varepsilon) > 0$ such that the following holds for any $(M, p) \in \mathcal{M}(n, -\delta)$ and $\Omega = \{x \in B_1(p) : B_r(x) \text{ is } (k, \varepsilon)\text{-splitting}\}$. If $B_5(p)$ is (k, δ)-splitting, then $\mathcal{H}^n(\Omega) \geq (1 - \varepsilon) \cdot \mathcal{H}^n(B_1(p))$.*

Remark 15.4.13   Under the same assumptions, it is not true that $B_r(x)$ is (k, ε)-splitting for every $x \in B_1(p)$. It is not true either that the $\mathbb{R}^k$ factors where $B_r(x)$ almost splits are "ε along" the same direction (see [11]) as the splitting directions of $B_r(x)$ and $B_5(p)$ may not close to each other. However, these are all true for Alexandrov spaces. See Corollary 15.4.17.

Definition 15.4.14 (Strong splitting maps)   Let $u_1, u_2, \ldots, u_k : B_R(p) \to \mathbb{R}$ be ε-concave functions. The map $u = (u_1, \ldots, u_k) : B_R(p) \to \mathbb{R}^k$ is called a (k, ε)-splitting map if the following are satisfied:

(1) $|\langle \nabla u_i, \nabla u_j \rangle - \delta_{ij}| \leq \varepsilon$.

(2) For any $x, y \in B_R(p)$ and any minimizing geodesic γ connecting x and y, it holds that

$$\langle \uparrow_x^y, \nabla_x u_i \rangle + \langle \uparrow_y^x, \nabla_y u_i \rangle \leq \varepsilon.$$

Here $\uparrow_x^y$ and $\uparrow_y^x$ denote the unit tangent directions of γ at x and y, respectively.

Remark 15.4.15 If X is a smooth Riemannian manifold, condition (2) in Definition 15.4.14, says that, on each geodesic, u has an integral lower Hessian bound.

It simply follows from Definition 15.4.14 that if $u : B_R(p) \to \mathbb{R}^k$ is a (k, ε)-splitting map, then $u|_{B_r}$ is also a (k, ε)-splitting map for any $B_r \subset B_R(p)$. Such a splitting map doesn't exist for the Ricci case. The following theorem is well known to the experts, and is usually stated using the language of (k, δ)-strainers. Here we state it using splitting functions. A proof is given in [16].

Theorem 15.4.16 *For any $n, \varepsilon > 0$, there exist $\delta = \delta(n, \varepsilon) > 0$ so that the following holds for any $X \in \text{Alex}^n(-\delta)$ and $R \in (0, 1]$:*

(1) *Let $u = (u_1, \dots, u_k) : B_{5R}(p) \to \mathbb{R}^k$ be a (k, δ)-splitting map. For any $B_r \subseteq B_R(p)$ and any $\xi \in u(B_r)$, there exists a map $\varphi : B_r \to u^{-1}(\xi)$ so that*

$$(u, \varphi) : B_r \to \mathbb{R}^k \times u^{-1}(\xi)$$

is an εr-isometry.

(2) *If $f : B_{5R}(p) \to B_{5R}(z)$, where $z \in \mathbb{R}^k \times Z$, is a δR-isometry, then there exists a (k, ε)-splitting map $u : B_R(p) \to \mathbb{R}^k$.*

(3) *If there is a (k, δ)-strainer $\{(a_i, b_i)\}$ with $d(p, a_i)$, $d(p, b_i) \geq 5R$ for every $1 \leq i \leq k$, then there exists a (k, ε)-splitting map $u : B_R(p) \to \mathbb{R}^k$.*

In words, the following corollary says that if some ball almost-splits off a Euclidean factor, then all sub-balls continue to almost-split off this factor.

Corollary 15.4.17 *For any $n, \varepsilon, r > 0$, there exists $\delta(n, \varepsilon) > 0$ such that if $X \in \text{Alex}^n(-\delta)$ and $B_5(p)$ is (k, δ)-splitting, then $B_r(x)$ is (k, ε)-splitting for any $x \in B_1(p)$ and $r \in (0, 1]$.*

15.4.3 Dimension Reduction

The dimension reduction technique is based on the observation that if a metric cone has two distinct cone points, then the cone splits off a line isometrically. In other words, every other point in a metric cone has a higher regularity than the cone points. In the study of quantitative stratification, various quantitative versions of the cone splitting principle are developed. For example, we have the following theorem in Alexandrov geometry.

Theorem 15.4.18 ([16])   *For any $n, k \in \mathbb{N}$ and $\varepsilon > 0$, there exist $\delta = \delta(n, \varepsilon)$ and $\varepsilon' = \varepsilon'(n, \varepsilon) > 0$ such that the following holds for any $p \in X \in \mathrm{Alex}^n(-\delta)$ and (k, δ)-splitting function $u = (u_1, \dots, u_k) :$ $B_{50}(p) \to \mathbb{R}^k$. Let $x \in B_1(p)$ and $y \in X$ with $d(x, y) = r > 0$. If $T_x^\delta(r, 2r) = 0$ and $\left| d(u(x), u(y)) - d(x, y) \right| > \varepsilon r$, then $B_s(y)$ is $(k+1, \varepsilon)$-splitting for every $0 < s \le \varepsilon' r$.*

If $k = 0$, we have that if $d(x, y) = r > 0$ and $T_x^\delta(r, 2r) = 0$, then $B_s(y)$ is $(1, \varepsilon)$-splitting for every $0 < s \le \varepsilon' r$.

This theorem helps us to find the distribution of the quantitative singular points, modulo a covering using almost cones and almost splitting sets. See the proof of Theorem 15.4.21 as an example. For the Ricci case, this leads to the neck decomposition theorem (see [8] and [13]), which requires many more concepts and terminologies to be able to explain it. In order to illustrate the main idea but avoid the special techniques, we will restrict the discussion to Alexandrov spaces from now on.

The following is the covering technique that will be combined with the cone splitting principle to locate the singular points. Theorem 15.3.14 is a different but more intuitive version of this.

Theorem 15.4.19 ([16])   *There exist $\delta(n, \varepsilon)$ and $\beta(n, \varepsilon) > 0$ so that if $u : B_{50}(p) \to \mathbb{R}^k$ is a (k, δ)-splitting function, and $\{B_{r_i}(x_i)\}$ is a disjoint collection with $x_i \in \mathcal{S}^k_{\varepsilon, \beta r_i}$, then for any $z \in \mathbb{R}^k$ we have*

$$\left| \left\{ i \in \mathbb{I} : B_{\beta r_i}(x_i) \cap u^{-1}(z) \ne \varnothing \right\} \right| < N(n, \varepsilon). \qquad (15.19)$$

Although the counterexample was not explicitly constructed in the literature, it is known to the experts that such a slice-wise result doesn't hold for the Ricci case. This is the primary simplification available in the lower sectional curvature case. Roughly speaking, in the Ricci case, the slice-wise statement is replaced by a measure/content estimate of such "nice" slices.

Theorem 15.4.19 implies that $\{B_{\frac{1}{2}\beta r_i}(u(x_i))\} = \cup_{j=1}^{N(n,\varepsilon)}\mathfrak{B}_j$, where each of the $\mathfrak{B}_j$ is a collection of disjoint balls in $B_2 \subset \mathbb{R}^k$, for which the packing estimate holds. Therefore, Theorem 15.3.7 holds if $B_1(p)$ is (k, ε)-splitting. The proofs of Theorem 15.3.7 is then finished by an induction on k.

15.4.4 Good-Scale Annuli Covering

We will give a proof of Theorem 15.3.14 which can be viewed as a version of Theorem 15.4.19 with $k = 0$ and a specially selected $\{B_{r_i}(x_i)\}$. The proofs of Theorem 15.3.14 and Theorem 15.4.19 are similar, though the latter requires the covering to be selected with respect to the previously given collection $\{B_{r_i}(x_i)\}$. Either Theorem 15.3.14 or Theorem 15.4.19 would imply Theorem 15.4.21.

Proof of Theorem 15.3.14 Let $N = N(n, \varepsilon, \lambda, \Lambda) > 0$ be a constant, which may vary line by line. In Step 1, we establish a covering technique. In Step 2, we construct the desired collection $\mathcal{C}$ inductively and show that $|\mathcal{C}| \le N$.

(Step 1) Let $W \subseteq X$ be a closed subset. We may write $W = W_r$ in order to remark that $W \subseteq \bar{B}_r$. Define the function

$$\sigma(x, \varepsilon, s) = \begin{cases} \inf\left\{\tau : T_x^\varepsilon(\tau s, 2s) = 0\right\}, & \text{if} \quad T_x^\varepsilon(s, 2s) = 0; \\ 1, & \text{otherwise.} \end{cases} \quad (15.20)$$

By Corollary 15.4.9, there exists $\eta = \eta(n, \varepsilon, \lambda, \Lambda) > 0$ such that, for any $x \in W_r$, there exists $\Lambda^{-2}r \ge r_x > \eta r$ such that $T_x^\varepsilon(\Lambda^{-1}\lambda r_x, \Lambda r_x) = 0$. Therefore, we have

$$\lambda_x \equiv \Lambda \cdot \sigma\left(x, \varepsilon, \Lambda r_x\right) \le \lambda. \quad (15.21)$$

Take a Vitali collection $\{B_{r_{x_i}}(x_i)\}_{i \in I}$ so that $\{B_{\frac{1}{5}r_{x_i}}(x_i)\}$ are disjoint and $\{B_{r_{x_i}}(x_i)\}_{i \in I}$ covers W_r. Note $\liminf_{z \to y} \sigma(z, \varepsilon, \Lambda r_x) \ge \sigma(y, \varepsilon, \Lambda r_x)$. For $i \in I$, pick $y_i \in \bar{B}_{\lambda_{x_i} r_{x_i}}(x_i)$, so $\sigma(y_i, \varepsilon, \Lambda r_{x_i}) = \inf\{\sigma(y, \varepsilon, \Lambda r_{x_i}) : y \in \bar{B}_{\lambda_{x_i} r_{x_i}}(x_i)\}$. Let $\bar{\lambda}_{y_i} = \Lambda \cdot \sigma(y_i, \varepsilon, \Lambda r_{x_i})$. It is clear that $\bar{\lambda}_{y_i} \le \lambda_{x_i} \le \lambda$. Define $\mathcal{F}(W_r) = \left\{A_{\lambda_{x_i} r_{x_i}}^{r_{x_i}}(x_i), A_{\bar{\lambda}_{y_i} r_{x_i}}^{r_{x_i}}(y_i)\right\}_{i \in I}$ and $\mathcal{D}(W_r) = \left\{\bar{B}_{\lambda_{x_i} r_{x_i}}(x_i) \cap \bar{B}_{\bar{\lambda}_{y_i} r_{x_i}}(y_i) = W_{\bar{\lambda}_{y_i} r_{x_i}}^i \subseteq \bar{B}_{\bar{\lambda}_{y_i} r_{x_i}}\right\}_{i \in I, \bar{\lambda}_{y_i} > 0}$. We now show

Claim 15.4.20 $\mathcal{F}(W_r)$ *and* $\mathcal{D}(W_r)$ *satisfy the following properties:*

(1) The annuli in $\mathcal{F}(W_r)$ *satisfy (A.1)–(A.3).*

(2) $\mathcal{F}(W_r) \cup \mathcal{D}(W_r)$ *is a covering of* W_r.

(3) $|\mathcal{D}(W_r)| \leq |\mathcal{F}(W_r)| \leq N$.

(4) *If* $\bar{\lambda}_{y_i} > 0$, *then, for every* $z \in W^i_{\bar{\lambda}_{y_i} r_{x_i}} = \bar{B}_{\bar{\lambda}_{y_i} r_{x_i}}(y_i) \cap \bar{B}_{\lambda_{x_i} r_{x_i}}(x_i) \in \mathcal{D}(W_r)$, *we have*

$$\left| Bad^\varepsilon(z) \cap [\Lambda^{-1}\bar{\lambda}_{y_i} r_{x_i}, 1] \right| \geq \left| Bad^\varepsilon(z) \cap [\Lambda^{-1}r, 1] \right| + 1. \quad (15.22)$$

Proof of Claim 15.4.20 (1) This follows from the construction.

(2) The following inclusion is obvious:

$$\bar{B}_{r_{x_i}}(x_i) \subseteq \left(A^{r_{x_i}}_{\lambda_{x_i} r_{x_i}}(x_i) \cup A^{r_{x_i}}_{\bar{\lambda}_{y_i} r_{x_i}}(y_i) \right) \cup \left(\bar{B}_{\lambda_{x_i} r_{x_i}}(x_i) \cap \bar{B}_{\bar{\lambda}_{y_i} r_{x_i}}(y_i) \right)$$
$$\cup \left(\bar{B}_{\lambda_{x_i} r_{x_i}}(x_i) \setminus \bar{B}_{r_{x_i}}(y_i) \right) \cup \left(\bar{B}_{\sigma_{x_i} r_{x_i}}(y_i) \setminus \bar{B}_{r_{x_i}}(x_i) \right).$$

$$(15.23)$$

It suffices to show that $\bar{B}_{\lambda_{x_i} r_{x_i}}(x_i) \subseteq \bar{B}_{r_{x_i}}(y_i)$ and $\bar{B}_{\bar{\lambda}_{y_i} r_{x_i}}(y_i) \subseteq \bar{B}_{r_{x_i}}(x_i)$. This is true because $d(x_i, y_i) \leq \lambda_{x_i} r_{x_i} \leq \bar{\lambda}_{y_i} r_{x_i} \leq r_{x_i}/10$.

(3) This follows from a standard volume comparison since $r_{x_i} > \eta r$.

(4) Let $z \in \bar{B}_{\bar{\lambda}_{y_i} r_{x_i}}(y_i) \cap \bar{B}_{\lambda_{x_i} r_{x_i}}(x_i) \in \mathcal{D}(W_r)$. By the definition of $\bar{\lambda}_{y_i}$, we have $\sigma(z, \varepsilon, \Lambda r_{x_i}) \geq \sigma(y_i, \varepsilon, \Lambda r_{x_i}) > 0$. So $T^\varepsilon_z \left(\frac{1}{2} \sigma(z, \varepsilon, r_x) r_{x_i}, \Lambda r_{x_i} \right) = 1$. The notion of bad scales implies $Bad^\varepsilon(z) \cap [\sigma(z, \varepsilon, r_x) r_{x_i}, \Lambda r_{x_i}] \neq \varnothing$. Then (4) follows because

$$[\Lambda^{-1}\bar{\lambda}_{y_i} r_{x_i}, \Lambda^{-1}r] = [\sigma(y_i, \varepsilon, \Lambda r_{x_i}) r_{x_i}, \Lambda^{-1}r] \supseteq [\sigma(z, \varepsilon, \Lambda r_{x_i}) r_{x_i}, \Lambda r_{x_i}].$$

$$(15.24)$$

$\square$

(Step 2) Let the decomposition functions $\mathcal{F}$ and D be defined as in Step 1. Begin with $W^{1,1}_1 = \Gamma \subseteq \bar{B}_1$. Let $\mathcal{C}_1 = \mathcal{F}(W^{1,1}_1)$ and $\mathcal{D}_1 = D(W^{1,1}_1)$, which satisfy (1)–(4) of Claim 15.4.20. Now we construct collections of annuli $\mathcal{C}_k$ and $\mathcal{D}_k = \{W^{k,j}_{r_j}\}$ inductively and show that they satisfy the following properties:

(A_k) The annuli in $\mathcal{C}_k$ satisfy (A.1)–(A.3), Theorem 15.3.14.

(B_k) $\mathcal{C}_k \cup \mathcal{D}_k$ is a covering of Γ.

(C_k) $|\mathcal{D}_k| \leq |\mathcal{C}_k| \leq (1+N)^k$.

(D_k) If $r_j > 0$, then $\left| Bad^\varepsilon(z) \cap [\Lambda^{-1}r_j, 1] \right| \geq k$ for every $z \in W^{k,j}_{r_j}$.

Clearly (A_1)–(D_1) hold by (1)–(4) of Claim 15.4.20. Suppose $\mathcal{C}_k$ and $\mathcal{D}_k = \left\{ W^{k,j}_{r_j} \right\}_{j \in I_k} \neq \varnothing$ have been constructed and satisfy (A_k)–(D_k). For each $j \in I_k$, apply the covering technique in Step 1 for $W = W^{k,j}_{r_j}$.

We get collections $\mathcal{F}(W_{r_j}^{k,j})$ and $\mathcal{D}_{k+1}^j = \mathcal{D}(W_{r_j}^{k,j}) = \left\{W_{\rho_m}^{k+1,m}\right\}_{m\in I_{k+1}^j}$ that satisfy (1)–(4) of Claim 15.4.20. Put $\mathcal{C}_{k+1} = \mathcal{C}_k \cup \left(\cup_{j\in I_k} \mathcal{F}(W_{r_j}^{k,j})\right)$ and $\mathcal{D}_{k+1} = \cup_{j\in I_k} \mathcal{D}_{k+1}^j$. It is clear that $\mathcal{C}_{k+1}$ and $\mathcal{D}_{k+1}$ satisfy (A_{k+1}) and (B_{k+1}). By (3), we have

$$|\mathcal{C}_{k+1}| \le |\mathcal{C}_k| + |I_k| \cdot N \le |\mathcal{C}_k| \cdot (1+N) \le (1+N)^{k+1}. \qquad (15.25)$$

Thus (C_{k+1}) holds. By (4) and the inductive hypothesis, for every $j \in I_k$, $m \in I_{k+1}^j$ and $z \in W_{\rho_m}^{k+1,m} \in \mathcal{D}_{k+1}^j$, we have

$$\left|Bad^\varepsilon(z) \cap [\Lambda^{-1}\rho_m, 1]\right| \ge \left|Bad^\varepsilon(z) \cap [\Lambda^{-1}r_j, 1]\right| + 1 \ge k+1. \qquad (15.26)$$

Now, by (D_k) and Theorem 15.4.8, there exists $K < N(n, \varepsilon)$ such that $\mathcal{D}_K = \varnothing$. Then $\mathcal{C} = \mathcal{C}_K$ is the desired covering. $\qquad \square$

We end this section by giving a proof for the following theorem.

Theorem 15.4.21 *If $(X, p) \in \mathrm{Alex}^n(-1)$, then $|\mathcal{S}_\varepsilon^0(X) \cap B_1(p)| < C(n, \varepsilon)$.*

Proof Let $\delta > 0$ and $\mathcal{C} = \left\{A_{\lambda_i r_i}^{r_i}(x_i), i \le N(n, \delta, 0.1, 0.1)\right\}$, be the good-scale annuli covering of $B_1(p)$, constructed as in Theorem 15.3.14. It suffices to prove that $\mathcal{S}_\varepsilon^0(X) \cap B_1(p) \subseteq \{x_i\}$. Let $y \in A_{\lambda_i r_i}^{r_i}(x_i)$ with $d(y, x_i) > 0$, then, by Theorem 15.4.18, the constant $\delta = \delta(n, \varepsilon) > 0$ can be chosen small so that $B_s(y)$ is $(1, \varepsilon)$-splitting for every $s > 0$ small. Therefore, $y \notin \mathcal{S}_\varepsilon^0(X)$. $\qquad \square$

15.5 Spaces whose Singular Sets are Cantor Sets

We begin with a simple example just to explain the idea, taken verbatim from [16]. This could help to build up some motivation with minimal technical details involved.

Let $Z = \bar{B}_1 \subset \mathbb{R}^2$ be a closed unit disk and $X_0 = Z \times [0, 1] \in \mathrm{Alex}^3(0)$ be a solid cylinder. For $\varepsilon > 0$ small, we have $\mathcal{S}^0(X_0) = \varnothing$, and $\mathcal{S}_\varepsilon^1(X_0) = \partial Z \times \{0, 1\}$ is a union of two unit circles. Now let $T \subseteq \partial Z$ be a closed subset, and thus $\partial Z \setminus T = \cup_\ell U_\ell$ is a collection of disjoint open intervals. Let p be the center of Z and define $C_\ell = \cup_{x\in U_\ell} \gamma_{px}$, where $\gamma_{x,y}$ denotes a line connecting x and y, to be the collection of sectors associated to the open sets U_ℓ. Let us observe for any $x \in \partial Z$ that the curvature at $(x, 1) \in X_0$ is $+\infty$ along the normal direction of $\partial Z \times \{1\}$ and strictly positive along its tangential direction. This will allow us to smoothly "sand off" each of $U_\ell \times \{1\}$ inside its convex hull $C_\ell \times [0,1]$, so that both the convexity of X_0 and the tangent cones at points in $X_0 \setminus (\cup C_\ell \times [0,1])$ are preserved. Let $X_1 \in \text{Alex}^3(0)$ be the resulting space. In particular, the tangent cones at the points of $T \times \{1\}$ are preserved, and thus we have that $\mathcal{S}_\varepsilon^1(X_1) = (T \times \{1\}) \cup (\partial Z \times \{0\})$. Similarly, we can smooth near $\partial Z \times \{0\}$ in order to construct X_2 with $\mathcal{S}_\varepsilon^1(X_2) = T \times \{1\}$. Now let Y_2 be the doubling of X_2, which is now a boundary free Alexandrov space $Y_2 \in \text{Alex}^3(0)$ for which $\mathcal{S}(Y_2) = \mathcal{S}_\varepsilon^1(Y_2) = T$ and $\mathcal{S}_\varepsilon^0(Y_2) = \varnothing$.

Now we start to prove Theorem 15.3.13. If $n \geq 4$, the difficulty lies not only in making $\mathcal{S}_\varepsilon^k$ a Cantor set (which is particularly technical when $k \leq n-3$), but also on the smoothing of every singular point in $\mathcal{S} \setminus \mathcal{S}_\varepsilon^k$ and $\mathcal{S}^{k-1}$. To begin with, let us state a lemma that will be used to smooth convex graphs.

Lemma 15.5.1 ([16])   *Let $\mathbb{T} \subset \mathbb{R}^n$ be a compact convex subset and $f : \mathbb{T} \to \mathbb{R}$ be a strictly convex function. Let $\Omega = \cup_{i=1}^\infty \Omega_i$, where Ω_i are disjoint open convex subsets in $\mathbb{T}$. For any $\delta > 0$, there exists a strictly convex function $F : \mathbb{T} \to \mathbb{R}$ such that the following hold:*

(1) $F|_\Omega$ is C^∞.
(2) $F|_{\mathbb{T}\setminus\Omega} = f|_{\mathbb{T}\setminus\Omega}$ and $|F - f| < \delta$ on Ω.
(3) For any $x \notin \Omega$ and any vector v, it holds that

$$\lim_{t \to 0+} \frac{F(x+tv) - F(x)}{t} = \lim_{t \to 0+} \frac{f(x+tv) - f(x)}{t}. \qquad (15.27)$$

In particular, if $Df(x)$ exists at $x \notin \Omega$, then $DF(x) = Df(x)$.

Inductively applying Lemma 15.5.1, we get the following corollary. It allows Ω_i to have non-empty intersections.

Corollary 15.5.2   *Lemma 15.5.1 holds if $\Omega = \cup_{i=1}^N \Omega_i$, $N < \infty$, where each of Ω_i is a countable union of disjoint open convex subsets.*

We now construct a family of base spaces Z_t. Given $1 \leq k \leq n-3$, choose points $p_1, \ldots, p_{n-1-k} \in \mathbb{R}^{n-1}$ so that the subspace

$$H = \left\{ x \in \mathbb{R}^{n-1} : d(x, p_1) = \cdots = d(x, p_{n-1-k}) \right\}$$

$$= \left\{ x \in \mathbb{R}^{n-1} : \left\langle x - \frac{p_i + p_j}{2}, p_i - p_j \right\rangle = 0 \right.$$

$$\left. \text{for all } 1 \leq i \neq j \leq n-1-k \right\} \qquad (15.28)$$

attains its minimal dimension $(n-1)-(n-1-k)+1 = k+1$. If $k = n-2$, we let $H = \mathbb{R}^{n-1}$. Let $B_r(p, H)$ denote the geodesic ball in H, centered at p with radius r. Let $\mathbb{S}_r^k(p, H) = \partial B_r(p, H)$ denote the k-dimensional sphere in H, centered at p with radius r.

Lemma 15.5.3   *Let* $W_{k,R} = \bigcap_{i=1}^{n-1-k} \bar{B}_R(p_i)$. *The following hold if* $W_{k,R} \neq \varnothing$:

(1) *There is a point* $\tilde{p} \in W_{k,R} \cap H$, *depending only on* $p_1, \ldots, p_{n-1-k}$, *and* $r > 0$ *so that*

$$\mathcal{S}_\varepsilon^k(W_{k,R}) = \partial W_{k,R} \cap H = \mathbb{S}_r^k(\tilde{p}, H). \qquad (15.29)$$

(2) *For* $\varepsilon = \varepsilon(p_1, \ldots, p_{n-1-k}) > 0$ *small, we have* $\mathcal{S}(W_{k,R}) = \partial W_{k,R}$ *and* $\mathcal{S}_\varepsilon^{k-1}(W_{k,R}) = \varnothing$.

(3) *Let* $m = (n-1-k)(n-2-k)$. *There exist open half spaces* T_i *in* $\mathbb{R}^{n-1}$, $i = 1, 2, \ldots, m$, *depending only on* $p_1, \ldots, p_{n-1-k}$, *so that*

$$W_{k,R} \setminus \mathcal{S}_\varepsilon^k(W_{k,R}) \subseteq \bigcup_{i=1}^{m} T_i. \qquad (15.30)$$

(4) *For every* $0 < t < R$, *we have that* $W_{k,t}$ *is the* $(R-t)$-*sub-level set of the distance function* $f : W_{k,R} \to \mathbb{R}$, $x \mapsto d(x, \partial W_{k,R})$. *Moreover, if* $W_{k,t} \neq \varnothing$ *then* $\tilde{p} \in W_{k,t}$ *is the unique point such that* $d(\tilde{p}, \partial W_{k,R}) = \sup\limits_{x \in W_{k,R}} \{d(x, \partial W_{k,R})\}$.

Proof (1) We observe that $\mathcal{S}_\varepsilon^k(W_{k,R}) = \bigcap_{i=1}^{n-1-k} \partial B_R(p_i) = \partial W_{k,R} \cap H$. Let $p' = \frac{1}{n-1-k} \sum_{i=1}^{n-1-k} p_i$ be the center of mass of $\{p_1, \ldots, p_{n-1-k}\}$. For any $x \in \mathbb{R}^{n-1}$, by direct computation we have that if $d(x, p_i) = R$ for all i, then

$$d(x, p') = \rho \equiv \sqrt{R^2 + |p'|^2 - \frac{1}{n-1-k} \sum_{i=1}^{n-1-k} |p_i|^2}. \qquad (15.31)$$

Thus we have

$$\bigcap_{i=1}^{n-1-k} \partial B_R(p_i) = \{x \in H : d(x, p_i) \equiv R \text{ for all } i\} = \partial B_\rho(\bar{p}) \cap H = \mathbb{S}_r^k(\tilde{p}, H),$$
$$\qquad (15.32)$$

where $\tilde{p} \in H$ with $d(p', \tilde{p}) = d(p', H)$ and $r = \sqrt{\rho^2 - d^2(p', H)} > 0$.

(2) This is obvious due to the construction.

(3) Let $H_{i,j} = \{x \in \mathbb{R}^{n-1} : d(x, p_i) = d(x, p_j)\}$, $1 \leq i \neq j \leq n-1-k$.

Let $H_{i,j}^{\pm}$ be the open half spaces in $\mathbb{R}^{n-1}$ divided by $H_{i,j}$. We will show that

$$W_{k,R} \setminus \mathcal{S}_\varepsilon^k(W_{k,R}) = W_{k,R} \setminus H \subseteq \bigcup_{i \neq j} (H_{i,j}^+ \cup H_{i,j}^-). \qquad (15.33)$$

The first equality follows from (1). For any $x \in W_{k,R}$, if $x \notin \bigcup_{i \neq j} (H_{i,j}^+ \cup H_{i,j}^-)$, then $x \in \bigcap_{i \neq j} H_{i,j} = H$ and the result follows.

(4) For $z \in W_{k,R}$, it suffices to prove the following chain of equivalences:

$$d(z, \partial W_{k,R}) \geq t \Leftrightarrow d(z, \partial B_R(p_i)) \geq t \text{ for every } i$$
$$\Leftrightarrow d(z, p_i) \leq R - t \text{ for every } i$$
$$\Leftrightarrow z \in \bar{B}_{R-t}(p_i) \text{ for every } i$$
$$\Leftrightarrow z \in W_{k,R-t}. \qquad (15.34)$$

We need only to prove the first equivalence, since the rest are obvious. Suppose $d(z, \partial W_{k,R}) \geq t$. Because $z \in W_{k,R}$ and $W_{k,R} = \bigcap_j \bar{B}_R(p_j) \subseteq \bigcap \bar{B}_R(p_i)$ for all i, we have $t \leq d(z, \partial W_{k,R}) \leq d(x, \partial B_R(p_i))$ for every i. Suppose $d(z, \partial B_R(p_i)) \geq t$ for every i. Note $\partial W_{k,R} \subseteq \bigcup_{i=1}^{n-1-k} \partial B_R(p_i)$. We have $d(z, \partial W_{k,R}) \geq t$. The second assertion in (4) follows from (1). $\qquad \square$

We have the following lemma for a convex body W in $\mathbb{R}^{n-1}$ which satisfies the property in Lemma 15.5.3(1). Given $p \in W$ we let $\Sigma_p(W)$ denote the unit tangent sphere of W at p. For $V \subseteq W$, we define

$$\Sigma_p(V) = \{\uparrow_p^x \colon x \in V, x \neq p\} \subseteq \Sigma_p(W). \qquad (15.35)$$

Let $C_p(V) = \{y \in \mathbb{R}^{n-1} \colon \uparrow_p^y \in \Sigma_p(V)\}$.

Lemma 15.5.4 *Let W be a convex body in $\mathbb{R}^{n-1}$. Suppose that there exist a $(k+1)$-dimensional plane H, a point $p \in H \setminus \partial W$ and $r > 0$, such that $H \cap W = \bar{B}_r(p, H)$. Then, for any open convex set $U \subseteq \partial \bar{B}_r(p, H) = \mathbb{S}_r^k(p, H)$, there exists an open convex subset $V \subseteq \Sigma_p(W)$ such that $U = C_p(V) \cap \mathbb{S}_r^k(p, H)$. Moreover, if $U_1, U_2 \subseteq \mathbb{S}_r^k(p, H)$ are open convex and disjoint, then the corresponding open convex sets $V_1, V_2 \subseteq \Sigma_p(W)$ can also be chosen so that $V_1 \cap V_2 = \varnothing$.*

Proof Obviously, $p \notin U$. Since U is open and convex in $\mathbb{S}_r^k(p, H)$, we have that $\Sigma_p(U)$ is open convex in $\Sigma_p(B_r(p, H)) = \mathbb{S}_1^k$. Because $\Sigma_p = \mathbb{S}_1^{n-2}$, we have that the spherical suspension $S^{n-2-k}(\Sigma_p(U))$, denoted by V, is open convex in Σ_p. It is easy to see that $U = C_p(V) \cap \mathbb{S}_r^k(p, H)$. The last assertion follows from an easy induction. $\qquad \square$

Theorem 15.3.13 follows from the following result.

Theorem 15.5.5 *Let $n \geq 3$, $1 \leq k \leq n - 2$ and $p \in \mathbb{R}^{k+1}$. Let K be a hyperplane passing through p and $\Gamma = \{x \in \partial B_1(p) : d(x, K) \geq 1/2\}$. For any $\mathcal{U} = \cup_{\ell=1}^{\infty} U_\ell$, where U_ℓ are disjoint open convex subsets in Γ, there exists $X \in \mathrm{Alex}^n(0)$ such that the following hold:*

1. *There is $\varepsilon > 0$, such that the singular sets $\mathcal{S}(X) = \mathcal{S}_\varepsilon^k(X)$ and $\mathcal{S}^{k-1}(X) = \varnothing$.*
2. *There is a bi-Lipschitz embedding $\varphi : \Gamma \to X$, such that $\mathcal{S}_\varepsilon^k(X) = \varphi(\Gamma \setminus \mathcal{U})$.*
3. *There is a sequence of n-dimensional manifolds M_i with $\sec_{M_i} \geq 0$ such that $\lim\limits_{i \to \infty} M_i = X$.*

Proof Given a convex body $W \subset \mathbb{R}^{n-1}$ and $f : W \to [0, \infty)$, we denote the subgraph of f by

$$G_{W,f} = \left\{(w, t) \in W \times \mathbb{R} : 0 \leq t \leq f(w)\right\}. \tag{15.36}$$

Select and fix points $p_1, \ldots, p_{n-1-k} \in \mathbb{R}^{n-1}$ so that the dimension of

$$H = \left\{x \in \mathbb{R}^{n-1} : d(x, p_1) = \cdots = d(x, p_{n-1-k})\right\} \tag{15.37}$$

is $k + 1$ and

$$Z = W_{k,1} = \bigcap_{i=1}^{n-1-k} \bar{B}_1(p_i) \neq \varnothing. \tag{15.38}$$

Let $\Lambda = d(\tilde{p}, \partial Z) = \max\limits_{z \in Z}\{d(z, \partial Z)\}$, where $\tilde{p}$ is the center point as in Lemma 15.5.3. Choose $\delta > 0$ small and define a strictly concave function on Z:

$$f_0(z) = \begin{cases} \sqrt{d(z, \partial Z)}, & \text{if} \quad d(z, \partial Z) \leq \frac{1}{2}\Lambda; \\ \delta \cdot \sqrt{d(z, \partial Z)} + (1 - \delta) \cdot \sqrt{\frac{1}{2}\Lambda}, & \text{if} \quad d(z, \partial Z) > \frac{1}{2}\Lambda. \end{cases} \tag{15.39}$$

Let $c = \sqrt{\frac{1}{2}\Lambda}$ and $Z_t = \{z \in Z : f_0(z) \geq t\}$ be the (f_0, t)-sub-level set. By Lemma 15.5.3(4), we have $Z_t = W_{k,1-t^2}$ for $t \leq c$ and $Z_t = W_{k,1-s^2}$ for $c < t \leq f_0(\tilde{p})$, where $s = \frac{t - (1 - \delta)c}{\delta}$. By Lemma 15.5.3(1), for each $0 \leq t \leq f_0(\tilde{p})$, there is $r_t > 0$, such that $Z_t \cap H = \bar{B}_{r_t}(\tilde{p}, H)$ and $\mathcal{S}_\varepsilon^k(Z_t) = \partial Z_t \cap H = \mathbb{S}_{r_t}^k(\tilde{p}, H)$.

If $n - 1 - k = 1$, $H = \mathbb{R}^{n-1}$, we let K be a hyperplane in $\mathbb{R}^{n-1}$ passing through p_1. For $n - 1 - k \geq 2$, $\dim(H) = k + 1 \leq n - 2$, let

K be a hyperplane so that $p_1, \ldots, p_{n-1-k} \in K$ and $K \perp H$. Let $K^{\pm}$ be the corresponding open half spaces in $\mathbb{R}^{n-1}$. Define convex sets $Z_t^+ = Z_t \cap \bar{K}^+$ and $Z^+ = Z_0^+$. We have $X_0 = G_{Z^+, f_0} \in \text{Alex}^n(0)$. See Figure 15.1.

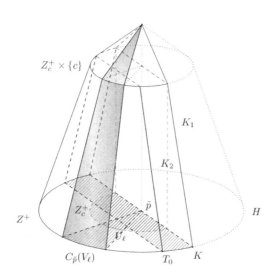

Figure 15.1 $X_0 = G_{Z^+, f_0} \in \text{Alex}^n(0)$.

Define

$$K_1 = \{(x, t) : x \in K, \, 0 \leq t \leq f_0(x)\}. \tag{15.40}$$

Let $\Delta = K_1 \cup (Z^+ \times \{0\})$. By Lemma 15.5.3(1) and (2), we have

(1) $\mathcal{S}(X_0) = \partial X_0$;
(2) $\mathcal{S}_\varepsilon^k(X_0) \setminus \Delta = \left(\mathcal{S}_\varepsilon^k(Z_c^+) \times \{c\}\right) \setminus \Delta = \left(\mathbb{S}_{r_c}^k(\tilde{p}, H) \cap K^+\right) \times \{c\}$ for some $r_c > 0$;
(3) $\mathcal{S}^{k-1}(X_0) \subseteq \Delta$.

Define half spaces $T_0^+ = \{x \in K^+ : d(x, K) > \frac{r_c}{2}\}$ and $T_0^- = \mathbb{R}^{n-1} \setminus \bar{T}_0^+$. Not losing generality, let $\left\{U_\ell \subseteq \mathbb{S}_{r_c}^k(\tilde{p}, H) \cap \bar{T}_0^+\right\}$ be a collection of disjoint open convex sets. It follows from Lemma 15.5.4 that there is a collection of open disjoint convex subsets $V_\ell \subseteq \Sigma_{\tilde{p}}(Z_c^+)$, such that $U_\ell = C_{\tilde{p}}(V_\ell) \cap \mathbb{S}_{r_c}^k(\tilde{p}, H)$ for each ℓ. Let T_i, $i = 1, \ldots, m$, be the open half spaces constructed in Lemma 15.5.3(3). Apply Corollary 15.5.2 to smooth f_0 over the finite collection

$$\left\{ \bigcup_\ell (Z^+ \cap C_{\tilde{p}}(V_\ell)), \quad Z_c^+ \setminus \partial Z_c^+, \quad Z^+ \cap T_0^-, \quad Z^+ \cap T_i, \quad i = 1, 2, \ldots, m \right\}. \tag{15.41}$$

Let f_1 be the strictly concave approximation of f_0 and $X_1 = G_{Z^+, f_1} \in$ Alex$^n(0)$. Because the tangent cones remain the same at the points outside the region where f_0 is smoothed, we have

(4) $\mathcal{S}^{n-2}(X_1) \setminus (\mathcal{S}^k_\varepsilon(X_1) \cup \Delta) \subseteq \{(z, f_1(z)) : f_1(z) < c\}$;
(5) $\mathcal{S}^k_\varepsilon(X_1) \setminus \Delta = \left(\mathcal{S}^k_\varepsilon(Z_c^+) \setminus (\cup_\ell U_\ell) \right) \times \{c\} = \left(\mathbb{S}^k_{r_c}(\tilde{p}, H) \cap T_0^+ \setminus \mathcal{U} \right) \times \{c\}$;
(6) $\mathcal{S}^{k-1}(X_1) \subseteq \Delta$.

Now we smooth out the singular points in (4). To do this, we want to view X_1 as a subgraph of a strictly concave function over K_1. For $0 < t < f_0(\tilde{p})$, let $\pi_t : \partial Z_t^+ \to Z_t \cap K$ be the projection map. We claim that for every t, the map π_t is a bijection; that is, for any $x \in Z_t \cap K$, there is a unique $z \in \partial Z_t^+$ so that $(z - x) \perp K$. First, we have the following fact by direct computation. If $z_1 \neq z_2$, $|z_1 - p_1| = |z_2 - p_2| = r$ and $|z_1 - p_2| = |z_2 - p_1| = t < r$, then $\langle z_1 - z_2, p_1 - p_2 \rangle < 0$. This implies that if $z_1, z_2 \in \pi_t^{-1}(x)$, then there are i and $r > 0$ so that $z_1, z_2 \in \partial B_r(p_i)$. This is not possible since $p_i \in K$, $z_1, z_2 \in \bar{B}_r(p_i) \cap K^+$ and $(z_1 - z_2) \perp K$. Note that for $z = \pi_t^{-1}(x)$, we have $d(z, x) = d(z, K)$. Therefore, Z_t^+ is a subgraph of function $x \mapsto d(x, \pi_t^{-1}(x))$ over $Z_t \cap K$. Because f_1 is a smoothing of f_0 that satisfies Corollary 15.5.2, the above property for (f_0, t)-sub-level sets Z_t also holds for the (f_1, t)-sub-level sets, which for simplicity we still denote by Z_t and define π_t accordingly.

Now let $f_2 : K_1 \to \mathbb{R}$, $(x, t) \mapsto d(x, \pi_t^{-1}(x))$. Then we have $X_1 = G_{f_2, K_1}$. It is easy to see that f_2 is strictly concave. Let convex set $K_2 = \{(z, t) : x \in K, 0 < t < \min\{c, f_1(z)\}\} \subset K_1$. It is clear that

$$G_{f_2, K_2} = \{(z, t) : z \in Z^+, t \leq f_1(z), 0 < t < c\} \supset \{(z, f_1(z)) : f_1(z) < c\}. \tag{15.42}$$

Smooth f_2 over $K_2 \subset K_1$. We get $X_2 \in$ Alex$^n(0)$ with

(7) $\mathcal{S}^{n-2}(X_2) \setminus \Delta = \mathcal{S}^k_\varepsilon(X_2) \setminus \Delta$;
(8) $\mathcal{S}^k_\varepsilon(X_2) \setminus \Delta = \left(\mathcal{S}^k_\varepsilon(Z_c^+) \setminus (\cup_\ell U_\ell) \right) \times \{c\} = \left(\mathbb{S}^k_{r_c}(\tilde{p}, H) \cap T_0^+ \setminus \mathcal{U} \right) \times \{c\}$;
(9) $\mathcal{S}^{k-1}(X_2) \subseteq \Delta$.

A similar but more straightforward smoothing procedure can be performed in a small neighborhood of Δ so that the resultant space $X_3 \in \text{Alex}^n(0)$ and

(10) $\mathcal{S}^{n-2}(X_3) = \mathcal{S}^k_\varepsilon(X_3)$;

(11) $\mathcal{S}^k_\varepsilon(X_3) = \left(\mathbb{S}^k_{r_c}(\widetilde{p}, H) \cap T_0^+ \setminus \mathcal{U}\right) \times \{c\}$;

(12) $\mathcal{S}^{k-1}(X_3) = \varnothing$.

Lastly, we double X_3 and arrive at a space $Y \in \text{Alex}^n(0)$ that satisfies $\mathcal{S}^{k-1}(Y) = \varnothing$ and $\mathcal{S}(Y) = \mathcal{S}^{n-2}(Y) = \mathcal{S}^k_\varepsilon(Y)$. Moreover, $\mathcal{S}^k_\varepsilon(Y)$ is bi-Lipschitz to $\mathbb{S}^k_{r_c}(\widetilde{p}, H) \cap T_0^+ \setminus \mathcal{U}$. It is easy to see that Y can be realized as a non-collapsed limit of n-dimensional manifolds M_i with $\sec_{M_i} \geq 0$. Moreover, Y is homeomorphic to a sphere. $\qquad \square$

References

[1] S. Alexander, V. Kapovitch, A. Petrunin, 2019. Alexandrov geometry. *arXiv: 1903.08539*.

[2] D. Burago, Y. Burago, S. Ivanov, 2001. A course in metric geometry. Graduate Studies in Mathematics., vol. 33, Amer. Math. Soc., Providence, RI.

[3] C. Breiner, T. Lamm, 2015. Quantitative stratification and higher regularity for biharmonic maps. *Manuscripta Math.*, **148**, 379–398.

[4] Y. Burago, M. Gromov, G. Perelman, 1992. A.D. Alexandrov spaces with curvature bounded below. *Uspekhi Mat. Nauk*, **47:2**, 3–51; translation in Russian Math. Surv., **47:2**, 1–58.

[5] J. Cheeger, T. Colding, 1996. Lower bounds on Ricci curvature and the almost rigidity of warped products. *Ann. of Math.*, **44**, 189–237.

[6] J. Cheeger, T. Colding, 1997. On the structure of spaces with Ricci curvature bounded below I. *J. Differ. Geom.*, **45**, 406–480.

[7] J. Cheeger, R. Haslhofer, A. Naber, 2013. Quantitative stratification and the regularity of mean curvature flow. *Geom. Funct. Anal.*, **23**, 828–847.

[8] J. Cheeger, W. Jiang, A. Naber, 2018. Rectifiability of singular sets in noncollapsed spaces. *arXiv:1805.07988*.

[9] J. Cheeger, A. Naber, 2013. Quantitative stratification and the regularity of harmonic maps and minimal currents. *Commun. Pure Appl. Math.*, **66**, 965–990.

[10] J. Cheeger, A. Naber, 2013. Lower bounds on Ricci curvature and quantitative behavior of singular sets. *Invent. Math.*, **191**, 321–339.

[11] J. Cheeger, A. Naber, 2015. Regularity of Einstein manifolds and the codimension 4 conjecture. *Ann. of Math.*, **182**, 1093–1165.

[12] J. Cheeger, A. Naber, D. Valtorta, 2015. Critical sets of elliptic equations. *Commun. Pure Appl. Math.*, **68**, 173–209.

[13] W. Jiang, A. Naber, 2016. L^2 curvature bounds on manifolds with bounded Ricci curvature. *arXiv:1605.05583*.

[14] V. Kapovitch, 2007. Perelman's stability theorem. *Surveys in Differential Geometry, Metric and Comparison Geometry*, International Press, Boston, MA, pp. 103–136.

[15] A. Naber, D. Valtorta, 2017. Rectifiable-Reifenberg and the regularity of stationary and minimizing harmonic maps. *Ann. of Math.*, **185**, no. 1, 131–227.

[16] N. Li, A. Naber, 2019. Quantitative estimates on the singular sets of Alexandrov spaces. *arXiv:1912.03615*.

[17] A. Naber, D. Valtorta, 2015. The singular structure and regularity of stationary and minimizing varifolds. *arXiv:1505.03428*.

[18] Y. Otsu, T. Shioya, 1994. The Riemannian structure of Alexandrov spaces. *J. Differ. Geom.*, **39**, 629–658.

[19] G. Perelman, 1991. Alexandrov spaces with curvatures bounded from below, II. *Preprint.*

16

Universal Covers of Ricci Limit and RCD Spaces

Jiayin Pan and Guofang Wei

Abstract

By Gromov's precompactness theorem, any sequence of n-dimensional manifolds with uniform Ricci curvature lower bound has a convergent subsequence. The limit spaces are referred to as Ricci limit spaces. Cheeger–Colding–Naber developed great regularity and geometric properties for Ricci limit spaces. However, unlike Alexandrov spaces, these spaces could locally have infinite topological type. Sormani and Wei [44, 46] gave the first topological result by showing that the universal cover of any Ricci limit space exists. Here, the universal cover is in the sense of the universal covering map (which does not need to be simply connected). This is extended to RCD spaces by Mondino–Wei [35]. Recently Pan–Wei [38] showed that the non-collapsing Ricci limit spaces are semi-locally simply connected and therefore the universal covers are simply connected. We give a survey of these results and pose some questions.

16.1 Introduction

In 1981 Gromov proved a precompactness result (see Theorem 16.2.3) for Gromov–Hausdorff topology, which revolutionized the field of Riemannian geometry. The Gromov–Hausdorff distance between two metric spaces, roughly speaking, is a coarse measure of their alikeness (see Definition 16.2.2). It provides a platform to study the set of all manifolds with curvature bounds and other geometric conditions (for example, diameter, volume) as a whole. For a Gromov–Hausdorff convergent sequence $M_i \xrightarrow{GH} X$, understanding the structure of the limit space X and the links between the geometrical and topological structures of X and M_i are crucial. When the sequence satisfies a lower sectional curvature bound, the limit spaces are the so called Alexandrov spaces [7]. Alexandrov spaces are locally contractible. On the other hand, the Ricci limit space may have infinite second homology even in the non-collapsing case, see [33]. Around 2000, Cheeger and Colding developed a rich theory on Ricci limit spaces [10, 11, 12, 13]. In recent years, Cheeger, Colding, and Naber further deepened this theory [18, 14, 15]. These results offer powerful tools in understanding Ricci curvature. On the other hand, very little is known about the topology of the Ricci limit spaces. Since Ricci curvature has control on the fundamental groups, it is natural to ask the following question:

Question 16.1.1   Is any Ricci limit space always semi-locally simply connected?

As a first step, Sormani and Wei proved the following.

Theorem 16.1.2 ([44, 46])   *If X is the Gromov–Hausdorff limit of a sequence of complete Riemannian manifolds M_i^n with Ricci curvature $\geq K$, then X has a universal cover.*

In [20, 21] descriptions of the universal cover in terms of the universal covers of the sequence are also given for Ricci limit spaces. Recently Theorem 16.1.2 has been generalized to RCD spaces.

Theorem 16.1.3 ([35])   *Any $\mathsf{RCD}^*(K, N)$ space $(X, \mathsf{d}, \mathsf{m})$ admits a universal cover $(\widetilde{X}, \widetilde{\mathsf{d}}, \widetilde{\mathsf{m}})$, which is itself $\mathsf{RCD}^*(K, N)$, where $K \in \mathbf{R}$, $N \in (1, +\infty)$.*

Note that the universal cover is not assumed to be simply connected, see Definition 16.3.2. Very recently, the authors were able to answer Question 16.1.1 positively in the non-collapsing case.

Theorem 16.1.4 ([38])   *Any non-collapsing Ricci limit space is semi-locally simply connected. Therefore the universal cover, is simply connected.*

With the existence of the universal cover, many results about the fundamental groups of manifolds with Ricci curvature bounded from below can be extended to the deck transformation groups on the universal of the Ricci limit and RCD spaces, see [44, 46, 35]. For the non-collapsing Ricci limit spaces, the results are extended to the fundamental group (see [38]).

In this chapter we present the ideas of these developments. In Section 16.2, we review some basic properties of Ricci limit and RCD spaces. In Section 16.3, we introduce the tools and sketch the ideas used in proving Theorems 16.1.2 and 16.1.3. In Section 16.4, we sketch the idea of the proof for Theorem 16.1.4.

Question 16.1.1 remains open for general Ricci limit spaces. One can also ask the same question for $\mathrm{RCD}(K, N)$ spaces. Some more questions are presented in Sections 16.3 and 16.4.

Acknowledgements   J.-Y.P. was supported by an AMS-Simon travel grant and G.-F.W. was supported by NSF Grant DMS 1811558.

16.2 Some Properties of Ricci Limit and RCD Spaces

In this section we review some basic properties for Ricci limit spaces developed by Cheeger–Colding [10, 11, 12, 13], many of which have been extended to RCD spaces by various authors.

First we recall Hausdorff and Gromov–Hausdorff distance, see [40, Chap. 11] for further details.

Definition 16.2.1   Let X and Y be two compact subsets in a metric space Z. We define the Hausdorff distance between X and Y as

$$d_H(X,Y) = \inf\{\varepsilon > 0 | X \subset B_\varepsilon(Y), Y \subset B_\varepsilon(X)\},$$

where $B_\varepsilon(X)$ is the ε-neighborhood of X in Z.

Definition 16.2.2   Let X and Y be two compact metric spaces. We define

$$d_{GH}(X,Y) = \inf_{Z,f,g}\{d_H(f(X),g(Y)) \mid f : X \to Z \text{ and } g : Y \to Z$$
$$\text{are isometric embeddings}\},$$

where the infimum is taken over all metric spaces Z and all isometric embeddings f, g.

So, d_{GH} defines a distance function on the set of all isometric classes of compact metric spaces. We say that a sequence of compact metric spaces X_i converges in the Gromov–Hausdorff topology to a limit space X, denoted as $X_i \xrightarrow{GH} X$, if $d_{GH}(X_i, X) \to 0$. For noncompact spaces, we say that a sequence of pointed metric spaces (X_i, x_i) converges to (X, x) in the pointed Gromov–Hausdorff sense if $B(x_i, R) \xrightarrow{GH} B(x, R)$

for all $R > 0$, where $B(x, R)$ is the closed ball centered at x with radius R with the restricted metric.

The starting point for this subject is Gromov's precompactness theorem (see [28]).

Theorem 16.2.3 *Let $\{(M_i, x_i)\}_i$ be a sequence of complete Riemannian n-manifolds of*

$$\mathrm{Ric}_{M_i} \geq -(n - 1),$$

then $\{(M_i, x_i)\}_i$ has a convergent subsequence with respect to the pointed Gromov–Hausdorff distance.

The Gromov–Hausdorff limits, referred as Ricci limit spaces, are length spaces. Recall that a length space is a metric space such that the distance between each pair of points equals the infimum of the length of the curves joining the points.

The Ricci limit space has the same Ricci lower bound as in the sequence in a weak sense. In fact they are $\mathrm{RCD}(-(n - 1), n)$-spaces (see below for the definition). Note that the sequence may collapse and that $d\mathrm{vol}_{M_i}$ goes to zero. Considering the renormalized measure $\mu_i = \frac{d\mathrm{vol}_{M_i}}{\mathrm{vol}B_1(x_i)}$, this converges uniformly to a limit measure μ_∞, called the renormalized limit measure. See [10].

$\mathrm{RCD}^*(K, N)$-spaces are the Ricci curvature analog of the celebrated Alexandrov spaces – the generalization of manifolds with Ricci curvature bounded from below by K and dimension bounded above by N to the (Riemannian) metric measure space. In order to define the notion of $\mathrm{RCD}^*(K, N)$-space we first recall the curvature dimension condition, the $\mathrm{CD}(K, N)$ condition, introduced in [49, 50, 31] using tools of optimal transport.

Definition 16.2.4 A metric measure space (X, d, μ) satisfies the (Ricci) curvature dimension condition $\mathrm{CD}(K, N)(K, N \in \mathbb{R}, N \geq 1)$ if the Renyi entropy H_N is K-convex on the space of probability measures.

Here the Renyi entropy, $H_{N,\mu} : P_2(X, \mu) = \{\nu | \nu = \rho\mu\} \to \mathbb{R}$, is

$$H_{N,\mu} = -\int_X \rho^{1-1/N} d\mu.$$

When $K = 0$, "K-convex" means the usual convex function.

There is also the notion of the $\mathrm{CD}^*(K, N)$ condition introduced in [4]. The $\mathrm{CD}^*(K, N)$ condition is a priori weaker than the $\mathrm{CD}(K, N)$ condition, and the two coincide for $K = 0$. The $\mathrm{CD}^*(K, N)$ condition is also equivalent to $\mathsf{CD}_{loc}(K, N)$, and (non-branching) $\mathsf{CD}^*(K, N)$ satisfies the local-to-global property. Recently, Cavalletti–Milman have shown the equivalence of the CD and CD^* conditions when the space is essentially non-branching and has finite measure [8, Cor. 13.6].

A natural version of the Bishop–Gromov volume growth estimate holds on $\mathsf{CD}^*(K, N)$-spaces (see [4, Thm. 6.2] and [9]). Here we only state a weaker version, which is enough for most applications:

Theorem 16.2.5   *Let* $K \in \mathbf{R}$, $N \geq 1$, *be fixed. Then there exists a function* $\Lambda_{K,N}(\cdot, \cdot) : \mathbf{R}_{>0} \times \mathbf{R}_{>0} \to \mathbf{R}_{>0}$ *such that if* $(X, \mathsf{d}, \mathfrak{m})$ *is a* $\mathsf{CD}^*(K, N)$-*space, for some* $K \in \mathbf{R}$, $N \geq 1$, *the following holds:*

$$\frac{\mathfrak{m}(B_r(x))}{\mathfrak{m}(B_R(x))} \geq \Lambda_{K,N}(r, R), \quad \forall 0 < r \leq R < \infty. \tag{16.1}$$

For $K = 0$ *the following more explicit bound holds:*

$$\frac{\mathfrak{m}(B_r(x))}{\mathfrak{m}(B_R(x))} \geq \left(\frac{r}{R}\right)^N, \quad \forall 0 < r \leq R < \infty. \tag{16.2}$$

Key features of both the CD and CD^* conditions are the compatibility with the smooth Riemannian case and the stability under measured Gromov–Hausdorff convergence of metric measure space. In particular, these classes include Ricci limit spaces. On the other hand, they also include Finsler manifolds. In order to rule out Finsler structures while retaining the crucial stability properties of Lott–Sturm–Villani spaces, the following stricter condition is introduced in [24].

Definition 16.2.6   For a metric measure space $(X, \mathsf{d}, \mathfrak{m})$ one says that the $\mathsf{RCD}(K, N)$ condition ($\mathsf{RCD}^*(K, N)$ condition) holds if it satisfies the $\mathsf{CD}(K, N)$ condition ($\mathsf{CD}^*(K, N)$ condition) and the Sobolev space $W^{1,2}(X, \mathfrak{m})$ is a Hilbert space.

The $\mathsf{RCD}^*(K, N)$ condition is also stable under convergence (see also [2], [26], and [22] for other important properties); therefore the class of $\mathsf{RCD}^*(K, N)$-spaces includes Ricci limit spaces (whether they are collapsed or not). Moreover, the class of $\mathsf{RCD}^*(K, N)$-spaces contains weighted manifolds satisfying Bakry–Émery lower curvature bounds as well as their non-smooth limits, cones, warped products, and Alexandrov spaces.

As $\mathsf{RCD}^*(K, N)$-spaces are essentially non-branching (see [41, Cor. 1.2], in particular under the assumption of finite measure, $\mathsf{RCD}(K, N)$ is equivalent to $\mathsf{RCD}^*(K, N)$. It is expected that $\mathsf{RCD}(K, N)$ is equivalent to $\mathsf{RCD}^*(K,N)$ without any further assumptions. As a result we sometimes abuse notation and do not distinguish between writing RCD and RCD^*.

The subject has been developed tremendously in the last few years. The Bochner inequality holds in a weak sense for $\mathsf{RCD}(K,N)$. As a result, many geometric and analytical results for manifolds with Ricci curvature bounded below, and Cheeger–Colding theory for Ricci limit spaces, have been generalized to $\mathsf{RCD}^*(K,N)$-spaces. Here we state a few fundamental properties of $\mathsf{RCD}^*(K,N)$-spaces.

A fundamental property of $\mathsf{RCD}^*(0,N)$-spaces is the extension of the celebrated Cheeger–Gromoll splitting theorem proved in [23]. For Ricci limit spaces, the splitting theorem was established in [10].

Theorem 16.2.7 (Splitting) *Let* $(X,\mathsf{d},\mathfrak{m})$ *be an* $\mathsf{RCD}^*(0,N)$-*space with* $1 \leq N < \infty$. *Suppose that* X *contains a line. Then* $(X,\mathsf{d},\mathfrak{m})$ *is isomorphic to* $(X' \times \mathbf{R}, \mathsf{d}' \times \mathsf{d}_E, \mathfrak{m}' \times \mathsf{L}_1)$, *where* d_E *is the Euclidean distance,* L_1 *is the Lebesgue measure, and* $(X',\mathsf{d}',\mathfrak{m}')$ *is an* $\mathsf{RCD}^*(0,N-1)$-*space if* $N \geq 2$ *and a singleton if* $N < 2$.

Definition 16.2.8 $x \in X$ *is called a* k-*regular point if there exists an integer* $k = k(x) \in [1,N] \cap \mathbb{N}$ *such that, for any sequence* $r_i \to \infty$, *the rescaled pointed metric spaces* $(X, r_i \mathsf{d}, x)$ *converge in the pointed Gromov–Hausdorff sense to the pointed Euclidean space* $(\mathbf{R}^k, \mathsf{d}_E, 0)$. *Otherwise* x *is called a singular point. We denote* $\mathcal{R}^k$ *as the set of* k-*regular points in* X, *and* $\mathcal{R} = \cup_k \mathcal{R}^k$ *as the set of all regular points.*

Note that the singular set could be dense in X.

Theorem 16.2.9 (Infinitesimal regularity of $\mathsf{RCD}^*(K,N)$-spaces, [34]) *Let* $(X,\mathsf{d},\mathfrak{m})$ *be an* $\mathsf{RCD}^*(K,N)$-*space for some* $K \in \mathbf{R}, N \in (1,\infty)$. *Then* $\mathfrak{m}$-*almost everywhere* $x \in X$ *is a regular point. In fact* (X,μ) *is* μ-*rectifiable; namely, up to measure zero, it is the union of sets which are bi-Lipschitz equivalent to subsets of Euclidean spaces.*

Some further regularity properties that follow include the existence of a unique integer k such that $\mathfrak{m}(\mathcal{R} \setminus \mathcal{R}_k) = 0$, which gives $\mathfrak{m}(X \setminus \mathcal{R}_k) = 0$; see [18] for Ricci limit spaces, [6] for RCD spaces. The isometry group of X is a Lie group. For Riemannian manifolds this goes back to [36]; for Ricci limit spaces, see [12, 18], and see [47] for RCD spaces.

A natural version of the Abresch–Gromoll inequality [1] holds on $\mathsf{RCD}^*(K,N)$-spaces (see [25]), which plays an important role in proving Theorems 16.1.2 and 16.1.3. This Abresch–Gromoll estimate has been recently improved by Mondino–Naber [34]. We will use the following simpler version, which is a particular case of [34, Cor. 3.8].

Theorem 16.2.10 (Abresch–Gromoll inequality)   *Given $K \in \mathbf{R}$ and $N \in (1, +\infty)$ there exist $\alpha(N) \in (0, 1)$ and $C(K, N) > 0$ with the following properties. Given $(X, \mathsf{d}, \mathfrak{m})$ an $\mathrm{RCD}^*(K, N)$-space, fix $p, q \in X$ with $\mathsf{d}_{p,q} := \mathsf{d}(p, q) \leq 1$ and let γ be a constant speed minimizing geodesic from p to q. Then*

$$e_{p,q}(x) \leq C(K, N) r^{1+\alpha(N)} \mathsf{d}_{p,q}, \quad \forall x \in B_{r \, \mathsf{d}_{p,q}}(\gamma(1/2)), \qquad (16.3)$$

where $e_{p,q}(x) := \mathsf{d}(p, x) + \mathsf{d}(x, q) - \mathsf{d}(p, q)$ is the so called excess function associated to p, q.

For non-collapsing Ricci limit spaces, we need the following volume convergence result to prove Theorem 16.1.4.

Theorem 16.2.11   [17, 11] *Let (M_i, p_i) be a sequence of complete n-manifolds converging to (X, p) in the pointed Gromov–Hausdorff topology. Suppose that*

$$\mathrm{Ric} \geq -(n - 1), \quad \mathrm{vol}(B_1(p_i)) \geq v > 0.$$

Then, for a sequence $q_i \in M_i$ converging to $q \in X$ and any $r > 0$, we have

$$\mathrm{vol}(B_r(q_i)) \to \mathcal{H}^n(B_r(q)),$$

where $\mathcal{H}^n$ is the n-dimensional Hausdorff measure of X.

A non-collapsing Ricci limit space (X, d, μ) has more regularity. For example, μ is a multiple of $\mathcal{H}^n$, the n-dim Hausdorff measure; the set of singular points has $\dim \mathcal{S} \leq n - 2$; every tangent cone of X is a metric cone $C(Y)$, Y is a length space with $\mathrm{diam}\,(Y) \leq \pi$, see [10, 11]. Non-collapsing RCD spaces are introduced in [19], and these results are mostly extended.

Definition 16.2.12 (Non-collapsed RCD spaces)   For $K \in \mathbb{R}$ and $N \in (1, +\infty)$, $(X, \mathsf{d}, \mathfrak{m})$ is called a non-collapsed $\mathrm{RCD}(K, N)$-space if it is an $\mathrm{RCD}(K, N)$-space and $\mathfrak{m} = \mathcal{H}^N$.

See [14, 16] for more structures on the singular set of non-collapsing Ricci limit spaces, and [15, 30] for the wonderful structure when the sequence is non-collapsing and has a two-sided Ricci curvature bound.

16.3 Universal and δ-Covers

In this section we first recall the definitions for various covering spaces and then indicate the proof of the existence of the universal covers for RCD spaces, namely Theorem 16.1.3.

Definition 16.3.1 We say that $\bar{X}$ is a *covering space* of X if there is a continuous map $\pi : \bar{X} \to X$ such that for all $x \in X$ there is an open neighborhood U such that $\pi^{-1}(U)$ is a disjoint union of open subsets of $\bar{X}$, each of which is mapped homeomorphically onto U by π (we say U is evenly covered by π).

When (X, d) is a locally compact length space, there is a unique length metric on any covering space $\bar{X}$ such that the covering map $\pi : \bar{X} \to X$ is distance nonincreasing and a local isometry. This is done by lifting the length structure on X to $\bar{X}$. See [42] for the study of the geometry and topology of length spaces. When X has a measure, we can also equip the covering space with a measure so that the covering map is locally measure preserving.

The universal cover is often defined as the simply connected cover. Here we do not assume it is simply connected, rather that the cover is universal in the sense of being the cover of all covers.

Definition 16.3.2 ([48, p. 82]) We say that $\widetilde{X}$ is a universal cover of a path connected space X if $\widetilde{X}$ is a cover of X such that, for any other cover $\bar{X}$ of X, there is a commutative triangle formed by a covering map $f : \widetilde{X} \to \bar{X}$ and the two covering projections:

$$\widetilde{X} \quad \xrightarrow{\ f\ } \quad \bar{X}$$
$$\searrow \qquad \swarrow$$
$$X$$

The universal cover may not exist, as can be seen by the Hawaiian earring (Figure 16.1). However, if it exists, then it is unique. Furthermore, if a space is locally path connected and semi-locally simply connected, then it has a universal cover and that cover is simply connected. On the other hand, the universal covering space of a locally path connected space may not be simply connected, as shown by the Griffiths twin cone (Figure 16.2) [27].[1]

---

[1] Figures reprinted with permission from J. Brazas' blog "Wild Topology" https://wildtopology.wordpress.com/2014/06/28/the-griffiths-twin-cone.

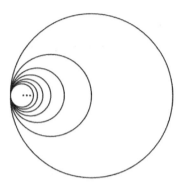

Figure 16.1 Hawaiian earring.

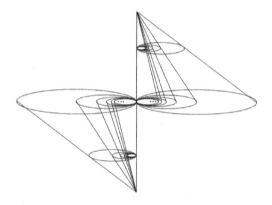

Figure 16.2 Griffiths twin cone.

Given $x \in X$, we denote the fundamental group of X based at x by $\pi_1(X, x)$. Recall that a locally path connected topological space X is said to be *semi-locally simply connected* (or semi-locally one connected) if for all $x \in X$ there is a neighborhood U_x of x such that any curve in U_x is contractible in X, i.e. $\pi_1(U_x, x) \to \pi_1(X, x)$ is trivial (see [48, p. 78], [32, p. 142]). This is weaker than saying that U_x is simply connected.

Let $\mathcal{U}$ be any open covering of X. For any $x \in X$, by [48, p. 81], there is a covering space $\widetilde{X}_{\mathcal{U}}$ of X with covering group $\pi_1(X, \mathcal{U}, p)$, where $\pi_1(X, \mathcal{U}, x)$ is a normal subgroup of $\pi_1(X, p)$ generated by homotopy classes of closed paths having a representative of the form $\alpha^{-1} \circ \beta \circ \alpha$, where β is a closed path lying in some element of $\mathcal{U}$ and α is a path from x to $\beta(0)$.

Now we recall the notion of δ-covers introduced in [44], which plays an important role in studying the existence of the universal cover.

Definition 16.3.3 Given $\delta > 0$, the δ-cover, denoted $\widetilde{X}^\delta$, of a length space X is defined to be $\widetilde{X}_{\mathcal{U}_\delta}$, where $\mathcal{U}_\delta$ is the open covering of X consisting of all balls of radius δ.

Intuitively, a δ-cover is the result of unwrapping all but the loops generated by small loops in X. Clearly $\widetilde{X}^{\delta_1}$ covers $\widetilde{X}^{\delta_2}$ when $\delta_1 \leq \delta_2$.

Note that the δ-cover is a regular or Galois cover. That is, the lift of any closed loop in X is either always closed or always open in the δ-cover.

Example 16.3.4 Let $T^2 = S_1^1 \times S_{1/2}^1$, the product of circles with radius 1 and $\frac{1}{2}$. Then the δ-cover of T^2 is the cylinder when $\pi \leq \delta < 2\pi$ and is $\mathbb{R}^2$ when $\delta < \pi$.

The basic properties of the δ-cover are well studied in the series of joint works of Sormani and Wei [44, 46, 45]. First we focus on the compact length spaces. The following theorem [44, Thm. 3.7] gives a way of proving the existence of the universal cover without showing that the space is semi-locally simply connected. We need only to show that the δ-covers eventually stabilize for sufficiently small δ.

Theorem 16.3.5 *For a compact metric space X, if there exists $\delta_0 > 0$ sufficiently small such that the covering map from $\widetilde{X}^\delta$ maps isometrically to $\widetilde{X}^{\delta_0}$ for all $0 < \delta \leq \delta_0$, then $\widetilde{X}^{\delta_0}$ is the universal cover of X.*

Unlike universal covers, δ-covers behave very well under the Gromov–Hausdorff convergence. The following are proved in [44, Thm. 3.6] and [45, Prop. 7.3].

- If a sequence of compact length spaces X_i converge to a compact length space X in the Gromov–Hausdorff topology, then for any $\delta > 0$ there is a subsequence of X_i such that their δ-covers also converge in the pointed Gromov–Hausdorff topology.
- If compact length spaces X_i converge in the Gromov–Hausdorff topology to a compact space X, and the δ-covering of X_i, $(\widetilde{X}_i^\delta, \widetilde{p}_i)$, converges in the pointed Gromov–Hausdorff topology to $(X^\delta, \widetilde{p}_\infty)$, then $(X^\delta, \widetilde{p}_\infty)$ is a covering space of X, which is almost the δ-cover of X. Namely, for all $\delta_1 > \delta$ we have the covering mapping

$$\widetilde{X}^\delta \to X^\delta \to \widetilde{X}^{\delta_1} \to X.$$

(Both results are not true in general for universal covers.)

Remark 16.3.6 X^δ may not be the δ-cover: take a sequence of flat tori of side lengths 1 by $(n-1)/(2n)$. Let $\delta = 1/2$. Then, δ-covers of these tori are cylinders since all loops of length $< \delta$ are not unraveled. However, they converge to a torus of side lengths 1 by $1/2$, whose δ-cover is Euclidean space.

Now we sketch the proof of Theorem 16.1.3 in the compact case. The proof is divided into two steps. The first step is to prove the stability of δ-covers for small balls at good points.

Theorem 16.3.7   *If X is an* $\mathrm{RCD}^*(K, N)$*-space and $x \in X$ is a regular point, then there exists $r_x > 0$ such that $B(x, r_x)$ lifts isometrically to $\widetilde{X}^\delta$ for all $\delta > 0$.*

Intuitively, if this is not true, there are shorter and shorter closed based geodesic loops in X shrinking towards x, so we can find corresponding closed curves in the tangent cone $\mathbb{R}^k$ which are "almost closed based geodesic loops." However, since $\mathbb{R}^k$ has no closed based geodesic loops, we get a contradiction. In order to make a rigorous proof, quantitative estimates are needed. Assuming the statement of Theorem 16.3.7 is false, then the orbit of $\widetilde{x}$ is getting closer and closer to $\widetilde{x}$ as $\delta \to 0$, where $\widetilde{x}$ in the δ-cover projects to $x \in X$. On each δ-cover, the segment between $\widetilde{x}$ and the closed orbit point projects to a geodesic loop based at x. The midpoint m of this geodesic loop is a cut point of x. In particular, for $y \in X$ with $d(x, y) > D$, where $D = d(x, m)$, we have

$$D + d(y, m) > d(x, y).$$

Using the Abresch–Gromoll inequality Theorem 16.2.10 on δ-covers, one can establish a quantitative version of this inequality: for all $D \leq \frac{1}{2}$, y with $d(x, y) \geq D + S_n D$, we have

$$D + d(y, m) - S_n D \geq d(x, y),$$

where $S_n > 0$ is a small constant. This estimate is called the *uniform cut* lemma [43]. Recall that there is a sequence of these geodesic loops based at x, thus we can pass the distance estimate to the tangent cone at x, which is $\mathbb{R}^k$. However, the uniform distance gap fails on $\mathbb{R}^k$, and this leads to a contradiction.

Theorem 16.2.9 states that the regular points have full measure. We need only one regular point. Then the second step is to use Theorem 16.3.7 at one regular point combined with *Bishop–Gromov's relative volume comparison theorem* on $\widetilde{X}^\delta$ and a *packing argument* to show that the $\widetilde{X}^\delta$ stabilizes everywhere.

When X is not compact, Theorem 16.3.5 is not true. A simple example is a cylinder with one side pinched to a cusp. For the noncompact case, a natural way is to consider bigger and bigger balls. In fact when (X, d) is semi-locally simply connected, the universal cover of X can be obtained as the Gromov–Hausdorff limit of the universal cover of larger and larger balls, see [21, Prop. 1.2]. Since we do not have this extra hypothesis, a different argument is required. Also, the universal cover of a ball may not exist, even if the universal cover of the whole space exists, and one would prefer to avoid the boundaries of balls. For this purpose, Wei and Sormani introduced the relative δ-covers in [46].

We use the following convention: open balls are denoted $B_R(x)$ and closed balls are denoted $B(x, R)$, all with intrinsic metric.

Definition 16.3.8 (Relative δ-cover)   Suppose X is a length space, $x \in X$ and $0 < r < R$. Let

$$\pi^\delta : \widetilde{B}_R(x)^\delta \to B_R(x)$$

be the δ-cover of the open ball $B_R(x)$. A connected component of

$$(\pi^\delta)^{-1}(B(x, r)),$$

where $B(x, r)$ is a closed ball, is called a relative δ-cover of $B(x, r)$ and is denoted $\widetilde{B}(x, r, R)^\delta$.

Instead of Theorem 16.3.5, which was the key to proving the existence of the universal cover for a compact length space, for noncompact spaces the key role will be played by the following result [46, Thm. 2.5].

Theorem 16.3.9   *Let (X, d) be a length space and assume that there is $x \in X$ with the following property: for all $r > 0$, there exists $R \geq r$, such that $\widetilde{B}(x, r, R)^\delta$ stabilizes for all δ sufficiently small. Then (X, d) admits a universal cover $\widetilde{X}$.*

The two steps in the proof of stabilizing the δ-cover for the compact case can be adapted to prove the stabilization of the relative δ-cover for the noncompact case. One subtle thing to note is that relative δ-covers are no longer RCD spaces, but volume comparison still holds on balls of controlled size. See [35] for details.

With the existence of the universal cover, results concerning the fundamental group of manifolds with Ricci curvature bounded from below can now be extended to the revised fundamental group for RCD spaces. Here the revised fundamental group is the group of deck transformations of the universal cover.

Note that the Abresch–Gromoll inequality (Theorem 16.2.10) plays an important role in the proof of the existence of universal cover here. The Abresch–Gromoll inequality does not hold for $CD(K, N)$-spaces. Therefore it is natural to ask the following question:

Question 16.3.10   Does the universal cover of a $CD(K, N)$-space always exist?

16.4 Non-Collapsing Ricci Limit Spaces

We outline a very recent work by the authors in this section, proving that non-collapsing Ricci limit spaces are semi-locally simply connected (Theorem 16.1.4). We point out that the techniques needed to prove Theorem 16.1.4 are independent of the material in Section 16.3.

Note that, by the regularity theory of non-collapsing Ricci limit spaces, local semi-simple connectedness holds at almost all points. In fact, any regular point has almost maximal local volume and the regular points form a subset of full measure in X; by the proof in [39], for a regular point $x \in X$ there exists $r > 0$ such that $B_r(x)$ is contractible in $B_{2r}(x)$. In particular, loops in $B_r(x)$ are contractible in $B_{2r}(x)$. However, this is far from the goal that X is semi-locally simply connected, since it is possible for a loop to go through or even be contained entirely in the singular set.

To attack the problem we quantify the points and make the following definition:

Definition 16.4.1   Let X be a metric space and let $x \in X$. We define the 1-contractibility radius at x as

$$\rho(t, x) = \inf\{\infty, \rho \geq t | \text{ any loop in } B_t(x) \text{ is contractible in } B_\rho(x)\}.$$

Note that X is semi-locally simply connected if for any $x \in X$ there is $T > 0$ such that $\rho(T, x) < \infty$; X is locally simply connected if for any $x \in X$ there is $t_i \to 0$ such that $\rho(t_i, x) = t_i$. For a Riemannian manifold M and $x \in M$, $\rho(t, x) = t$ for t smaller than the injectivity radius at x.

We state the local version of Theorem 16.1.4 with an estimate on the 1-contractibility radius.

Theorem 16.4.2   *Let (M_i, p_i) be a sequence of Riemannian n-manifolds (not necessarily complete) converging to (X, p) such that, for all i,*

(1) $B_2(p_i) \cap \partial M_i = \emptyset$ and the closure of $B_2(p_i)$ is compact;

(2) Ric $\geq -(n-1)$ *on* $B_2(p_i)$, $\mathrm{vol}(B_1(p_i)) \geq v > 0$.

Then $\lim\limits_{t \to 0} \rho(t,x)/t = 1$ *holds for any* $x \in B_1(p)$.

The key step in proving Theorem 16.4.2 is showing that $\lim \rho(t,x) = 0$, which is sometimes called *essentially locally simply connected*. After obtaining this, we can further improve the result to $\lim \rho(t,x)/t = 1$ by using the structure of tangent cones and Sormani's uniform cut techniques.

We classify the points in X according to the local 1-contractibility radius on manifolds:

Definition 16.4.3 For $x \in X$, let x_i in M_i converge to x. We say

- x is of *type I* if there exists $r > 0$ such that the family of functions $\{\rho(q,t) | q \in B_r(x_i), i \in \mathbb{N}\}$ is equi-continuous at $t = 0$;
- x is of *type II* if $\{\rho(x_i, t)\}_{i \in \mathbb{N}}$ is not equi-continuous at $t = 0$;
- x is of *type III* if it is neither of type I nor type II.

Definition 16.4.3 is well defined: the type of x does not depend on the choice of x_i. Type I requires control at every point around x_i, not just at x_i. Type II points may exist due to a positive Ricci curvature example by Otsu [37, p. 262, Rem. 2] (see also the Eguchi–Hanson metric on the tangent bundle of $\mathbb{R}P^2$ for a Ricci flat example). By definition, x being of type III implies that $\{\rho(x_i, t)\}_i$ is equi-continuous at $t = 0$.

For a family of functions $\{\rho_\alpha(t)\}_{\alpha \in A}$ with $\rho_\alpha(0) = 0$, the family is equi-continuous at $t = 0$ if and only if there is a continuous function $\lambda(t)$ defined on $[0,T]$ with $\lambda(0) = 0$ such that $\rho_\alpha(t) \leq \lambda(t)$ for all $t \in [0,T]$ and all $\alpha \in A$. For type I points, we can pass the local 1-contractibility control on local balls around x_i to that around x in the limit space:

Theorem 16.4.4 *Let* (X_i, x_i) *be a sequence of length spaces converging to* (X, x) *satisfying the following conditions:*

(1) the closure of $B_1(x_i)$ *is compact;*

(2) there exists a continuous function λ *on* $[0,T)$ *with* $\lambda(0) = 0$ *such that, for all* i *and all* $q \in B_2(x_i)$, $\rho(t,q) \leq \lambda(t) < 1/2$ *holds on* $[0,T)$.

Then $\rho(t,q) \leq \lambda(t)$ *for all* $t \in [0,T)$ *and all* $q \in B_{1/2}(x)$. *In particular,* $\lim_{t \to 0} \rho(t,q) = 0$.

Note that Theorem 16.4.4 does not require any curvature conditions (X_i may not even be manifolds). We briefly explain two different proofs of Theorem 16.4.4. For a loop c in the limit space, we can find a sequence of loops c_i in M_i that converges uniformly to c. For c in a sufficiently small ball, we can contract c_i for all sufficiently large i. The first proof is related to the method given in [5]: we transfer the null homotopy of c_i along the sequence and pass it to the limit space by uniform convergence; in this method, we need to control the distance between null homotopies to assure uniform convergence. The second proof constructs the null homotopy gradually in the limit space through a sequence of refining skeletons (see Figure 16.3). By controlling the extensions on the new skeletons at each step, these maps on the skeletons converge uniformly to a continuous map defined on the disk.

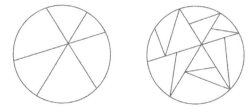

Figure 16.3 Constructing a null homotopy by refining 1-skeletons.

For a type II point x, there are $\varepsilon > 0$ and $t_i \to 0$ such that $\rho(t_i, x_i) \geq \varepsilon$. Let U_i be a small ball centered at y_i ($\pi_i(y_i) = x_i$) in the universal covering space of $B_\varepsilon(x_i)$. The local fundamental group Γ_i has a subgroup generated by these non-contractible short loops H_i. By a result of [3], each H_i is a finite group whose order is controlled by some constant $C(n, v)$. We consider the equivariant Gromov–Hausdorff convergence:

$$
\begin{array}{ccc}
(U_i, y_i, \Gamma_i, H_i) & \xrightarrow{\;GH\;} & (Y, y, G, H) \\
\big\downarrow{\scriptstyle \pi_i} & & \big\downarrow{\scriptstyle \pi} \\
(B_\varepsilon(x_i), x_i) & \xrightarrow{\;GH\;} & (B_\varepsilon(x) = Y/G, x),
\end{array}
$$

Figure 16.4 Equivariant Gromov–Hausdorff convergence.

Roughly speaking, the covering group H_i increases the volume around y_i when compared with the volume around x_i. Together with volume convergence (see Theorem 16.2.11) the local volume around y is at least twice of that around x: $\mathcal{H}^n(B_s(y)) \geq 2 \cdot \mathcal{H}^n(B_s(x))$. Recall that volume comparison states that $\mathcal{H}^n(B_s(y)) \leq \text{vol}(B_s^n(-1))$, where $B_s^n(-1)$ is the s-ball in the n-dimensional space form of constant curvature -1. Therefore, if x has $\mathcal{H}^n(B_s(x)) > \text{vol}(B_s^n(-1))/2$, then Figure 16.4 cannot occur. In fact, this implies that such a point x must be of type I, and we have a linear estimate of the 1-contractibility radius at these points.

Theorem 16.4.5 *Given $n \geq 2$, $\kappa \geq 0$, and $\omega > 1/2$, there exist positive constants $\varepsilon(n, \kappa, \omega)$ and $C(n, \kappa, \omega)$ such that the following holds. Let (M, p) be a complete n-manifold satisfying*

$$\text{Ric} \geq -(n-1)\kappa, \quad \text{vol}(B_1(p)) \geq \omega \cdot \text{vol}(B_1^n(-\kappa)).$$

Then every loop in $B_r(p)$ is contractible in $B_{Cr}(p)$, where $r \in [0, \varepsilon)$.

Note that by Otsu's example [37], the half volume lower bound cannot be weakened to a non-collapsing condition. One may compare the 1-contractibility radius estimate with Theorem 16.4.6 below on sectional curvature and volume lower bounds from [29]:

Theorem 16.4.6 (Grove–Petersen) *Given $n \geq 2, \kappa \geq 0$, and $v > 0$, there exist positive constants $\varepsilon(n, \kappa, v)$ and $C(n, \kappa, v)$ such that, for any complete n-manifold (M, p) of*

$$\sec_M \geq -\kappa, \quad \text{vol}(B_1(p)) \geq v,$$

$B_r(p)$ *is contractible in* $B_{Cr}(p)$, *where* $r \in [0, \varepsilon)$.

If $B_s(x)$ does not have a half volume lower bound, then the volume of $B_s(y)$ in Y will be doubled. This inspires us to use an induction argument on the local volume.

We define $\omega(x) = \lim_{r \to 0} \frac{\mathcal{H}^n(B_r(x))}{\text{vol}(B_r^n(0))}$. Note that, by relative volume comparison, the limit always exists and $0 < \omega(x) \leq 1$; the equality holds if and only if x is a regular point. Also, $\omega(x)$ has a uniform lower bound for all $x \in B_1(p)$.

If $\omega(x) > 1/2$ for all $x \in B_1(p)$, then every point $x \in B_1(p)$ is of type I and we can apply Theorem 16.4.4 directly. Next we consider the case that $\omega(x) > 1/4$ for all $x \in B_1(p)$. If x is type I, then we again apply Theorem 16.4.4. If x is of type II, we can take the convergence of the local universal covers (U_i, y_i), then the corresponding limit point y has $\omega(y) > 1/2$. For s small, we can lift the loop in $B_s(x)$ to a loop in

$B_s(y)$, which we know is contractible. Projecting the homotopy down, we obtain the desired homotopy in X. We need the following technical result for type III points.

Theorem 16.4.7   *Let (X_i, x_i) be a sequence of length spaces converging to (X, x) with the closure of each $B_2(x_i)$ being compact. Suppose that $\lim\limits_{t \to 0} \rho(t, y) = 0$ holds for all points y of type II in $B_{3/2}(x)$, then it holds for all points of type III in $B_1(x)$. Consequently, $\lim\limits_{t \to 0} \rho(t, y) = 0$ holds for all $y \in B_1(x)$.*

Similar to the statement of Theorem 16.4.4, Theorem 16.4.7 does not require any curvature conditions. Assuming Theorem 16.4.7, we clear the step $\omega(x) > 1/4$ and continue the induction argument.

We briefly illustrate the proof of Theorem 16.4.7. We observe that, with the assumptions of Theorem 16.4.7, for any $y \in B_1(x)$, either $\{\rho(t, y_i)\}$ is equi-continuous at $t = 0$ (case y is type I or III), or $\lim \rho(t, y) = 0$ (case y is type I or II). The starting step is the same as the second proof of Theorem 16.4.4, using the homotopy on a manifold from the sequence M_i to extend the loop on a 1-skeleton. Then we apply two different procedures to the sub-triangles. Roughly speaking, if a sub-triangle is displaced from a point of type III, we can directly contract this sub-triangle; if not, we will use the local 1-contractibility from the sequence to extend the map on a finer 1-skeleton (see Figure 16.5). The actual proof is quite technical since we need to assure uniform convergence and control the size of the eventual null homotopy.

Unlike Theorem 16.1.2, which is generalized to RCD spaces, a generalization of Theorem 16.1.4 to RCD spaces seems very difficult. Note that the proof relies heavily on the sequence, which consists of manifolds. More specifically, investigating the convergence of universal covers of local balls $B_\varepsilon(x_i)$ is a key step in the study of type II points. On the other hand, if we are assuming that a space is locally essentially simply connected, then we can always take the universal cover of any local ball. Using the same arguments, it can be shown that for a convergent sequence of uniformly non-collapsing RCD spaces (see Definition 16.2.12), if each space of the sequence is essentially locally simply connected, then the limit space is as well.

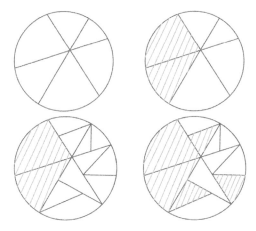

Figure 16.5 Refining 1-skeletons with partial extensions.

In closing, we ask a question about the stability of fundamental groups.

Question 16.4.8   Is the following statement true?

Given $n, D, v > 0$, there is a constant $\varepsilon(n, D, v) > 0$ such that for any two compact Riemannian n-manifolds M_i $(i = 1, 2)$ satisfying:

$$\mathrm{Ric}_{M_i} \geq -(n-1), \quad \mathrm{diam}(M_i) \leq D, \quad \mathrm{vol}(M_i) \geq v, \quad d_{GH}(M_1, M_2) \leq \varepsilon,$$

then $\pi_1(M_1)$ and $\pi_1(M_2)$ must be isomorphic.

From [3] it is known that the fundamental groups of the spaces in the class above have only finitely many isomorphism types. It is known that the fundamental group of X may be different from that of M_i for all i: in Otsu's example, a sequence of Riemannian metrics on $S^3 \times \mathbb{R}P^2$ converges to a simply connected Ricci limit space.

References

[1] Abresch, Uwe, and Gromoll, Detlef. 1990. On complete manifolds with nonnegative Ricci curvature. *J. Amer. Math. Soc.*, **3**(2), 355–374.

[2] Ambrosio, Luigi, Gigli, Nicola, and Savaré, Giuseppe. 2014. Metric measure spaces with Riemannian Ricci curvature bounded from below. *Duke Math. J.*, **163**(7), 1405–1490.

[3] Anderson, Michael T. 1990. Short geodesics and gravitational instantons. *J. Differ. Geom.*, **31**(1), 265–275.

[4] Bacher, Kathrin, and Sturm, Karl-Theodor. 2010. Localization and tensorization properties of the curvature-dimension condition for metric measure spaces. *J. Funct. Anal.*, **259**(1), 28–56.

[5] Borsuk, Karol. 1955. On some metrizations of the hyperspace of compact sets. *Fund. Math.*, **41**, 168–202.

[6] Brué, Ella, and Semola, Daniele. 2019. Constancy of the dimension for RCD(K,N) spaces via regularity of Lagrangian flows. *Commun. Pure Appl. Math. https://doi.org/10.1002/cpa.21849.*

[7] Burago, Yuri., Gromov, Misha., and Perelman, Grisha. 1992. A. D. Aleksandrov spaces with curvatures bounded below. *Uspekhi Mat. Nauk*, **47**(2(284)), 3–51, 222. Translation in Russian Math. Surveys **47**(2), 1—58.

[8] Cavalletti, Fabio, and Milman, Emanuel. The globalization theorem for the curvature-dimension condition. *arXiv:1612.07623.*

[9] Cavalletti, Fabio, and Sturm, Karl-Theodor. 2012. Local curvature-dimension condition implies measure-contraction property. *J. Funct. Anal.*, **262**(12), 5110–5127.

[10] Cheeger, Jeff, and Colding, Tobias H. 1996. Lower bounds on Ricci curvature and the almost rigidity of warped products. *Ann. of Math. (2)*, **144**(1), 189–237.

[11] Cheeger, Jeff, and Colding, Tobias H. 1997. On the structure of spaces with Ricci curvature bounded below. I. *J. Differ. Geom.*, **46**(3), 406–480.

[12] Cheeger, Jeff, and Colding, Tobias H. 2000. On the structure of spaces with Ricci curvature bounded below. II. *J. Differ. Geom.*, **54**(1), 13–35.

[13] Cheeger, Jeff, and Colding, Tobias H. 2000. On the structure of spaces with Ricci curvature bounded below. III. *J. Differ. Geom.*, **54**(1), 37–74.

[14] Cheeger, Jeff, and Naber, Aaron. 2013. Lower bounds on Ricci curvature and quantitative behavior of singular sets. *Invent. Math.*, **191**(2), 321–339.

[15] Cheeger, Jeff, and Naber, Aaron. 2015. Regularity of Einstein manifolds and the codimension 4 conjecture. *Ann. of Math. (2)*, **182**(3), 1093–1165.

[16] Cheeger, Jeff, Jiang, Wenshuai, and Naber, Aaron. 2018. Rectifiability of singular sets in noncollapsed spaces with Ricci curvature bounded below. *arXiv:1805.07988.*

[17] Colding, Tobias H. 1997. Ricci curvature and volume convergence. *Ann. of Math. (2)*, **145**(3), 477–501.

[18] Colding, Tobias H., and Naber, Aaron. 2012. Sharp Hölder continuity of tangent cones for spaces with a lower Ricci curvature bound and applications. *Ann. of Math. (2)*, **176**(2), 1173–1229.

[19] De Philippis, Guido, and Gigli, Nicola. 2018. Non-collapsed spaces with Ricci curvature bounded from below. *J. Éc. Polytech. Math.*, **5**, 613–650.

[20] Ennis, John, and Wei, Guofang. 2006. Describing the universal cover of a compact limit. *Differ. Geom. Appl.*, **24**(5), 554–562.

[21] Ennis, John, and Wei, Guofang. 2010. Describing the universal cover of a noncompact limit. *Geom. Topol.*, **14**(4), 2479–2496.

[22] Erbar, Matthias, Kuwada, Kazumasa, and Sturm, Karl-Theodor. 2015. On the equivalence of the entropic curvature-dimension condition and Bochner's inequality on metric measure spaces. *Invent. Math.*, **201**(3), 993–1071.

[23] Gigli, Nicola. 2013. The splitting theorem in non-smooth context. *arXiv:1302.5555.*

[24] Gigli, Nicola. 2015. *On the differential structure of metric measure spaces and applications*. Memiors of the American Mathematical Soceity, vol. 236, no. 1113. Amer. Math. Soc., Providence, RI.

[25] Gigli, Nicola, and Mosconi, Sunra. 2014. The Abresch-Gromoll inequality in a non-smooth setting. *Discrete Contin. Dyn. Syst.*, **34**(4), 1481–1509.

[26] Gigli, Nicola, Mondino, Andrea, and Savaré, Giuseppe. 2015. Convergence of pointed non-compact metric measure spaces and stability of Ricci curvature bounds and heat flows. *Proc. Lond. Math. Soc. (3)*, **111**(5), 1071–1129.

[27] Griffiths, H. Brian. 1954. The fundamental group of two spaces with a common point. *Quart. J. Math. Oxford Ser. (2)*, **5**, 175–190.

[28] Gromov, Misha. 2007. *Metric structures for Riemannian and non-Riemannian spaces*. (English edn.), Modern Birkhäuser Classics. Birkhäuser Boston, Inc., Boston, MA. Based on the 1981 French original. With appendices by M. Katz, P. Pansu, and S. Semmes. Translated from the French by Sean Michael Bates.

[29] Grove, Karsten, and Petersen, Peter. 1988. Bounding homotopy types by geometry. *Ann. of Math. (2)*, **128**(1), 195–206.

[30] Jiang, Wenshuai, and Naber, Aaron. 2016. L^2 curvature bounds on manifolds with bounded Ricci curvature. *arXiv:1605.05583*.

[31] Lott, John, and Villani, Cédric. 2009. Ricci curvature for metric-measure spaces via optimal transport. *Ann. of Math. (2)*, **169**(3), 903–991.

[32] Massey, William S. 1991. *A basic course in algebraic topology*. Graduate Texts in Mathematics, vol. 127. Springer-Verlag, New York.

[33] Menguy, Xavier. 2000. Noncollapsing examples with positive Ricci curvature and infinite topological type. *Geom. Funct. Anal.*, **10**(3), 600–627.

[34] Mondino, Andrea, and Naber, Aaron. 2019. Structure theory of metric measure spaces with lower Ricci curvature bounds. *J. Eur. Math. Soc. (JEMS)*, **21**(6), 1809–1854.

[35] Mondino, Andrea, and Wei, Guofang. 2019. On the universal cover and the fundamental group of an RCD$^*(K, N)$-space. *J. Reine Angew. Math.*, **753**, 211–237.

[36] Myers, Sumner B., and Steenrod, Norman. E. 1939. The group of isometries of a Riemannian manifold. *Ann. of Math. (2)*, **40**(2), 400–416.

[37] Otsu, Yukio. 1991. On manifolds of positive Ricci curvature with large diameter. *Math. Z.*, **206**(2), 255–264.

[38] Pan, Jiayin, and Wei, Guofang. 2019. Semi-local simple connectedness of non-collapsing Ricci limit spaces. To appear in *J. Eur. Math. Soc. arXiv:1904.06877*.

[39] Perelman, Grisha. 1994. Manifolds of positive Ricci curvature with almost maximal volume. *J. Amer. Math. Soc.*, **7**(2), 299–305.

[40] Petersen, Peter. 2016. *Riemannian geometry*. 3rd edn. Graduate Texts in Mathematics, vol. 171. Springer, Cham.

[41] Rajala, Tapio, and Sturm, Karl-Theodor. 2014. Non-branching geodesics and optimal maps in strong $CD(K, \infty)$-spaces. *Calc. Var. Partial Differ. Equ.*, **50**(3–4), 831–846.

[42] Rinow, Willi. 1961. *Die innere Geometrie der metrischen Räume.* Die Grundlehren der mathematischen Wissenschaften, Bd. 105. Springer-Verlag, Berlin.

[43] Sormani, Christina. 2000. Nonnegative Ricci curvature, small linear diameter growth and finite generation of fundamental groups. *J. Differ. Geom.*, **54**(3), 547–559.

[44] Sormani, Christina, and Wei, Guofang. 2001. Hausdorff convergence and universal covers. *Trans. Amer. Math. Soc.*, **353**(9), 3585–3602.

[45] Sormani, Christina, and Wei, Guofang. 2004. The covering spectrum of a compact length space. *J. Differ. Geom.*, **67**(1), 35–77.

[46] Sormani, Christina, and Wei, Guofang. 2004. Universal covers for Hausdorff limits of noncompact spaces. *Trans. Amer. Math. Soc.*, **356**(3), 1233–1270.

[47] Sosa, Gerardo. 2018. The isometry group of an RCD^* space is Lie. *Potential Anal.*, **49**(2), 267–286.

[48] Spanier, Edwin H. 1966. *Algebraic topology.* McGraw-Hill Book Co., New York.

[49] Sturm, Karl-Theodor. 2006. On the geometry of metric measure spaces. I. *Acta Math.*, **196**(1), 65–131.

[50] Sturm, Karl-Theodor. 2006. On the geometry of metric measure spaces. II. *Acta Math.*, **196**(1), 133–177.

17

Local and Global Homogeneity for Manifolds of Positive Curvature

Joseph A. Wolf

Abstract

We study globally homogeneous Riemannian quotients $\Gamma \backslash (M, ds^2)$ of homogeneous Riemannian manifolds (M, ds^2). The homogeneity conjecture is that $\Gamma \backslash (M, ds^2)$ is (globally) homogeneous if and only if (M, ds^2) is homogeneous and every $\gamma \in \Gamma$ is of constant displacement on (M, ds^2). We provide further evidence for that conjecture by (i) verifying it for normal homogeneous Riemannian manifolds of positive curvature and (ii) showing that in most cases the normality condition can be dropped.

17.1 Introduction

In this chapter we show that a certain conjecture, concerning global homogeneity for locally homogeneous Riemannian manifolds, holds for positively curved manifolds whose universal Riemannian cover is a normal homogeneous space of strictly positive curvature. That is Theorem 17.4.5. It depends on certain classifications (see Table 17.1 on p. 376) and on earlier results [32] concerning isotropy-split fibrations. We then explore the possibility of dropping the normality requirement. Theorem 17.5.3 eliminates the normality condition on the Riemannian metric for most of the positively curved Riemannian manifolds. That requires a modification (17.2) of the isotropy-splitting condition of [32].

We start by describing the homogeneity conjecture. Let (M, ds^2) be a connected, simply connected Riemannian homogeneous space. Let $\pi : M \to M'$ be a Riemannian covering. In other words, $\pi : M \to M'$ is a topological covering space that is a local isometry. Then the base of the covering must have form $M' = \Gamma \backslash M$, where Γ is a discontinuous group of isometries of M such that only the identity element has a fixed point. Clearly M', with the induced Riemannian metric ds'^2 from $\pi : M \to M'$, is locally homogeneous. We ask when (M', ds'^2) is globally homogeneous.

If $M' = \Gamma \backslash M$ is homogeneous then [27] every element $\gamma \in \Gamma$ is of constant displacement $\delta_\gamma(x) = dist(x, \gamma x)$ on M. For the isometry group $G' = \mathbf{I}(\Gamma \backslash M, ds'^2) = N_G(\Gamma)/\Gamma$, where $N_G(\Gamma)$ is the normalizer of Γ in the isometry group $G = \mathbf{I}(M, ds^2)$. Since Γ is discrete, the identity component $N_G(\Gamma)^0$ centralizes Γ in G. This centralizer $Z_G(\Gamma)$ is transitive on M. If $x, y \in M$ and $\gamma \in \Gamma$ we write $y = g(x)$ with $g \in Z_G(\Gamma)$, and we see that $\delta_\gamma(y) = dist(y, \gamma y) = dist(gx, \gamma gx) = dist(gx, g\gamma x) = dist(x, \gamma x) = \delta_\gamma(x)$. That is the easy half of the

Homogeneity conjecture Let M be a connected, simply connected Riemannian homogeneous manifold and $M \to \Gamma \backslash M$ a Riemannian covering. Then $\Gamma \backslash M$ is homogeneous if and only if every $\gamma \in \Gamma$ is an isometry of constant displacement on M.

The first case is implicit in the thesis of Georges Vincent [25, Sect. 10.5]; he noted that the linear transformations $\mathrm{diag}\{R(\theta), \ldots, R(\theta)\}$, $R(\theta) = \begin{pmatrix} \cos(\theta) & \sin(\theta) \\ -\sin(\theta) & \cos(\theta) \end{pmatrix}$, are of constant displacement on the sphere S^{2n-1}. If Γ is the cyclic group of order k, generated by $\mathrm{diag}\{R(2\pi/k), \ldots, R(2\pi/k)\}$, then it has centralizer $U(n)$ in $SO(2n)$ for $k > 2$, all of $SO(2n)$ for $k \leq 2$, so its centralizer is transitive on S^{2n-1}, and $\Gamma \backslash S^{2n-1}$ is homogeneous. Vincent did not consider homogeneity, but he referred to such linear transformations as *Clifford translations* ("translation au sens de Clifford") and examined space forms $\Gamma \backslash S^{2n-1}$ where Γ is a cyclic group $\langle \mathrm{diag}\{R(2\pi/k), \ldots, R(2\pi/k)\} \rangle$.

This was extended to a proof of the homogeneity conjecture, first for spherical space forms [28] and then for locally symmetric Riemannian manifolds [29] by the author. The proof in [29] used classification and case by case checking. This was partially improved by Freudenthal [19] and Ozols ([20], [21], [22]), who gave direct proofs for the case where Γ is contained in the identity component of $\mathbf{I}(M, ds^2)$.

Since then a number of special cases of the homogeneity conjecture have been verified. The ones known by the author are the case [30] where M is of non-positive sectional curvature, extended by Druetta [18] to the case where M has no focal points; Cámpoli's work ([7], [8]) on the case where M is a Stieffel manifold and ds^2 is the normal Riemannian metric;

the case [17] where M admits a transitive semisimple group of isometries that has no compact factor; the case [31] where M admits a transitive solvable group of isometries; the case [32] where M has a fibration such as that of Stieffel manifolds over Grassmann manifolds; and the case [34] where every element of Γ is close to the identity and M belongs to a certain class of Riemannian normal homogeneous spaces. Incidently, the homogeneity conjecture is valid for locally symmetric Finsler manifolds as well [9].

There has also been a lot of work on the infinitesimal version of constant displacement isometries. These are the Killing vector fields of constant length. For example, see papers ([3], [4], [5]) of Berestovskii and Nikonorov, and, in the Finsler manifold setting, of Deng and Xu ([10], [11], [12], [13], [14], [15], [16]).

In this chapter we verify the homogeneity conjecture for (i) the case where (M, ds^2) is a normal homogeneous Riemannian manifold of strictly positive curvature and (ii) most cases where (M, ds^2) is a homogeneous Riemannian manifold, not necessarily normal, of strictly positive curvature. The three cases from which we have not yet eliminated the normality requirement are the odd dimensional spheres $M = G/H = SU(m + 1)/SU(m), Sp(m + 1)/Sp(m), SU(2)/\{1\}$ with possibly nonstandard metrics. In effect, this is a progress report.

Added in proof: the author now has eliminated the normality requirement for these three cases, see [33].

Acknowledgements Research partially supported by a Simons Foundation grant and by hospitality from the MATRIX math research institute at the Creswick campus of the University of Melbourne.

17.2 The Classification for Positive Curvature

The connected, simply connected homogeneous Riemannian manifolds of positive sectional curvature were classified by Marcel Berger [6], Nolan Wallach [26], Simon Aloff and Nolan Wallach [1], and Lionel Bérard-Bergery [2]. Their isometry groups were worked out by Krishnan Shankar [23]. The spaces and the isometry groups are listed in the first two columns of Table 17.1. When there is a fibration that will be relevant to our verification of the homogeneity conjecture, it will also be listed in the first column.

Table 17.1. Isometry groups of csc homogeneous spaces of positive curvature and fibrations over symmetric spaces

	$M = G/H$	$\mathbf{I}(M, ds^2)$
(1)	$S^n = SO(n+1)/SO(n)$	$O(n+1)$
(2)	$P^m(\mathbb{C}) = SU(m+1)/U(m)$	$PSU(m+1) \rtimes \mathbb{Z}_2$
(3)	$P^k(\mathbb{H}) = Sp(k+1)/(Sp(k) \times Sp(1))$	$Sp(k+1)/\mathbb{Z}_2)$
(4)	$P^2(\mathbb{O}) = F_4/Spin(9)$	F_4
(5)	$S^6 = G_2/SU(3)$	$O(7)$
(6)	$P^{2m+1}(\mathbb{C}) = Sp(m+1)/(Sp(m) \times U(1))$ $P^{2m+1}(\mathbb{C}) \to P^m(\mathbb{H})$	$(Sp(m+1)/\mathbb{Z}_2) \times \mathbb{Z}_2$
(7)	$F^6 = SU(3)/T^2$ $F^6 \to P^2(\mathbb{C})$	$(PSU(3) \rtimes \mathbb{Z}_2) \times \mathbb{Z}_2$
(8)	$F^{12} = Sp(3)/(Sp(1) \times Sp(1) \times Sp(1))$ $F^{12} \to P^2(\mathbb{H})$	$(Sp(3)/\mathbb{Z}_2) \times \mathbb{Z}_2$
(9)	$F^{24} = F_4/Spin(8)$ $F^{24} \to P^2(\mathbb{O})$	F_4
(10)	$M^7 = SO(5)/SO(3)$	$SO(5)$
(11)	$M^{13} = SU(5)/(Sp(2) \times_{\mathbb{Z}_2} U(1)$ $M^{13} \to P^4(\mathbb{C})$	$PSU(5) \rtimes \mathbb{Z}_2$
(12)	$N_{1,1} = (SU(3) \times SO(3))/U^*(2)$	$(PSU(3) \rtimes \mathbb{Z}_2) \times SO(3)$
(13)	$N_{k,\ell} = SU(3)/U(1)_{k,\ell}$ $(k,\ell) \neq (1,1), 3|(k^2 + \ell^2 + k\ell)$ $N_{k,\ell} \to P^2(\mathbb{C})$	$(PSU(3) \rtimes \mathbb{Z}_2) \times (U(1) \rtimes \mathbb{Z}_2)$
(14)	$N_{k,\ell} = SU(3)/U(1)_{k,\ell}$ $(k,\ell) \neq (1,1), 3 \nmid (k^2 + \ell^2 + k\ell)$ $N_{k,\ell} \to P^2(\mathbb{C})$	$U(3) \rtimes \mathbb{Z}_2$
(15)	$S^{2m+1} = SU(m+1)/SU(m)$ $S^{2m+1} \to P^m(\mathbb{C})$	$U(m+1) \rtimes \mathbb{Z}_2$
(16)	$S^{4m+3} = Sp(m+1)/Sp(m)$ $S^{4m+3} \to P^m(\mathbb{H})$	$Sp(m+1) \rtimes_{\mathbb{Z}_2} Sp(1)$
(17)	$S^3 = SU(2)$ $S^3 \to P^1(\mathbb{C}) = S^2$	$O(4)$
(18)	$S^7 = Spin(7)/G_2$	$O(8)$
(19)	$S^{15} = Spin(9)/Spin(7)$ $S^{15} \to S^8$	$Spin(9)$

Most of the embeddings $H \hookrightarrow G$ in Table 17.1 are obvious, but a few might need explanation. For (9), $Spin(8) \hookrightarrow Spin(9) \hookrightarrow F_4$. For (10), $SO(3) \hookrightarrow SO(5)$ is the irreducible representation of highest weight 4λ, where λ is the fundamental highest weight; the tangent space representation is the irreducible representation of highest weight 6λ. For (11), $Sp(2) \hookrightarrow SU(4)$ so $Sp(2) \times U(1)$ maps to $U(5)$ with kernel $\{\pm(I, 1)\}$ and image in $SU(5)$. The $\mathbb{Z}_2$ for the non-identity component of $\mathbf{I}(M, ds^2)$ here corresponds to complex conjugation on $SU(5)$. For (12), $U^*(2)$ is the image of $U(2) \hookrightarrow (SU(3) \times SO(3))$, given by $h \mapsto (\alpha(h), \beta(h))$, where $\alpha(h) = \begin{pmatrix} h & 0 \\ 0 & 1/\det(h) \end{pmatrix}$ and β is the projection $U(2) \to U(2)/(center) \cong SO(3)$. For (13) and (14),

$$U(1) = H \hookrightarrow G = SU(3) \text{ is } e^{i\theta} \mapsto \operatorname{diag}\{e^{ik\theta}, e^{i\ell\theta}, e^{-i(k+\ell)\theta}\}.$$

For (19), $Spin(7) \hookrightarrow Spin(8) \hookrightarrow Spin(9)$.

Note that the first four spaces $M = G/H$ of Table 17.1 are Riemannian symmetric spaces with $G = \mathbf{I}(M)^0$. The fifth space is $S^6 = G_2/SU(3)$, where the isotropy group $SU(3)$ is irreducible on the tangent space, so the only invariant metric is the one of constant positive curvature; thus it is isometric to a Riemannian symmetric space. In view of [29] we have

Proposition 17.2.1 *The homogeneity conjecture is valid for the entries (1) through (5) of* Table 17.1.

17.3 Positive Curvature and Isotropy Splitting

Some of the entries of Table 17.1 are of positive Euler characteristic, i.e. have rank $H = \operatorname{rank} G$. In those cases every element of $G = \mathbf{I}(M, ds^2)^0$ is conjugate to an element of H, hence has a fixed point on M. Those are the table entries (6), (7), (8) and (9). Each is isotropy-split with fibration over a projective (thus Riemannian symmetric) space, as defined in [32, (1.1)]. The homogeneity conjecture follows for these (M, ds^2) where ds^2 is the normal Riemannian metric [32, Cor. 5.7]:

Proposition 17.3.1 *The homogeneity conjecture is valid for the entries (6) through (9) of* Table 17.1, *where* ds^2 *is the normal Riemannian metric on* M.

The argument of Proposition 17.3.1 applies with only obvious changes to a number of other table entries, using [32, Thm. 6.1] instead of [32, Cor. 5.7]. That gives us

Proposition 17.3.2 *The homogeneity conjecture is valid for the entries* (11), (13), (14), (15), (16), (17) *and* (19) *of Table 17.1, where* ds^2 *is the normal Riemannian metric on* M.

17.4 The Three Remaining Positive Curvature Cases

In positive curvature it remains only to verify the homogeneity conjecture for table entries $M = G/H$ given by (10) $M^7 = SO(5)/SO(3)$, (12) $N_{1,1} = (SU(3) \times SO(3))/U^*(2)$, and (18) $S^7 = Spin(7)/G_2$. For (10) and (18), H is irreducible on the tangent space of M. In particular for (18) ds^2 must be the constant positive curvature metric on S^7, where the homogeneity conjecture is known:

Lemma 17.4.1 *The homogeneity conjecture is valid for entry* (18) *of Table 17.1, where* ds^2 *is the invariant Riemannian metric on* M.

Now consider the case $M^7 = G/H = SO(5)/SO(3)$. There H acts irreducibly on the tangent space, so the only invariant metric is a normal one. Let $\mathfrak{g} = \mathfrak{h} + \mathfrak{m}$ where $\mathfrak{h} \perp \mathfrak{m}$ and $\mathfrak{m}$ represents the tangent space at $x_0 = 1H \in G/H$. If $\eta \in \mathfrak{m}$ then $t \mapsto \exp(t\eta)$ is a geodesic based at x_0.

Suppose that we have an isometry γ of some constant displacement $d > 0$. As $\mathbf{I}(M, ds^2) = SO(5)$ is connected, we have $\xi \in \mathfrak{m}$ such that $\sigma(t) = \exp(t\xi)x_0$, $0 \leq t \leq 1$, is a minimizing geodesic in M from x_0 to $\gamma(x_0)$. Let X denote the Killing vector field on M corresponding to ξ. Note that $||X_{x_0}|| = ||\xi|| = d$. Let $g \in G$ and $y = gx_0 \in M$. Then $t \mapsto g\sigma(t) = g\exp(t\xi)x_0$ is a minimizing geodesic in M from y to $g\gamma(x_0) = \mathrm{Ad}(g)(\gamma)(y)$. Since $\mathrm{Ad}(g)(\gamma)$ has the same constant displacement d as γ we have $||(g_*X)_{x_0}|| = d$. But $||(g_*X)_{x_0}|| = ||X_y||$, in other words $||X_y|| = ||X_{x_0}||$. Thus X is a Killing vector field of constant length on $SO(5)/SO(3)$. There is no such nonzero vector field [34], so γ does not exist. We have proved

Lemma 17.4.2 *There is no isometry* $\neq 1$ *of constant displacement on the manifold* M^7 *with normal Riemannian metric. In particular the homogeneity conjecture is valid for entry* (10) *of Table 17.1, where* ds^2 *is the normal Riemannian metric on* M^7.

Now consider the case (12) of $N_{1,1} = G/H = (SU(3) \times SO(3))/U^*(2)$. Let γ be an isometry of constant displacement $d > 0$ on $N_{1,1}$. Suppose that γ^2 is also an isometry of constant displacement. The argument of Lemma 17.4.1 shows that γ cannot belong to the identity component $G = SU(3) \times SO(3)$ of $\mathbf{I}(N_{1,1})$. Further, γ^2 belongs to that identity component, so the argument of Lemma 17.4.1 shows that $\gamma^2 = 1$.

Now $\gamma = (g_1, g_2)\nu$, where $g_1 \in SU(3)$, $g_2 \in SO(3)$, $\nu^2 = 1$, $\mathrm{Ad}(\nu)$ is complex conjugation on $SU(3)$, and $\mathrm{Ad}(\nu)$ is the identity on $SO(3)$. The centralizer of ν is $K := ((SO(3) \times SO(3)) \cup (SO(3) \times SO(3))\nu$. Let T_1 (resp. T_2) be a maximal torus of the first (resp. second) $SO(3)$. Following de Siebenthal [24] we may assume $g_i \in T_i$, where we replace γ by a conjugate. Compute $\gamma^2 = (g_1, g_2)(\overline{g_1}, g_2) = (g_1^2, g_2^2)$. We have reduced our considerations to the cases where g_1 is either the identity matrix I_3 or the matrix $I_3' := \left(\begin{smallmatrix} -1 & 0 & 0 \\ 0 & -1 & 0 \\ 0 & 0 & +1 \end{smallmatrix} \right)$, and also g_2 is either I_3 or I_3'.

Recall that $H = U^*(2)$ is the image of $U(2) \hookrightarrow (SU(3) \times SO(3))$, given by $h \mapsto (\alpha(h), \beta(h))$, where $\alpha(h) = \left(\begin{smallmatrix} h & 0 \\ 0 & 1/\det(h) \end{smallmatrix} \right)$ and β is the projection $U(2) \to U(2)/(center) \cong SO(3)$. Further, $\mathbf{I}(L_{1,1}) = G \cup G\nu$ and its isotropy subgroup is $H \cup H\nu$. Observe that

$$\text{if } (g_1, g_2) = (I_3, I_3) \text{ then } \gamma = (\alpha(I_2), \beta(I_2))\nu \in H\nu, \text{ and}$$
$$\text{if } (g_1, g_2) = (I_3', I_3) \text{ then } \gamma = (\alpha(-I_2), \beta(-I_2))\nu \in H\nu.$$

We can replace I_3' by its $SO(3)$ conjugate $I_3'' = \left(\begin{smallmatrix} -1 & 0 & 0 \\ 0 & +1 & 0 \\ 0 & 0 & -1 \end{smallmatrix} \right)$. Set $I_2'' = \left(\begin{smallmatrix} -1 & 0 \\ 0 & +1 \end{smallmatrix} \right)$. Note that

$$\text{if } (g_1, g_2) = (I_3'', I_3'') \text{ then } \gamma = (\alpha(I_2''), \beta(I_2''))\nu \in H\nu.$$

When $\gamma \in H\nu$ it cannot be of nonzero constant displacement. We have reduced our considerations to the case $\gamma = (I_3, I_3')\nu$, or equivalently to one of its conjugates. Compute

$$\left(\left(\begin{pmatrix} i & 0 & 0 \\ 0 & i & 0 \\ 0 & 0 & -1 \end{pmatrix}, I_3 \right) \cdot (I_3, I_3')\nu \cdot \left(\begin{pmatrix} i & 0 & 0 \\ 0 & i & 0 \\ 0 & 0 & -1 \end{pmatrix}, I_3 \right)^{-1} \right.$$

$$= \left(\begin{pmatrix} i & 0 & 0 \\ 0 & i & 0 \\ 0 & 0 & -1 \end{pmatrix}, I_3 \right) \cdot (I_3, I_3') \cdot \left(\begin{pmatrix} i & 0 & 0 \\ 0 & i & 0 \\ 0 & 0 & -1 \end{pmatrix}, I_3 \right) \nu$$

$$= (I_3', I_3')\nu \in H\nu,$$

so again γ cannot be of nonzero constant displacement. We have proved

Lemma 17.4.3 *There is no isometry $\gamma \neq 1$ of constant displacement on the manifold $N_{1,1}$ with normal Riemannian metric, for which γ^2 is also of constant displacement. In particular the homogeneity conjecture is valid for entry (12) of Table 17.1 with normal Riemannian metric on $N_{1,1}$.*

Combining Lemmas 17.4.1, 17.4.2 and 17.4.3 we have

Proposition 17.4.4 *The homogeneity conjecture is valid for entries (10), (12) and (18) of Table 17.1, where ds^2 is the normal Riemannian metric on M.*

Finally, combine Propositions 17.2.1, 17.3.1, 17.3.2 and 17.4.4 to obtain our main result:

Theorem 17.4.5 *Let (M, ds^2) be a connected, simply connected normal homogeneous Riemannian manifold of strictly positive sectional curvature. Then the homogeneity conjecture is valid for (M, ds^2).*

17.5 Dropping Normality in Positive Curvature

In several cases one can eliminate the normality requirement of Theorem 17.4.5. Of course this is automatic when the adjoint action of H on the tangent space $\mathfrak{g}/\mathfrak{h}$ is irreducible; there, every invariant Riemannian metric on $M = G/H$ is normal, so Theorem 17.4.5 applies. Those are the spaces given by entries (1), (2), (3), (4), (5), (10), (11), (12) and (18) of Table 17.1. Some other cases require an extension of certain results from [32] concerning normal Riemannian homogeneous spaces.

The homogeneity conjecture was verified by Wolf [32] for a class of normal Riemannian homogeneous spaces $M = G/H$ that fiber over homogeneous spaces $M' = G/K$, where H is a local direct factor of K. We are going to weaken the normality conditions in such a way that the results still apply to some of the homogeneous Riemannian manifolds of positive sectional curvature. In [32] the metrics on M and M' were required to be the normal Riemannian metrics defined by the Killing form of G. Instead, we look at Riemannian surjections $\pi : M \to M'$, with fiber F, as in (17.2).

G is a compact, connected, simply connected Lie group.

$H \subset K$ be closed, connected subgroups of G such that

(i) $M = G/H$, $M' = G/K$, and $F = H\backslash K$,
(ii) $\pi : M \to M'$ by $\pi(gH) = gK$, right action of K,
(iii) M' and F are Riemannian symmetric spaces, and   (17.2)
(iv) the tangent spaces $\mathfrak{m}'$ for M', $\mathfrak{m}''$ for F and $(\mathfrak{m}'+\mathfrak{m}'')$
    for M satisfy $\mathfrak{m}' \perp \mathfrak{m}''$.

We first modify [32, Lem. 5.2]:

Lemma 17.5.1   *Assume* (17.2). *Then the fiber F of $M \to M'$ is totally geodesic in M In particular it is a geodesic orbit space, and any geodesic of M tangent to F at some point is of the form $t \mapsto \exp(t\xi)x$ with $x \in F$ and $\xi \in \mathfrak{m}''$.*

Proof   Consider the restriction $\mathrm{Ad}_{\mathfrak{g}}|_{\mathfrak{h}}$. Then $\mathfrak{g} = \mathfrak{h} + \mathfrak{m}' + \mathfrak{m}''$, where $\mathfrak{h}$ acts by its adjoint representation (on itself), by the restriction of the isotropy representation of $\mathfrak{k}$ on $\mathfrak{m}'$, and by its isotropy representation on $\mathfrak{m}''$. If $\xi \in \mathfrak{m}''(\subset \mathfrak{k})$ and $\eta \in \mathfrak{m}'(\subset \mathfrak{g})$ then $[\xi, \eta] \in \mathfrak{m}'$, so $\langle [\xi, \eta]_{\mathfrak{m}'+\mathfrak{m}''}, \xi \rangle = 0$ because $\langle \xi, \mathfrak{m}' \rangle = 0$. If $\xi, \eta \in \mathfrak{m}''$ then $\mathrm{ad}(\eta)$ is antisymmetric so $[\xi, \eta] \perp \xi$, i.e. $\langle [\xi, \eta]_{\mathfrak{m}'+\mathfrak{m}''}, \xi \rangle = 0$. We have just shown that if $\xi \in \mathfrak{m}''$ then $t \mapsto \exp(t\xi)H$ is a geodesic in M based at $1H$. Since it is a typical geodesic in the symmetric space $F = H\backslash K$, we conclude that F is totally geodesic in M.   □

Now we set up the conditions for applying (17.2) in the possible absence of normality.

Lemma 17.5.2   *Suppose that (M, ds^2) is one of the entries of Table 17.1 for which $G = \mathbf{I}(M, ds^2)^0$, that (M, ds^2) satisfies* (17.2), *and that $\pi : (M, ds^2) \to (M', ds'^2)$ is a Riemannian submersion. Then the homogeneity conjecture holds for (M, ds^2).*

Proof   Running through the last column of Table 17.1 we see something surprising: in each case, $\mathrm{rank}\, G = \mathrm{rank}\, K$, in other words $\chi(M') > 0$. That noted, we use the starting point of the proof of [32, Prop. 5.4]. There one takes $\gamma \in G$ of constant displacement on M to be of the form $(g, r(k))$, with $g \in G$ acting on the left on $M = G/H$ and $r(k)$ given by the right action of the normalizer of H in G. The metric ds'^2 is the usual one in each case because it is G-invariant. The argument of [32, Prop. 5.4] adapts to show that $\Gamma \cap G$ is central in G. Further, the proof of [32,

Lem. 5.5] carries through for the intersection of Γ with other components of $\mathbf{I}(M, ds^2)$. Thus, if Γ is a group of isometries of constant displacement on (M, ds^2), then Γ centralizes G. The homogeneity conjecture follows for (M, ds^2).   □

The argument of Lemma 17.5.2 does not apply directly to entry (6) of Table 17.1, so we give another argument specific to that case. There rank H = rank G, so every element of G has a fixed point on M. If $\Gamma \neq \{1\}$ is a subgroup of $\mathbf{I}(M, ds^2)$ acting freely on M then $\Gamma = \{1, g\nu\}$, where $g \in G$ and ν denotes complex conjugation. As $(g\nu)^2 = 1$ we have $g \cdot {}^t g^{-1} = g\bar{g} = \pm I$ in terms of matrices. Then ${}^t g = cg$ for some $c \in \mathbb{C}$ and $g = c^2 g$ so $c = \pm 1$. If $c = 1$ then $g = {}^t g$. In that case g is diagonalized by a real matrix and $g\nu$ has a fixed point. Thus $c = -1$ and we may assume $g = {}_\bullet \left(\begin{smallmatrix} 0 & I \\ -I & 0 \end{smallmatrix} \right)$. Thus $g\nu$ centralizes G, and that proves the homogeneity conjecture for case (6) of Table 17.1.

A small modification of the argument of Lemma 17.5.2 applies to entry (19), where $G = \mathbf{I}(M, ds^2)$. We simply note that the proof of [32, Thm. 6.1] proves the homogeneity conjecture for case (19) of Table 17.1.

Now we run through the entries of Table 17.1. As noted above, in cases (1), (2), (3), (4), (5), (10), (11), (12) and (18) the group H is irreducible on the tangent space of M, so the metric is the normal Riemannian metric, and the homogeneity conjecture follows from the normal metric case, Theorem 17.4.5. And we just dealt with entries (6) and (19). Now we run through the other entries.

In table entries (7), (8) and (9), $G = \mathbf{I}(M, ds^2)^0$, and in entries (13) and (14) we may assume $G = U(3) = \mathbf{I}(M, ds^2)^0$. For those entries, $\mathfrak{m}' \perp \mathfrak{m}''$ because the representations of H on those spaces have no common summand, and $\pi : (M, ds^2) \to (M', ds'^2)$ is a Riemannian submersion because (M', ds'^2) is an irreducible Riemannian symmetric space. Thus Lemma 17.5.2 applies to (7), (8), (9), (13) and (14).

Consider the table entries for which M is an odd sphere. Cases (1), (18) and (19) have already been dealt with, leaving (15), (16) and (17). Combining all the results of this section we have

Theorem 17.5.3 *Let (M, ds^2) be a connected, simply connected homogeneous Riemannian manifold of strictly positive curvature, where ds^2 is not required to be the normal Riemannian metric. Suppose that (M, ds^2) is not one of the entries (15), (16) or (17) of Table 17.1. Then the homogeneity conjecture is valid for (M, ds^2).*

Added in proof: the author recently proved the homogeneity conjecture for the entries (15), (16) and (17) of Table 17.1. So those entries need not be excluded in Theorem 17.5.3, see [33]

References

[1] S. Aloff & N. R. Wallach, An infinite family of distinct 7-manifolds admitting positively curved Riemannian structure, Bull. Amer. Math. Soc. **81** (1975), 93–97.

[2] L. Bérard-Bergery, Les variétés riemanniennes homogènes simplement connexes de dimension impaire à courbure strictement positive, J. Math. Pures Appl. **55** (1976), 47–67.

[3] V. N. Berestovskii & Y. G. Nikonorov, Killing vector fields of constant length on locally symmetric Riemannian manifolds. Transform. Group. **13** (2008), 25–45.

[4] V. N. Berestovskii & Y. G. Nikonorov, On Clifford–Wolf homogeneous Riemannian manifolds, Doklady Math. **78** (2008), 807–810.

[5] V. N. Berestovskii & Y. G. Nikonorov, Clifford–Wolf homogeneous Riemannian manifolds. J. Differ. Geom. **82** (2009), 467–500.

[6] M. Berger, Les variétés riemanniennes homogènes normales simplement connexes à courbure strictement positive, Ann. Scuola Norm. Sup. Pisa **15** (1961), 179–246.

[7] O. Cámpoli, Clifford isometries of compact homogeneous Riemannian manifolds, Revista Unión Math. Argentina **31** (1983), 44–49.

[8] O. Cámpoli, Clifford isometries of the real Stieffel manifolds, Proc. Amer. Math. Soc. **97** (1986), 307–310.

[9] S. Deng & J. A. Wolf, Locally symmetric homogeneous Finsler spaces, Intern. Mat. Res. Not. (IMRN) **2013** (2013), 4223–4242.

[10] S. Deng & M. Xu, Clifford–Wolf homogeneous Randers spaces, J. Lie Theory **23** (2013), 837–845.

[11] S. Deng & M. Xu, Clifford–Wolf homogeneous Randers spheres, Israel J. Math. **199** (2014), 507–525.

[12] S. Deng & M. Xu, Clifford–Wolf translations of left invariant Randers metrics on compact Lie groups, Quart. J. Math. **65** (2014), 133–148.

[13] S. Deng & M. Xu, Clifford–Wolf translations of Finsler spaces, Forum Math. **26** (2014), 1413–1428.

[14] S. Deng & M. Xu, Left invariant Clifford–Wolf homogeneous (α, β)-metrics on compact semisimple Lie groups, Transform. Group. **20** (2015), 395–416.

[15] S. Deng & M. Xu, Clifford–Wolf homogeneous Finsler metrics on spheres, Ann. Mat. Pura Appl. **194** (2015), 759–766.

[16] S. Deng & M. Xu, (α_1, α_2)–metrics and Clifford–Wolf homogeneity. J. Geom. Anal. **26** (2016), 2282–2321.

[17] I. Dotti Miatello, R. J. Miatello & J. A. Wolf, Bounded isometries and homogeneous Riemannian quotient manifolds, Geom. Dedicata **21** (1986), 21–28.

[18] M. J. Druetta, Clifford translations in manifolds without focal points, Geom. Dedicata **14** (1983), 95–103.

[19] H. Freudenthal, Clifford–Wolf-Isometrien symmetrischer Räume, Math. Ann. **150** (1963), 136–149.

[20] V. Ozols, Critical points of the displacement function of an isometry, J. Differ. Geom. **3** (1969), 411–432.

[21] V. Ozols, Critical sets of isometries, Proc. Sympos. Pure Math. **27** (1973), 375–378.

[22] V. Ozols, Clifford translations of symmetric spaces, Proc. Amer. Math. Soc. **44** (1974), 169–175.

[23] K. Shankar, Isometry groups of homogeneous spaces with positive sectional curvature, Differ. Geom. Appl. **14** (2001), 57–78.

[24] J. de Siebenthal, Sur les groupes de Lie compacts non connexes, Comment. Math. Helv. **31** (1956), 41–89.

[25] G. Vincent, Les groupes linéaires finis sans points fixes, Comment. Math. Helv. **20** (1947), 117–171.

[26] N. R. Wallach, Compact homogeneous Riemannian manifolds with strictly positive curvature, Ann. Math. **96** (1972), 277–295.

[27] J. A. Wolf, Sur la classification des variétés riemanniènnes homogènes à courbure constante, C. R. Acad. Sci. Paris, **250** (1960), 3443–3445.

[28] J. A. Wolf, Vincent's conjecture on Clifford translations of the sphere, Comment. Math. Helv. **36** (1961), 33–41.

[29] J. A. Wolf, Locally symmetric homogeneous spaces. Comment. Math. Helv. **37** (1962), 65–101.

[30] J. A. Wolf, Homogeneity and bounded isometries in manifolds of negative curvature, Ill. J. Math. **8** (1964), 14–18.

[31] J. A. Wolf, Bounded isometries and homogeneous quotients, J. Geom. Anal. **26** (2016), 1–9.

[32] J. A. Wolf, Homogeneity for a class of Riemannian quotient manifolds, Differ. Geom. Appl. **56** (2018), 355–372.

[33] J. A. Wolf, Local and Global Homogeneity for Three Obstinate Spheres, preprint, arXiv:2005.09702 [math.DG], 2020.

[34] M. Xu & J. A. Wolf, Killing vector fields of constant length on Riemannian normal homogeneous spaces, Transform. Group. **21** (2016), 871–902.